Modern Optimal Control

LECTURE NOTES

IN PURE AND APPLIED MATHEMATICS

1. *N. Jacobson,* Exceptional Lie Algebras
2. *L. -Å. Lindahl and F. Poulsen,* Thin Sets in Harmonic Analysis
3. *I. Satake,* Classification Theory of Semi-Simple Algebraic Groups
4. *F. Hirzebruch, W. D. Newmann, and S. S. Koh,* Differentiable Manifolds and Quadratic Forms (out of print)
5. *I. Chavel,* Riemannian Symmetric Spaces of Rank One (out of print)
6. *R. B. Burckel,* Characterization of C(X) Among Its Subalgebras
7. *B. R. McDonald, A. R. Magid, and K. C. Smith,* Ring Theory: Proceedings of the Oklahoma Conference
8. *Y.-T. Siu,* Techniques of Extension on Analytic Objects
9. *S. R. Caradus, W. E. Pfaffenberger, and B. Yood,* Calkin Algebras and Algebras of Operators on Banach Spaces
10. *E. O. Roxin, P.-T. Liu, and R. L. Sternberg,* Differential Games and Control Theory
11. *M. Orzech and C. Small,* The Brauer Group of Commutative Rings
12. *S. Thomeier,* Topology and Its Applications
13. *J. M. Lopez and K. A. Ross,* Sidon Sets
14. *W. W. Comfort and S. Negrepontis,* Continuous Pseudometrics
15. *K. McKennon and J. M. Robertson,* Locally Convex Spaces
16. *M. Carmeli and S. Malin,* Representations of the Rotation and Lorentz Groups: An Introduction
17. *G. B. Seligman,* Rational Methods in Lie Algebras
18. *D. G. de Figueiredo,* Functional Analysis: Proceedings of the Brazilian Mathematical Society Symposium
19. *L. Cesari, R. Kannan, and J. D. Schuur,* Nonlinear Functional Analysis and Differential Equations: Proceedings of the Michigan State University Conference
20. *J. J. Schäffer,* Geometry of Spheres in Normed Spaces
21. *K. Yano and M. Kon,* Anti-Invariant Submanifolds
22. *W. V. Vasconcelos,* The Rings of Dimension Two
23. *R. E. Chandler,* Hausdorff Compactifications
24. *S. P. Franklin and B. V. S. Thomas,* Topology: Proceedings of the Memphis State University Conference
25. *S. K. Jain,* Ring Theory: Proceedings of the Ohio University Conference
26. *B. R. McDonald and R. A. Morris,* Ring Theory II: Proceedings of the Second Oklahoma Conference
27. *R. B. Mura and A. Rhemtulla,* Orderable Groups
28. *J. R. Graef,* Stability of Dynamical Systems: Theory and Applications
29. *H.-C. Wang,* Homogeneous Branch Algebras
30. *E. O. Roxin, P.-T. Liu, and R. L. Sternberg,* Differential Games and Control Theory II
31. *R. D. Porter,* Introduction to Fibre Bundles
32. *M. Altman,* Contractors and Contractor Directions Theory and Applications
33. *J. S. Golan,* Decomposition and Dimension in Module Categories
34. *G. Fairweather,* Finite Element Galerkin Methods for Differential Equations
35. *J. D. Sally,* Numbers of Generators of Ideals in Local Rings
36. *S. S. Miller,* Complex Analysis: Proceedings of the S.U.N.Y. Brockport Conference
37. *R. Gordon,* Representation Theory of Algebras: Proceedings of the Philadelphia Conference
38. *M. Goto and F. D. Grosshans,* Semisimple Lie Algebras
39. *A. I. Arruda, N. C. A. da Costa, and R. Chuaqui,* Mathematical Logic: Proceedings of the First Brazilian Conference

40. *F. Van Oystaeyen,* Ring Theory: Proceedings of the 1977 Antwerp Conference
41. *F. Van Oystaeyen and A. Verschoren,* Reflectors and Localization: Application to Sheaf Theory
42. *M. Satyanarayana,* Positively Ordered Semigroups
43. *D. L. Russell,* Mathematics of Finite-Dimensional Control Systems
44. *P.-T. Liu and E. Roxin,* Differential Games and Control Theory III: Proceedings of the Third Kingston Conference, Part A
45. *A. Geramita and J. Seberry,* Orthogonal Designs: Quadratic Forms and Hadamard Matrices
46. *J. Cigler, V. Losert, and P. Michor,* Banach Modules and Functors on Categories of Banach Spaces
47. *P.-T. Liu and J. G. Sutinen,* Control Theory in Mathematical Economics: Proceedings of the Third Kingston Conference, Part B
48. *C. Byrnes,* Partial Differential Equations and Geometry
49. *G. Klambauer,* Problems and Propositions in Analysis
50. *J. Knopfmacher,* Analytic Arithmetic of Algebraic Function Fields
51. *F. Van Oystaeyen,* Ring Theory: Proceedings of the 1978 Antwerp Conference
52. *B. Kedem,* Binary Time Series
53. *J. Barros-Neto and R. A. Artino,* Hypoelliptic Boundary-Value Problems
54. *R. L. Sternberg, A. J. Kalinowski, and J. S. Papadakis,* Nonlinear Partial Differential Equations in Engineering and Applied Science
55. *B. R. McDonald,* Ring Theory and Algebra III: Proceedings of the Third Oklahoma Conference
56. *J. S. Golan,* Structure Sheaves over a Noncommutative Ring
57. *T. V. Narayana, J. G. Williams, and R. M. Mathsen,* Combinatorics, Representation Theory and Statistical Methods in Groups: YOUNG DAY Proceedings
58. *T. A. Burton,* Modeling and Differential Equations in Biology
59. *K. H. Kim and F. W. Roush,* Introduction to Mathematical Consensus Theory
60. *J. Banas and K. Goebel,* Measures of Noncompactness in Banach Spaces
61. *O. A. Nielson,* Direct Integral Theory
62. *J. E. Smith, G. O. Kenny, and R. N. Ball,* Ordered Groups: Proceedings of the Boise State Conference
63. *J. Cronin,* Mathematics of Cell Electrophysiology
64. *J. W. Brewer,* Power Series Over Commutative Rings
65. *P. K. Kamthan and M. Gupta,* Sequence Spaces and Series
66. *T. G. McLaughlin,* Regressive Sets and the Theory of Isols
67. *T. L. Herdman, S. M. Rankin, III, and H. W. Stech,* Integral and Functional Differential Equations
68. *R. Draper,* Commutative Algebra: Analytic Methods
69. *W. G. McKay and J. Patera,* Tables of Dimensions, Indices, and Branching Rules for Representations of Simple Lie Algebras
70. *R. L. Devaney and Z. H. Nitecki,* Classical Mechanics and Dynamical Systems
71. *J. Van Geel,* Places and Valuations in Noncommutative Ring Theory
72. *C. Faith,* Injective Modules and Injective Quotient Rings
73. *A. Fiacco,* Mathematical Programming with Data Perturbations I
74. *P. Schultz, C. Praeger, and R. Sullivan,* Algebraic Structures and Applications Proceedings of the First Western Australian Conference on Algebra
75. *L. Bican, T. Kepka, and P. Nemec,* Rings, Modules, and Preradicals
76. *D. C. Kay and M. Breen,* Convexity and Related Combinatorial Geometry: Proceedings of the Second University of Oklahoma Conference
77. *P. Fletcher and W. F. Lindgren,* Quasi-Uniform Spaces
78. *C.-C. Yang,* Factorization Theory of Meromorphic Functions
79. *O. Taussky,* Ternary Quadratic Forms and Norms
80. *S. P. Singh and J. H. Burry,* Nonlinear Analysis and Applications
81. *K. B. Hannsgen, T. L. Herdman, H. W. Stech, and R. L. Wheeler,* Volterra and Functional Differential Equations

82. *N. L. Johnson, M. J. Kallaher, and C. T. Long,* Finite Geometries: Proceedings of a Conference in Honor of T. G. Ostrom
83. *G. I. Zapata,* Functional Analysis, Holomorphy, and Approximation Theory
84. *S. Greco and G. Valla,* Commutative Algebra: Proceedings of the Trento Conference
85. *A. V. Fiacco,* Mathematical Programming with Data Perturbations II
86. *J.-B. Hiriart-Urruty, W. Oettli, and J. Stoer,* Optimization: Theory and Algorithms
87. *A. Figa Talamanca and M. A. Picardello,* Harmonic Analysis on Free Groups
88. *M. Harada,* Factor Categories with Applications to Direct Decomposition of Modules
89. *V. I. Istrăţescu,* Strict Convexity and Complex Strict Convexity: Theory and Applications
90. *V. Lakshmikantham,* Trends in Theory and Practice of Nonlinear Differential Equations
91. *H. L. Manocha and J. B. Srivastava,* Algebra and Its Applications
92. *D. V. Chudnovsky and G. V. Chudnovsky,* Classical and Quantum Models and Arithmetic Problems
93. *J. W. Longley,* Least Squares Computations Using Orthogonalization Methods
94. *L. P. de Alcantara,* Mathematical Logic and Formal Systems
95. *C. E. Aull,* Rings of Continuous Functions
96. *R. Chuaqui,* Analysis, Geometry, and Probability
97. *L. Fuchs and L. Salce,* Modules Over Valuation Domains
98. *P. Fischer and W. R. Smith,* Chaos, Fractals, and Dynamics
99. *W. B. Powell and C. Tsinakis,* Ordered Algebraic Structures
100. *G. M. Rassias and T. M. Rassias,* Differential Geometry, Calculus of Variations, and Their Applications
101. *R.-E. Hoffmann and K. H. Hofmann,* Continuous Lattices and Their Applications
102. *J. H. Lightbourne, III, and S. M. Rankin, III,* Physical Mathematics and Nonlinear Partial Differential Equations
103. *C. A. Baker and L. M. Batten,* Finite Geometries
104. *J. W. Brewer, J. W. Bunce, and F. S. Van Vleck,* Linear Systems Over Commutative Rings
105. *C. McCrory and T. Shifrin,* Geometry and Topology: Manifolds, Varieties, and Knots
106. *D. W. Kueker, E. G. K. Lopez-Escobar, and C. H. Smith,* Mathematical Logic and Theoretical Computer Science
107. *B.-L. Lin and S. Simons,* Nonlinear and Convex Analysis: Proceedings in Honor of Ky Fan
108. *S. J. Lee,* Operator Methods for Optimal Control Problems
109. *V. Lakshmikantham,* Nonlinear Analysis and Applications
110. *S. F. McCormick,* Multigrid Methods: Theory, Applications, and Supercomputing
111. *M. C. Tangora,* Computers in Algebra
112. *D. V. Chudnovsky and G. V. Chudnovsky,* Search Theory: Some Recent Developments
113. *D. V. Chudnovsky and R. D. Jenks,* Computer Algebra
114. *M. C. Tangora,* Computers in Geometry and Topology
115. *P. Nelson, V. Faber, T. A. Manteuffel, D. L. Seth, and A. B. White, Jr.* Transport Theory, Invariant Imbedding, and Integral Equations: Proceedings in Honor of G. M. Wing's 65th Birthday
116. *P. Clément, S. Invernizzi, E. Mitidieri, and I. I. Vrabie,* Semigroup Theory and Applications
117. *J. Vinuesa,* Orthogonal Polynomials and Their Applications: Proceedings of the International Congress
118. *C. M. Dafermos, G. Ladas, and G. Papanicolaou,* Differential Equations: Proceedings of the EQUADIFF Conference
119. *E. O. Roxin,* Modern Optimal Control: A Conference in Honor of Solomon Lefschetz and Joseph P. Lasalle
120. *J. C. Díaz,* Mathematics for Large Scale Computing

Other Volumes in Preparation

Solomon Lefschetz (1884–1972)

Joseph Pierre LaSalle (1916–1983)

Modern Optimal Control

A Conference in Honor of Solomon Lefschetz and Joseph P. LaSalle

edited by

Emilio O. Roxin
The University of Rhode Island
Kingston, Rhode Island

Marcel Dekker, Inc. New York • Basel

Library of Congress Cataloging-in-Publication Data

Modern optimal control : a conference in honor of Solomon Lefschetz
and Joseph P. LaSalle / edited by Emilio O. Roxin
p. cm. -- (Lecture notes in pure and applied mathematics ; v. 116)
Bibliography: p.
ISBN 0-8247-8168-6
1. Control theory--Congresses. 2. Mathematical optimization--Congresses. 3. Lefschetz, Solomon, 1884-1972--Congresses.
4. LaSalle, Joseph P.. --Congresses. I. Lefschetz, Solomon, 1884-1972. II. LaSalle, Joseph P. III. Roxin, Emilio O.
IV. Series.
QA402.3.M62 1989
629.8'312--dc20 89-11888
CIP

This book is printed on acid-free paper.

MARCEL DEKKER, INC.
270 Madison Avenue, New York, New York 10016

Current printing (last digit):
10 9 8 7 6 5 4 3 2 1

PRINTED IN THE UNITED STATES OF AMERICA

Foreword

Solomon Lefschetz (1884-1972) and Joseph Pierre La Salle (1916-1983)

It is indeed an honor to make a few remarks about Solomon Lefschetz and Joseph P. LaSalle. It is impossible for me to cover their accomplishments and overall influence on science. Therefore, I will concentrate on a part of their role in differential equations and control theory, not dwelling on specific mathematical research.

Lefschetz was born in Moscow in 1884 and died in Princeton, New Jersey, in 1972. LaSalle was born in State College, Pennsylvania, in 1916 and died in Little Compton, Rhode Island, in 1983.

Already famous for his contributions to algebraic geometry and topology, Lefschetz, at age 60, became interested in differential equations through his contacts with engineering departments at Princeton. This interest remained throughout his life. To further develop the subject he formed a project on differential equations at Princeton from 1946–1953, under the sponsorship of ONR (which lasted until 1958 but phased out after his retirement in 1953). As a result of this project, the subject of differential equations began to take on a new meaning. Lefschetz himself said that there emanated a number of capable fellows; also, a book: *Differential Equations—Geometric Theory* (1957). In the western world, this book continues to serve as the model for the presentation of the geometric theory of differential equations.

During this period, Lefschetz consistently monitored the Russian literature, translated, and encouraged the translations of significant works. I mention in particular the books of Nemitskii and Stepanov, Krylov and Bogolinbov, and the republication at Princeton of the 1891 thesis of Liapunov.

After beginning his career in functional analysis, LaSalle shifted his interest to differential equations and control theory during World War II. He was an instructor at the Radar School at MIT and worked with physicists at Cornell designing a magnetron for a naval communications system.

In 1947–48, LaSalle was at Princeton on Lefschetz's project in differential equations. They developed a lasting friendship, which led to mutual cooperation on ventures that were most influential in the development of control theory and dynamical systems as we know it today. They were instrumental in the creation of research institutes, journals, and the timely organization of international meetings. Both were leaders in the scientific community, unselfish with their time and always concerned with the general welfare of their colleagues. They were especially influential on young people, stimulating them to perform at their maximum potential.

In 1958, Lefschetz, at age 74, with LaSalle as codirector, formed the Mathematical Center at the Research Institute for Advanced Studies (RIAS), a Division of the Martin Company. The emphasis at this center was on differential equations and control theory with a small core of permanent people (I was fortunate to be one of these). The thrust of the activity came from visitors from around the world—many were present at the Kingston conference. This center gave new life to differential equations and control theory, and the origins of many accomplishments in these fields in the last 30 years can be traced to this group. Other influential activities of the center at RIAS were international meetings in Mexico in 1959 and Colorado Springs in 1961. (In the west, meetings in those days were infrequent or nonexistent.)

After six years of prosperity at RIAS, financial restrictions on non-Project-oriented research forced us to find a new home. In September 1964, the Center for Dynamical Systems at Brown was formed with LaSalle as Director and Lefschetz, at the age of 80, visiting one day a week from Princeton. In Baltimore, he came for two days, but his wife said he had to slow down. The program continued in the same spirit as RIAS, but it was now possible to train young people through the undergraduate, graduate, and postdoctoral programs.

After the death of Lefschetz in 1972, the center at Brown was renamed the Lefschetz Center for Dynamical Systems, in 1974. In the words of LaSalle, it was hoped that "this center will be a lasting memorial to him and that it will remind everyone who works and visits here, or knows of us, that we reside eternally in his debt."

Lefschetz and LaSalle certainly will be remembered for their scientific contributions; however, their influence goes far beyond that. They were able to recognize new areas of basic research, they provided a unique opportunity for the training of young people, they insisted on academic scholarship, they exhibited moral integrity, and they were an inspiration to everyone that knew them. The fields of differential equations and control theory will be indebted eternally to them.

Jack K. Hale

Preface

Thirty Years of Modern Optimal Control

A Historical Perspective

In 1958 Donald Bushaw's paper on the controlled harmonic oscillator [3] appeared in the Princeton University series *Contributions to the Theory of Nonlinear Oscillations*, vol. IV. In this paper, Bushaw presented a main result from his dissertation written under the guidance of the great Solomon Lefschetz. The paper can be taken as the starting point of the somewhat explosive development of modern optimal control theory (at least in the western world) that took place in the 1950s and 1960s. As any date of this kind, 1958 is to some degree artificial; Bushaw's result had been obtained earlier (dissertation finished in 1952 [4]), and significant contributions had appeared already [1,2,5–7]. And this reflects only the western literature, leaving out the important contributions of the Russian school [11]. Nevertheless, it is a good date, also coming exactly 30 years before the Kingston Conference.

Of course, related problems were posed and solved much earlier [8–10,12–14]. The classical calculus of variations, of which optimal control is an extension, is again much older, going back to Newton, Euler, the Bernoulli's, and, to some degree, to antiquity, with the "Problem of Dido" [13]. Still the recognition of optimal control by this name and with its peculiar methods developed mainly (at least to my knowledge) in the 1950s and 1960s.

I was extremely fortunate to enter this field in the mid-1950s myself. I was already interested in nonlinear processes when I discovered the paper by Bellman et al., which led me to a lot of new and surprising results. I was particularly fascinated by the funny shapes of reachable (or attainable) sets and wrote my dissertation about them in Argentina in 1956–58; later I found out that many other people were working on similar problems and that this subject was called "optimal control," a subject I have been devoted to ever since.

As I said before, I consider myself extremely fortunate to have lived through the development of this branch of mathematics. I have seen its making from the inside; getting to know personally many of the great contributors to this field. It is to them, and especially to Lefschetz and LaSalle, that this volume is dedicated.

Finally, I express my thanks to the Office of Naval Research, which with its financial support made the Kingston Conference possible. My thanks go also to The University of Rhode Island for additional support and for hosting the Kingston Conference.

Emilio O. Roxin

REFERENCES

1. Bellman, R., On the application of dynamic programming to the study of control processes, Proc. Symposium Nonlinear Circuit Analysis, Polytech. Inst., Brooklyn (1956), 199–213.
2. Bellman, R., I. Glicksberg and O. Gross, On the "bang-bang" control problem, *Quart. Appl. Math., 14* (1956), 11–18.
3. Bushaw, Donald, Optimal discontinuous forcing terms, in S. Lefschetz (ed.) *Contributions to the Theory of Nonlinear Oscillations*, vol. IV, Princeton Univ. Press, Princeton NJ, (1958), 29–52.
4. Bushaw, Donald, Stevens Institute of Technology, Report 469, Jan. 1953.
5. Flügge-Lotz, Irmgard, *Discontinuous Automatic Control*, Princeton University Press, 1953.
6. Isaacs, Rufus, Rand Reports RM-1391, 1411, 1486 (1954, 1955).
7. LaSalle, J. P., Basic principle of the "bang-bang" servo, *Bull. Amer. Math. Soc., 60* (1954), 154.
8. Marchaud, M. A., Sur les champs des demi-droites et les équations differentielles de premier ordre, *Bull. Soc. Math. France, 63* (1934), 1–38.
9. Marchaud, M. A., Sur les champs continus des demicones convexes et leur integrales, *Compositio Math., 3* (1936), 89–127.
10. McShane, E. J., Generalized curves, *Duke Math. J., 6* (1940), 513–536.
11. Pontryagin, L. S., Optimal Control Processes (in Russian), *Usp. Mat. Nauk, 14,* No. 3 (1959), 3–20, transl. in *Amer. Math. Soc. Translations, 18* (1961), 321–339.
12. Young, L. C., On approximation by polygons in the calculus of variations, *Proc. Royal Soc., A, 141* (1933), 325–341.
13. Young, L. C., *Lectures on the Calculus of Variations and Optimal Control Theory,* W. B. Saunders Co., Philadelphia, London, Toronto, 1969, p. 215.
14. Zaremba, M. S. C., Sur les equations au paratingent, *Bull. Sci. Math., I, 60* (1936), 139–160.

Contents

Contributors

Leonard D. Berkovitz, Professor, Department of Mathematics, Purdue University, West Lafayette, Indiana

Dean Carlson, Professor, Department of Mathematics, Southern Illinois University at Carbondale, Carbondale, Illinois

Lamberto Cesari, Professor, Department of Mathematics, University of Michigan, Ann Arbor, Michigan

Constantin Corduneanu, Professor, Department of Mathematics, The University of Texas at Arlington, Arlington, Texas

M. Corless, Professor, School of Aeronautics & Astronautics, Purdue University, West Lafayette, Indiana

Hector Fattorini, Professor, Department of Mathematics, University of California at Los Angeles, Los Angeles, California

Halina Frankowska, Professor, CNRS-Centre National de la Recherche Scientifique, CEREMADE, Universite Paris-Dauphine, Paris, France

Andrzej Fryszkowski, Professor, Institute of Mathematics, Technical University of Warsaw, Warsaw, Poland

John Gregory, Professor, Department of Mathematics, Southern Illinois University at Carbondale, Carbondale, Illinois

Istvan Györi, Professor, Department of Mathematics, The University of Rhode Island, Kingston, Rhode Island

Jack K. Hale, Division of Applied Mathematics, Brown University, Providence, Rhode Island

Sherry E. Hammel, Naval Underwater Systems Center, Newport, Rhode Island

Henry Hermes, Professor, Department of Mathematics, University of Colorado, Boulder, Colorado

Bronislaw Jakubczyk, Professor, Institute of Mathematics, Polish Academy of Sciences, Warsaw, Poland

Velimir Jurdjevic, Professor, Department of Mathematics, University of Toronto, Toronto, Ontario, Canada

Efthimios Kappos, Professor, Department of Electrical and Computer Engineering, Northeastern University, Boston, Massachusetts

Charles Kaufman, Professor, Physics Department, The University of Rhode Island, Kingston, Rhode Island

Jacob Kogan, Professor, Department of Mathematics, University of Maryland, Baltimore County Campus, Baltimore, Maryland

G. S. Ladde, Professor, Department of Mathematics, The University of Texas at Arlington, Arlington, Texas

Urszula Ledzewicz-Kowalewska, Professor, Department of Mathematics and Statistics, Southern Illinois University at Edwardsville, Edwardsville, Illinois

S. J. Lee, Department of Mathematics, University of South Florida, Tampa, Florida

George Leitmann, Professor, Department of Mechanical Engineering, University of California at Berkeley, Berkeley, California

Cantian Lin, Department of Mathematics, Southern Illinois University, Carbondale, Illinois

Pan-Tai Liu, Professor, Department of Mathematics, University of Rhode Island, Kingston, Rhode Island

Stanislaw Lojasiewicz, Jr., Professor, Department of Mathematics, University of Rhode Island, Kingston, Rhode Island

Lawrence Markus, Professor, Department of Mathematics, Institute of Technology, University of Minnesota, Minneapolis, Minnesota and Department of Mathematics, University of Warwick, Coventry, England

G. Müller, Physics Department, The University of Rhode Island, Kingston, Rhode Island

M. Z. Nashed, Department of Mathematical Sciences, University of Delaware, Newark, Delaware

O. Sirisaengtaksin, Professor, Department of Applied Mathematical Sciences, University of Houston-Downtown, Houston, Texas

Eduardo D. Sontag, Professor, Department of Mathematics, Rutgers University, New Brunswick, New Jersey

Pablo Spinadel, Flexible Automation Group, Seibersdorff, Austria, Technical University, Vienna, Austria, and University of Buenos Aires, Buenos Aires, Argentina

Vera W. de Spinadel, Professor, University of Buenos Aires, Buenos Aires, Argentina

N. Srivastava, Physics Department, The University of Rhode Island, Kingston, Rhode Island

Hector J. Sussmann, Professor,Department of Mathematics,Rutgers University, New Brunswick, New Jersey

W. Szpunar, Department of Mathematics, Rochester Institute of Technology, Rochester, New York

Roberto Triggiani, Professor, Applied Mathematics Department, University of Virginia, Charlottesville, Virginia

Jack Warga, Professor, Department of Mathematics, Northeastern University, Boston, Massachusetts

George Yin, Professor, Department of Mathematics, Wayne State University, Detroit, Michigan

Laurence C. Young, Professor, Center for the Mathematical Sciences, University of Wisconsin at Madison, Madison, Wisconsin

Modern Optimal Control

Thirty Years of Differential Games

Leonard D. Berkovitz, Professor, Department of Mathematics, Purdue University, West Lafayette, Indiana

1 INTRODUCTION

In this paper we sketch those developments in differential games over the past thirty years or so that have interested us and that are related to optimal control theory. We do not claim to present a complete historical survey. Work that is not discussed should not be interpreted necessarily as being of limited interest or value.

An intuitive formulation of a zero-sum two person differential games is the following. The state $x(t)$ of the game at time t is a vector in R^n determined by a system of differential equations

$$\frac{dx}{dt} = f(t, x, y, z) \qquad x(t_0) = x_0, \tag{1.1}$$

where $y = y(t)$ is a vector in R^r chosen from some set Y at each instant of time by Player I and $z = z(t)$ is a vector in R^s chosen from some set Z at each instant of time by Player II. The choice of $y(t)$ is governed by a "strategy" U selected by Player I prior to the start of play. The choice of $z(t)$ is governed by a strategy V selected by Player II prior to the start of play. Play proceeds from an initial point (t_0, x_0) until the point $(t, \phi(t))$, where ϕ is the solution of (1.1), reaches some preassigned terminal set $\mathcal{T}$. The point at which $(t, \phi(t))$ reaches $\mathcal{T}$ is called the terminal point and is denoted by $(t_f, \phi(t_f))$ or (t_f, x_f). The payoff is

$$P(t_0, x_0, U, V) = g(t_f, x_f) + \int_{t_0}^{t_f} f^0(s, \phi(s), y(s), z(s))ds. \tag{1.2}$$

This paper was prepared under National Science Foundation Grant No. DMS 8700813

Player I wishes to choose U so as to maximize P while Player II wishes to choose V so as to minimize P. Such games are called games of survival. If $g \equiv 0$ they are called games of generalized pursuit and evasion. If the terminal set $\mathfrak{T}$ is the hyperplane $t = T$, the game is said to be a game of fixed duration. The payoff (1.2) then becomes

$$P(t_0, x_0, U, V) = g(x_f) + \int_{t_0}^{T} f^0(s, \phi(s), y(s), z(s))ds. \tag{1.3}$$

If

$$\sup_U \inf_V P(t_0, x_0, U, V) = \inf_V \sup_U P(t_0, x_0, U, V), \tag{1.4}$$

then we denote this number by $W(t_0, x_0)$ and say that the game has value equal to $W(t_0, x_0)$. A pair of strategies U^*, V^* is said to be a saddle point if for all U and V over which the sup and inf are taken,

$$P(t_0, x_0, U, V^*) \leq P(t_0, x_0, U^*, V^*) \leq P(t_0, x_0, U^*, V).$$

The strategies U^* and V^* are called optimal strategies.

Differential games as described above are models of zero–sum conflicts in which the state of the system is governed by differential equations controlled by two antagonists. Pursuit and evasion problems, which arise in military applications, are examples of problems that can be modelled as differential games. In a pursuit and evasion problem an evader E wishes to escape capture, or interception, by a pursuer P. Capture can be defined as coincidence of the position vectors of P and E, or by the attainment of a preassigned distance between P and E. The game terminates if capture occurs prior to some preassigned time T or at time T, if capture has not occurred prior to T. The evader wishes to maximize the time of capture, the pursuer wishes to minimize. The time T is chosen so as to reflect the situation that if T units of time have elapsed, the evader has reached his destination or target.

Military tactical problems which are concerned with the allocation of resources among different tasks by each of two combatants are further examples of problems that can be modelled as differential games [**3**].

Differential games can also be used to study either optimal control problems or problems of system design in which the system is subject to unknown disturbances, and the disturbances are considered to be chosen by a malevolent Nature, who is assigned the role of Player II. In the optimal control problem Player I wishes to choose controls that guarantee the best possible performance in the event of the worst possible disturbance. In the design problem, Player I chooses the design parameters to give the best possible system performance in the event of the worst possible disturbance.

A satisfactory theory of differential games should address the following problems:

(1) Formulate the differential game as a two–person zero–sum game in the sense of game theory. This involves defining the notion of strategy precisely and defining the payoff as a function of the strategies.

(2) For the given definitions of strategy and payoff does the game have a value?

(3) If the game has a value, does it have a saddle point?

(4) If the game has a saddle point, determine at least one pair of optimal strategies U^* and V^* that constitute the saddle point.

2 EARLY YEARS

The study of differential games was initiated by Rufus Isaacs in the early 1950's, and the first results of his work were published in a series of RAND Corporation memoranda [**22**]. These were later expanded into a book [**23**]. In his pioneering work, Isaacs tacitly assumed that the notion of strategy had been defined, that a saddle point existed and that the game had a continuously differentiable value function. To paraphrase the essentials of his approach most economically, we assume that the payoff in (1.2) has $f^0 \equiv 0$. There is no loss of generality in this assumption, since (1.2) can be written as a payoff with $f^0 \equiv 0$ by introducing a new coordinate x^0, a new equation $dx^0/dt = f^0(t,x,y,z)$, $x^0(t_0) = 0$, and by taking the payoff to be $g(t_f, x_f) + x^0_f$.

Let (t,x) be a point that is reached in the course of play, and let $W(t,x)$ denote the value of the game with initial point (t,x). Then, assuming requisite smoothness, along a trajectory resulting from choices y and z by the Players at (t,x) we have

$$dW = W_t(t,x) + \langle W_t(t,x), f(t,x,y,z)\rangle, \tag{2.1}$$

where $\langle \ , \rangle$ denotes inner product. Clearly, Player I wishes to choose y so as to maximize (2.1) while Player II wishes to choose z so as to minimize (2.1). It follows from the Principle of Optimality, which Isaacs called the Tenet of Transition, and the assumption that $f^0 \equiv 0$, that for optimal choices (y^*, z^*) at this point, $dW = 0$. Thus

$$W_t(t,x) + \langle W_x(t,x), f(t,x,y^*,z^*)\rangle = 0. \tag{2.2}$$

It then further follows from (2.1) and the Principle of Optimality that

$$W_t(t,x) + \langle W_x(t,x), f(t,x,y^*,z)\rangle \geq 0 \tag{2.3.i}$$

for all z in Z and

$$W_t(t,x) + \langle W_x(t,x), f(t,x,y,z^*)\rangle \leq 0 \tag{2.3.ii}$$

for all y in Y. From these relations and from (2.2) we get that (y^*, z^*) is a saddle point of the game over $Y \times Z$ with payoff $\langle W_x(t,x), f(t,x,y,z)\rangle$.

The last conclusion can be rewritten as follows. Let

$$H^+(t,x,p) = \min_{z \in Z} \max_{y \in Y} \langle p, f(t,x,y,z)\rangle$$
$$H^-(t,x,p) = \max_{y \in Y} \min_{z \in Z} \langle p, f(t,x,y,z)\rangle.$$

We assume that for all (t,x,p) in $\mathbf{R} \times \mathbf{R}^n \times \mathbf{R}^n$

$$H^+(t,x,p) = H^-(t,x,p), \tag{2.4}$$

and denote the common value by $H(t,x,p)$. Condition (2.4) is now known as the Isaacs condition. The arguments of the previous paragraph show that if a saddle point and a continuously differentiable value exist, then the value is a solution of

$$\begin{aligned} W_t + H(t,x,W_x) &= 0 \\ W(t_f, x_f) &= g(t_f, x_f). \end{aligned} \tag{2.5}$$

The equation and boundary condition (2.5) are known as the Isaacs equation.

Most of Isaacs' work was concerned with solving (2.5) in specific examples, essentially by the method of characteristics. He also attempted to classify the structures of the fields of optimal trajectories that arise. Although many of the interesting examples that he and others considered failed to have continuously differentiable values, they were still able to treat some of the examples by continuing across manifolds of discontinuity.

In [**2**], W. Fleming and this writer attempted to establish rigorously, in the case of differential games of fixed duration in the plane, some of the formulas and methods used by Isaacs. The saddle point condition

$$P(t,x,U,V^*) \leq P(t,x,U^*,V^*) \leq P(t,x,U^*,V) \tag{2.6}$$

suggested that if the game had a value for every initial point (t,x) and if a saddle point existed, then the optimal trajectories would be solutions of a minimum and a maximum problem in the calculus of variations. The presence of constraints on the controls seemed to put the problem outside of the standard theory. The introduction of slack variables, as was done by Valentine [**27**], enabled us to apply known necessary conditions in the calculus of variations to this problem. This approach was extended to problems in $\mathbf{R}^n$ in [**5**] and led to this writer's paper linking the classical calculus of variations and the Pontryagin maximum principle [**4**]. A more definitive statement of necessary conditions conditions that saddle points must satisfy was given in [**6**] in control formulation. In all of this work it was assumed that an equilibrium point (U^*,V^*) existed as in (2.6), that the strategies U and V were piecewise C' functions of (t,x) and that the field of trajectories resulting from (U^*,V^*) had a special structure.

3 MIDDLE YEARS

In the early 1960's Fleming, in [**18**] [**19**], addressed the problem of game formulation and existence of value. In the late 1960's and early 1970's Friedman also addressed these problems, as did Elliott and Kalton. Friedman's contributions are summarized in [**20**] and [**21**], and those of Elliott and Kalton in [**16**]. Another paper addressing this problem which was influential is that of Varaiya and Lin [**28**].

In the work of Fleming, Friedman and Elliott and Kalton, the direct definition of a game as is indicated by the description in the introduction is abandoned. Instead, indirect definitions are used which involve choices of y and z at discrete times, the introduction of small noise terms and passages to the limit, first as the lengths of times between choices tend to zero and then as the noise tends to zero. The arguments involved are extremely technical and non–elementary.

The principal results of all of these investigations were the following. In games of fixed duration, if the Isaacs condition (2.4) holds, then the games have values and saddle points. Under reasonable assumptions on the data of the problem, the value is uniformly Lipschitz continuous on bounded sets and satisfies the Isaacs equation (2.5) at points of differentiability. In games of generalized pursuit and evasion and in games of survival, if the terminal set has a certain smooth structure and if a capturability condition, which essentially says that near the boundary either player can force termination, holds, then all of the above results hold, with the exception of the existence of a saddle point.

The definitions of strategy and value in the approaches discussed above are such that the results obtained generally do not seem to lead to solution methods. Friedman in **[20]** does consider problems in which the value function is $C^{(2)}$ and treats some examples under this assumption. If the value function is $C^{(2)}$, however, the Isaacs approach can be justified rigorously.

In the Soviet Union, parallel to these developments and independent of them, Krasovskii and Subbotin and their students also developed a theory of differential games. Their work is described in **[25]**. For a translation into English, which incorporates later developments and improvements, see **[26]**. This work also involves choices at discrete times and passage to the limit, but the approach is less technical, more direct and more geometric than the Western work. We shall say more about this work in the next section.

From the mid 1960's onward, Pontryagin and his collaborators actively studied pursuit and evasion problems. Since their work is not in the framework of a zero–sum two person game, we do not include it in this survey.

3 RECENT PERIOD

From the late 1970's to the early 1980's, the theoretical development of differential games was dormant in the U.S., although it was pursued actively in the Soviet Union by Krasovskii and Subbotin and their collaborators. Interest in the theoretical aspects of differential games was revived in this country in the early 1980's as a result of the introduction of the notion of viscosity solution of a first order partial differential equation.

See [14], [15]. It became a relatively easily matter to show that in games of fixed duration, the upper values as defined by Fleming, Friedman and Elliott and Kalton are viscosity solutions of the equation (2.5) with H replaced by H^+. Also, all of the lower values of these games are viscosity solutions of the equation (2.5) with H replaced by H^-. If the Isaacs condition is assumed to hold, then $H = H^+ = H^-$, and both the upper values and the lower values are viscosity solutions of the Isaacs equation and boundary condition (2.5). From uniqueness theorems for viscosity solutions of first order partial differential equations, it follows that the upper and lower values are equal. Hence the values exist in the Fleming sense, the Friedman sense and the Elliott and Kalton sense. Moreover, all of these values are equal. See [1], [15], [17], [24]. We emphasize that the use of the uniqueness theory of viscosity solution has done away with the need for very complicated and technical arguments involving partial differential equations and stochastic processes in establishing that the various values exist, are solutions of the Isaacs equation in some sense, and are equal to each other.

In the early 1980's, this writer again became interested in differential games. His objective was to develop a simpler theory than those of Fleming, Friedman or Elliott and Kalton. This theory also involves choices at discrete times and passages to the limit, but in a more direct way than that of the other authors. It is hoped that this more direct approach will make it amenable to application in specific examples. Our definition of strategy is a combination of "K strategies" discussed by Isaacs [23] and Friedman's lower strategies [20]. Our definition of payoff follows that of Krasovskii and Subbotin [25]

To simplify our description of the game formulation, we suppose that it is a game of fixed duration with terminal payoff g; i.e. $f^0 \equiv 0$. A strategy Γ for Player I is a sequence of instructions $\{\Gamma_n\}$ and a strategy Δ for Player II is a sequence of instructions $\{\Delta_n\}$. For each $n = 1, 2, 3, \ldots,$ Γ_n consists, first, of a choice of a partition of the interval $[\tau, T]$, say $\tau = \tau_0 < \tau_1 < \cdots < \tau_{p(n)} = T$, where $p(n) = 0(n)$ and $\max[\tau_i - \tau_{i-1}; i = 1, \ldots, p] = 0(\frac{1}{n})$. These are the times at which Player I will make his decisions. Similarly, Δ_n consists first of a similar partition $\tau = \sigma_0 < \sigma_1 < \cdots < \sigma_{q(n)} = T$ of $[\tau, T]$. The instruction Γ_n also choose a control $u_{n,1}$ on $[\tau_0, \tau_1)$, and the instruction Δ_n chooses a control $v_{n,1}$ on $[\tau_0, \sigma_1)$. Suppose now for definitions that $\tau_1 < \sigma_1$. The instruction Γ_n then chooses a control $u_{n,2}$ for Player I on $[\tau_1, \tau_2)$ on the basis of the choices $u_{n,1}, v_{n,1}$ restricted to $[\tau_0, \tau_1)$. (If

$\sigma_1 < \tau_1$, then the instruction Δ_n chooses a control $v_{n,2}$ for Player II on $[\sigma_1, \sigma_2)$ based on the choices $u_{n,1}, v_{n,1}$ restricted to $[\tau_0, \sigma_1)$.) Again if, say, $\tau_2 < \sigma_1$, then Γ_n chooses a control $u_{n,3}$ on $[\tau_2, \tau_3)$ based on the choices $u_{n,1}, u_{n,2}, v_{n,1}$ restricted to $[\tau_0, \tau_2)$. If now, say, $\sigma_1 < \tau_3$, then Δ_n chooses $v_{n,2}$ on $[\sigma_1, \sigma_2)$ based on the choices $u_{n,1}, u_{n,2}, u_{n,3}, v_{n,1}$ restricted to $[\tau_0, \sigma_1)$. In other words, at time $\tau_i, i = 1, \ldots, p-1$ the instruction Γ_n chooses a control for Player I on $[\sigma_i, \sigma_{i+1})$ based on the concatenations of the choices of controls for Player I on $[\tau_0, \tau_i)$ and the concatenation of the choices of controls for Player II on $[\tau_0, \tau_i)$. At times σ_i, $i = 1, \ldots, q, \Delta_n$ chooses a control for Player II on $[\sigma_i, \sigma_{i+1})$ on the basis of the controls chosen for the two players on $[\tau_0, \sigma_i)$. By concatenating the $u_{n,i}$'s $i = 1, \ldots, p$ we obtain a control u_n for Player I on $[\tau, T]$. By concatenating the $v_{n,i}$'s $i = 1, \ldots, q$ we obtain a control v_n on $[\tau, T]$.

Under reasonable hypotheses on the data of the problem, the differential equation

$$x' = f(t, x, u_n(t), v_n(t)) \qquad x(\tau) = \xi$$

has a unique solution $\varphi_n(\cdot, \tau, \xi, u_n, v_n)$ on $[\tau, T]$. We call the sequence $\{\varphi_n\}$ so obtained the sequence of n–th stage trajectories resulting from (Γ, Δ). Under the usual assumptions on the data of the problem, the sequence $\{\varphi_n\}$ is compact in the uniform topology on $[\tau, T]$. We denote a uniform limit of a subsequence by $\varphi[\ \ , \tau, \xi, \Gamma, \Delta]$ and call it a motion resulting from (Γ, Δ), following Krasovskii and Subbotin [**25**]. Also following [**25**], we define the payoff to be the set valued function $g(\varphi[T; \tau, \xi, \Gamma, \Delta])$; here $\varphi[\ \ , \tau, \xi, \Gamma, \Delta]$ ranges over all motions resulting from (Γ, Δ).

As in the other theories, the principal results are the following. In games of fixed duration, if the Isaacs condition is satisfied, then the games have values and saddle points. The value is uniformly Lipschitz continuous on bounded sets and is a viscosity solution of the Isaacs equation (2.5). In games of generalized pursuit and evasion and in games of survival in which the terminal set has a certain structure and a capturability condition holds, similar statements hold for the value function. The existence of saddle points, however, cannot be asserted. If the Isaacs condition does not hold, then the results described can be obtained if instead of choosing ordinary controls, the Players choose relaxed controls. For a summary of our recent results, see [**13**]. For the details see [**7**], [**8**], [**9**], [**10**], [**11**], [**12**].

REFERENCES

[1] E. N. BARRON, L. C. EVANS, R. JENSEN, *Viscosity solutions of Isaacs' equations and differential games with Lipschitz controls*, J. Differential Equations **53** (1984), 213–233.

[2] L. D. BERKOVITZ AND W. H. FLEMING, *On differential games with integral payoff*, *Contributions to the Theory of Games Vol. III*, Annals of Math Study No. 39 (1957), 413–435, Princeton University Press, Princeton, NJ.

[3] L. D. BERKOVITZ AND M. DRESHER, *A game theory analysis of tactical air war*, Opns. Res. **7** (1959), 599–620.

[4] L. D. BERKOVITZ, *Variational methods in problems of control and programming*, J. Math. Anal. Appl. **3** (1961), 145–169.

[5] L. D. BERKOVITZ, *A variational approach to differential games*, Annals of Math Study No. 52 (1964), 127–174, Princeton University Press, Princeton, NJ.

[6] L. D. BERKOVITZ, *Necessary conditions for optimal strategies in a class of differential games and control problems*, SIAM J. Control **5** (1967), 1–24.

[7] L. D. BERKOVITZ, *The existence of value and saddle point in games of fixed duration*, SIAM J. Control **23** (1985), 172–196.

[8] L. D. BERKOVITZ, *Differential games of generalized pursuit and evasion*, SIAM J. Control **24** (1986), 361–373.

[9] L. D. BERKOVITZ, *Differential games without the Isaacs condition*, in "Recent Advances in Communication and Control Theory," R. E. Kalman, G. I. Marchuk, A. E. Ruberti and A. J. Viterbi, eds., Optimization Software, New York, 1987, pp. 308–336.

[10] L. D. BERKOVITZ, *Characterizations of the values of differential games*, Appl. Math. Optim. **17** (1988), 177–183.

[11] L. D. BERKOVITZ, *Differential games of survival*, J. Math. Anal. Appl. **129** (1988), 493–504.

[12] L. D. Berkovitz, *Correction to 'The existence of value and saddle point in games of fixed duration*, SIAM J. Control **26** (1988), 740–742.

[13] L. D. Berkovitz, *A survey of recent results in differential games*, in "Differential Games and Applications," T. Basar and P. Bernhard, eds., Lecture Notes in Control and Information Sciences, Springer–Verlag, Berlin, Heidelberg, New York, 1988, pp. 35–50.

[14] M. G. Crandall and P. L. Lions, *Viscosity solutions of Hamilton–Jacobi equations*, Trans. Amer. Math. Soc. **277** (1983), 1–42.

[15] M. G. Crandall, L. C. Evans and P. L. Lions, *Some properties of viscosity solutions of Hamilton–Jacobi equations*, Trans. Amer. Math. Soc. **282** (1984), 487–502.

[16] R. J. Elliott and N. J. Kalton, *The existence of value in differential games*, Mem. Amer. Math. Soc. **126** (1972).

[17] L. C. Evans and P. E. Souganidis, *Differential games and representation formulas for solutions of Hamilton–Jacobi–Isaacs equations*, Indiana Univ. Math. J. **33** (1984), 773–795.

[18] W. H. Fleming, *The convergence problem for differential games*, J. Math. Anal. Appl. **3** (1961), 102–116.

[19] W. H. Fleming, *The convergence problem for differential games II*, Annals of Mathematics Study No. 52 (1964), 195–210, Princeton University Press, Princeton, NJ.

[20] A. Friedman, "Differential Games," Wiley–Interscience, New York, 1971.

[21] A. Friedman, Regional Conference Series in Mathematics No. 18 (1974), "Differential games," American Mathematical Society, Providence, RI,.

[22] R. Isaacs, *Differential Games I, II, III, IV*, The RAND Corporation, Research Memoranda RM–1391, RM–1399, RM–1411, RM–1486 (1954–1955).

[23] R. Isaacs, "Differential Games," John Wiley and Sons, New York, London, Sydney, 1965.

[24] H. Ishi, *Uniqueness of unbounded viscosity solutions of Hamilton–Jacobi equations*, Indiana Univ. Math. J. **33** (1984), 721–748.

[25] N. N. Krasovskii and A. I. Subbotin, "Positional Differential Games," Nauka, Moscow, 1974. (Russian)

[26] N. N. Krasovskii and A. I. Subbotin, "Game–Theoretical Control Problems," Springer–Verlag, New York, Berlin, Heidelberg, 1988. Translation and revision of "Positional Differential Games", Nauka, Moscow 1974, Samuel Kotz, translator.

[27] F. A. Valentine, *The problem of Lagrange with differential inequalities as added side conditions*, Contributions to the Calculus of Variations, 1933–37, University of Chicago Press, Department of Mathematics, University of Chicago, Chicago.

[28] P. Varaiya and J. Lin, *Existence of saddle points in differential games*, SIAM J. Control **7** (1969), 141–157.

Some Concepts of Optimality for Infinite Horizon Optimal Control and Their Interrelationships

Dean Carlson, Professor, Department of Mathematics, Southern Illinois University at Carbondale, Carbondale, Illinois

I. INTRODUCTION

The study of optimal control problems defined on unbounded time intervals, commonly called infinite horizon optimal control problems, has its birth in the 1928 paper of F. Ramsey [12]. In that work, he proposed a model of economic growth consisting of maximizing the accumulated utility of a nation, described by an improper integral $\int_0^\infty f^0(u(t))dt$, over all arcs satisfying the

---

This research was supported by a grant from the National Science Foundation (Grant No. DMS-8700706) and by the Southern Illinois University at Carbondale Summer Research Fellowship Program.

ordinary differential equation $\dot{x}(t) = f(x(t)) - \lambda x(t) - u(t)$ and initial condition $x(0) = x_0$.

The control u represents consumer consumption, the trajectory x represents capital stock, $f:[0,\infty) \to [0,\infty)$ is a production function, and $\lambda > 0$ denotes the rate of depreciation of capital.

The possible divergence of the objective functional did not go unnoticed by Ramsey, and he recognized that this presents a severe difficulty. To overcome this hardship he introduced the notion of the optimal steady state, which he called the maximum sustainable rate of enjoyment or "Bliss". Specifically he let $B = \max_{k \geq 0}\{f^0(u) : u = f(k) - \lambda k\}$, and replaced the original functional by $J(u) = \int_0^\infty [f^0(u(t)) - B]dt$. He then considered only those control-trajectory pairs $\{x, u\}$ for which J(u) is finite. The rest of his paper is devoted to developing necessary conditions for optimality, through the Euler-Lagrange equations, obtaining what is now referred to as Ramsey's Rule.

After the advent of optimal control theory it was shown in 1962 by Chakravarty [3] that Ramsey's rule was not sufficient. In particular, he presents an example in which Ramsey's rule provides the "minimum" rather than the "maximum" and moreover the accumulated utility minus "bliss", J(u), diverges. This discovery was surprising in that if one considers the finite horizon analog of Ramsey's model, the convexity hypotheses typically assumed are strong enough to insure that the necessary conditions are sufficient as well. As a consequence of this result, researchers were led to reevaluate Ramsey's method. This led to a heirarchy of new concepts of optimality. These first new concepts of optimality appear in a 1965 paper of von Weizäcker [14] where he introduces the notion of overtaking or catching up optimality. Since 1965 a variety of similar concepts of optimality have been introduced into the literature and they have been investigated extensively. For a detailed survey of these results the reader may consult the monograph of Carlson and Haurie [2]. (see also Stern [13]).

The concepts of optimality alluded to above were introduced to circumvent the divergence of the objective functional, and as a consequence they investigate the limiting behavior of the objective functional, $J_T(u) = \int_0^T f^0(u(t))dt$, as $T \to \infty$. For brevity we will refer to these concepts as Limiting-Objective Type or more briefly (LOF) types of optimality. In the late 1950's, the works of Madigliani and Hohn [10], Charnes, Cooper, and Mellon [4], and Wagner and Whitin [15] introduced the concepts of decision and forecast horizons in applications to production planning. These ideas lead naturally to different concepts of optimality (known as agreeable plans) which, instead of examining the limiting behavior of the objective, seek a comparison between the optimal value of corresponding finite horizon problems. For brevity we refer to these concepts as Finite-Horizon Solution or (FHS) types of optimality. For a discussion of the existing literature in this area we refer the reader to the paper of Morton [11] as well as the 1973 paper of Hammond and Mirrlees [6] (see also [9]).

A natural question which now arises is the relationships between (LOF) and (FHS) types of optimality. This question has been treated by Hammond and Kennan [7] in a discrete time setting as well as by Haurie and Sethi [9] in a continuous time control theory setting. In [7] a link is established between the two classes and examples are given which show these two classes are distinct. Further in [9] it is demonstrated that both classes of optimality enjoy a common "principle of optimality", and as a consequence any optimal solution must satisfy the infinite horizon Pontryagin Maximum principle due to Halkin [5]. In [9] however no link is established between the (LOF) and (FHS) classes and it is remarked that a direct comparison cannot be made. It is the belief of the present author that this lack of comparison is attributable to the lack of uniformity with respect to feasible elements in the (LOF) definitions whereas in (FHS) such a uniformity is satisfied. Indeed in [1] it is shown that by imposing this uniformity in the (LOF) types of optimality a comparison can be made.

With the above remarks, the object of the remainder of this paper is to briefly describe the concepts of optimality discussed above and present the relationships between them. The plan of our paper is as follows. In Section 2 we describe the general infinite horizon model. Section 3 is devoted to defining the various concepts of optimality and their interrelationships. Finally we given some concluding remarks in Section 4.

II. THE MODEL

We begin by considering a control system of the form

$$\dot{x}(t) = f(t,x(t),u(t)), \text{ a.e. } 0 \leq t \tag{2.1a}$$

$$x(0) = x_0, \tag{2.1b}$$

$$(t,x(t)) \in X, \text{ for } 0 \leq t, \tag{2.1c}$$

$$u(t) \in U(t,x(t)), \text{ a.e. } 0 \leq t; \tag{2.1d}$$

in which $x_0 \in \mathbb{R}^n$ is fixed, $X \subset [0,\infty) \times \mathbb{R}^n$ is closed, $U: X \to 2^{\mathbb{R}^m}$ has a closed graph $M = \{(t,x,u): (t,x) \in X,\ u \in U(t,x)\}$ and $f: M \to \mathbb{R}^n$ is continuous and satisfies the growth condition

$$|f(t,x,u)| \leq K(t) + N(t)|x|, \tag{2.2}$$

with K and N locally Lebesgue integrable functions on $[0,\infty)$, $(t,x,u) \in M$. We let S_∞ denote the set of all pairs of functions $\{x,u\}: [0,\infty) \to \mathbb{R}^{n+m}$, with x locally absolutely continuous and u Lebesgue measurable, which satisfy the control system (2.1).

The performance of the system at any finite time $T > 0$ is measured by the integral functional

$$J_T(x,u) = \int_0^T f^0(t,x(t),u(t))dt, \tag{2.3}$$

where $f^0: M \to \mathbb{R}$ is a given lower semicontinuous function. This functional is well defined for all $T > 0$ whenever $\{x,u\} \in S_\infty$ is such that $t \to f^0(t,x(t),u(t))$ is locally Lebesgue integrable. We let $A_\infty \subseteq S_\infty$ denote the set of all such pairs and refer to these as admissible. The optimization problem is to "minimize" $J_\infty(x,u) = \lim_{T\to\infty} J_T(x,u)$ over $\{x,u\} \in A_\infty$.

III. CONCEPTS OF OPTIMALITY

With the notations of the previous section we give the following definitions.

DEFINITION 2.1 ((LOF)-types of optimality). A pair $\{x^*,u^*\} \in A_\infty$ is called

(i) *Strongly optimal* (SO) if $J_\infty(x^*,u^*) = \lim_{T\to\infty} J_T(x,u)$ is finite and if for every $\{x,u\} \in A_\infty$, there exists $\hat{T} = \hat{T}(x,u) \geq 0$ so that for all $T \geq \hat{T}$,

$$J_T(x^*,u^*) \leq J_T(x,u), \tag{2.4}$$

(ii) *Overtaking (or catching-up) optimal* (OO) if for each $\epsilon > 0$ and $\{x,u\} \in A_\infty$, there exists $\hat{T} = \hat{T}(\epsilon,x,u) \geq 0$ such that for all $T \geq \hat{T}$

$$J_T(x^*,u^*) \leq J_T(x,u) + \epsilon, \tag{2.5}$$

(iii) *Weakly overtaking (sporadically catching up) optimal* (SC) if for each $\epsilon > 0$, $\{x,u\} \in A_\infty$ and $\hat{T} \geq 0$, there exists $T = T(\epsilon,x,u) \geq \hat{T}$ such that (2.5) above holds,

(iv) *Uniformly strongly optimal* (USO) if $J_\infty(x^*,u^*)$ is finite and if there exists $\hat{T} \geq 0$ such that for all $T \geq \hat{T}$ and all $\{x,u\} \in A_\infty$ the inequality (2.4) holds,

(v) *Uniformly overtaking optimal* (UOO) if for each $\epsilon > 0$ there exists $\hat{T} = \hat{T}(\epsilon) \geq 0$ such that for all $T \geq \hat{T}$ and all $\{x,u\} \in A_\infty$ the inequality (2.5) holds,

(vi) *Uniformly weakly overtaking optimal* (USC) if for each $\epsilon > 0$ and $\hat{T} \geq 0$ there exists $T = T(\epsilon) \geq \hat{T}$ such that for all $\{x,u\} \in A_\infty$ the inequality (2.5) holds.

The original ideas for the above definitions appear in von Weizäcker [14] with the uniformity notions appearing later in Hammond and Kennan [7] and Stern [13]. The relationships between these various notions is fully treated in [13] and we give a summary in the following proposition.

PROPOSITION 2.1. (a) If $\{x^*,u^*\}$ is optimal with respect to the optimality concept (i) or (ii) above it is optimal with respect to the succeeding concepts (ii) and (iii) (i.e., (i) implies (ii) and (ii) implies (iii)).

(b) If $\{x^*,u^*\}$ is optimal with respect to the concepts (iv) or (v) it is optimal with respect to the succeeding concepts (v) and (vi).

(c) The uniform concepts of optimality imply their nonuniform counterparts (e.g., (iv) implies (i)).

PROOF. See Stern [13].

To define the (FHS) types of optimality we introduce the following notation. Given $\{x,u\} \in A_\infty$ and $\tau > 0$ we let $\{x^\tau,u^\tau\}$ denote the restriction of $\{x,u\}$ to $[0,\tau]$, and let $A(x,\tau) = \{\{y,v\} \in A_\infty\colon y^\tau = x^\tau\}$. For $T > 0$ let

$$W(T) = \inf\{J_T(x,u)\colon \{x,u\} \in A_\infty\}$$

and for $\{x^*,u^*\} \in A_\infty$ and $\tau > 0$ let

$$W(T;x^*,\tau) = \inf\{J_T(x,u)\colon \{x,u\} \in A(x^*,\tau)\}.$$

Clearly, since $A(x^*,\tau) \subseteq A_\infty$, we have $W(T) \leq W(T;x^*,\tau)$ allowing us to give the following definitions.

DEFINITION 3.2. ((FHS)-types of optimality). A pair $\{x^*,u^*\} \in A_\infty$ is called

(i) *Decision horizon optimal* (DH) if for all $\tau > 0$ there exists $\hat{T} = \hat{T}(\tau) \geq \tau$ such that for all $T \geq \hat{T}$,

$$W(T) = W(T;x^*,\tau) \tag{2.6}$$

(ii) *Agreeable* (A) if for all $\tau \geq 0$

$$\lim_{T\to\infty} [W(T;x^*,\tau)-W(T)] = 0 \tag{2.8}$$

(iii) *Weakly Agreeable* (WA) if for all $\tau \geq 0$

$$\liminf_{T\to\infty} [W(T;x^*,\tau)-W(T)] = 0. \tag{2.9}$$

The notion of an agreeable pair is apparently due to Hammond and Mirrlees [6] while decision horizon optimality appears in [9]. The remaining concept of weakly agreeable is apparently new but arises naturally when one considers the heirarchy in Definition 3.1. The relationship between the (FHS) optimality concepts is easily established and given below.

PROPOSITION 2.2. If $\{x^*,u^*\}$ is DH-optimal it is agreeable. Further, if $\{x^*,u^*\}$ is agreeable, then it is weakly agreeable.

PROOF. See Haurie and Sethi [9].

We are now ready to establish a link between Definitions (2.1) and (2.2).

PROPOSITION 2.3. Let $\{x^*,u^*\} \in A_\infty$. Then

(a) if $\{x^*,u^*\}$ is (USO)-optimal it is (DH)-optimal,
(b) if $\{x^*,u^*\}$ is (UCO)-optimal it is (A)-optimal,
(c) if $\{x^*,u^*\}$ is (USC)-optimal it is (WA)-optimal.

PROOF. See Carlson [1].

The above propositions establish a connection between (LOF) type optimality and (FHS) type optimality. While our results do not establish a direct link between say (SO) and (DH) we do see that they are related.

IV. CONCLUDING REMARKS

Another link that has been established between the above conncepts of optimality is that they all enjoy the following "principle of optimality".

DEFINITION 2.3. A pair $\{x^*,u^*\}$ satisfies the *Principle of Optimality* if for each T > 0 one has $J_T(x^*,u^*) \leq J_T(x,u)$ for all $\{x,u\} \in A_\infty$ satisfying $x(T) = x^*(T)$.

For a proof of the above fact we refer the reader to Carlson [1; Proposition 3.4]. The significance of this last observation is realized through the fact that any pair $\{x^*,u^*\}$ enjoying the principle of optimality also satisfies the Maximum Principle given by Halkin [5]. Unfortunately there are no known neccessary conditions which distinguish one type of optimality from another. On the other hand, sufficient conditions for optimality have been given (see e.g. Carlson and Haurie [2] or Hartl [8]) which enable one to distinguish between (SC) optimality and (CO) optimality. These results take the form of a strengthened maximum principle. In [9] it is remarked that similar results are unknown for (DH)-optimality and (A)-optimality. Recently, it has been shown in Carlson [1] that one can strengthen the known sufficiennt conditions for (SC) and (CO) optimality to determine if a given pair is (USC) or (UCO) optimality. In this way one obtains sufficient conditions for a pair $\{x^*,u^*\}$ to be both overtaking (resp. weakly overtaking) optimal and agreeable (resp. weakly agreeable). It should be remarked that all these results further rely on an asymptotic transversality condition which involves the map $t \to p(t)^T(x(t)-x^*(t))$, as $t \to \infty$ in which $p\colon [0,\infty) \to \mathbb{R}^n$ arises out of the maximum principle and $\{x,u\}$ and $\{x^*,u^*\}$ are admissible pairs.

Other results yet to be investigated concern the existence of uniform (LOF) solutions and as a consequence the existence of DH-optimal and agreeable pairs. It has been conjectured that such results can be obtained through a strengthening of the hypotheses in theorems concerning the existence of nonuniform (LOF) type solutions. We refer the reader to [2] for a variety of such results.

REFERENCES

1. Carlson, D. A., Uniform overtaking and weakly overtaking optimal solutions in infinite horizon optimal control: When optimal solutions are agreeable, Journal of Optimization Theory and Applications, to appear.
2. Carlson, D. A. and Haurie, A (1987). *Infinite horizon optimal control, theory and application*, Lecture notes in economics and mathematical systems, 290, Springer-Verlag, New York, New York.
3. Chakravarty, S. (1962). The existence of an optimum savings program, *Econometrica*, 30: 178-187
4. Charnes, A., Cooper, W. W., and Mellon, B. (1955). A model for optimizing production by reference to cost surrogates, *Econometrica*, 23: 307-323.
5. Halkin, H. (1974). Necessary conditions for optimal control problems with infinite horizon, *Econometrica*, 42: 267-272.
6. Hammond, P. J. and Mirrlees, J. A. (1973). Agreeable plans, in *Models of economic growth*, (J. A. Mirrlees and N. H. Stern, eds.), MacMillan, New York, New York, pp. 283-299.
7. Hammond, P. J. and Kennan, J. (1979). Uniformly optimal infinite horizon plans, *International Economic Review*, 20: 283-296.
8. Hartl, R. F. (1986). Arrow-type sufficient optimality conditions for nondifferentiable optimal control problems with state constraints, *Applied Mathematics and Optimization*, 14: 229-247.
9. Haurie, A. and Sethi, S. (1984). Decision and forecast horizons, agreeable plans, and the maximum principle for infinite horizon control problems, *Operations Research Letters*, 3: 261-265.
10. Modigliani, F. and Hohn, F. (1955). Production planning over time and the nature of the expectations and planning horizons, *Econometrica*, 23: 46-66.
11. Morton, T. (1978). Universal planning horizons for generalized convex production scheduling, *Operations Research*, 26: 1046-1058.
12. Ramsey, F. P. (1928). A mathematical theory of saving, *Economic Journal*, 38: 543-559.
13. Stern, L. J. (1984). Criteria for optimality in the infinite-time optimal control problem, Journal of *Optimization Theory and Applications*, 44: pp. 497-5088.
14. Von Weizäcker, C. C. (1965). Existence of optimal programs of accumulation for an infinite time horizoon, *Review of Economic Studies*, 32: 85-103.

15. Wagner, H. M. and Whitin, T. M. (1958). Dynamic version of the economic lot size of models, *Management Science*, 5: 89–96.

Discontinuous Solutions of Bounded Variation to Problems of the Calculus of Variations and of Quasi-Linear Hyperbolic Differential Equations

Integrals of Serrin and Weierstrass

Lamberto Cesari, Professor, Department of Mathematics, University of Michigan, Ann Arbor, Michigan

I. BV FUNCTIONS OF $\nu \geq 1$ INDEPENDENT VARIABLES.

In 1936 Cesari [5] introduced a concept of real functions $z : G \rightarrow \mathbf{R}$, or $z(t)$, or $z(t^1, \ldots, t^\nu)$, of bounded variation (BV) from a domain G of $\mathbf{R}^\nu$. For the case $\nu = 2$, G the rectangle $(a, b; c, d)$, the definition is very simple: we say that z if BV in $G = (a, b; c, d)$ provided $z \in L_1(G)$ and there is a set E of measure zero in G such that the total variation $V_x(y)$ of $z(\cdot, y)$ in (a, b) is of class $L_1(c, d)$, and the total variation $V_y(x)$ of $z(x, \cdot)$ in (c, d) is of class $L_1(a, b)$, where these total variations are computed completely disregarding the values taken by z in E. The number

$$V_0 = V_0(z, G) = \int_a^b V_y(x)\, dx + \int_c^d V_x(y)\, dy$$

may well be taken as a definition of total variation of z in $G = (a, b; c, d)$, (with respect to such a set $E \subset G$ of measure zero). Analogous definitions hold for BV functions $z(t^1, \dots, t^\nu)$ in an interval G of $\mathbf{R}^\nu$.

We omit here the more involved definition of BV functions in a general bounded domain G of $\mathbf{R}^\nu$.

If z is continuous in G, then no set E need be considered and the concept reduces to Tonelli's concept of BV continuous functions. For discontinuous functions, examples show how essential it is to disregard sets E of measure zero in G. On the other hand, the concept obviously concerns equivalent classes in $L_1(G)$.

We may think of $z(t)$, $t \in G \subset \mathbf{R}^\nu$, as defining a nonparametric possibly discontinuous surface, $S : z = z(t)$, $t \in G$, in $\mathbf{R}^{\nu+1}$, and we may take as generalized Lebesgue area $L(S)$ of S the lower limit of the elementary areas $a(\sum)$ of the polyhedral surfaces $\sum : z = Z(t)$, $t \in G$, converging to z pointwise a.e. in G (or in $L_1(G)$). More precisely, if $(\sum_k)$ denotes any sequence of polyhedral surfaces $\sum_k : z = z_k(t)$, $t \in G$, converging to z pointwise a.e. in G (or in $L_1(G)$),we take for $L(S)$ the number, $0 \leq L(S) \leq +\infty$, defined by

$$L(S) = \operatorname*{Inf}_{(\sum_k)} \varliminf_{k \to \infty} a(\textstyle\sum_k).$$

Cesari proved [5] that $L(S)$ is finite if and only if z is BV in G. This shows that the concept of BV functions is independent of the direction of the axes in $\mathbf{R}^\nu$. More than that, the concept of BV functions is actually invariant with respect to $1-1$ continuous transformations in $\mathbf{R}^\nu$ which are Lipschitzian in both directions.

In 1937 Cesari [6a] proved that for $\nu = 2$, $G = (0, 2\pi; 0, 2\pi)$ and z BV in G, then the double Fourier series of z converges to z (by rectangles, by lines and by columns) a.e. in G. Comparable, though weaker, results hold for BV functions of $\nu > 2$ independent variables and their multiple Fourier series [6b].

In 1950 Cafiero [4] and later in 1957 Fleming [15] proved the relevant compactness theorem: any sequence (z_k) of BV functions with equibounded total variations, say $V_0(z_k, G) \leq C$, and equibounded mean values in G, possesses a subsequence z_{k_s} which is pointwise convergent a.e. in G as well as strongly convergent in $L_1(G)$ toward a BV function z.

In 1967 Conway and Smoller [12] used these BV functions in connection with the weak solutions (shock waves) of conservation laws, a class of nonlinear hyperbolic partial differential equations in $\mathbf{R}^+ \times \mathbf{R}^\nu$. Indeed they proved that, if the Cauchy data on $(0) \times \mathbf{R}^\nu$ are locally BV, then there is a unique solution on $\mathbf{R}^+ \times \mathbf{R}^\nu$, also locally BV and satisfying an entropy condition. Without any entropy condition there are in general infinitely many weak solutions. Analogous results for $\nu = 1$ had been obtained before by Oleinik [18]. Later, Dafermos [13] and Di Perna [14] characterized the properties of the BV weak solutions of conservation laws.

Meanwhile, in the fifties, distribution theory became known, and in 1957 Krickeberg [17] proved that the BV functions are exactly those $L_1(G)$ functions whose first order partial derivatives in the sense of distributions are finite measures in G.

Thus a BV function $z(t)$, $t \in G$, G a bounded domain in $\mathbf{R}^\nu$, possesses first order partial derivatives in the sense of distributions which are finite measures μ_j, $j = 1, \dots, \nu$. On the other hand, if we think of the initial definition of z, we see that the set E of measure zero in G has intersection $E \cap \ell$ of linear measure zero on almost all lines ℓ parallel to the axes. Hence z is BV on almost all such straight lines when we disregard the values taken by z on E, and therefore has "usual" partial derivatives $D^j z$ a.e. in G, and these derivatives are functions in G of class $L_1(G)$. We call these $D^j z(t)$, $t \in G$, $j = 1, \dots, \nu$, computed by usual incremental quotients disregarding the values taken by z on E, the generalized first order partial derivatives of z in G.

Much work followed on BV functions in terms of the new definition, that is, thought of as those $L_1(G)$ functions whose first order derivatives are finite measures. We mention here Fleming [15], Volpert [22], Gagliardo [16], Anzellotti and Giaquinta ([1]), and also De Giorgi, Da Prato, Ferro, Caligaris, Oliva, Fusco, Temam. However, there are advantages in using both view points.

Great many properties of BV functions have been proved. To begin with, a "total variation" $V(z, G)$ can be defined globally in terms of functional analysis,

$$V(z, G) = \operatorname{Sup}\left[\left(\int_G f_1 d\mu_1\right)^2 + \dots + \left(\int_G f_\nu d\mu_\nu\right)^2\right]^{1/2},$$

where the Sup is taken for all $f_1, \ldots, f_\nu \in C(\bar{G})$ with $f_1^2 + \ldots + f_\nu^2 \leq 1$.

If (z_k) is a sequence of BV functions on G with equibounded total variations, say $V(z_k, G) \leq C$, and $z_k \to z$ in $L_1(G)$, then z is BV and $V(z, G) \leq \underline{\lim}_{k\to\infty} V(z_k, G)$.

The question of the existence of traces $\gamma : \partial G \to \mathbf{R}$ for BV functions $z : G \to \mathbf{R}$ has been discussed under both view points. Note that for a BV function z in an interval $G = (a, b; c, d)$ it is trivial that the generalized limits $z(a+, y)$ and $z(b-, y)$, $z(x; c+)$ and $z(x, d-)$, exist a.e. and are L_1 functions, i.e., the trace $\gamma(z)$ of z on ∂G exists and is $L_1(\partial G)$. For general domains G in $\mathbf{R}^\nu$ possessing the cone property everywhere on ∂G, a theorem of Gagliardo [16] characterizes the properties of ∂G, and one can prove that any BV function z in a bounded domain G with the cone property and $\mathcal{H}^{\nu-1}(\partial G) < \infty$ possesses a trace $\gamma(z)$ on ∂G with $\gamma(z) \in L_1(\partial G)$.

We mention here the following theorem by Gagliardo on bounded domains G with the cone property: If G is a bounded open domain in $\mathbf{R}^\nu$ having the cone property, then there is a finite system $(G_1, \ldots, G_m)$ of open subsets of G with max diam G_s as small as we want, each G_s has the cone property and has locally Lipschitzian boundary ∂G_s.

From this result, and trace properties for Lipschitzian domains, it is possible to define the traces $\gamma(z)$ of a BV function on ∂G, for G bounded and with the cone property. An equivalent definition of traces of BV functions in terms of the distributional definition is also well known.

We come now to the delicate question of the continuity of the traces of $\gamma(z)$ of BV functions z in a domain G, in other words whether $z_k \to z$, say in $L_1(G)$, may actually imply—under assumptions—that $\gamma(z_k) \to \gamma(z)$ in $L_1(\partial G)$. A number of devices have been proposed to this effect. For instance, Anzellotti and Giaquinta ([1]) have recently proved the following statement in terms of the distributional definition of BV functions: If G has the cone property at every point of ∂G, if $\mathcal{H}^{\nu-1}(\partial G) < \infty$, if the functions z_k are BV with $V(z_k) \leq C$, if $z_k \to z$ in $L_1(G)$ with $V(z_k) \to V(z)$, then $\gamma(z_k) \to \gamma(z)$ in $L_1(\partial G)$. A parallel proof of this statement is available in terms of the original definition of BV functions. We mention here that it is well known that any BV function $z(t)$, $t \in G \subset \mathbf{R}^\nu$, can be approximated in $L_1(G)$ by BV smooth functions z_k with $V(z_k) \to V(z)$, hence $V(z_k) \leq C$.

We shall see now how these ideas have been used in questions of optimization.

II. CALCULUS OF VARIATIONS IN CLASSES OF BV FUNCTIONS.

When the state variable z, or $z(t) = (z^1, \ldots, z^m)$, $t \in G \subset \mathbf{R}^\nu$, is only BV, that is, each component z^i is BV in G, the usual Lebesgue integral of the calculus of variations

$$I(z) = \int_G f_0(t, z(t), Dz(t))\, dt,$$

$$t = (t^1, \ldots, t^\nu) \in G \subset \mathbf{R}^\nu,\ z(t) = (z^1, \ldots, z^m),\ \nu \geq 1,\ m \geq 1,$$

may not give a true, or stable value for the functional of interest. There are two basic processes to determine a true, or stable value for the underlying functional, and both have generated a great deal of recent work.

One is the limit process already proposed by *Weierstrass*, leading to a functional, or Weierstrass integral, W(z). Tonelli made use of it in his early work (1914) on the direct method in the calculus of variations on parametric continuous curves C, or $z(t) = (z^1, \ldots, z^m)$, $a \leq t \leq b$, of finite length, hence all z^i are BV and continuous. Recently, Cesari [8ab] presented an abstract formulation of the Weierstrass integral as a Burkill-type limit on "quasi additive" set functions $\phi(I) = (\phi_1, \ldots, \phi_n)$, $I \subset G$, and state functions $z(t) = (z^1, \ldots, z^m)$, $t \in G \subset \mathbf{R}^\nu$. Cesari also proved [8b] that $W(z)$ has a representation as a Lebesgue-Stieltjes integral in terms of measures and Radon-Nikodym derivatives derived from the set function ϕ. Warner [23] then proved lower semicontinuity theorems for continuous varieties, and very recently Brandi and Salvadori [1defg] extended further the abstract formulation, proved further representation properties, and lower semicontinuity theorems, both in the parametric and in the non-parametric case, and for vector functions z (state variables) only BV, possibly discontinuous, possibly not Sobolev (see Section VII below).

Another approach was proposed by *Serrin* [20a] leading to a functional, or Serrin integral, $\mathcal{T}(z)$, in classes of BV vector functions $z(t)$, $t \in G \subset \mathbf{R}^{\nu}$. The Serrin functional $\mathcal{T}(z)$ is obtained by taking lower limits on the values of I on AC, or $W^{1,1}(G)$ functions, a process which is similar to the one with which Lebesgue area is defined. Recently, Cesari, Brandi and Salvadori [10ab] proved closure and lower closure theorems, hence theorems of lower semicontinuity in the L_1-topology, and finally theorems of existence of the absolute minimum of $\mathcal{T}(z)$ in classes of BV vector functions whose total variations $V(z)$ are equibounded [10ab]. We proved also that $I(z) \leq \mathcal{T}(z)$, and that $\mathcal{T}$ is a proper extension of I in the sense that $\mathcal{T}(z) = I(z)$ for all z which are AC, or $W^{1,1}(G)$ (see Sections III,IV below). A number of applications of this approach has been announced [9abc, 11ab].

III. PROBLEMS OF OPTIMIZATION FOR SIMPLE INTEGRALS, $\nu = 1$, BY THE USE OF SERRIN'S FUNCTIONAL.

We may be interested either in problems of the classical calculus of variations involving a vector valued state variable $z(t) = (z^1, \dots, z^n)$, $t_1 \leq t \leq t_2$, or in problems of optimal control involving an analogous state variable $z(t) = (z^1, \dots, z^n)$ and a control variable $u(t) = (u^1, \dots, u^m)$, $t_1 \leq t \leq t_2$, with given control space $U(t, z)$ and constraint $u(t) \in U(t, z(t))$.

It is more general and more satisfactory (cfr. [7b]), to deparametrize the problems of optimal control, and concern ourselves exclusively with generalized problems of the calculus of variations with constraints on the derivatives, say

$$I(z) = \int_{t_1}^{t_2} f_0(t, z(t), z'(t))\, dt = \text{minimum},$$

$$(t, z(t)) \in A \subset \mathbf{R}^{n+1},\ z'(t) \in Q(t, z(t)), \tag{1}$$

where $t \in [t_1, t_2] \subset \mathbf{R}$ (a.e.), where A is a subset of $\mathbf{R}^{n+1}$ whose projection on the t-axis contains $[t_1, t_2]$, and where, for every $(t, z) \in A$, a set $Q(t, z)$ is given constraining the direction $z'(t)$ of the tangent to the state variable

z a.e. in $[t_1, t_2]$. The process of deparametrization has been discussed in detail in [7b].

For what concerns boundary conditions for problem (1), we restrict ourselves here to Dirichlet type boundary conditions

$$z(t_1) = z_1, \; z(t_2) = z_2. \tag{2}$$

Above, let M denote the set $M = [(t, z, \xi)]|(t, z) \in A, \xi \in Q(t, z)] \subset \mathbf{R}^{1+2n}$, and let $f_0(t, z, \xi)$ be a real valued function on M. Let Ω be a class of admissible functions, i.e., functions $z : [t_1, t_2] \to \mathbf{R}^n$, or $z(t) = (z^1, \dots, z^n)$, such that (i) z is BV in $[t_1, t_2]$; (ii) $(t, z(t)) \in A$, $z'(t) \in Q(t, z(t))$ a.e. in $[t_1, t_2]$; (iii) $f_0(\cdot, z(\cdot), z'(\cdot)) \in L_1[t_1, t_2]$.

It is easy to see that the Lebesgue integral definition (1) of the functional (I) does not yield stable and realistic values for I, and one may use a Serrin type integral. To this effect, for every $z \in \Omega$ we denote by $\Gamma(z)$ the class of all sequences (z_k) of elements $z_k \in \Omega$ with (a) z_k is AC in $[t_1, t_2]$; (b) $z_k \to z$ pointwise a.e. in $[t_1, t_2]$.

If $\Gamma(z)$ is empty we take $\mathcal{T}(z) = +\infty$. If $\Gamma(z)$ is not empty, then we take

$$\mathcal{T}(z) = \inf \underline{\lim} \int_{t_1}^{t_2} f_0(t, z_k(t), z_k'(t))\, dt = \inf_{\Gamma(z)} \underline{\lim}_{k\to\infty} I(z_k). \tag{3}$$

This is a Serrin type definition of the functional which was inspired to the Lebesguc area.

If problem (1) has assigned boundary conditions, say of the Dirichlet type 2, then let $\Gamma(z)$ denote the class of all sequences (z_k) of elements z_k in Ω with (a) z_k is AC and satisfies the boundary conditions; (b$'$) $z_k \to z$ pointwise a.e. in $[t_1, t_2]$, in particular $z_k(t_i) \to z(t_i)$, $i = 1, 2$. Then the analogous integral defined by (3) could be denoted by $\mathcal{T}^*$ and obviously $\mathcal{T} \le \mathcal{T}^*$.

We can now state a lower semicontinuity theorem and an existence theorem for the integrals I and $\mathcal{T}$ on BV functions. To this purpose we have first to define as usual the "augmented" sets $\tilde{Q}(t, z)$ as follows:

$$\tilde{Q}(t, z) = [(\tau, \xi)|\tau \ge f_0(t, z, \xi), \; \xi \in Q(t, z)] \subset \mathbf{R}^{n+1}.$$

A lower semicontinuity theorem. Let us assume that (i) A is closed; (ii) the sets $\tilde{Q}(t,z)$ are closed, convex, and satisfy property (Q) with respect to (t,z) at every $(t,z) \in A$; (iii) $f_0(t,z,\xi)$ is lower semicontinuous in M, and there exists some function $\lambda \in L_1[t_1,t_2]$ such that $f_0(t,z,\xi) \geq \lambda(t)$ for all $(t,z,\xi) \in M$. Let $z(t)$, $t \in [t_1,t_2]$, be BV, and let $z_k(t)$, $t \in [t_1,t_2]$, $k = 1,2,\ldots$, be a sequence of AC functions z_k such that $z_k \to z$ pointwise a.e. in $[t_1,t_2]$, $(t,z_k(t)) \in A$, $z_k'(t) \in Q(t,z_k(t))$ a.e. in $[t_1,t_2]$, and $V(z_k) \leq C$. Then, $(t,z(t)) \in A, z'(t) \in Q(t,z(t))$ a.e. in $[t_1,t_2]$, and $I(z) \leq \underline{\lim}_{k\to\infty} I(z_k)$.

A fundamental consequence of this lower semicontinuity theorem is that if z_k is any of the sequences of AC elements in $\Gamma(z)$, with $V(z_k) \leq C$, and we take $j = \underline{\lim}_{k\to\infty} I(z_k)$, then

$$I(z) \leq \mathcal{T}(z) \leq j = \underline{\lim}_{k\to\infty} I(z_k).$$

Furthermore, the Serrin integral $\mathcal{T}$ is actually an extension of the integral I. Indeed, if $z \in \Omega \cap AC$, then, by taking $z_k = z$ we conclude that $I(z) \leq \mathcal{T}(z) \leq \underline{\lim}_{k\to\infty} I(z_k) = I(z)$.

Note that for sequences (z_k) as above with $V(z_k)$ unbounded, it may well occur that $\mathcal{T}(z) < I(z)$ as it has been proved by examples (cfr. [10a]).

We mention here that Kuratowski's property (K) at a point (t_0,z_0) is expressed by the relation

$$\tilde{Q}(t_0,z_0) = \cap_{\delta>0}\, cl\, [\cup\tilde{Q}(t,z)\, (t-t_0)^2 + |z-z_0|^2 \leq \delta^2].$$

The analogous condition (Q) at the point (t_0,z_0) is expressed by the relation

$$\tilde{Q}(t_0,z_0) = \cap_{\delta>0}\, cl\, co\, [\cup\tilde{Q}(t,z),\, (t-t_0)^2 + |z-z_0|^2 \leq \delta^2].$$

If problem (1) has assigned boundary conditions of the type (2), then in the theorem above we assume that $z_k \to z$ a.e. in $[t_1,t_2]$, in particular $z_k(t_i) = z(t_i)$, $i = 1,2$, and the same statement holds for $\mathcal{T}^*$.

An existence theorem for the integral $\mathcal{T}$. Let us assume that (i) A is compact and M is closed; (ii) the sets $\tilde{Q}(t,z)$ are closed, convex, and satisfy property (Q) with respect to (t,z) at every point (t,z) of A; (iii) $f_0(t,z,\xi)$ is lower semicontinuous in M. Assume that the class Ω is nonempty and closed, $V(z) \leq C$ for all $z \in \Omega$, and $\Gamma(z)$ is nonempty for at least one z. Then the functional T has an absolute minimum $z \in BV$ in Ω [10a].

In other words, let i denote the infimum of $I(z)$ for $z \in AC \cap \Omega$, let (z_k) denote a sequence of elements $z_k \in AC \cap \Omega$ with $I(z_k) \to i$. Then, there is an element $z \in \Omega$, $z \in BV$, such that $I(z) \leq T(z) = i$.

Examples have been given in [10a].

IV. PROBLEMS OF OPTIMIZATION FOR MULTIPLE INTEGRALS AND BV DISCONTINUOUS FUNCTIONS, $\nu > 1$, BY THE USE OF SERRIN'S FUNCTIONAL.

Let $\nu > 1$, $n \geq 1$, and let $G \subset \mathbf{R}^\nu$ be a bounded domain in the t-space $\mathbf{R}^\nu$, $t = (t^1, \dots, t^\nu)$, possessing the cone property at every point of its boundary ∂G. Let $A \subset \mathbf{R}^{\nu+n}$ be a compact subset of the tz-space $\mathbf{R}^{\nu+n}$, whose projection on the t-space contains G.

We shall deal with vector valued functions $z(t) = (z^1, \dots, z^n)$, z^i BV in G, therefore possessing first order partial derivatives in the sense of distributions which are measures $\mu_{ij}, j = 1, \dots, \nu$, $i = 1, \dots, n$, and in addition also generalized first order derivatives $D^j z^i$ a.e. in G, as functions of class $L_1(G)$, which are obtained as limits of incremental quotients when we disregard the values taken by the functions in suitable sets E of measure zero in G. We may need only a subset of such derivatives $D^j z^i$ as follows.

For every $i = 1, \dots, n$, let $\{j\}_i$ be a system of indices $1 \leq j_1 < \dots < j_s \leq \nu$, let $D^j z^i, j \in \{j\}_i$ denote the corresponding system of first order partial derivatives of the function z^i, and let N be their total number. Then by Dz we denoted the N-vector function $Dz(t) = (D^j z^i, j \in \{j\}_i, i = 1, \dots, n)$, $t \in G$ (a.e.).

For every $(t, z) \in A$ let $Q(t, z)$ be a given subset of $\mathbf{R}^N$. Let $M \subset \mathbf{R}^{\nu+n+N}$ denote the set $M = [(t, z, \xi) | (t, z) \in A,\ \xi \in Q(t, z)]$, and let $f_0(t, z, \xi)$ be a given real-valued function in M. We are interested in the multiple integral problem of the calculus of variations with constraints on the derivatives

$$I(z) = \int_G f_0(t, z(t), Dz(t))\, dt = \text{minimum},$$

$$(t, z(t)) \in A,\ Dz(t) \in Q(t, z(t)),\ t \in G\ (a.e.), \tag{1}$$

and possible Dirichlet type boundary conditions of the form $z(t) = \phi(t), t \in \partial G(\mathcal{H}^{\nu-1} - a.e.)$ on ∂G. Again we introduce a Serrin type integral.

Let Ω be a class of admissible functions $z(t) = (z^1, \dots, z^n)$, $t \in G$, such that (i) z is BV in G; (ii) $(t, z(t)) \in A, Dz(t) \in Q(t, z(t))$, $t \in G(a.e.)$; (iii) $f_0(\cdot, z(\cdot), Dz(\cdot)) \in L_1(G)$.

To simplify notations, let AC, or $AC(G)$, denote the class of functions $z(t) = (z^1, \dots, z^n)$, $t \in G$, whose components z^i are of Sobolev class $W^{1,1}(G)$, or briefly, Beppo Levi functions.

For any element $z \in \Omega$ let $\Gamma(z)$ denote the class of all sequences (z_k) of elements z_k in Ω with (a) z_k is AC in G; (b) $z_k \to z$ strongly in $L_1(G)$.

If $\Gamma(z)$ is empty we take $\mathcal{T}(z) = +\infty$. If $\Gamma(z)$ is not empty we take

$$\mathcal{T}(z) = \inf \underline{\lim} \int_G f_0(t, z_k(t), Dz_k(t))\, dt = \inf_{\Gamma(z)} \underline{\lim}_{k\to\infty} I(z_k). \tag{2}$$

To state an existence theorem we introduce, as usual, the augmented sets $\tilde{Q}(t, z) \subset \mathbf{R}^{N+1}$ as follows

$$\tilde{Q}(t, z) = [(\tau, \xi) | \tau \geq f_0(t, z, \xi),\ \xi \in Q(t, z)].$$

Beside property (Q) we shall require on the sets $\tilde{Q}(t, z)$ another property, or property $\tilde{F}_1$.

We say that the sets $Q(t, z)$, $(t, z) \in A$, have property $\tilde{F}_1$ with respect to z at the point $(t_0, z_0) \in A$ provided, given any number $\sigma > 0$, there are constants $C > 0$, $\delta > 0$ which depend on t_0, z_0, σ, such that for any set of measurable vector functions $\eta(t), z(t), \xi(t), t \in H$, on a measurable subset H of points t of G with $(t, z(t)) \in A$, $|z(t) - z_0| > \sigma$, $(\eta(t), \xi(t)) \in \tilde{Q}(t, z(t))$ for $t \in H$, $|t - t_0| \leq \delta$, there are other measurable vector functions $\bar{\eta}(t)$, $\bar{z}(t)$, $\bar{\xi}(t)$, $t \in H$, such that

$$(t, \bar{z}(t)) \in A,\ |\bar{z}(t) - z_0| \leq \sigma,\ (\bar{\eta}(t), \bar{\xi}(t)) \in \tilde{Q}(t, z(t)),$$
$$|\xi(t) - \bar{\xi}(t)| \leq C[|z(t) - \bar{z}(t)| + |t - t_0|],$$
$$\bar{\eta}(t) \leq \eta(t) + C[|z(t) - \bar{z}(t)| + |t - t_0|] \text{ for } t \in H$$

We denote by $\tilde{F}_2$ the same condition with $\bar{z}(t) = z_0$. These conditions are inspired to analogous ones proposed by Rothe, Berkovitz, Browder (cfr. Cesari [7], Section 13).

An existence theorem. Let us assume that (i) A is compact and M is closed; (ii) the sets $\tilde{Q}(t,z)$ are closed, convex and satisfy properties (Q) and $(\tilde{F}_1)$ at every point $(t,z) \in A$; (iii) $f_0(t,z,\xi)$ is bounded below and lower semicontinuous in (t,z,ξ). Also, let us assume that the class Ω is nonempty and closed, and $\Gamma(z)$ is nonempty for at least one $z \in \Omega$. Then the functional $\mathcal{T}$ has an absolute minimum z in Ω, $z \in BV$ in G [10b].

In other words, let i denote the infimum of $I(z)$ for $z \in AC \cap \Omega$, let (z_k) denote any sequence of elements $z_k \in AC \cap \Omega$ with $I(z_k) \to i$. Then there is at least one element $z \in \Omega$, $z \in BV$, such that $I(z) \le \mathcal{T}(z) = i$.

V. EXISTENCE OF BV POSSIBLY DISCONTINUOUS ABSOLUTE MINIMA FOR CERTAIN INTEGRALS WITHOUT GROWTH PROPERTIES.

Recently I considered [9a] multiple integrals of the form

$$I(z) = \int_G \sum_{i=1}^{m} \left| \sum_{j=1}^{\nu} [U_{ij}(t,z)]_{t_j} + V_i(t,z) \right| dt,$$
$$z(t) = (z_1, \ldots, z_m),\ t = (t_1, \ldots, t_\nu) \in G \subset \mathbf{R}^\nu,$$
$$z(t) = w(t),\ t \in B \subset \partial G, \tag{1}$$

and associated Serrin functionals $\mathcal{T}(z)$. I studied these integrals in classes of BV vector functions $z(t) = (z_1, \ldots, z_m)$, $t \in G$, with equibounded total variations. Here the U_{ij} are given functions of class C^1 and the V_i are given locally Lipschitzian functions. The existence theorems we mentioned in Section IV above, and we had proved in [10c], do not apply directly to the integrals (1). However, I proved in [9a] that the same integrals $I(z)$ and $\mathcal{T}(z)$ can be transformed into integrals $H(v)$ and $\mathcal{H}(v)$ to which the existence theorems in [10c] apply. Thus, I could obtain the expected existence theorems for the absolute minimum of $\mathcal{T}(z)$ for BV possibly discontinuous vector functions z, and of course $0 \le I(z) \le \mathcal{T}(z)$.

In [9b] I also studied a number of variants of the Serrin functional $\mathcal{T}(z)$ associated to the integral $I(z)$, namely, functionals $\mathcal{T}^*(z)$, $\mathcal{T}^{**}(z)$. I proved the needed properties of lower semicontinuity in the topology of L_1, and the basic relation $0 \le I(z) \le \mathcal{T}^{**}(z) \le \mathcal{T}^*(z) \le \mathcal{T}(z)$. It is clear that whenever we can prove that for the optimal solution z we have $I(z) = 0$, then z is a solution of the differential system $\sum_{j=1}^{\nu}[U_{ij}(t,z)]_{t_j} + V_i(t,z) = 0$, $i = 1,\ldots,m$, $t \in G$ (a.e).

VI. RANKINE-HUGONIOT TYPE PROPERTIES IN TERMS OF THE CALCULUS OF VARIATIONS AND BV SOLUTIONS.

In [9c] I investigated in more details integrals $I(z)$ of the form

$$I(z) = \int_G |z_x + (F(z))_y|\,dxdy \tag{2}$$

and corresponding Serrin type functionals, say $0 \le I(z) \le \mathcal{T}^{**}(z) \le \mathcal{T}^*(z) \le \mathcal{T}(z)$.

For $m = 1$, $\nu = 1$, we are dealing with the original integral

$$I(z) = \int_G |z_x + (F(z))_y|\,dxdy,\ G \subset \mathbf{R}^2,$$

$$x, y, z, F \text{ scalars}. \tag{3}$$

If z has a line $\Gamma : y = \ell(x), a \le x \le b$, of class C^1 and of jump discontinuity, say

$$z_2(x) = z(x, \ell(x)+),\ z_1(x) = z(x, \ell(x)-),\ a \le x \le b,$$

then, under mild assumptions, the contribution of Γ on the value of the Serrin type functional $\mathcal{T}^*$ is ≥ 0, and such a contribution is zero if and only if

$$[z_2(x) - z_1(x)]\ell'(x) = F(z_2(x)) - F(z_1(x)),\ a \le x \le b,$$

along the line Γ (Cesari [9c]). This is the same relation which is well known for weak solutions of conservation laws (cf. Oleinik [18], Conway and Smoller [12]).

For $m = 1, \nu \ge 1$, we are dealing with the original integral

$$I(z) = \int_G |z_x + \sum_{j=1}^{\nu}(F_j(z))_{y_j}|\,dxdy,\ G \subset \mathbf{R}^{\nu+1},\ dy = dy_1 \ldots dy_\nu,$$

$$x, z \text{ scalar},\ z(x,y) = z(x, y_1, \ldots, y_\nu),\ F(z) = (F_1, \ldots, F_\nu). \tag{4}$$

If $z(x,y)$ has a surface $\Gamma : x = L(y) = L(y_1, \ldots, y_\nu)$, $y \in D$, of class C^1 and of jump discontinuity for z, say

$$z_2(y) = z(L(y)+, y),\ z_1(y) = z(L(y)-, y),\ y = (y_1, \ldots, y_\nu) \in D,$$

then under mild assumptions the contribution of Γ on the value of the same Serrin type functional $\mathcal{T}^*$ is ≥ 0, and such a contribution is zero if and only if

$$z_2(y) - z_1(y) = \sum_{j=1}^{\nu} (L(y))_{y_j} [F_j(z_2(y)) - F_j(z_1(y))],\ y \in D,$$

on the surface Γ (Cesari [9c]).

For $m > 1$, $\nu = 1$ we are dealing with the original integral

$$I(u) = \int_G \sum_{i=1}^{m} |z_{ix} + (F_i(z))_y|\, dxdy,\ G \subset \mathbf{R}^2,$$

$$x, y,\ \text{scalars},\ z(x,y) = (z_1, \ldots, z_m),\ (F(z) = (F_1, \ldots, F_m), (5)$$

and in this situation we must use the Serrin type integral $\mathcal{T}^{**}$. Let us assume that for a given $i = 1, \ldots, m$, the component $z_i(x, y)$ of z has a line $\Gamma : y = \ell(x)$, $a \leq x \leq b$, of class C^1 and of jump discontinuity for z_i, say

$$z_{i2}(x) = z_i(x, \ell(x)+),\ z_{i1}(x) = z_i(x, \ell(x)-),\ a \leq x \leq b,$$

while the remaining components $z_h(x, y)$, $h = 1, \ldots, m$, $h \neq i$, are continuous in a neighborhood of Γ. In this situation, let us take

$$z^{i,2}(x) = (z_i(x, \ell(x)+); z_h(x, \ell(x)),\ h \neq i,\ h = 1, \ldots, m),$$
$$z^{i,1}(x) - (z_i(x, \ell(x)-); z_h(x, \ell(x)),\ h \neq i,\ h = 1, \ldots, m),\ a \leq x \leq b,$$

I proved in [9c], under mild assumptions, that the contribution of Γ on the value of the Serrin type integral $\mathcal{T}^{**}$ is ≥ 0, and such a contribution is zero if and only if

$$[z_{i2}(x) - z_{i1}(x)]\ell'(x) = F_i(z^{(i,2)}(x)) - F_i(z^{(i,1)}(x)),\ a \leq x \leq b,$$

along Γ (Cesari [9c]).

VII. THE WEIERSTRASS INTEGRAL W(Z).

In [8ab] Cesari established a very general axiomatization concerning extensions of Burkill's integral on set functions. Namely, Cesari [8a] introduced a concept of quasi-additivity for set functions, guaranteeing the existence of a limit, now called the Burkill-Cesari integral. About the non-linear integral $I = \int_T F(p,q)$ over a variety T, Cesari considered the set function $\Phi(I) = F(T(\omega(I)), \phi(I))$, where $\omega(I)$ is a choice function, i.e. $\omega(I) \in I$, and ϕ is a set function. Cesari proved [8a] that if T is any continuous parametric mapping and ϕ is quasi-additive and BV, then also Φ is quasi-additive and BV. In other words, the non-linear transformation F preserves quasi-additivity and bounded variation. Then the integral $W(z)$ is defined by the Burkill-Cesari process on the function Φ, and is thus defined as a Weierstrass-type integral.

Later, many authors studied this integral, both in the parametric and in the non-parametric case, for curves and for varieties, and framed in this theory many of their properties (see [21] for a survey). Note that if F does not depend on the variety, i.e., it is of the type $F(q)$, then the sole concept of quasi-additivity permits the extension of $W(z)$ over BV curves and surfaces, not necessarily continuous nor Sobolev's.

In the last years Brandi and Salvadori [2defg] have extended the definition of $W(z)$ over BV curves or varieties, not necessarily continuous nor Sobolev's, for complete integrand $F(p,q)$.

First the term $T(\omega(I))$, in the definition of $\Phi(I)$ was replaced [2d] by a set function $P(I)$ whose values are in a metric space K, while $\phi(I)$ a set function whose values are in a uniformly convex Banach space X and $F : K \times X \to E$, with E a real Banach space. In order to guarantee the existence of the integral $W(z)$ for BV transformations T, a condition on the pair of set functions (P, ϕ) was proposed in [2d], which is of the quasi-additivity-type, and was called Γ-quasi-additivity. This condition reduces to the quasi-additivity on ϕ when P is the usual set function $T(\omega(I))$ and T is continuous. In this new situation, Brandi and Salvadori proved that, if (P, ϕ) is Γ-quasi-additive and ϕ is BV, then still $\Phi(I) = F(P(I), \phi(I))$ is quasi-additive and BV. Thus the integral $W(z)$ is still defined by the Burkill-Cesari process on the set function Φ, and $W(z)$ is still a Weierstrass-type integral even for T only BV, possibly discontinuous.

Note that the new condition on (P, ϕ) is weaker than the couple of assumptions: continuity on T and quasi-additivity on ϕ. Moreover, it takes advantage of the power of the quasi-additivity-type properties to extend

I over BV curves and varieties, for integrands of the type $F(p, q)$, both in the parametric and in the non-parametric case (see many applications in [2def]).

Even in this more general setting, the integral $W(z)$ admits of a Lebesgue-Stieltjes integral representation ([2d])

$$W(z) = \int_G F(\pi(s), d\mu/d\|\mu\|)(s))d\|\mu\|,$$

in terms of a vectorial measure μ related to ϕ, its total variation $\|\mu\|$, and Radon-Nikodym derivative $d\mu/d\|\mu\|$, as in the previous work of Cesari [8b] in Euclidean spaces and in the successive extensions to abstract spaces, always for continuous varieties T (see [21] for a survey).

In the non-parametric case (see [2e]) Brandi and Salvadori transform the integral $I = \int_T f(t, p, q)$ into a suitable parametric integral in the manner of McShane, with integrand $F(t, p; \ell, q)$ defined by $F(t, p; \ell, q) = \ell f((t, p, q/\ell)$ for $\ell > 0$ and $F(t, p; 0, q) = \lim_{\ell \to 0+} F(t, p; \ell, q)$. Then the set function Φ becomes $\Phi(I) = \lambda(I) f(p(I), \phi(I)/\lambda(I)) = F(p(I); \lambda(I), \phi(I))$, and the existence result is still given in terms of Γ-quasi-additivity. Now the representation of I in terms of Lebesgue-Stieltjes integral becomes

$$W(z) = \int_G f(\pi(s), (d(\nu, \mu)/d\|(\nu, \mu)\|)(s))\, d\|(\nu, \mu)\|,$$

where μ is the vectorial measure related to ϕ, ν is the real measure related to λ and $\|(\nu, \mu)\|$ is the total variation of the measure (ν, μ). Furthermore, in this non-parametric situation Brandi and Salvadori proved a Tonelli-type inequality in [2e] relating $W(z)$ to a corresponding Lebesgue-Stieltjes integral, namely,

$$W(z) \geq \int_G f(\pi(s), (\delta\mu/\delta\nu)(s)\, d\nu,$$

where $\delta\mu/\delta\nu$ is a derivative of the Radon-Nikodym type, and the equality sign holds if and only if the set function ϕ is absolutely continuous with respect to the set function λ. If ϕ is absolutely continuous with respect to λ, then $\delta\mu/\delta\nu$ reduces to the usual Radon-Nikodym derivative $\partial\mu/\partial\nu$. In proving this last result, as in the proof of the representation theorem, Brandi and Salvadori used a connection between the Burkill-Cesari process and the convergence of martingales, a connection which was already made in previous papers (see [21] and also in the quoted papers [2def]).

Finally, in [2f] Brandi and Salvadori dealt with the problem of the lower semicontinuity for the integral $W(z)$, both in the parametric and in the non-parametric case. A first abstract lower semicontinuity theorem was proved in terms of a suitable global convergence on the sequence (P_n, ϕ_n), defined in the same spirit of the Γ-quasi-additivity and therefore again inspired to Cesari's concept of quasi-additivity. In a number of applications this convergence is implied by the L_1-convergence of equiBV varieties.

REFERENCES

1. G. Anzellotti and M. Giaquinta, Funzioni BV e tracce, Rend. Sem. Mat. Padova 60 (1978), 1–21.
2. P. Brandi - A. Salvadori, (a) "Sull'integrale debole alla Burkill-Cesari", Atti Sem. Mat. Fis. Univ. Modena *27* (1978) 14–38; (b) "Existence, semi-continuity and representation for the integrals of the Calculus of Variations. The BV case", Atti Convegno celebrativo I centenario Circolo Matematico di Palermo (1984), 447–462; (c) "L'integrale del Calcolo delle Variazioni alla Weierstrass lungo curve BV e confronto con i funzionali integrali di Lebesgue e Serrin", Atti Sem. Mat. Fis. Univ. Modena *35* (1987); (d) "A quasi-additivity type condition and the integral over a BV variety", to appear; (e) "On the non-parametric integral over a BV surface", to appear; (f) "On the lower semicontinuity of certain integrals of the Calculus of Variations", to appear; (g) "On Weierstrass-type variational integrals over BV varieties", Rend. Accad. Naz. Lincei Roma, to appear.
3. J. C. Breckenridge, "Burkill-Cesari integrals of quasi additive interval functions", Pacific. J. Math. *37* (1971), 635–654.
4. F. Cafiero, Criteri di compattezza per le successioni di funzioni generalmente a variazione limitata, Atti Accad. Naz. Lincei Roma *8* (1950), 305–310.
5. L. Cesari, Sulle funzioni a variazione limitata, Annali Scuola Norm. Sup. Pisa (2) *5*, 1936, 299–313.
6. L. Cesari, (a) Sulle funzioni di due variabili a variazione limitata e sulla convergenza delle serie doppie di Fourier, Rend. Sem. Mat. Univ. Roma *1*, 1937, 277–294; (b) Sulle funzioni di più variabili a variazione limitata e sulla convergenza delle relative serie multiple di Fourier, Pontificia Accad. Scienze, Commentationes *3*, 1939, 171–197.

7. L. Cesari, (a) Surface Area, Princeton University Press, 1956; (b) Optimization-Theory and Applications, Springer Verlag, 1983.

8. L. Cesari, (a) "Quasi additive set functions and the concept of integral over a variety", Trans. Amer. Math. Soc. *102* (1962), 94–113; (b) "Extension problem for quasi additive set functions and Radon-Nikodym derivatives", Trans. Amer. Math. Soc. *102* (1962), 114–146.

9. L. Cesari, (a) Existence of discontinuous absolute minima for certain multiple integrals without growth properties, Rend. Accad. Naz. Lincei Roma, to appear; (b) Existence of discontinuous absolute minima for modified multiple integrals of the calculus of variations, to appear; (c) Rankine-Hugoniot type properties in terms of the calculus of variations for BV solutions, to appear.

10. L. Cesari, P. Brandi, and A. Salvadori, (a) Discontinuous solutions in problems of optimization, Annali Scuola Normale Sup. Pisa, to appear; (b) Existence theorems concerning simple integrals of the calculus of variations for discontinuous solutions, Archive Rat. Mech. Anal. *98*, 1987, 307–328; (c) Existence theorems for multiple integrals of the calculus of variations for discontinuous solutions, Annali Mat. Pura Appl., to appear.

11. L. Cesari and P. Pucci, (a) Remarks on discontinuous optimal solutions for simple integrals of the calculus of variations, Atti Sem. Mat.-Fis. Univ. Modena, to appear; (b) Existence of BV discontinuous solutions of the Cauchy problem for conservation laws, to appear.

12. E. Conway and J. Smoller, Global solutions of the Cauchy problem for quasi linear first order equations in several space variables, Comm. Pure Appl. Math. *19*, 1966, 95–105.

13. C. M. Dafermos, Generalized characteristics and the structure of solutions of hyperbolic conservation laws, Indiana Univ. Math. J. *26*, 1977, 1097–1119.

14. R. J. Di Perna, Singularities of solutions of nonlinear hyperbolic sytems of conservation laws, Archive Rat. Mech. Anal. *60*, 1974, 75–100.

15. W. H. Fleming, Functions with generalized gradient and generalized surfaces, Annali Mat. Pura Appl. *44*, 1957, 93–103.

16. E. Gagliardo, Proprietà di alcune classi di funzioni di più variabili, Ricerche Mat. *7*, 1959, 24–51.

17. K. Krickeberg, Distributionen, Funktionen beschränkter Variation und Lebesguescher Inhalt nichtparametrischer Flächen, Annali Mat. Pura Appl. *44*, 1957, 105–133.

18. O. A. Oleinik, On discontinuous solutions of nonlinear differential equations, Uspekhi Mat. Nauk *12*, 1957, 3–73. English translation, Amer. Math. Soc. Transl. (2) *26*, 95–172.

19. A. Salvadori, Theorems of lower semicontinuity, to appear.

20. J. Serrin, On the definition and properties of certain variational integrals, Trans. Amer. Math. Soc. *101*, 1961, 139–167; (b) On the differentiability of functions of several variables, Archive Rat. Mech. Anal. *7*, 1961, 359-372.

21. C. Vinti, Non linear integration and Weierstrass integral over a manifold. Connection with theorems on martingales, Journ. of Optimization Theory Applications *41*, 1983, 213–237.

22. A. L. Volpert, The space BV and quasi linear equations, Mat. Sb. *73*, 1967, 225–267.

23. G. Warner, (a) The Burkill-Cesari integral, Duke Math. J. *35*, 1968, 61–78; (b) The generalized Weierstrass-type integral $\int f(\xi, \phi)$. Annali Scuola Norm. Sup. Pisa *22*, 1968, 163–192.

Control Problems for Abstract Volterra Functional-Differential Equations

Constantin Corduneanu, Professor, Department of Mathematics,
The University of Texas at Arlington, Arlington, Texas

INTRODUCTION

The control systems we want to investigate can be described by functional-differential equations involving Volterra operators, such as

$$\dot{x}(t) = (Lx)(t) + u(t), \quad t \in [0, T], \tag{1}$$

where x is a map from $[0, T]$ into R^n, L is an abstract Volterra linear operator acting on a convenient function space, and $u(t)$ stands for the controller. Of course, more general systems could be considered, for instance

$$\dot{x}(t) = (Lx)(t) + (Au)(t),$$

where A is another operator defined on the function space to which the control belongs to.

In order to cover the cases of various types of differential-delay equations, it will be necessary to consider also the systems

$$\dot{x}(t) = (Lx)(t) + (B\phi)(t) + u(t), \quad t \in [0, T], \tag{2}$$

where ϕ is a functional parameter (belonging to the space of initial functions in the particular case we have in mind), and B stands for an operator taking the function space to which ϕ belongs to, into a convenient function space whose elements are defined on [0, T].

In either case (1) or (2), the initial condition

$$x(0) = x^0 \in R^n \tag{3}$$

will be attached.

The functional to be minimized will be taken, for illustration, to be the usual quadratic functional

$$C(x, u) = \int_0^T \left\{ < Px, x > + < Qu, u > \right\} dt, \tag{4}$$

where $P = P^*$ and $Q = Q^*$ are adequately chosen. In (4), $x = x(t)$ stands for the (unique) solution of the problem (1), (3), or of the problem (2), (3). The control u will be subject to restrictions to be specified later.

AN INTEGRAL REPRESENTATION FORMULA

In order to be able to discuss the optimal control problem described in the Introduction (as well as other control problems related to the systems (1) or (2)), it is necessary to provide a representation formula for the solution of the problems (1), (3). Let us notice the fact that existence and uniqueness has been proved in [3], where the integral representation formula

$$x(t) = X(t, 0)\, x^0 + \int_0^t X(t, s)\, u(s)\, ds, \quad t \in [0, T], \tag{5}$$

has been derived.

To be more specific, the validity of the formula (5) for the solution of the problems (1), (3) is assured by the following conditions:

1) L is a continuous linear Volterra operator from the space $L^2([0, T], R^n)$ into itself

2) $u \in L^2([0, T], R^n)$.

The properties of the kernel X(t, s) are given in [3]. We will notice only the fact that X(t, 0) is absolutely continuous, and X(t, s), $0 \le s \le t \le T$, is regular enough, such that the integral in the right hand side of (5) represents also an absolutely continuous function (in t), for any $u \in L^2([0, T], R^n)$.

In particular, the Volterra integral operator in (5) is compact from $L^2([0, T], R^n)$ into the space $AC_2([0, T], R^n)$ of absolutely continuous functions whose derivative is in $L^2([0, T], R^n)$. This fact implies [2] some properties for the kernel $X(t, s)$, such as

$$\int_0^t |X(t, s)|^2 \, ds \leq M, \quad t \in [0, T], \tag{6}$$

and

$$\lim_{t \to t_0} \int_0^T |X(t, s) - X(t_0, s)|^2 \, ds = 0, \tag{7}$$

with the agreement that $X(t, s) = 0$ for $s > t$. These properties will be useful in what follows.

THE OPTIMAL CONTROL PROBLEM

Let us consider now the system (7), under initial condition (3), with the cost functional $C(x, u)$ given by (4). Our aim is to minimize $C(x, u)$ with respect to all controllers $u(t)$, $t \in [0, T]$, which are subject to the constraint

$$\int_0^T |u(t)|^2 \, dt \leq r. \tag{8}$$

Of course, we have in mind measurable controls, which are square integrable on $[0, T]$: $u \in L^2([0, T], R^n)$.

Before we proceed to the statement of the existence of optimal controls, let us point out that the initial value condition (3) can be easily reduced to the simpler form

$$x(0) = \theta \in R^n, \tag{9}$$

if we change accordingly the constraint (8). Indeed, by means of the change of variable $x(t) - x^0 = y(t)$, the equation (1) becomes

$$\dot{y}(t) = (Ly)(t) + (Lx^0)(t) + u(t), \tag{10}$$

while the initial condition is $y(0) = \theta \in R^n$. Since $(Lx^0)(t)$ is a fixed element in the space $L^2([0, T], R^n)$, it is obvious that the substitution $(Lx^0)(t) + u(t) = v(t)$ reduces equation (10) to

$$\dot{y}(t) = (Ly)(t) + v(t), \tag{11}$$

with the constant for v

$$\int_0^T \left[v(t) - (Lx^0)(t) \right]^2 dt \leq r. \tag{12}$$

Consequently, choosing the initial condition in the special form (9), implies a change of the constraint (8): instead of assuming that the control vector u belongs to the ball of radius r in $L^2([0, T], R^n)$, centered at the zero element θ, we will assume that it belongs to the ball of same radius r, but centered at the element $(Lx^0)(t)$.

The following existence result can be stated in regard to the problem: find the control $u \in L^2([0, T], R^n)$, such that the cost functional (4) is minimized, where x(t) stands for the (unique) solution of the abstract Volterra equation (1), such that $x(0) = \theta \in R^n$, the constraint on the control being of the form

$$\int_0^T \left[u(t) - u_0(t) \right]^2 dt \leq r. \tag{13}$$

It is understood that $u_0(t) \in L^2([0, T], R^n)$ is a fixed element.

Theorem. The optimal control problem formulated above has a unique solution $u(t) \in L^2([0, T], R^n)$, satisfying the constraint (13), if the following conditions hold true:

a) $L : L^2([0, T], R^n) \to L^2([0, T], R^n)$ is a linear continuous operator of Volterra type (i.e., nonanticipative, or causal);

b) the matrix $P = P^*$ is non-negative, while the matrix $Q = Q^*$ is positive (i.e., $P \geq 0$ and $Q > 0$).

Remark. It is possible to consider more general cost functionals, in which $P = P(t)$ and $Q = Q(t)$. We leave to the reader the task to formulate the conditions that will allow the same treatment of the existence and uniqueness of the optimal control.

PROOF OF THE THEOREM

A rather simple proof of the existence can be based on the following well-known result from Hilbert space theory:

Let H be a real Hilbert space, and $C \subset H$ a closed convex set. Then C contains a unique element x, such that

$$\| x \| = \min \{ \| y \| ; y \in C \}.$$

We consider the space $L^2([0, T], R^n)$ as the Hilbert space, while C will denote the ball of radius r, centered at u_0. In order to apply the above result on the existence of the element of minimum norm, we shall endow the space $L^2([0, T], R^n)$ with another inner product, instead of the usual one

$$(u, v) \to \int_0^T < u(t), v(t) > dt,$$

where $< \cdot, \cdot >$ signifies the Euclidean scalar product of the vectors $u(t)$ and $v(t)$. Namely, if $x(t)$ represents the solution of (1), (9), and $y(t)$ represents the solution of (11), such that $y(0) = \theta \in R^n$, then we denote

$$\ll u, v \gg = \int_0^T \Big\{< Px(t), y(t) > + < Qu(t), v(t) >\Big\} dt. \tag{14}$$

The condition b) of the Theorem allows us to conclude that the new inner product verifies the usual conditions: symmetry, non-negativeness, the distributivity with respect to addition in $L^2([0, T], R^n)$, as well as the fact that $\ll u, u \gg = 0$ implies $u = \theta$ (of course, a.e. on [0, T]). Indeed, from $\ll u, u \gg = 0$ one obtains a.e. on [0, T] $< Qu(t), u(t) > = 0$, if we take into account that $< Px(t), x(t) > \geq 0$. Since the matrix Q is positive, there results $u(t) = 0$ a.e. on [0, T]. It is useful to keep in mind that

$$x(t) = \int_0^t X(t, s)\, u(s)\, ds, \tag{15}$$

which results from the formula (15), while $y(t)$ is given by a similar formula, with $u(t)$ replaced by $v(t)$. One sees immediately that $u(t) = 0$ a.e. on [0, T] implies $x(t) = 0$ on the same interval.

In order to conclude that $L^2([0, T], R^n)$ is also a Hilbert space with respect to the inner product (14), it suffices to prove that it is also complete in the norm derived from (14). In other words, it suffices to show that there are two positive constants α and β, such that

$$\alpha \int_0^T |u(t)|^2\, dt \leq C(x, u) \leq \beta \int_0^T |u(t)|^2\, dt, \tag{16}$$

for any $u \in L^2([0, T], R^n)$. Of course, x in (16) is determined from (1), (9).

Since the matrix Q is positive, one can obviously take α to be the smallest eigenvalue of Q. The number β is determined as follows: since

$$C(x, u) \le \lambda \int_0^T \left\{ |x(t)|^2 + |u(t)|^2 \right\} dt, \tag{17}$$

with λ greater or equal to the largest eigenvalues of matrices P and Q, we need to show only that

$$\int_0^T |x(t)|^2 \, dt \le K \int_0^T |u(t)|^2 \, dt, \tag{18}$$

for some $K > 0$. The inequality (18) follows immediately from (15), if we take into account (6). Hence, one can choose $\beta = \lambda\,(K+1)$ in (16).

The inequalities (16) prove that with the new norm, the convergence is the same as in $L^2([0, T], R^n)$.

Therefore, $L^2([0, T], R^n)$ is still a Hilbert space with the inner product defined by (14), which ends the proof of the Theorem.

SOME GENERALIZATIONS

1. From the proof of the Theorem, one can see that the constraint on the control (13) has been used only to the extent to conclude that u belongs to a closed and convex set of the Hilbert space $L^2([0, T], R^n)$. Therefore, instead of (13), one can ask only

$$u \in C, \tag{19}$$

with $C \subset H$ an arbitrary closed convex set. Of course, (19) is a much weaker constraint than (13).

2. Another generalization of the Theorem is related to the case of equations of the form (2). This case is no different than the case in which we deal with a "nonhomogeneous" system of the form

$$\dot{x}(t) = (Lx)(t) + f(t) + u(t), \tag{20}$$

in which $f \in L^2([0, T], R^n)$ is a fixed element. The discussion of this case is very simple, due to the fact that one can treat $f(t) + u(t) = v(t)$ as a new control variable, the only condition to be changed being the constraint on the controller. For instance, if we consider the problem (20), (9), with cost functional $C(x, u)$ given by (4), and constraint (8), then it is easily reduced to (1), (9), (4), with the constraint

$$\int_0^T [u(t) - f(t)]^2 \, dt \le r.$$

This generalization is significant for the case of linear differential-delay equations such as

$$\dot{x}(t) = L(t, x_t) + u(t), \tag{21}$$

under the initial condition

$$x(0) = x^0 \in R^n, \quad x_0 = \phi, \tag{22}$$

where $x_t(s) = x(t+s)$, $-h \leq s < 0$. The function ϕ belongs to some function space $S([-h, 0], R^n)$, the case $S = L^2$ being a convenient choice.

The operator $x \to L(t, x_t)$ is a Volterra type linear operator if L is linear, and the application of the Theorem is a routine matter.

3. Another generalization of the optimal control problem discussed above is the following one: find (or prove the existence of) an optimal control $u \in L^2([0, T], R^n)$, such that the pair (x, u), with x determined from

$$\dot{x}(t) = (Lx)(t) + B(t)\,u(t), \quad t \in [0, T], \tag{23}$$

and (9), minimizes the cost functional $C(x, u)$ given by (4), where $B(t)$ is a matrix (say, $B \in L^\infty$). The constraint on the control could be also (13), or $u \in C$, with C a convex closed set in $L^2([0, T], R^n)$.

The equation (23) will allow us to deal with controls of different dimensions than the state vector x. Of course, more general situations can be considered, as the one mentioned in the Introduction.

4. Finally, another generalization that requires a different approach than that used above is related to a cost functional of the form

$$C(x, u) = \int_0^T F(t, x(t); u(t))\,dt,$$

under convenient conditions on F. The formula of representation (15), for $x(t)$ in terms of $u(t)$, provides a good tool in handling this more general case. Convexity conditions will certainly occur.

OTHER CONTROL PROBLEMS (NON-OPTIMAL)

Using the theory of linear functional-differential equations involving Volterra operators, as sketched in our book [3], it is possible to approach various control problems, so far considered only in regard to ordinary differential equations, or in certain special cases of Volterra equations (particularly, differential-delay equations).

Let us consider now the system (1), under initial condition (9). Maintaining our assumption on the control, $u \in L^2([0, T], R^n)$, but dropping the constraint, let us ask the following question: is it possible to determine the control $u \in L^2([0, T], R^n)$ such that the trajectory of the system (1), (9) is preassigned?

First, let us notice the fact that the trajectory $x(t)$, $0 \leq t \leq T$, given by (15), is in the space $AC_2([0, T], R^n)$, and vanishes at $t \geq 0$. Hence, our problem can be precised in the sense that the trajectory belongs to $AC_2([0, T], R^n)$, and satisfies (9). If we want to reformulate the problem, then we should ask whether the integral operator given by (15) is invertible, as an operator from $L^2([0, T], R^n)$ into the subspace of $AC_2([0, T], R^n)$, consisting of the elements satisfying (9).

It turns out that the problem has always a unique solution in $L^2([0, T], R^n)$, namely $u(t) = \dot{x}(t) - (Lx)(t)$, a.e. on $[0, T]$. In other words, the operator defined by (15) is invertible.

Another problem that is usually investigated in regard to control systems defined by (1), (3), (8) is related to the (geometric) properties of the "attainability set". Using the representation formula (5), one can prove various properties of this set, such as the convexity or the compactness (as a subset of R^n).

Further properties of the control systems defined by (1), (3), (8), or (1), (3) and $u \in C \subset L^2([0, T], R^n)$, can be obtained by using methods as those emphasized in [1], [4], [5]. We take the opportunity to mention that the interest for control problems described by equations involving abstract Volterra operators has been illustrated for the first time in the work of L. Neustadt, the reference [4] being the most comprehensive source in this regard.

REFERENCES

1. R. Conti, *Infinite dimensional linear autonomous controllability*, School of Mathematics, University of Minnesota (1982).

2. C. Corduneanu, *Integral Equations and Stability of Feedback Systems.* Academic Press, New York (1973).

3. C. Corduneanu, *Integral Equations and Applications* (in print).

4. L. Neustadt, *Optimization: A Theory of Necessary Conditions.* Princeton University Press (1976).

5. J. Warga, *Optimal Control for Differential and Functional Equations.* Academic Press, New York (1972).

Deterministic Control of Uncertain Systems

M. Corless, Professor, School of Aeronautics & Astronautics, Purdue University, West Lafayette, Indiana

George Leitmann, Professor, Department of Mechanical Engineering, University of California at Berkeley, Berkeley, California

ABSTRACT

We present a controller design methodology for uncertain systems which is based on the constructive use of Lyapunov functions. The uncertainties, which are deterministic, are characterized by certain structural conditions and known bounds. As a consequence of the Lyapunov approach, the methodology is not restricted to linear systems. The robustness of these controllers in the presence of singular perturbations is considered. The situation in which the full state of the system is not available for measurement is also considered. These controllers are illustrated by application to the tracking control of a Manutec r3 robot which has an uncertain payload.

1 INTRODUCTION

A fundamental feedback control problem is that of obtaining some specified desired behavior from a system about which there is incomplete or uncertain information. Here we consider systems whose uncertainties are characterized deterministically rather than stochastically; for a stochastic approach, see, e.g. (9).

Our model of an uncertain system is of the form

$$\dot{x}(t) = F(t, x(t), u(t), \omega) \tag{1.1}$$

Supported by the U.S. National Science Foundation and the U.S. Air Force Office of Scientific Research under grants MSM-8706927 and ECS-8602524.

where $t \in \mathbb{R}$ is the "time" variable, $x(t) \in \mathbb{R}^n$ is the state, and $u(t) \in \mathbb{R}^m$ is the control input. All the uncertainty in the system is represented by the lumped uncertain element ω. It could be an element of $\mathbb{R}^q$ representing unknown constant parameters and inputs; it could be a function from $\mathbb{R}$ into $\mathbb{R}^q$ representing unknown time varying parameters and inputs; it could also be a function from $\mathbb{R} \times \mathbb{R}^n \times \mathbb{R}^m \to \mathbb{R}^q$ representing nonlinear elements which are difficult to characterize exactly; it could be an index. The only information assumed about ω is the knowledge of a non-empty set Ω to which it belongs.

We consider control to be given by a memoryless state feedback controller, i.e.,

$$u(t) = p(t, x(t)) \,. \tag{1.2}$$

Ideally we wish to choose p so that the feedback controlled system

$$\dot{x}(t) = f(t, x(t), \omega) \,, \tag{1.3}$$

where

$$f(t, x, \omega) \triangleq F(t, x, p(t,x), \omega) \,,$$

has the property of g.u.a.s. (global uniform asymptotic stability) about the zero state for all $\omega \in \Omega$. However to assure g.u.a.s. of an uncertain system one sometimes has to resort to controllers which are discontinuous in the state; see e.g. (33, 34, 47, 49, 50, 51, 70, 71, 96). To avoid such discontinuous controllers, we relax the problem to that of obtaining a family of controllers which assure that the behavior of (1.3) can be made arbitrarily close to g.u.a.s.; such a family is called a *practically stabilizing* family; see (33, 34).

2 A LYAPUNOV THEORY APPROACH

The following theorem (see (33) for a proof) is useful in the construction of practically stabilizing sets of controllers.

Theorem 2.1. Consider an uncertain system described by (1.1) with $\omega \in \Omega$ and suppose that P is a collection of feedback control functions $p: \mathbb{R} \times \mathbb{R}^n \to \mathbb{R}^m$. If there exists a candidate Lyapunov function[(1)] $V: \mathbb{R} \times \mathbb{R}^n \to R_+$ and a class K function[(1)] $\gamma: \mathbb{R}_+ \to \mathbb{R}_+$ such that for each $\varepsilon > 0$ there exists $p^\varepsilon \in P$ which assures that for all $\omega \in \Omega$

$$\dot{x}(t) = F(t, x(t), p^\varepsilon(t, x(t)), \omega) \tag{2.1}$$

has existence and indefinite extension of solutions and

$$\frac{\partial V}{\partial t}(t, x) + \frac{\partial V}{\partial x}(t, x) F(t, x, p^\varepsilon(t, x), \omega) \le -\gamma(\|x\|) + \varepsilon \tag{2.2}$$

for all $t \in \mathbb{R}$, $x \in \mathbb{R}^n$, then P is a practically stabilizing family.

2.1 A Specific Class of Uncertain Systems

An uncertain system under consideration here is described by (1.1) and satisfies the following assumption.

Assumption A1. There exist a continuous function $B: \mathbb{R} \times \mathbb{R}^n \to \mathbb{R}^{n \times m}$, a candidate Lyapunov function $V: \mathbb{R} \times \mathbb{R}^n \to \mathbb{R}_+$, a class K function $\gamma: \mathbb{R}_+ \to \mathbb{R}_+$ and continuous functions $\kappa, \rho: \mathbb{R} \times \mathbb{R}^n \to \mathbb{R}_+$ such that

(1) A definition is given in (33, 34).

$$F(t, x, u, \omega) = f^s(t, x, \omega) + B(t, x)g(t, x, u, \omega) \tag{2.3}$$

for some functions f^s and g which satisfy:

1) For each $\omega \varepsilon \Omega$, $f^s(\cdot, \omega)$ is continuous and

$$\frac{\partial V}{\partial t}(t, x) + \frac{\partial V}{\partial x}(t, x)f^s(t, x, \omega) \leq -\gamma(\|x\|) \tag{2.4}$$

for all $t \in \mathbb{R}$, $x \in \mathbb{R}^n$.

2) For each $\omega \in \Omega$, $g(\cdot, \omega)$ is continuous and

$$u^T g(t, x, u, \omega) \geq -\beta_1(t, x, \omega)\|u\| + \beta_2(t, x, \omega)\|u\|^2 \tag{2.5}$$

where $\beta_1(t, x, \omega)$, $\beta_2(t, x, \omega) \geq 0$ and

$$\beta_1(t, x, \omega) \leq \beta_2(t, x, \omega)\, \rho(t, x)\,, \tag{2.6}$$
$$\beta_1(t, x, \omega) \leq \kappa(t, x) \tag{2.7}$$

for all $t \in \mathbb{R}$, $x \in \mathbb{R}^n$, $u \in \mathbb{R}^m$.

2.2 Proposed Controllers

Here we present some practically stabilizing controller sets for the systems considered in the previous section. These controllers can be regarded as continuous approximations of those presented in (47, 50). Their construction is based on satisfying the hypotheses of Theorem 2.1; see (34).

Consider any uncertain system described in sec. 2.1 and let $(B, V, \gamma, \rho, \kappa)$ be any quintuple which assures the satisfaction of Assumption A1. Choose any continuous functions ρ^c, κ^c which satisfy

$$\rho^c(t, x) \geq \rho(t, x)\,, \quad \kappa^c(t, x) \geq \kappa(t, x) \tag{2.8}$$

and define

$$\alpha(t, x) \triangleq B(t, x)^T \frac{\partial V}{\partial x}(t, x)^T\,, \tag{2.9}$$

$$\eta(t, x) \triangleq \kappa^c(t, x)\, \alpha(t, x)\,. \tag{2.10}$$

A practically stabilizing family of controllers is the set

$$P = \{p^\varepsilon : \varepsilon > 0\} \tag{2.11}$$

where p^ε is any continuous function which satisfies

$$\|\alpha(t, x)\|\, p^\varepsilon(t, x) = -\|p^\varepsilon(t, x)\|\, \alpha(t, x)\,, \tag{2.12}$$

i.e., $p^\varepsilon(t, x)$ is opposite in direction to $\alpha(t, x)$, and

$$\|\eta(t, x)\| > 0 \Rightarrow \|p^\varepsilon(t, x)\| \geq \rho^c(t, x)[1 - \|\eta(t, x)\|^{-1}\varepsilon]\,. \tag{2.13}$$

As an example of a function satisfying the above requirements on p^ε, consider

$$p^\varepsilon(t, x) = \begin{cases} -\rho^c(t, x)\varepsilon^{-1}\eta(t, x) & \text{if } \|\eta(t, x)\| \leq \varepsilon \\ -\rho^c(t, x)\,\|\eta(t, x)\|^{-1}\,\eta(t, x) & \text{if } \|\eta(t, x)\| > \varepsilon\,; \end{cases} \tag{2.14}$$

see (26, 33, 34).

As another example, consider

$$p^\varepsilon(t, x) = -\rho^c(t, x)[\|\eta(t, x)\| + \varepsilon]^{-1}\eta(t, x)\,;$$

see (1).

2.3 Matching Conditions

Given a system described by (1.1), the choice of B, f^s, g (see (2.3)) to assure satisfaction of Assumption A1 (if possible) may not be obvious. This choice is usually easier if the uncertainties are *matched* in the sense that there exist functions f^o, B, g, with $B(t, x) \in \mathbb{R}^{n \times m}$, such that

$$F(t, x, u, \omega) = f^o(t, x) + B(t, x)g(t, x, u, \omega) ;$$

i.e., the uncertainty and the control $u \in \mathbb{R}^m$ enter the system description via the same nxm matrix B(t, x).

Much of the literature concerns systems in which the uncertainties are matched. Refs. (6, 19, 111, 113) consider systems with *unmatched* uncertainties. In (6,19) the norm of the unmatched portion of the uncertain term must be smaller than a certain *threshold* value. In (113) linear systems are considered in which the uncertainty satisfies *generalized matching conditions,* i.e., structural conditions which are less restrictive than the matching condition. There (as in the matched case) the norm bounds on the uncertain terms can be arbitrarily large. Linear time-invariant systems with scalar control input are treated in (111).

2.4 Other Problems

Refs. (28, 29, 32) deal with systems in which the uncertainty bounds (e.g., $\rho(t, x)$ in sec. 2.1) are not known exactly but depend on unknown constants; the controllers presented there are parameter adaptive controllers.

Problems in which one wishes to keep the system state within or outside a prescribed region of the state space are considered in (20, 31, 37, 73, 80, 81).

Systems with delays are considered in (114, 116).

Discrete time systems, i.e., systems described by difference equations, are treated in (21, 38, 59).

3 ROBUSTNESS IN THE PRESENCE OF SINGULAR PERTURBATIONS

Consider an uncertain singularly perturbed system described

$$\dot{x} = F(t, x, y, u, \mu, \omega) \tag{3.1a}$$
$$\mu\dot{y} = G(t, x, y, u, \mu, \omega) \tag{3.1b}$$

where $(x, y) \in \mathbb{R}^n \times \mathbb{R}^l$ describe the state of the system, $\mu \in (0, \infty)$ is the *singular perturbation parameter,* and all the other variables are as described in sec. 1. Here one wants to obtain memoryless feedback controllers (generating u) which assure that, for all $\omega \in \Omega$ and for all sufficiently small μ, the behavior of the feedback-controlled system is close to that of g.u.a.s.

Assuming that, for each x, u, ω, there exists a unique vector $H(x, u, \omega) \in \mathbb{R}^l$ such that

$$G(t, x, H(x, u, \omega), u, 0, \omega) = 0 \tag{3.2}$$

for all t, the *reduced order system* associated with (3.1) (let $\mu = 0$ in (3.1)) is given by

$$\dot{x} = \bar{F}(t, x, u, \omega) \tag{3.3a}$$

where

$$\bar{F}(t, x, u, \omega) \triangleq F(t, x, H(x, u, \omega), u, 0, \omega) . \tag{3.3b}$$

For each t, x, u, ω, the *boundary layer system* associated with (3.1) is given by

$$\frac{dy}{d\tau}(\tau) = G(t, x, y(\tau), u, 0, \omega) . \tag{3.4}$$

Refs. (22, 36, 78, 79) require that the boundary layer system satisfies g.u.a.s. about its equilibrium state H(x, u, ω) and present "stabilizing" controllers whose designs are based on the reduced order system.

In (42, 43) the boundary layer system is not required to be stable. The "stabilizing" controllers presented there are composite controllers in the sense that they consist of two parts; one part is utilized to stabilize the boundary layer system and the other part is based on a nominal reduced order system.

4 OUTPUT FEEDBACK

In secs. 1 and 2, it is implicitly assumed that the complete state of the system is available for feedback. Consider now the more general situation in which the output $y(t) \in \mathbb{R}^s$ available for feedback is related to the state by

$$y(t) = c(t, x(t), \omega) \tag{4.1}$$

for some function $c: \mathbb{R} \times \mathbb{R}^n \times \Omega \rightarrow \mathbb{R}^s$.

Memoryless output feedback controllers are treated in (14, 41, 112, 119).

The dynamic output feedback controllers in the literature utilize state estimators. The estimated state is fed to a memoryless controller whose design is based on having the complete state available for feedback.

Full order observers are utilized in (5, 117). In (5) the bounds on the uncertain terms cannot be arbitrarily large. In (117) the bound on the uncertain term can only depend on t and y.

Refs. (10, 11) utilize reduced order observers. There the uncertain terms must satisfy certain structural conditions and the differential equation describing the evolution of the state estimation error is decoupled from the system state; see also (15, 99).

5 TRACKING CONTROLLERS FOR UNCERTAIN MECHANICAL SYSTEMS

In this section we consider uncertain mechanical systems, i.e., systems which are modelled using the laws of mechanics. Examples include robots and spacecraft. The systems treated here have a finite number of degrees of freedom (d.o.f.'s) and their main characterization is that the number of independent scalar control inputs is the same as the number of d.o.f.'s of the system.

Utilizing the results of sec. 2, we obtain a set of controllers which assure that the trajectory of the controlled system can track a prespecified desired trajectory to within any desired degree of accuracy.

In sec. 7 the controllers are illustrated by application to a 3 d.o.f. model of a Manutec r3 robot which has an uncertain payload.

5.1 Systems Under Consideration and Problem Statement

Consider an uncertain mechanical system with N d.o.f.'s described by

$$J(t, q(t), \omega)\ddot{q}(t) = U(t, q(t), \dot{q}(t), \omega) + W(t, q(t), \dot{q}(t))\hat{u}(t) \tag{5.1}$$

$$\omega \in \Omega \tag{5.2}$$

where $t \in \mathbb{R}$ represents time; $q(t) \in \mathbb{R}^N$ is a vector of generalized coordinates which describes the configuration of the system; $\hat{u}(t) \in \mathbb{R}^N$ is a vector of control inputs; and ω is the lumped uncertain element.

The set Ω and the functions $J: \mathbb{R} \times \mathbb{R}^N \times \Omega \to \mathbb{R}^{N\times N}$, $U: \mathbb{R} \times \mathbb{R}^N \times \mathbb{R}^N \times \Omega \to \mathbb{R}^N$, and $W: \mathbb{R} \times \mathbb{R}^N \times \mathbb{R}^N \to \mathbb{R}^{N\times N}$ are known.

The following assumption is satisfied.

Assumption M1

1) For each $\omega \in \Omega$, the functions $J(\cdot, \omega): \mathbb{R} \times \mathbb{R}^N \to \mathbb{R}^{N\times N}$ and $U(\cdot, \omega): \mathbb{R} \times \mathbb{R}^N \times \mathbb{R}^N \to \mathbb{R}^N$ are continuous.

2) For each $(t, q, \dot{q}, \omega) \in \mathbb{R} \times \mathbb{R}^N \times \mathbb{R}^N \times \Omega$, $J(t, q, \omega)$ is symmetric and $W(t, q, \dot{q})$ is non-singular.

3) There exist continuous functions $\hat{\beta}_o: \mathbb{R} \times \mathbb{R}^N \times \mathbb{R}^N \to \mathbb{R}_+$ and $\hat{\beta}_1, \hat{\beta}_2: \mathbb{R} \times \mathbb{R}^N \to \mathbb{R}_+$ such that for all $(t, q, \dot{q}) \in \mathbb{R} \times \mathbb{R}^N \times \mathbb{R}^N$

$$\|U(t, q, \dot{q}, \omega)\| \leq \hat{\beta}_o(t, q, \dot{q}), \tag{5.3}$$

$$\lambda_{max}[J(t, q, \omega)] \leq \hat{\beta}_1(t, q), \tag{5.4}$$

$$\lambda_{min}[J(t, q, \omega)] \geq \hat{\beta}_2(t, q) > 0 \tag{5.5}$$

for(2) all $\omega \in \Omega$.

For a prescribed, desired, C^2 motion $\bar{q}(\cdot): \mathbb{R} \to \mathbb{R}^N$, we consider the problem of obtaining a set P of feedback control functions with the following property. For each $\delta > 0$, there exists $\hat{p} \in P$ such that by letting

$$\hat{u}(t) = \hat{p}(t, q(t), \dot{q}(t))$$

in (5.1) the resulting controlled system has the property that for any initial condition $(t_o, q^o, \dot{q}^o) \in \mathbb{R} \times \mathbb{R}^N \times \mathbb{R}^N$ and any $\omega \in \Omega$, $\|q(\cdot)-\bar{q}(\cdot)\|$, $\|\dot{q}(\cdot)-\dot{\bar{q}}(\cdot)\|$ are bounded and

$$\limsup_{t\to\infty} \|q(t)-\bar{q}(t)\| \leq \delta, \tag{5.6}$$

where $q(t_o) = q^o$ and $\dot{q}(t_o) = \dot{q}^o$. Roughly speaking, we wish to assure asymptotic tracking of $\bar{q}(\cdot)$ by $q(\cdot)$ to within any desired degree of accuracy.

5.2 Proposed Controllers

First we reformulate the problem into one already considered in sec. 2. Towards that end introduce the *state tracking error*

$$x(t) = \begin{bmatrix} q(t) - \bar{q}(t) \\ \dot{q}(t) - \dot{\bar{q}}(t) \end{bmatrix} ; \tag{5.7}$$

also choose any nonsingular matrix $T \in \mathbb{R}^{N\times N}$ and define a new control input

$$\tilde{u}(t) = T^T W(t, q(t), \dot{q}(t))\hat{u}(t) . \tag{5.8}$$

Thus,

$$\hat{u}(t) = [T^T W(t, q(t), \dot{q}(t))]^{-1}\tilde{u}(t) \tag{5.9}$$

and the evolution of $x(\cdot)$ is described by

$$\dot{x}(t) = Ax(t) + B[h(t, x(t), \omega) + G(t, x(t), \omega)\tilde{u}(t)] \tag{5.10}$$

with

(2) If $M \in \mathbb{R}^{N\times N}$ has all real eigenvalues, $\lambda_{max}[M]$ ($\lambda_{min}[M]$) is the maximum (minimum) eigenvalue of M.

$$A = \begin{bmatrix} 0 & I \\ 0 & 0 \end{bmatrix}, \quad B = \begin{bmatrix} 0 \\ T \end{bmatrix}, \tag{5.11}$$

where 0 and I are the zero and identity matrices in $\mathbb{R}^{N\times N}$, respectively, and[3]

$$h = T^{-1}J^{-1}U - T^{-1}\ddot{\bar{q}}, \tag{5.12}$$

$$G = \tilde{J}^{-1}, \quad \tilde{J} = T^TJT. \tag{5.13}$$

The original problem can be solved now by obtaining a zero-state practically stabilizing set P of controllers (which generate $\tilde{u}$) for (5.10).

To obtain a system description which satisfies Assumption A1, let

$$\tilde{u}(t) = p^1(t, x(t)) + p^2(t, x(t)) + u(t) \tag{5.14}$$

which yields

$$\dot{x}(t) = f^s(t, x(t), \omega) + Bg(t, x(t), u(t), \omega) \tag{5.15}$$

where

$$f^s(t, x, \omega) = Ax + BG(t, x, \omega)p^1(t, x) \tag{5.16}$$

$$g(t, x, u, \omega) = h(t, x, \omega) + G(t, x, \omega)p^2(t, x) + G(t, x, \omega)u \tag{5.17}$$

The function $p^1: \mathbb{R} \times \mathbb{R}^n \to \mathbb{R}^N$ $(n = 2N)$ is chosen so that the system

$$\dot{x}(t) = f^s(t, x(t), \omega) \tag{5.18}$$

is g.u.a.s. about 0. Specifically, choose any positive definite, symmetric matrix $Q \in \mathbb{R}^{n\times n}$, any positive scalar $\sigma > 0$, and any continuous function $\gamma_1 : \mathbb{R} \times \mathbb{R}^n \to \mathbb{R}_+$ which satisfies

$$\gamma_1(t, x) \geq \frac{1}{2}\sigma\lambda_{max}[\tilde{J}(t, x, \omega)] \tag{5.19}$$

for all $(t, x, \omega) \in \mathbb{R} \times \mathbb{R}^n \times \Omega$; the existence of γ_1 is guaranteed by Assumption M1(3). Then

$$p^1(t, x) = -\gamma_1(t, x)B^TPx \tag{5.20}$$

where $P \in \mathbb{R}^{n\times n}$ is the unique positive definite, symmetric solution of the algebraic Riccati equation

$$PA + A^TP - \sigma PBB^TP + 2Q = 0. \tag{5.21}$$

With p^1 chosen as indicated above, consider the function $V: \mathbb{R} \times \mathbb{R}^n \to \mathbb{R}_+$, given by

$$V(t, x) = \frac{1}{2}x^TPx, \tag{5.22}$$

as a candidate Lyapunov function for (5.18). Then,

$$\frac{\partial V}{\partial t}(t, x) + \frac{\partial V}{\partial x}(t, x)f^s(t, x, \omega) \leq -x^TQx;$$

hence inequality (2.4) of Assumption A1 is satisfied with

$$\gamma(\|x\|) = \lambda_{min}[Q]\|x\|^2. \tag{5.23}$$

The function $p^2: \mathbb{R} \times \mathbb{R}^n \to \mathbb{R}^N$ is continuous and is chosen to reduce the magnitude of the uncertain term $h + Gp^2$.

Utilizing (5.17) we note that

$$u^Tg(t, x, u, \omega) \geq -\beta_1(t, x, \omega)\|u\| + \beta_2(t, x, \omega)\|u\|^2$$

(3) For ease of exposition, arguments have been omitted in the following expressions.

where

$$\beta_1(t, x, \omega) = \|h(t, x, \omega) + G(t, x, \omega)p^2(t, x)\|, \tag{5.24}$$

$$\beta_2(t, x, \omega) = \lambda_{min}[G(t, x, \omega)] = [\lambda_{max}[\tilde{J}(t, x, \omega)]]^{-1} > 0. \tag{5.25}$$

It follows from Assumption M1(3) that there exist continuous functions κ, ρ: $\mathbb{R} \times \mathbb{R}^n \to \mathbb{R}_+$ such that for all $(t, x) \in \mathbb{R} \times \mathbb{R}^n$

$$\beta_1(t, x, \omega) \le \kappa(t, x), \tag{5.26}$$

$$\beta_1(t, x, \omega)/\beta_2(t, x, \omega) \le \rho(t, x) \tag{5.27}$$

for all $\omega \in \Omega$; thus (2.5)-(2.7) of A1 are satisfied and hence A1 is satisfied.

Utilizing the results of sec. 2, a practically stabilizing set of controllers for (5.15) is the set

$$P = \{p^\varepsilon: \varepsilon > 0\} \tag{5.28}$$

where p^ε: $\mathbb{R}\times\mathbb{R}^n \to \mathbb{R}^N$ is any continuous function which satisfies (2.12),(2.13) where

$$\alpha(t, x) = B^T P x \tag{5.29}$$

and κ^c, ρ^c, η are chosen to satisfy (2.8), (2.10).

Thus, a practically stabilizing set of controllers for (5.10) is

$$\tilde{P} = \{\tilde{p}: \tilde{p} = p^1 + p^2 + p^\varepsilon, \ p^\varepsilon \in P\}$$

where p^1, p^2 are as specified previously. The controllers generating $\hat{u}(t)$ are given by (5.9) with

$$\tilde{u}(t) = \tilde{p}(t, x(t))$$

and $\tilde{p} \in \tilde{P}$.

6 APPLICATIONS

Controllers whose designs are based on Lyapunov theory have been applied to a variety of engineering control problems including the tracking control of robotic manipulators (2, 12, 17, 23, 35, 52, 56, 95, 97, 100-102), the suspension control of a magnetically levitated vehicle (10, 11), the control of structures in the presence of seismic excitations (50, 60), and aerospace control problems (103-105, 110); see also (53). Experimental results are contained in (56, 95, 100, 102).

Applications to economic systems may be found in (39, 82-84). Harvesting problems are treated in (28, 60) and river pollution control problems are considered in (18, 63, 64, 75-77).

7 EXAMPLE: APPLICATION TO A MANUTEC r3 ROBOT

In this section, we apply the results of sec. 5 to a 3 d.o.f. model of a Manutec r3 robot as illustrated in Figure 1. A complete description of the robot is given in (92). Torques are applied to the arms by three electric motors, one for each arm.

Letting

$$q = [q_1 \ q_2 \ q_3]^T,$$

$$\hat{u} = [\hat{u}_1 \ \hat{u}_2 \ \hat{u}_3]^T$$

where $\hat{u}_i$ is the control voltage applied to motor i, the robot can be modelled by (5.1). In this model, ω is the mass of an uncertain payload modelled as a point mass located at P. The

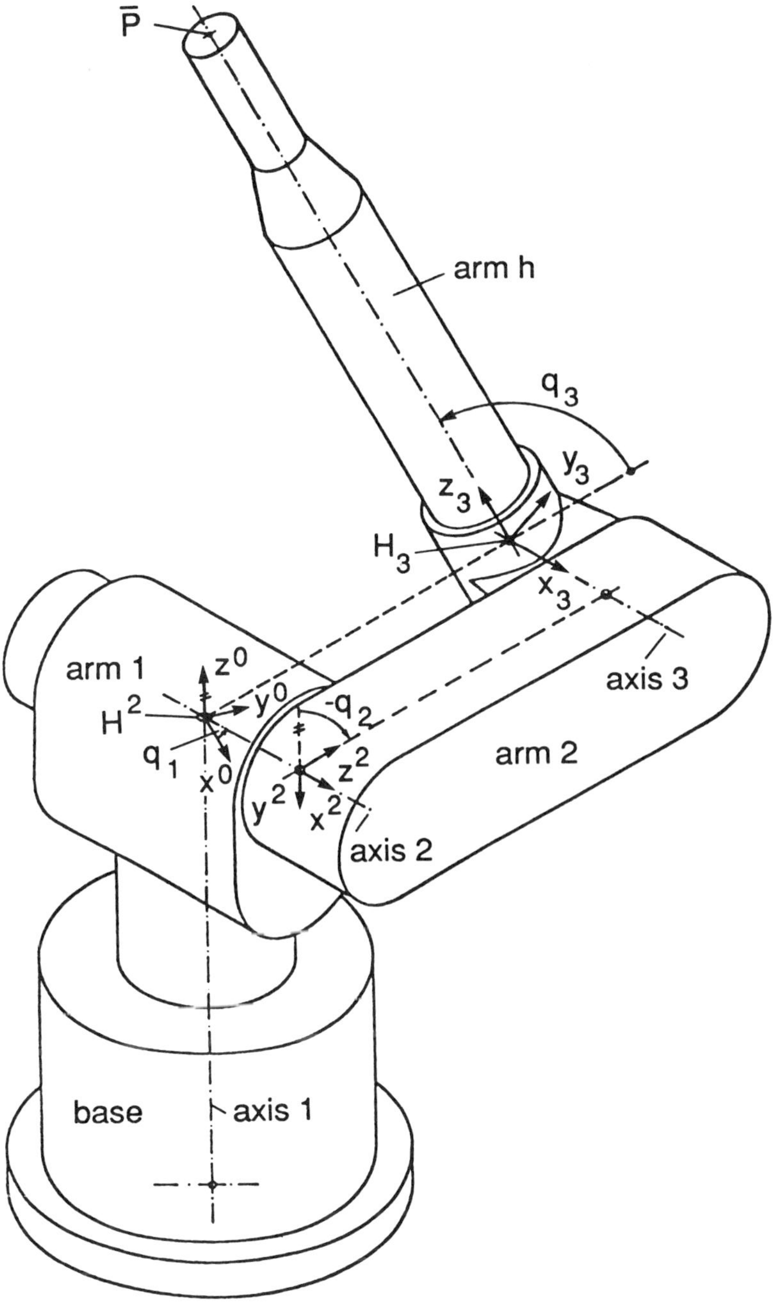

Figure 1. A 3 d.o.f. model of a Manutec r3 robot (from (92)).

only information available on ω is an upper bound $\overline{m}$. Thus,

$$\omega \in \Omega = [0, \overline{m}].$$

The mass matrix J(t, q, ω) is symmetric and W is constant, diagonal, and non-singular. The generalized force term U(t, q, $\dot{q}$, ω) is of the form

$$U(t, q, \dot{q}, \omega) = \chi^g(q, \omega) + \chi^d(q, \dot{q}, \omega).$$

Utilizing the expressions given in (92) for J, χ^g, χ^d, one can show that there exist scalars c_1, c_2, c_3, c_4, $\in \mathbb{R}_+$ such that

$$\|\chi^g(q, \omega)\| \le c_1, \qquad \|\chi^d(q, \dot{q}, \omega)\| \le c_2 \|\dot{q}\|^2,$$

$$\lambda_{max}[J(t, q, \omega)] \le c_3, \qquad \lambda_{min}[J(t, q, \omega)] \ge c_4 > 0$$

for all (t, q, $\dot{q}$) $\in \mathbb{R}\times\mathbb{R}^3\times\mathbb{R}^3$ and $\omega \in \Omega$. Thus Assumption M1 is satisfied; let

$$\hat{\beta}_0(t, q, \dot{q}) = c_1 + c_2 \|\dot{q}\|^2, \qquad \hat{\beta}_1(t, q) = c_3, \qquad \hat{\beta}_2(t, q) = c_4.$$

Suppose $\overline{q}(\cdot)$: $\mathbb{R} \to \mathbb{R}^3$ is a given desired trajectory and define

$$x = \begin{bmatrix} q - \overline{q}(t) \\ \dot{q} - \dot{\overline{q}}(t) \end{bmatrix}.$$

Now choose any nonsingular matrix $T \in \mathbb{R}^{3\times 3}$, any positive definite symmetric matrix $Q \in \mathbb{R}^{6\times 6}$, any positive scalar $\sigma > 0$, and let $P \in \mathbb{R}^{6\times 6}$ be the unique positive definite solution of (5.21).

Then, utilizing the results of sec. 5, a set of controllers which yield tracking of $\overline{q}(\cdot)$ by q($\cdot$) to within any desired degree of accuracy is given by

$$\hat{u}(t) = (T^T W)^{-1} \tilde{p}^{\varepsilon}(t, q(t), \dot{q}(t))$$

where, for any $\varepsilon > 0$,

$$\tilde{p}^{\varepsilon}(t, q, \dot{q}) = p^1(t, x) + p^2(t, x) + p^{\varepsilon}(t, x)$$

with

$$p^1(t, x) = -\gamma_1 \alpha(t, x)$$

$$p^2(t, x) = \gamma_2 T^{-1}\ddot{\overline{q}}(t)$$

$$p^{\varepsilon}(t, x) = \rho^c(t, x)[\varepsilon + \|\eta(t, x)\|]^{-1}\eta(t, x)$$

where

$$\alpha(t, x) = B^T P x$$

$$\eta(t, x) = \kappa^c(t, x)\alpha(t, x);$$

the bounding functions κ^c and ρ^c are chosen to be of the form

$$\kappa^c(t, x) = \kappa_1 + \kappa_2 \|\dot{q}\|^2 + \kappa_3 \|T^{-1}\ddot{\overline{q}}(t)\|,$$

$$\rho^c(t, x) = \rho_1 + \rho_2 \|\dot{q}\|^2 + \rho_3 \|T^{-1}\ddot{\overline{q}}(t)\|.$$

The scalars γ_1, γ_2, κ_i, ρ_i, i=1,2,3, are chosen to satisfy the inequalities given in Table 7.1 for all (t, q, $\dot{q}$, ω) $\in \mathbb{R} \times \mathbb{R}^3 \times \mathbb{R}^3 \times \Omega$.

$\tilde{J} = T^T J T$	$\lambda_{min}[\tilde{J}] \geq \underline{\sigma} > 0$	$\lambda_{max}[\tilde{J}] \leq \bar{\sigma}$
$\gamma_1 = \frac{1}{2}\,\underline{\sigma}\bar{\sigma}$	$\gamma_2 = 2(\underline{\sigma}^{-1} + \bar{\sigma}^{-1})^{-1}$	
$\|T^{-1}J^{-1}\chi^g\| \leq \kappa_1$	$\|T^{-1}J^{-1}\chi^d\| \leq \kappa_2\|\dot{q}\|^2$	$\|\gamma_2\tilde{J}^{-1} - I\| \leq \kappa_3$
$\|\tilde{J}\|\,\|T^{-1}J^{-1}\chi^g\| \leq \rho_1$	$\|\tilde{J}\|\,\|T^{-1}J^{-1}\chi^d\| \leq \rho_2\|\dot{q}\|^2$	$\|\tilde{J}\|\,\|\gamma_2\tilde{J}^{-1} - I\| \leq \rho_3$

Table 7.1

In numerical simulations we used the system parameter values given in (92); in particular, we chose

$$\bar{m} = 15 \text{ kg} \quad .$$

We let

$$T = \begin{bmatrix} 1.4142 & 0 & 0 \\ 0 & 1 & 0 \\ 0 & 0 & 1.8708 \end{bmatrix} \quad , \quad \sigma = 0.5 \ .$$

Q was chosen so that

$$B^T P = -2\sigma^{-1}\, T^{-1}\, \underline{A}$$

where

$$\underline{A} = \begin{bmatrix} -a_1 & 0 & 0 & -a_2 & 0 & 0 \\ 0 & -a_1 & 0 & 0 & -a_2 & 0 \\ 0 & 0 & -a_1 & 0 & 0 & -a_2 \end{bmatrix} \quad , \quad a_1 = 72, \ \ a_2 = 12 \ .$$

The rest of the controller parameters were as follows.

$$\varepsilon = 5, \ \ \sigma = 33.1, \ \ \bar{\sigma} = 182,$$
$$\kappa_1 = 7.29, \ \kappa_2 = 0.744, \ \ \kappa_3 = 1.25,$$
$$\rho_1 = 941, \ \ \rho_2 = 136, \ \ \rho_3 = 226 \ .$$

Figures 2-5 illustrate the results obtained for desired motion

$$\bar{q}_1(t) = \begin{cases} 2.5t & , \ t < 0.5 \\ 2.5(1-t) & , \ t \geq 0.5 \end{cases}$$

$$\bar{q}_2(t) \equiv 0 \cdot 1$$

$$\bar{q}_3(t) = \begin{cases} -4t & , \ t < 0.5 \\ -4(1-t) & , \ t \geq 0.5 \end{cases}$$

and initial conditions

$$q_1(0), \, q_3(0) = 0 \quad , \quad q_2(0) = \pi/4,$$

$$\dot{q}_i(0) \quad = 0 \qquad i = 1, 2, 3 \ .$$

For Figures 2 and 4, $\omega = 0$; for Figures 3 and 5, $\omega = 15$ kg.

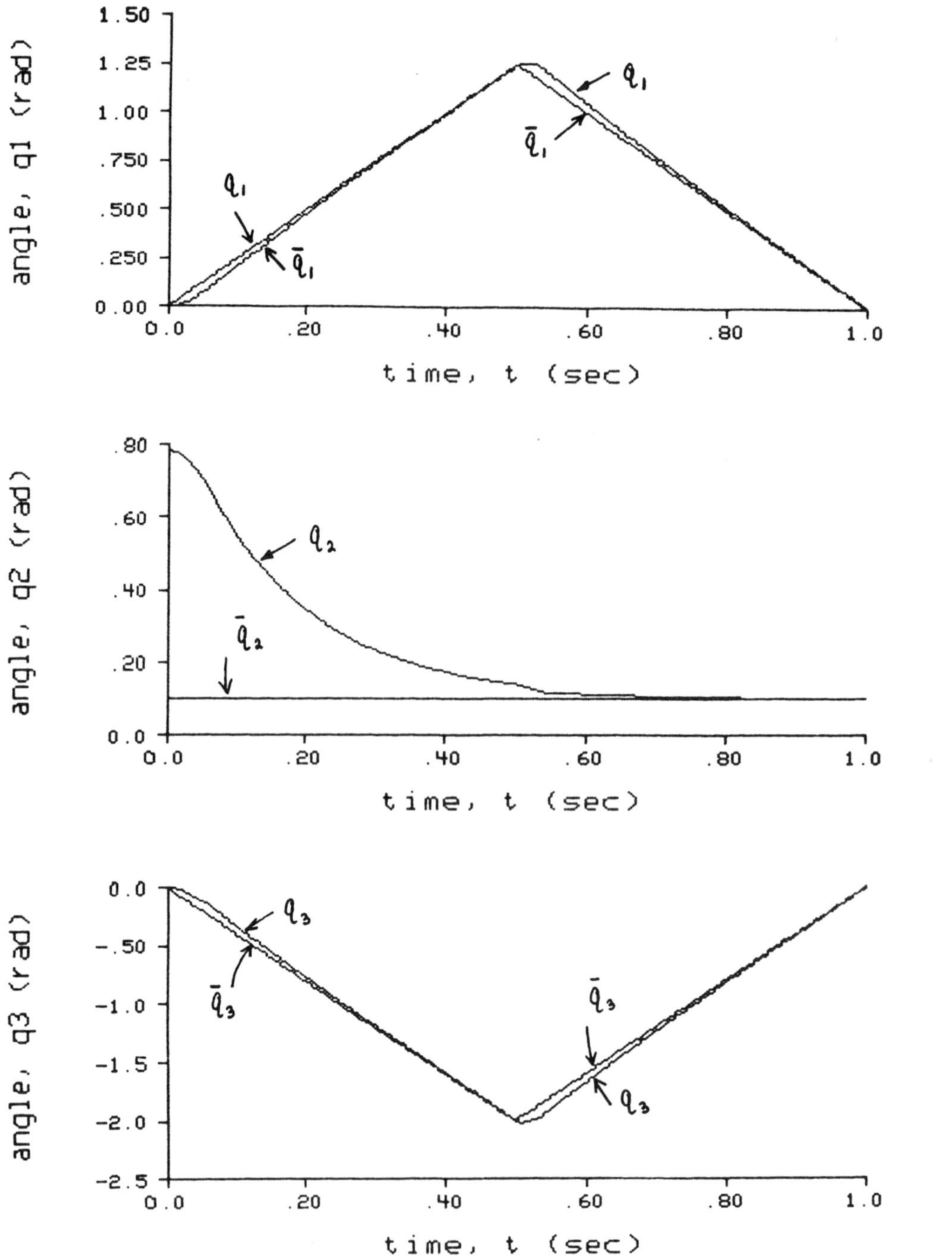

Figure 2. Angle responses for $\omega = 0\,\mathrm{kg}$.

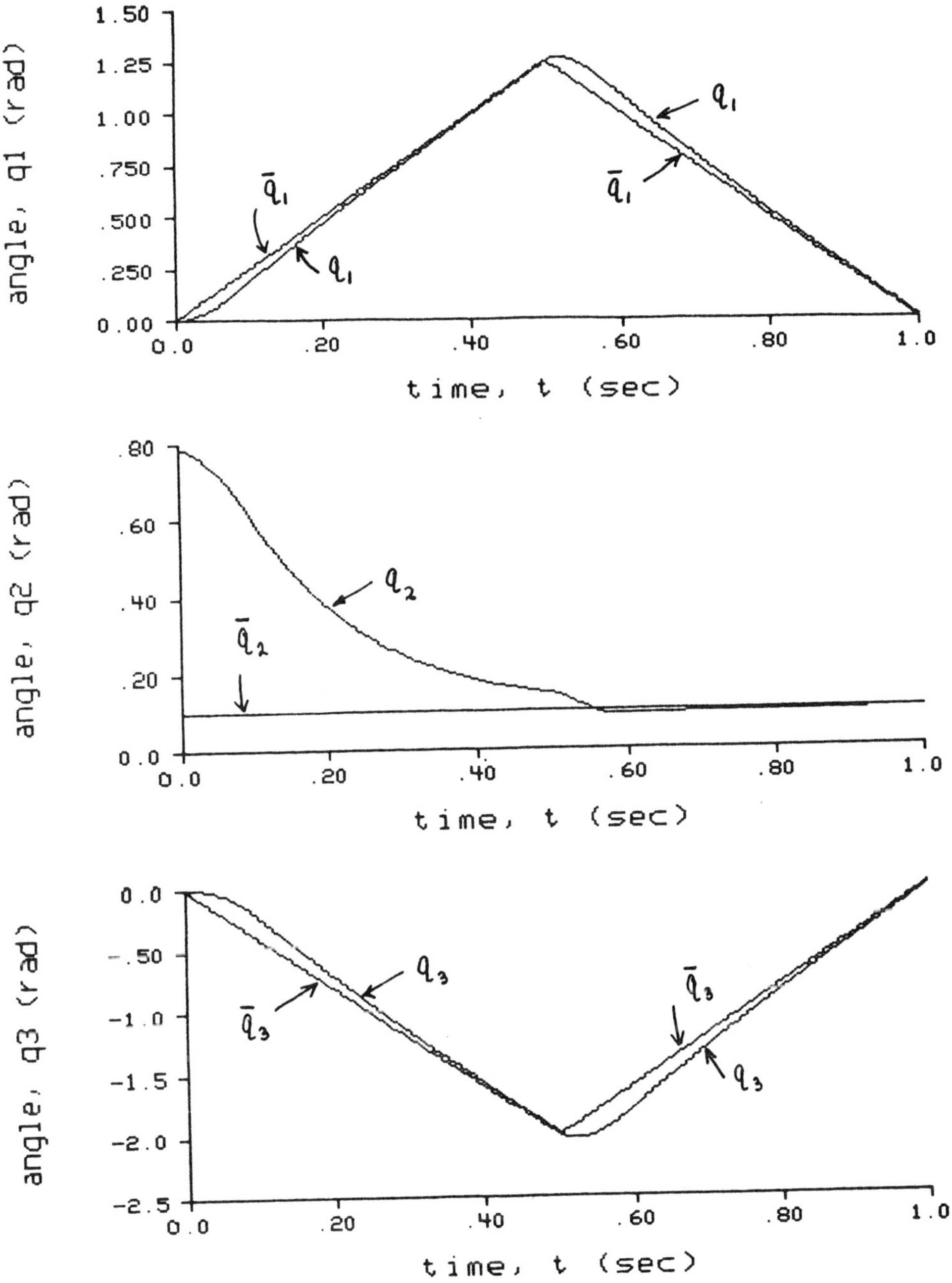

Figure 3. Angle responses for ω = 15 kg.

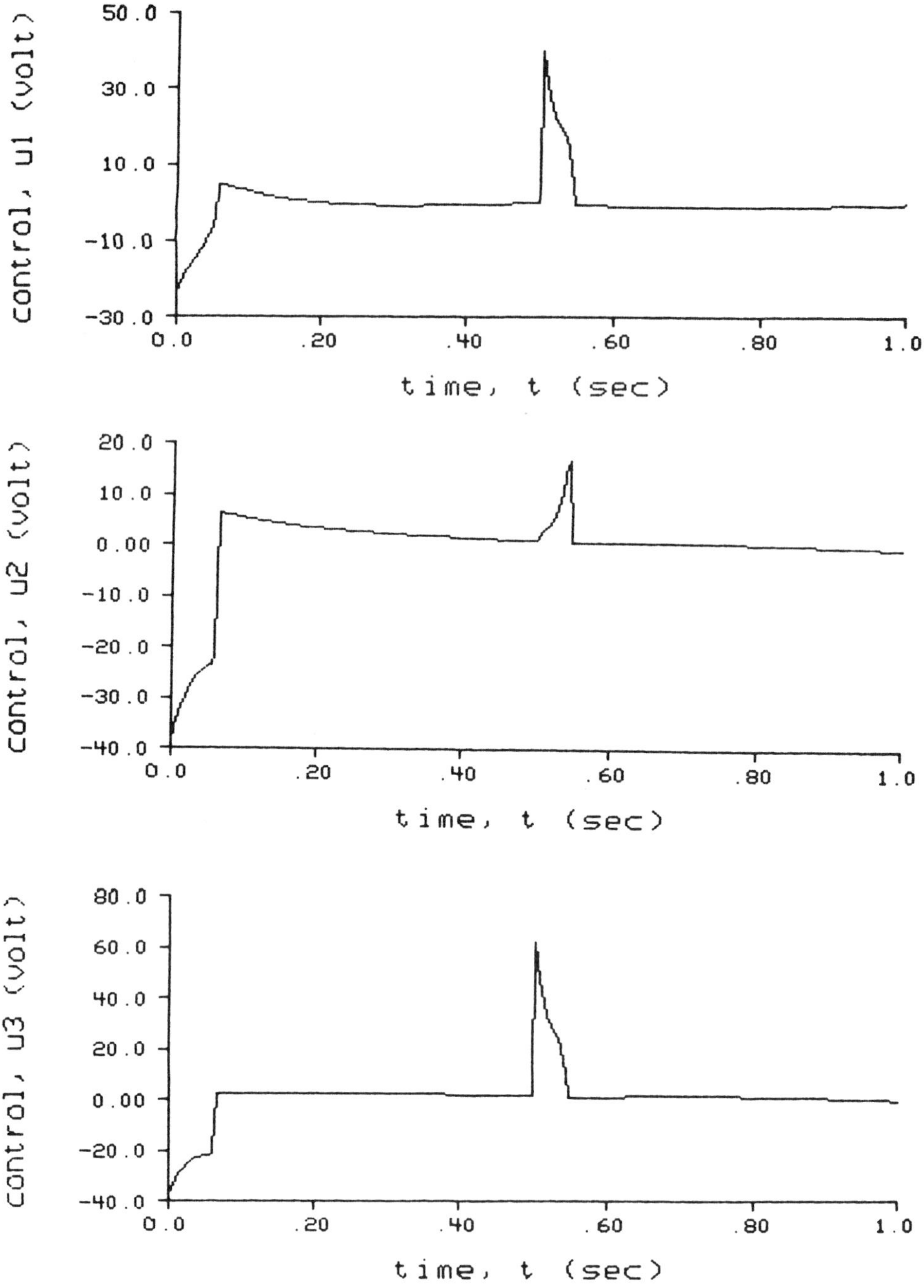

Figure 4. Control voltage histories for $\omega = 0$ kg.

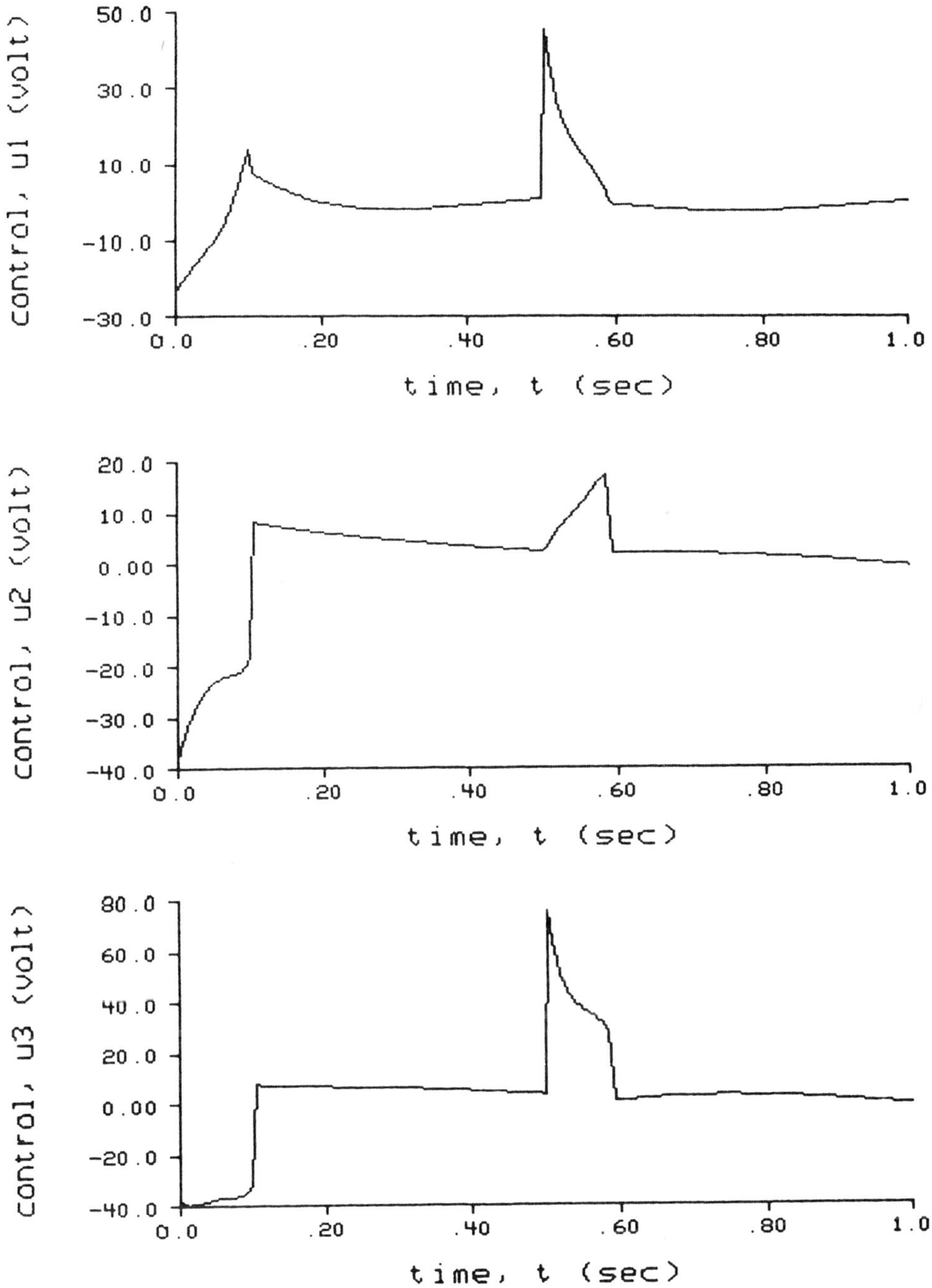

Figure 5. Control voltage histories for ω = 15 kg.

REFERENCES

[1] Ambrosino, G., G. Celentano, and F. Garofalo, Robust Model Tracking Control for a Class of Nonlinear Plants, *IEEE Trans. Automatic Control,* **AC-30,** 275, 1985.

[2] Ambrosino, G., G. Celentano, and F. Garofalo, Tracking Control of High-Performance Robots via Stabilizing Controllers for Uncertain Systems, *J. Optimiz. Theory Appl.,* **2,** 239, 1986.

[3] Barmish, B.R., Necessary and Sufficient Conditions for Quadratic Stabilizability of Uncertain Linear Systems, *J. Optimiz. Theory Appl.,* **46,** 399, 1985.

[4] Barmish, B. R., M. Corless, and G. Leitmann, A New Class of Stabilizing Controllers for Uncertain Dynamical Systems, *SIAM J. Contr. Optimiz.,* **21,** 246, 1983.

[5] Barmish, B. R., and A. R. Galimidi, Robustness of Luenberger Observers: Linear System Stabilized via Nonlinear Control, *Automatica,* **22,** 413, 1986.

[6] Barmish, B. R., and G. Leitmann, On Ultimate Boundedness Control of Uncertain Systems in the Absence of Matching Conditions, *IEEE Trans. Automatic Control,* **AC-27,** 1253, 1982.

[7] Barmish, B. R., I. R. Petersen and A. Feuer, Linear Ultimate Boundedness Control of Uncertain Dynamical Systems, *Automatica,* **19,** 523, 1983.

[8] Bass, R. W., Discussion of "Die Stabilität von Regelsystemen mit nachgebender Rückführung" by A. M. Letov, *Proc. Heidelberg Conf. Automatic Control,* 209, 1957.

[9] Bernstein, D. S., Robust Static and Dynamic Output-Feedback Stabilization: Deterministic and Stochastic Perspectives, *IEEE Trans. Automatic Control,* **AC-32,** 1076, 1987.

[10] Breinl, W., and G. Leitmann, Zustandsrückführung fur dynamische Systeme mit Parameterunsicherheiten, *Regelungstechnik,* **31,** 95, 1983.

[11] Breinl, W., and G. Leitmann, State Feedback for Uncertain Dynamical Systems, *Applied Mathematics and Computation,* to appear.

[12] Bremer, H., and A. Truckenbrodt, Robust Control for Industrial Robots, *Proc. of RoManSy'84, Theory and Practice of Robots and Manipulators,* MIT Press, 1985.

[13] Chen, Y. H., On the Deterministic Performance of Uncertain Dynamical Systems, *Int. J. Control,* **43,** 1557, 1986.

[14] Chen, Y. H., Robust Output Feedback Controller: Direct Design, *Int. J. Control,* **46,** 1083, 1987.

[15] Chen, Y. H., Robust Output Feedback Controller: Indirect Design, *Int. J. Control,* **46,** 1093, 1987.

[16] Chen, Y. H., Deterministic Control of Large-Scale Uncertain Dynamical Systems, *J. Franklin Inst.,* to appear.

[17] Chen, Y. H., Robust Control of Mechanical Manipulators, *J. Dynam. Syst. Meas. Control,* submitted.

[18] Chen, Y. H., and C. S. Lee, On the Control of an Uncertain Water Quality System, *Optimal Control Appl. Methods,* **8,** 279, 1987.

[19] Chen, Y. H., and G. Leitmann, Robustness of Uncertain Systems in the Absence of Matching Assumptions, *Int. J. Control,* **45,** 1527, 1987.

[20] Corless, M., Controlling Uncertain Systems Within a Subset of the State Space, *Proc. American Control Conf.,* Boston, 1985.

[21] Corless, M., Stabilization of Uncertain Discrete-Time Systems, *Proc. IFAC Workshop on Model Error Concepts and Compensation,* Boston, 1985.

[22] Corless, M., Robustness of a Class of Feedback-Controlled Uncertain Nonlinear Systems in the Presence of Singular Perturbations, *Proc. American Control Conf.,* Minneapolis, 1987.

[23] Corless, M., Tracking Controllers for Uncertain Systems: Application to a Manutec r3 Robot, *J. Dynam. Syst. Meas. Contr.,* to appear.

[24] Corless, M., Guaranteed Rates of Exponential Convergence for Uncertain Systems, *J. Optimiz. Theory Applic.,* to appear.

[25] Corless, M., D. P. Goddall, G. Leitmann, and E. P. Ryan, Model-Following Controls for a Class of Uncertain Dynamical Systems, *Proc. IFAC Conf. on Identification and System Parameter Estimation,* York University, England, 1985.

[26] Corless, M., and G. Leitmann, Continuous State Feedback Guaranteeing Uniform Ultimate Boundedness for Uncertain Dynamic Systems, *IEEE Trans. Automatic Control,* **AC-26,** 1139, 1981.

[27] Corless, M., and G. Leitmann, Erratum to 'Continuous State Feedback Guaranteeing Uniform Ultimate Boundedness for Uncertain Dynamic Systems,' *IEEE Trans. Automatic Control,* **AC-28,** 249, 1983.

[28] Corless, M., and G. Leitmann, Adaptive Control of Systems Containing Uncertain Functions and Unknown Functions with Uncertain Bounds, *J. Optimiz. Theory Applic.,* **41,** 155, 1983.

[29] Corless, M., and G. Leitmann, Adaptive Control for Uncertain Dynamical Systems, in *Dynamical Systems and Microphysics: Control Theory and Mechanics,* (A. Blaquiere and G. Leitmann, eds.), Academic Press, New York, 1984.

[30] Corless, M., and G. Leitmann, Adaptive Long-Term Management of Some Ecological Systems Subject to Uncertain Disturbances, in *Optimal Control Theory and Economic Analysis 2,* (G. Feichtinger, ed.), Elsevier Science Publishers, Amsterdam, Holland, 1985.

[31] Corless, M., and G. Leitmann, Adaptive Controllers for Avoidance or Evasion in an Uncertain Environment: Some Examples, *Computers and Mathematics with Applications,* to appear.

[32] Corless, M., and G. Leitmann, Adaptive Controllers for Uncertain Dynamical Systems, *Proc. Conf. on Modeling and Adaptive Control,* Sopron, Hungary, Lecture Notes in Control and Information Sciences, Springer Verlag, to appear.

[33] Corless, M., and G. Leitmann, Deterministic Control for Uncertain Dynamical Systems, *Proc. Conf. on Modeling and Adaptive Control,* Sopron, Hungary, Lecture Notes in Control and Information Sciences, Springer Verlag, to appear.

[34] Corless, M., and Leitmann, G., Deterministic Control of Uncertain Systems: A Lyapunov Theory Approach, in *Deterministic Nonlinear Control of Uncertain Systems: Variable Structure and Lyapunov Control,* (A. Zinober, ed.), IEE Publishers, to appear.

[35] Corless, M., G. Leitmann and E. P. Ryan, Tracking in the Presence of Bounded Uncertainties, *Proc. 4th IMA Int. Conf. Control Theory,* Cambridge University, England.

[36] Corless, M., G. Leitmann, and E. P. Ryan, Control of Uncertain Systems with Neglected Dynamics, in *Deterministic Nonlinear Control of Uncertain Systems: Variable Structure and Lyapunov Control,* (A. Zinober, ed.), IEE Publishers, to appear.

[37] Corless, M., G. Leitmann, and J. M. Skowronski, Adaptive Control for Avoidance or Evasion in an Uncertain Environment, *Computers and Mathematics with*

Applications, **13,** 1, 1987.

[38] Corless, M. and J. Manela, Control of Uncertain Discrete-Time Systems, *Proc. American Control Conf.,* Seattle, Washington, 1986.

[39] Deissenberg, C., Long-run Macroeconomic Stabilization Under Bounded Uncertainty, *Proc. Optimiz. Days,* 1986.

[40] Eslami, M., and D. L. Russell, On Stability with Large Parameter Variations: Stemming from the Direct Method of Lyapunov, *IEEE Trans. Automatic Control,* **AC-25,** 1980.

[41] Galimidi, A. R., and B. R. Barmish, The Constrained Lyapunov Problem and its Application to Robust Output Feedback Stabilization, *IEEE Trans. Automatic Control,* **AC-31,** 410, 1986.

[42] Garofalo, F., and G. Leitmann, Nonlinear Composite Control of a Nominally Linear Singularly Perturbed Uncertain System, *Proc. 12th IMACS World Congress,* Paris, 1988.

[43] Garofalo, F., and G. Leitmann, Nonlinear Composite Control of a Class of Nominally Linear Singularly Perturbed Uncertain Systems, in *Variable Structure Control Systems,* (A. Zinober, ed.), IEE Publishers, London, to appear.

[44] Grayson, L. P., Two Theorems on the Second Method, *IEEE Trans. Automatic Control,* **AC-9,** 587, 1964.

[45] Grayson, L. P., The Status of Synthesis Using Lyapunov's Method, *Automatica,* **3,** 91, 1965.

[46] Gutman, S., *Uncertain Dynamical Systems - A Differential Game Approach,* NASA TMX-73, 135, 1976.

[47] Gutman, S., Uncertain Dynamical Systems--Lyapunov Min-Max Approach, *IEEE Trans. Automatic Control,* **AC-24,** 437, 1979.

[48] Gutman, S., and G. Leitmann, On a Class of Linear Differential Games, *J. Optimiz. Theory Appl.,* **17,** 511, 1975.

[49] Gutman, S., and G. Leitmann, Stabilizing Control for Linear Systems with Bounded Parameter and Input Uncertainty, *Proc. 2nd IFIP Conf. Optimiz. Techniques,* Springer-Verlag, Berlin, 1975.

[50] Gutman, S., and G. Leitmann, Stabilizing Feedback Control for Dynamical Systems with Bounded Uncertainty, *Proc. IEEE Conf. Decision Control,* 1976.

[51] Gutman, S., and Z. Palmor, Properties of Min-Max Controllers in Uncertain Dynamical Systems, *SIAM J. Control Optimiz.,* **20,** 850, 1982.

[52] Ha, I. J., and E. G. Gilbert, Robust Tracking in Nonlinear Systems, *IEEE Trans. Automatic Control,* **AC-32,** 763, 1987.

[53] Hagedorn, P., On the Active Control of Rotors with Uncertain Parameters, *Int. Conf. on Rotor Dynamics,* Tokyo, 1986.

[54] Hollot, C. V., Bounded Invariant Lyapunov Functions: a Means for Enlarging the Class of Stabilizable Uncertain Systems, *Int. J. Control,* **46,** 161, 1987.

[55] Hollot, C. V., and B. R. Barmish, Optimal Quadratic Stabilizability of Uncertain Linear Systems, *Proc. 18th Allerton Conf. Communications Contr. Computing,* 1980.

[56] Horowitz, R., H. I. Stephens, and G. Leitmann, Experimental Verification of a Deterministic Controller for a D. C. Motor with Uncertain Dynamics, *Proc. American Control Conf.,* Minneapolis, 1987.

[57] Johnson, G. W., Synthesis of Control Systems with Stability Constraints via the Direct Method of Lyapunov, *IEEE Trans. Automatic Control,* **AC-9,** 270, 1964.

[58] Kalman, R. E., and J. E. Bertram, Control System Analysis and Design via the "Second Method" of Lyapunov, I: Continuous-Time Systems, *J. Basic Engineering,* **82,** 371, 1960.

[59] Kalman, R. E., and J. E. Bertram, Control System Analysis and Design via the "Second Method" of Lyapunov, II: Discrete-Time Systems, *J. Basic Engineering,* **82,** 394, 1960.

[60] Kelly, J., G. Leitmann, and A. Soldatos, Robust Control of Base-Isolated Structures Under Earthquake Excitation, *J. Optimiz. Theory Appl.,* **52,** 3, 1987.

[61] LaSalle, J., Stability and Control, *SIAM J. Control,* **1,** 3, 1962.

[62] Lee, C. S. and G. Leitmann, On Optimal Long-Term Management of Some Ecological Systems Subject to Uncertain Disturbances, *Int. J. Syst. Sci.,* **14,** 979, 1983.

[63] Lee, C. S., and G. Leitmann, Deterministic Control of an Uncertain Single Reach River: a Discrete Model, *Proc. of SICE '87,* Hiroshima, Japan, 1987.

[64] Lee, C. S., and G. Leitmann, Uncertain Dynamical Systems: An Application to River Pollution Control, *Proc. Modeling and Management of Resources under Uncertainty,* Honolulu 1985, Lecture Notes in Biomathematics, **72,** 167, Springer-Verlag, 1987.

[65] Leitmann, G., A Simple Differential Game, *J. Optimiz. Theory Appl.,* **2,** 220, 1968.

[66] Leitmann, G., On Stabilizing a Linear System with Bounded State Uncertainty, in *Topics in Contemporary Mechanics,* Springer-Verlag, Vienna, 1974.

[67] Leitmann, G., Stabilization of Dynamical Systems under Bounded Input Disturbance and Parameter Uncertainty, in *Differential Games and Control Theory II,* (E. O. Roxin, P.-T. Liu, and R. L. Sternberg, eds.), 47, Marcel Dekker, New York, 1976.

[68] Leitmann, G., Guaranteed Ultimate Boundedness for a Class of Uncertain Linear Dynamical Systems, in *Differential Games and Control Theory III,* (P.-T. Liu and E. Roxin, eds.), 29, Marcel Dekker, New York, 1978.

[69] Leitmann, G., Guaranteed Ultimate Boundedness for a Class of Uncertain Linear Dynamical Systems, *IEEE Trans. Automatic Control,* **AC-23,** 1109, 1978.

[70] Leitmann, G., Guaranteed Asymptotic Stability for a Class of Uncertain Linear Dynamical Systems, *J. Optimiz. Theory Appl.,* **27,** 99, 1979.

[71] Leitmann, G., Guaranteed Asymptotic Stability for Some Linear Systems with Bounded Uncertainties, *J. Dynam. Syst. Meas. Contr.,* **101,** 212, 1979.

[72] Leitmann, G., Deterministic Control of Uncertain Systems, *Acta Astronautica,* **7,** 1457, 1980.

[73] Leitmann, G., Guaranteed Avoidance Strategies, *J. Optimiz. Theory Appl.,* **32,** 569, 1980.

[74] Leitmann, G., On the Efficacy of Nonlinear Control in Uncertain Linear Systems, *J. Dynam. Syst. Meas. Control,* **102,** 95, 1981.

[75] Leitmann, G., and C. S. Lee, Continuous Feedback Guaranteeing Uniform Ultimate Boundedness for Uncertain Linear Delay Systems: An Application to River Pollution Control, *J. Comput. and Math. with Appl.,* to appear.

[76] Leitmann, G., C. S. Lee, and Y. H. Chen, Decentralized Control for a Large Scale Uncertain River System, *Proc. IFAC Workshop on Modelling, Decisions and Games for Social Phenomena,* Beijing, China, 1986.

[77] Leitmann, G., C. S. Lee, and Y. H. Chen, Decentralized Control for an Uncertain Multi-Reach River System, *Proc. Conf. on Optimal Control and Variational Calculus,* Oberwolfach, 1986, West Germany, Lecture Notes in Control and Information Sciences, **95,** Springer-Verlag, 1987.

[78] Leitmann, G., and E. P. Ryan, Output Feedback Control of a Class of Singularly Perturbed Uncertain Dynamical Systems, *Proc. American Control Conf.*, Minneapolis, 1987.

[79] Leitmann, G., E. P. Ryan, and A. Steinberg, Feedback Control of Uncertain Systems: Robustness with Respect to Neglected Actuator and Sensor Dynamics, *Int. J. Control,* **43,** 1243, 1986.

[80] Leitmann, G., and J. Skowronski, Avoidance Control, *J. Optimiz. Theory Appl.,* **581,** 581, 1977.

[81] Leitmann, G., and J. Skowronski, A Note on Avoidance Control, *Optimal Control Appl. Methods,* **4,** 335, 1983.

[82] Leitmann, G., and H. Y. Wan, Jr., A Stabilization Policy for an Economy with Some Unknown Characteristics, *J. Franklin Institute,* **306,** 23, 1978.

[83] Leitmann, G., and H. Y. Wan, Jr., Macro-Economic Stabilization Policy for an Uncertain Dynamic Economy, in *New Trends in Dynamic System Theory and Economics,* Academic Press, New York, 1979.

[84] Leitmann, G., and H. Y. Wan, Jr., Performance Improvement of Uncertain Macroeconomic Systems, in *Dynamic Optimization and Mathematical Economics,* (P.-T. Liu, ed.), Plenum Press, New York, 1979.

[85] Letov, A. M., *Stability of Nonlinear Regulating Systems* (in Russian), Izdatel'stvo Technichesko-Teoreticheskoi Literatury, Moscow, 1955.

[86] Lur'e, A. I., *Some Nonlinear Problems in the Theory of Automatic Control,* (in Russian), Gostekhizdat, Moscow, 1951; German translation, Akademie Verlag, 1957; English translation, Her Majesty's Stationery Office, 1957.

[87] Madani-Esfahani, S. M., R. A. DeCarlo, M. J. Corless, and S. H. Zak, On Deterministic Control of Uncertain Nonlinear Systems, *Proc. American Control Conf.,* Seattle, Washington, 1986.

[88] Monopoli, R. V., Discussion on 'Two Theorems on the Second Method,' *IEEE Trans. Automatic Control,* **AC-10,** 369, 1965.

[89] Monopoli, R. V., Synthesis Techniques Employing the Direct Method, *IEEE Trans. Automatic Control,* **AC-10,** 369, 1965.

[90] Monopoli, R. V., Corrections to: Synthesis Techniques Employing the Direct Method, *IEEE Trans. Automatic Control,* **AC-11,** 631, 1966.

[91] Monopoli, R. V., *Engineering Aspects of Control System Design Via the 'Direct Method' of Lyapunov,* CR-654, NASA, 1966.

[92] Otter, M., and S. Turk, Mathematical Model of the Manutec r3 Robot, Institüt für Dynamik der Flugsysteme, DFVLR, West Germany, 1987.

[93] Petersen, I. R., Structural Stabilization of Uncertain Systems: Necessity of the Matching Condition, *SIAM J. Control Optimiz.,* **23,** 286, 1985.

[94] Petersen, I. R., and B. R. Barmish, Control Effort Considerations in the Stabilization of Uncertain Dynamical Systems, *Proc. American Control Conf,* 1984.

[95] Riedl, M., Regelung eines nichtlinearen, elastischen, Antriebssystems auf der Grundlage eines unsicheren Modells, *Studienarbeit,* **STUD-30,** University of Stuttgart, Germany, 1986.

[96] Ryan, E. P., and M. Corless, Ultimate Boundedness and Asymptotic Stability of a Class of Uncertain Dynamical Systems via Continuous and Discontinuous Feedback Control, *IMA J. Math. Contr. Inf.,* **1,** 223, 1984.

[97] Ryan, E. P., G. Leitmann, and M. Corless, Practical Stabilizability of Uncertain Dynamical Systems, Application to Robotic Tracking, *J. Optimiz. Theory Applic.,* **47,**

235, 1985.

[98] Schmitendorf, W. E., A Design Methodology for Robust Stabilization, *J. Guidance, Control, Dynamics,* **10,** 250, 1987.

[99] Schmitendorf, W. E., Design of Observer-Based Robust Stabilizing Controllers, Applied Mathematics Tech. Report 8714, Northwestern University, 1987.

[100] Settelmeyer, E., Kraftregelung eines zweigliedrigen Roboterarmes, *Diplomarbeit,* **DIPL-18,** University of Stuttgart, Germany, 1987.

[101] Shoureshi, R., M. J. Corless, and M. D. Roesler, Control of Industrial Manipulators with Bounded Uncertainties, *J. Dynam. Syst. Meas. Contr.,* **109,** 53, 1987.

[102] Shoureshi, R., M. E. Momot, and M. D. Roesler, Robust Control for Manipulators with Uncertain Dynamics, *Proc. IFAC Congress,* Munich, Germany, 1987.

[103] Singh, S. N., Attitude Control of a Three Rotor Gyrostat in the Presence of Uncertainty, *J. Astronautical Sciences,* **35,** 1987.

[104] Singh, S. N., Nonlinear Adaptive Attitude Control of Spacecraft, *IEEE Trans. Aerospace and Electronic Systems,* **AES-23,** 371, 1987.

[105] Singh, S. N., and A. A. R. Coelho, Nonlinear State Feedback Control of Uncertain Linear Systems and Application to Aircraft Control, *Proc. Conf. Decision Control,* 1983.

[106] Singh, S. N., and A. A. R Coelho, Ultimate Boundedness Control of Set Points of Mismatched Uncertain Systems, *Int. J. Syst. Sci.,* **14,** 1983.

[107] Skowronski, J. M., *Control Dynamics of Robotic Manipulators,* Academic Press, 1986.

[108] Spong, M. W., and H. Sira-Ramirez, Robust Control Design Techniques for a Class of Nonlinear Systems, *Proc. American Control Conf.,* Seattle, 1986.

[109] Spong, M. W., J. S. Thorp, and J. M. Kleinwaks, The Control of Robot Manipulators with Bounded Input, Part II: Robustness and Disturbance Rejection, *Proc. IEEE Conf. Decision Control, 1984.*

[110] Stalford, H. L., On Robust Control of Wing Rock Using Nonlinear Control, *Proc. American Control Conf.,* Minneapolis, 1987.

[111] Stalford, H. L., Robust Control of Uncertain Systems in the Absence of Matching Conditions: Scalar Input, *Proc. Conf. Decision and Control,* 1987.

[112] Steinberg, A., and M. Corless, Output Feedback Stabilization of Uncertain Dynamical Systems, *IEEE Trans. Automatic Control,* **AC-30,** 1025, 1985.

[113] Thorp, J. S. and B. R. Barmish, On Guaranteed Stability of Uncertain Systems via Linear Control, *J. Optimiz. Theory Appl.,* **35,** 559, 1981.

[114] Thowsen, A., Uniform Ultimate Boundedness of the Solutions of Uncertain Dynamic Delay Systems with State-Dependent and Memoryless Feedback Control, *Int. J. Control,* **37,** 1135, 1983.

[115] Yedavalli, R. K., and Z. Liang, Reduced Conservatism in the Ultimate Boundedness Control of Mismatched Uncertain Linear Systems, *J. Dynam. Syst. Meas. Contr.,* **109,** 1, 1987.

[116] Yu, Y., On Stabilizing Uncertain Linear Delay Systems, *J. Optimiz. Theory Appl.,* **3,** 503, 1983.

[117] Walcott, B. L., and S. H. Zak, State Observation of Nonlinear Uncertain Dynamical Systems, *IEEE Trans. Automatic Control,* **AC-32,** 166, 1987.

[118] Zak, S. H., J. D. Brehove, and M. J. Corless, Control of Uncertain Systems with Unmodeled Actuator and Sensor Dynamics and Incomplete State Information, *IEEE*

Trans. Syst. Man Cyber, to appear.

[119] Zeheb, E., A Sufficient Condition for Output Feedback Stabilization of Uncertain Systems, *IEEE Trans. Automatic Control,* **AC-31,** 1055, 1986.

[120] Zhou, K., and P. P. Khargonekar, On the Stabilization of Linear Uncertain Systems via Bound Invariant Lyapunov Functions, submitted.

Explicit Estimates for Convergence of Suboptimal Controls

Hector Fattorini,* Professor, Department of Mathematics, University of California at Los Angeles, Los Angeles, California

Halina Frankowska, Professor, CNRS-Centre National de la Recherche Scientifique, CEREMADE, Universite Paris-Dauphine, Paris, France

1. INTRODUCTION

This paper is an announcement of some of the results in [12]. It deals mostly with systems governed by quasilinear initial value problems

$$y'(t) = Ay(t) + f(t,y(t),u(t)) \quad (0 \leq t \leq T), \tag{1.1}$$

$$y(0) = \xi, \tag{1.2}$$

where A is the infinitesimal generator of a strongly continuous

*The work of this author was supported in part by the NSF under grant DMS-8701877

semigroup $\{S(t);t \geq 0\}$ in a Hilbert space E (although results for boundary control systems are also obtained). Controls $u(t)$ are strongly measurable functions defined in $0 \leq t \leq \bar{t}$ $(\bar{t} \leq T)$ taking values in a subset U (called the *control set*) of a second Hilbert space F; the set of all these controls, called the *control space* is denoted by $W(0,\bar{t};U)$. Precise conditions on the control set U and on $f(t,y,u)$ will be given in next section. Assuming that (1.1)-(1.2) has a unique solution $y(t) = y(t,\xi,u)$ for each ξ and each control u we call $y(t,\xi,u)$ the *trajectory* corresponding to ξ, u. The *cost functional* or *performance index* of the problem is

$$y_0(t,\xi,u) = \int_0^t f_0(\sigma,y(\sigma,\xi,u),u(\sigma))d\sigma , \tag{1.3}$$

where $f_0(t,y,u)$ is a real valued function; in particular, $f_0 \equiv 1$ for the time optimal problem. The *target condition* is

$$y(t,\xi,u) = \bar{y} , \tag{1.4}$$

where $\bar{y}$ is the *target point*. Both $\bar{y}$ and the initial condition $\bar{\xi}$ are fixed in advance.

The control problem is standard: a control $\bar{u}(\cdot)$ in $W(0,\bar{t};U)$ is *optimal* if

$$y_0(\bar{t},\bar{\xi},\bar{u}) = m = \inf y_0(t,\bar{\xi},u) , \tag{1.5}$$

among all controls $u = u(\cdot)$ in $W(0,t;U)$ whose corresponding trajectories $y(t) = y(t,\bar{\xi},u)$ satisfy the target condition (1.4) at the time t (the same as in (1.5)). The initial condition $\bar{\xi}$ is the same for all trajectories.

We investigate in [12] the *robustness* of this optimal control problem, understood in the following sense. Let $\{t_n\}$ be a numerical sequence, $\{u^n\}$ a sequence of controls with u^n in the control space $W(0,t_n;U)$, $\{\xi^n\}$ a sequence in E. Assume that

$$\lim_{n\to\infty} \xi^n = \bar{\xi} , \quad \lim_{n\to\infty} y(t_n,\xi^n,u^n) = \bar{y} ,$$

$$\limsup_{n\to\infty} y_0(t_n,\xi^n,u^n) \leq m , \tag{1.6}$$

where m is the minimum in (1.5). Does it follow from these conditions that

$$\lim_{n \to \infty} u^n(\cdot) = \bar{u}(\cdot) , \tag{1.7}$$

where $\bar{u}(\cdot)$ is an optimal control?

The answer to this problem must in general be negative due to lack of uniqueness of the optimal control; if several optimal controls exist, (1.7) may be expected to hold at best for suitable subsequences. Also, the problem depends on what kind of convergence is expected in (1.7); if only weak convergence is at play, elementary affirmative results (for subsequences) exist.

The results in [12] reported here are on *strong* weighted L^2 convergence; in fact, explicit bounds for the weighted L^2 norm $\|u^n - \bar{u}\|$ (or for associated norms) are obtained in terms of the norms $\|\xi^n - \xi\|$, $\|y(t_n,\xi^n,u^n) - \bar{y}\|$ and of the deviation from the minimum $y_o(t_n,\xi^n,u^n) - m$. These estimates are based on a (suitable version of) the maximum principle and only suceed completely when the maximum principle is nontrivial on the whole control interval; otherwise, the weight in the L^2 norm above may be zero in certain subsets of the control interval $0 \leq t \leq \bar{t}$. Estimates of the required form are actually obtained only for linear systems; for nonlinear equations similar bounds exist, but depend as well on the L^1 norm of the difference $y(\cdot,\xi^n,u^n) - y(\cdot,\bar{\xi},\bar{u})$ of the optimal and the approximating trajectory.

Due to notorious difficulties in establishing the maximum principle with point target condition (1.4) in infinite dimensional spaces, the resuls have a limited range of applicability in that setting. Nevertheless, many important examples are amenable to the theory; we consider in Section 7 a system described by a quasilinear wave equation. In the case of parabolic systems the maximum principle may fail, but modified versions are available and give rise to weaker estimates (Sections 8 and 9). Finally, we mention in Section 10 a result in the same spirit for boundary control systems. In all cases the estimates are explicit (modulo a weight function) and can be used as a basis for obtaining explicit convergence rates for approximation of optimal controls.

For answers to the robustness problem obtained by other means, see [5], [6], [7], [8], [9], [10] and [11].

2. ASSUMPTIONS ON THE INITIAL VALUE PROBLEM

By definition, a *solution* of (1.1)-(1.2) is a solution of

$$y(t) = S(t)\xi + \int_0^t S(t - \sigma)f(\sigma,y(\sigma),u(\sigma))d\sigma \quad (0 \leq t \leq T) \tag{2.1}$$

where $S(t)$ is the semigroup generated by A. The control set U is assumed *bounded*. The conditions on f are standard (although not the most general possible) for local existence and uniqueness, as well as for insuring the necessary properties of the linearized equation (3.2) below:

(F) $f(t,y,u)$ has a Fréchet derivative $\partial_y f(t,y,u)$ with respect to y and $f(t,y,u)$ (resp. $\partial_y f(t,y,u)$) is continuous (resp. strongly continuous) in $[0,T] \times E \times U$ and bounded on bounded subsets.

Under these assumptions on A and $f(t,y,u)$ (1.1)-(1.2) (or rather (2.1)) has a unique local solution. To be able to extend solutions to a fixed interval $0 \leq t \leq T$ it is enough to have an *a priori* bound

$$\|y(t)\| \leq C \quad (0 \leq t \leq T_o) \tag{2.2}$$

for any solution of (2.1) in an interval $0 \leq t \leq T_o$. For information on how to obtain such bounds see [5], [6].

The conditions on f imply that, for every $C > 0$ there exists a constant $K = K(C)$ such that

$$\|\partial_y f(t,y,u)\| \leq K \quad (0 \leq t \leq T, \|y\| \leq C, u \in U). \tag{2.3}$$

This implies the local Lipschitz condition

$$\|f(t,y,u) - f(t,y',u)\| \leq K\|y - y'\|$$
$$(0 \leq t \leq T, \|y\|, \|y'\| \leq C, u \in U) \tag{2.4}$$

where K is the same constant in (2.3).

3. THE MAXIMUM PRINCIPLE

Although [12] deals with a general performance index y_0, we treat mostly the time optimal problem here. The maximum principle gives a vector $z \neq 0$ and a set e of full measure in $0 \leq s \leq \bar{t}$ such that

$$(z,\ S(\bar{t},s;\bar{u})\{f(s,y(s,\bar{\xi},\bar{u}),v) - f(s,y(s,\bar{\xi},\bar{u}),\bar{u}(s))\}) \leq 0 \tag{3.1}$$

for all $s \in e$, $v \in U$, where $S(t,s;u)$ is the solution operator of the linearized equation

$$z'(s) = (A + \partial_y f(s,y(s,\bar{\xi},\bar{u}),\bar{u}(s)))z(s)\ . \tag{3.2}$$

However, the maximum principle holds only under exact controllability conditions on the linearized system

$$z'(s) = (A + \partial_y f(s,y(s,\bar{\xi},\bar{u}),\bar{u}(s)))z(s) + w(s) \tag{3.3}$$

$$z(0) = 0 \tag{3.4}$$

where $w(s) = f(s,y(s,\bar{\xi},\bar{u}),v(s)) - f(s,y(s,y(s,\bar{\xi},\bar{u}),u(s))$ with $v(\cdot) \in W(0,t;U)$. See [3], [4], [10], [11] for precise statements and proofs, as well as for the treatment of general cost functionals. The maximum principle is studied in [10], [11] as (a particular case of) an infinite dimensional nonlinear programming problem in Hilbert space. This approach has been generalized in [13] to certain Banach spaces.

4. GEOMETRY OF THE APPROXIMATION PROBLEM

Let E be an arbitrary Hilbert space and V a subset of E. A vector z in E is *normal to* V *at* $\bar{v} \in Cl(V)$ ($Cl(V)$ is the closure of V) if and only if

$$(z,\ \bar{v} - v) \geq 0 \quad (v \in V)\ . \tag{4.1}$$

The set V is said to be *R-spherical* if and only if there exists a constant R such that, for each $\bar{v}$ in $Cl(V)$ and each z normal to V at $\bar{v}$ (if any) we have

$$\|z\|\,\|\bar{v} - v\|^2 \leq 2R(\bar{v} - v,\ z) \quad (v \in V)\ . \tag{4.2}$$

Geometrically, (4.2) means that for every $\bar{v}$ in $Cl(V)$, if z is normal to V at $\bar{v}$ then there exists a ball $B = B(\bar{v})$ of radius R such that $V \in B(\bar{v}) \subseteq H(\bar{v})$, where $H(\bar{v})$ is the half-space defined by the inequality $(\bar{v} - v, z) \geq 0$.

Convex R-spherical sets were introduced in [16] under the name R-*convex* and used there and in [14], [15] in connection with finite dimensional control problems.

We show how these geometric concepts apply to the robustness problem in the particular case where

$$f(t,y,u) = B(t)u .$$

Let $\bar{u}(\cdot)$ be an optimal control or more generally an *extremal* control (that is, a control that satisfies (3.1) almost everywhere in $0 \leq t \leq \bar{t}$ for some vector $z \neq 0$). We may assume that $\| z \| = 1$. Rewrite (3.1) in the form

$$(B^*(s)S(\bar{t} - s)^*z, \bar{u}(s) - v) \geq 0 . \tag{4.3}$$

This inequality simply means that for each s in the set e the vector $B(s)^*S(\bar{t} - s)^*z$ is normal to the control set U at $\bar{u}(s)$. Assume now that U is R-spherical and let $u(\cdot)$ be arbitrary, $y(t,\xi,u)$ the trajectory of (1.1) corresponding to $u(\cdot)$ and to an arbitrary initial condition ξ. Then it follows from (4.2) that

$$\begin{aligned}
&\int_0^{\bar{t}} \| B(\sigma)^*S(\bar{t} - \sigma)^*z \| \, \| \bar{u}(\sigma) - u(\sigma) \|^2 d\sigma \leq \\
&\leq 2R \int_0^{\bar{t}} (B(\sigma)^*S(\bar{t} - \sigma)^*z, \bar{u}(\sigma) - u(\sigma))d\sigma = \\
&= 2R\left(z , \int_0^{\bar{t}} S(\bar{t} - \sigma)B(\sigma)(\bar{u}(\sigma) - u(\sigma))d\sigma \right) \leq \\
&\leq 2R \left\| \int_0^{\bar{t}} S(\bar{t} - \sigma)B(\sigma)(\bar{u}(\sigma) - u(\sigma))d\sigma \right\| \leq \\
&\leq 2R \, \| y(\bar{t},\bar{\xi},\bar{u}) - y(\bar{t},\xi,u) \| + 2RM \, \| \bar{\xi} - \xi \| .
\end{aligned} \tag{4.4}$$

In the nonlinear case, a similar result can be obtained under the assumption that, for almost all s in $0 < s < \bar{t}$ the set

$$f(s,y(s,\bar{\xi},\bar{u}),U)$$

is R-spherical. The final estimate is

$$\int_0^{\bar{t}} \|S(\bar{t},\sigma;\bar{u})^*z\| \, \| f(\sigma,y(\sigma,\bar{\xi},\bar{u}),\bar{u}(\sigma)) - f(\sigma,y(\sigma,\bar{\xi},\bar{u}),u(\sigma))\|^2 d\sigma \leq$$

$$\leq 2R \| y(\bar{t},\bar{\xi},\bar{u}) - y(\bar{t},\xi,u)\| + 2RM \|\bar{\xi} - \xi\| +$$

$$+ 4RMK \int_0^{\bar{t}} \|y(\sigma,\bar{\xi},\bar{u}) - y(\sigma,\xi,u)\| d\sigma \tag{4.5}$$

which is obtained linearizing (1.1) about the optimal trajectory $y(t,\bar{\xi},\bar{u})$. The constant M in (4.4) and (4.5) is a bound for the norm of $S(\bar{t},s;u)$ in $0 \leq s \leq \bar{t}$; the constant K in (4.5) is the local Lipschitz constant in (2.4).

5. COMPUTATIONAL APPLICATIONS

An estimate like (4.4) can be used to justify the convergence of suboptimal (i. e. approximately optimal) controls obtained with the help of penalty methods. For instance, consider the functional

$$F(t,u) = t + \varepsilon^{-1}\|y(t,\bar{\xi},u) - \bar{y}\|$$

($\bar{y}$ the target point) defined on pairs (t,u), where $u \in W(0,\bar{t};U)$. If $\bar{u}$ is the optimal control and $\bar{t}$ is the optimal time then $F(\bar{t},\bar{u}) = \bar{t}$. Accordingly, if (t,u) is an ε-solution of the problem of minimizing $F(t,u)$ (that is, if $F(t,u) \leq \min F(t,u) + \varepsilon$) we have as well $F(t,u) \leq \bar{t} + \varepsilon$, so that, in particular, $t \leq \bar{t} + \varepsilon$, $\| y(t,\bar{\xi},\bar{u}) - \bar{y} \| \leq C\varepsilon$. If

$$B(\sigma)^*S(\bar{t} - \sigma)^*z \neq 0 \quad \text{a. e. in} \quad 0 \leq \sigma \leq \bar{t} , \tag{5.1}$$

weighted L^2-convergence of u to $\bar{u}$ results from (4.6). We obtain a rate of convergence $O(\sqrt{\varepsilon})$ if the norm of the left hand side of (5.1) can be bounded below away from zero and if we have $t = \bar{t}$, but the last relation assumes knowledge of the optimal time $\bar{t}$.

6. GENERAL OPTIMAL PROBLEMS

Estimates similar to (4.8) exist also for general cost functionals y_0 given by (1.3). As in Section 4, let V be an arbitrary subset of a Hilbert space E, and let Z be a *cone* in E (that is, a set such that $cZ \subseteq Z$ for $c \geq 0$). We say that V is *R-spherical with respect to* Z if (4.2) holds for each $\bar{v}$ in $Cl(V)$ and each z in Z normal to V at $\bar{v}$ (if any such z exists).

Assuming that the set of all vectors of the form

$$\{f_o(s,y(s,\bar{\xi},\bar{u}),U),\ f(s,y(s,\bar{\xi},\bar{u}),U)\}$$

in $\mathbb{R} \times E$ is R-spherical with respect to the cone Z of all vectors of the form $\{z_o,z\}$ with $z_o \leq 0$ and that f_o satisfies smoothness conditions similar to those for f then an estimate of the form (4.5) holds with a different weight function, different constants and an additional term of the form $C_o|y_o(\bar{t},\bar{\xi},\bar{u}) - y_o(\bar{t},\xi,u)|$ on the right hand side.

Estimates valid in intervals $0 \leq s \leq t$ where t is different from the optimal time $\bar{t}$ can be obtained as well for the time optimal problem and for the general optimal problem, with terms of the form $C\,\omega(\bar{t},t)$ on the right hand side (ω the modulus of continuity of the approximating trajectory).

7. APPLICATION: A HYPERBOLIC DISTRIBUTED PARAMETER SYSTEM

This system is described by the quasilinear wave equation

$$y_{tt}(t,x) = \sum\sum D^j(a_{jk}(x)D^k y(t,x)) - \phi(y(t,x)) + u(t,x) \tag{7.1}$$

in a bounded domain Ω with boundary Γ of class $C^{(2)}$ in m-dimensional Euclidean space $\mathbb{R}^m$; here $x = (x_1,x_2,\ldots,x_m)$, $D^j = \partial/\partial x_j$. The boundary condition is $y(t,x) = 0$ on Γ, but the treatment of other boundary conditions is similar. The a_{jk} satisfy $a_{jk} = a_{kj}$, are continuously differentiable on $Cl(\Omega)$ and satisfy an uniform ellipticity condition there. The control set is the ball of radius R in $E = L^2(\Omega)$ centered at the origin, so that $W(0,\bar{t};U)$ is the corresponding ball in $L^2(0,\bar{t};L^2(\Omega))$. Assumption (F) in Section 2 for the function $f(t,y,u)(x) = \phi(y(x)) + u(t,x)$ will be satisfied if ϕ is continuously differentiable and satisfies estimates at infinity depending on the dimension m (see [6] for full details). In order to establish the *a priori* bound (2.2) we need to assume that the nonlinearity in (7.1) is a *restoring* force, namely

$$y\phi(y) > 0 \quad (-\infty < y < \infty). \tag{7.2}$$

The final estimate in the time optimal case is

$$\begin{aligned}
&\int_0^{\bar{t}} \|z_t(\sigma,\cdot)\| \, \|\bar{u}(\sigma,\cdot) - u(\sigma,\cdot)\|^2 d\sigma \\
&\leq 2R|||y(\bar{t},\cdot,\bar{\xi}_o,\bar{\xi}_1,\bar{u}) - y(\bar{t},\cdot,\xi_o,\xi_1,u)||| + \\
&+ 2R\| y_t(\bar{t},\cdot,\bar{\xi}_o,\bar{\xi}_1,\bar{u}) - y_t(\bar{t},\cdot,\xi_o,\xi_1,u)\| + \\
&+ 2R|||\bar{\xi}_o - \xi_o||| + 2R\|\bar{\xi}_1 - \xi_1\| + \\
&+ 4RK\int_0^{\bar{t}} |||y(\sigma,\cdot,\bar{\xi}_o,\bar{\xi}_1,\bar{u}) - y(\sigma,\cdot,\xi_o,\xi_1,u)|||d\sigma + \\
&+ 4RK\int_0^{\bar{t}} \|y_t(\sigma,\cdot,\bar{\xi}_o,\bar{\xi}_1,\bar{u}) - y_t(\sigma,\cdot,\xi_o,\xi_1,u)\|d\sigma .
\end{aligned} \tag{7.3}$$

Here $|||\ |||$ (resp. $\|\ \|$) is the norm in $H^1(\Omega)$ (resp. $L^2(\Omega)$) and we denote by $y(t,x,\xi_o,\xi_1,u)$ the solution $y(t,x)$ of (7.1) corresponding to $u = u(t,x)$ with initial conditions $y(0,x) = \xi_o(x)$, $y_t(0,x) = \xi_1(x)$. In (7.3), $z(t,x)$ is a solution of the linearized equation

$$z_{tt}(t,x) = \sum\sum D^j(a_{jk}(x)D^k z(t,x)) - \phi'(y(t,x,\bar{\xi}_o,\bar{\xi}_1,\bar{u}))z(t,x) .$$

We can show that $z(s,\cdot) \not\equiv 0$ except at a finite number of points in the control interval $0 < t < \bar{t}$.

8. APPLICATION: A PARABOLIC DISTRIBUTED PARAMETER SYSTEM IN L^2

The system is described by the quasilinear parabolic equation

$$\begin{aligned}
y_t(t,x) &= \sum\sum D^j(a_{jk}(x)D^k y(t,x)) + \sum b_j(x)D^j y(t,x) + \\
&+ \phi(t,y(t,x)) + c(x)y(t,x) + u(t,x)
\end{aligned} \tag{8.1}$$

in a domain Ω in m-dimensional Euclidean space. The assumptions on Ω, the boundary Γ and the coefficients a_{jk} are the same as in Section 7; the b_j and c are assumed continuous in $Cl(\Omega)$. Again we assume that $y(t,x)$ satisfies the Dirichlet boundary condition, although other boundary conditions are similarly treated. The linear part of the operator (8.1),

$$Ay = \sum \sum D^j(a_{jk}(x)D^k y) + \sum b_j(x)D^j y + c(x)y \tag{8.2}$$

(defined, say, on all test functions with support in Ω) admits an extension (by definition, associated with the Dirichlet boundary condition) that we denote by A_2; this operator generates an analytic compact semigroup $S_2(t;A)$ in $E = L^2(\Omega)$. The domain of the operator A_2 is $H^2(\Omega) \cap H_0^1(\Omega)$.

We describe below the treatment of the time optimal problem. The nonlinear case is apparently not amenable to the methods in [12], thus we limit ourselves to the linear case. For parabolic systems of the present type the maximum principle does not hold in the form requires in Section 3 and must be modified (see [1] for details). Based on this version of the maximum principle, the following estimate results under the assumption that the control set U is R-spherical:

$$\int_0^{\bar{t}} \|A_2^* S_2(\bar{t} - \sigma)^* z\| \, \|\bar{u}(\sigma,\cdot) - u(\sigma,\cdot)\|^2 d\sigma \le$$

$$2R\|A_2(y(\bar{t},\bar{\xi},\bar{u}) - y(\bar{t},\xi,u))\| + 2R\|A_2 S_2(\bar{t};A)(\bar{\xi} - \xi)\|. \tag{8.3}$$

However, we must assume that

$$y(\bar{t},\bar{\xi},\bar{u}) - y(\bar{t},\xi,u) \in D(A_2) . \tag{8.4}$$

Obviously, (8.3) is weaker than the basic estimate (4.4): (8.4) must be required and the final deviation of the trajectories must be measured in a more demanding norm. We note that if A_2 has a bounded inverse (which can be insured via a translation) then $\|A_2 u\|$ is (equivalent to) the A_2-graph norm of u and also equivalent to the norm of u in the space $H^2(\Omega)$.

The weight function on the integral in the estimate (8.3) is $(A_2^* S(\bar{t} - s)^* z)(x) = z(s,x)$ where the adjoint A_2^* of the infinitesimal generator A_2 is given by $A_2^* = (A^*)_2$, A^* the formal adjoint of A. Accordingly, the function $z(s, x)$ is a solution of the backwards equation

$$-z_t(s,x) = \sum\sum D^j(a_{jk}(x)D^k z(s,x)) + \sum b_j D^j y(t,x) + + c(x)z(s,x) \tag{8.5}$$

in $[0,T] \times \Omega$ satisfying the Dirichlet boundary condition on Γ and the "final condition"

$$z(\bar{t},\cdot) = \bar{z}(\cdot) \neq 0 \tag{8.6}$$

in the sense of distributions. We can guarantee that the estimate (8.3) is nontrivial; in fact, since the function $S_2(t;A)^*$ is analytic in $s > 0$, the L^2-valued function $z(s,\cdot)$ cannot vanish for any t in $0 < t < \bar{t}$, because if it did, solving the equation backwards, $z(s,\cdot) = 0$ in $0 \leq s \leq t$ and then, by analyticity, $z(s,\cdot) = 0$ in the whole interval $0 \leq s \leq \bar{t}$. This would contradict the final condition (8.6).

The treatment of the nonlinear case of (8.1) is simpler for a general cost functional y_0 (under adequate conditions on the nonlinearity). The maximum principle for this case can be found in [4], and the full treatment of the present approximation problem in [12].

9. APPLICATION: A PARABOLIC DISTRIBUTED PARAMETER SYSTEM IN L^∞

The system is again described by (8.1)(same assumptions on the coefficients). This time, however, we have pointwise bounds on the controls; the control space consists of all measurable functions $u = u(t,x)$ defined in $[0,T] \times \Omega$ and such that

$$|u(t,x)| \leq r \quad (0 \leq t \leq T,\ x \in \Omega) \tag{9.1}$$

Call this space W_r. We might try to incorporate this problem in our previous framework by using the control set

$$U = \{u \in L^2(\Omega);\ |u(x)| \leq r \quad \text{a.e. in } \Omega\}.$$

However, U is not R-spherical in $L^2(\Omega)$. Worse yet, although

$$W(0,T;U) \subset W_r \tag{9.2}$$

the inclusion is strict (for instance, if $\Omega = (0,1)$ and $u(t,x)$ is the characteristic function of the triangle $0 < t < x < T$ the function $t \to u(t,\cdot)$ is not strongly measurable as a $L^\infty(\Omega)$-valued function). Thus, a different approach is necessary. Observe first that, in view of the definition of W_r, the "right" space must be $L^\infty(\Omega)$. However, A does not admit any reasonable extension that generates a semigroup in $L^\infty(\Omega)$, so an indirect method must be used. The treatment of the maximum principle, which follows the abstract scheme in [1], is actually carried out in the space $E = C_o(Cl(\Omega))$ of continuous functions in $Cl(\Omega)$ vanishing on Γ (endowed with the supremum norm), where A admits an extension A_C which generates a compact analytic semigroup $S_C(t;A)$. The system under study will be

$$y'(t) = A_C y(t) + u(t) \tag{9.3}$$

in the space $L^\infty(\Omega)$ (where, of course, A_C is not densely defined). Taking into account the strict inclusion (9.2), the control space will be $W_{wm}(0,T;U)$, consisting of all weakly measurable functions $u(\cdot)$ defined in $0 < t < T$ and taking values in the ball of center 0 and radius R in $L^\infty(\Omega)$. Here, "weakly" is understood in the $(L^\infty(\Omega), L^1(\Omega))$ sense. We check easily that $W_{wm}(0,T;U) = W_r$. If we use this set of controls, trajectories have to be redefined; the integral

$$y(t) = \int_0^t S_\infty(t - \sigma)u(\sigma)d\sigma \tag{9.4}$$

(where $S_\infty(t)$ is the canonical extension of $S_C(t)$ to $L^\infty(\Omega)$) has to be interpreted as a Pettis rather than a Bochner integral. Using the maximum principle as proved in [1], we obtain the following analogue of (8.3), where $\bar{u}$ is an extremal control in the interval $0 < t < \bar{t}$ and u is an arbitrary control in $W_{wm}(0,\bar{t};U)$:

$$\int_{(0,\bar{t}) \times \Omega} |z(t,x)||\bar{u}(t,x) - u(t,x)|dtdx \leq$$

$$\leq K \, \|A_C(y(\bar{t},\cdot,\bar{\xi},\bar{u}) - y(\bar{t},\cdot,\xi,u))\|_{C_0(Cl(\Omega))} +$$

$$+ K \, \|A_C S_C(\bar{t};A)(\bar{\xi} - \xi)\|_{C_0(Cl(\Omega))} \tag{9.5}$$

under the assumption that

$$y(\bar{t},\bar{\xi},\bar{u}) - y(\bar{t},\xi,u) \in D(A_C) \ . \tag{9.6}$$

We note that the estimate (9.5) (unlike 8.3) involves the difference $|u(\bar{t},x) - u(t,x)|$ thus is of linear type.

10. APPLICATION: A BOUNDARY CONTROL SYSTEM

Although boundary control systems cannot be naturally forced in the form (1.1), modifications of the methods in the above sections provide results. As in the last three sections, Ω is a bounded domain in Euclidean space $\mathbb{R}^m$, but we assume now that the boundary Γ is *analytic*. The system is described by the heat equation

$$y_t(t,x) = \Delta y(t,x) \quad (t > 0, \ x \in \Omega) \tag{10.1}$$

with control of Dirichlet type,

$$y(t,x) = u(t,x) \quad (t > 0, \ x \in \Gamma) \ . \tag{10.2}$$

The control space $W(0,T;U)$ is the 1-sphere of $L^\infty((0,\infty) \times \Gamma)$, that is, the set of all measurable functions in $(0,\infty) \times \Gamma$ satisfying the estimate

$$|u(t,x)| \leq 1 \quad (t > 0, \ x \in \Gamma) \tag{10.3}$$

Denote by $\{-\lambda_n; \ n \geq 1\}$ the eigenvalues of the Laplacian Δ in Ω with Dirichlet boundary condition and by $\{\phi_n; \ n \geq 1\}$ the corresponding eigenfunctions. Given constants c, t with c arbitrary and $t \geq 0$ we define a Hilbert space $X_{c,t}(\Omega)$ of formal Fourier series as follows: the function

$$y = \sum c_n \phi_n$$

belongs to $X_{c,t}(\Omega)$ if and only if

$$\|y\|^2_{c,t} = \sum |c_n|^2 \exp\{2(c\sqrt{\lambda_n} - \lambda_n t)\} < \infty \tag{10.4}$$

The spaces $X_{c,t}(\Omega)$ were introduced by Russell [17] in relation to *controllability* of the boundary control system (10.1)-(10.2). The pertinent controllability result in [17] can be summarized as follows: if $\xi = \xi(x)$ is an arbitrary element of $L^2(\Omega)$ and $\bar{y} = \bar{y}(x)$ belongs to $X_{c,0}(\Omega)$ (where the value of c only depends on Ω) then for every $\bar{t} > 0$ there exists a bounded measurable control $u(t,x)$ such that the solution of (10.1)-(10.2) with initial condition

$$y(0,x) = \xi(x) \quad (x \in \Omega) \tag{10.5}$$

satisfies

$$y(t,x) = y(x) \quad (x \in \Omega) \tag{10.6}$$

Again, we only consider the time optimal problem for the system (10.1)-(10.2). A version of the maximum principle was proved in [2] and we summarize the result below. Assume the target point $\bar{y}(\cdot)$ satisfies

$$\bar{y} \in X_{c,0}(\Omega) \tag{10.7}$$

(c the constant in Russell's controllability result). Then, if $\bar{u}(t,x)$ is an optimal control, it admits the representation

$$\bar{u}(t,x) = \operatorname{sign} D_\eta h(t,x) \quad \text{a.e. in } (0,\bar{t}) \times \Gamma , \tag{10.8}$$

where D_η indicates outer normal derivative on Γ. Here, $h(t,x)$ is a classical solution of the backwards heat equation

$$h_t(t,x) = -\Delta h(t,x) \quad (t < \bar{t},\ x \in \Omega) \tag{10.9}$$

with null boundary condition

$$h(t,x) = 0 \quad (t < \bar{t},\ x \in \Omega) \tag{10.10}$$

and final value $h(\bar{t},\cdot) \in X_{-c,0}(\Omega)$, $h(\bar{t},\cdot) \neq 0$. Moreover,

$$D_\eta h(t,x) \neq 0 \quad \text{a.e. on} \quad (0,\bar{t}) \times \Gamma , \tag{10.11}$$

and

$$\int_{(0,\bar{t}) \times \Gamma} |D_\eta h(t,x)| dt d\Sigma < \infty \tag{10.12}$$

($d\Sigma$ the hyperarea differential on Γ). Using this maximum principle in the style of the last three sections we obtain the following estimate for an extremal control $\bar{u}$ (that is, a control that satisfies the maximum principle) and an arbitrary control in the control space:

$$\begin{aligned} &\int_{(0,\bar{t}) \times \Gamma} |D_\eta h(t,x)| |\bar{u}(t,x) - u(t,x)| dt d\Sigma \leq \\ &\leq \|h(\bar{t},\cdot)\|_{-c,0} \|y(\bar{t},\cdot,\bar{\xi},\bar{u}) - y(\bar{t},\cdot,\xi,u)\|_{c,0} + \\ &+ \|h(0,\cdot)\|_{-c,-\bar{t}} \|\xi(\cdot) - \xi(\cdot)\|_{c,\bar{t}} \end{aligned} \tag{10.13}$$

We assume here that

$$y(\bar{t},\cdot,\bar{\xi},\bar{u}) - y(\bar{t},\cdot,\xi,u) \in X_{c,0}(\Omega) . \tag{10.14}$$

For details, see [12]. It has been proved by G. Schmidt that for the pertinent version of the maximum principle for (10.1)-(10.2) we only need to assume that the boundary Γ is C^∞.

REFERENCES

[1] H. O. Fattorini, The time optimal problem in Banach spaces, *Appl. Math. Optimization* 1 (1974) 163-188

[2] H. O. Fattorini, The time optimal problem for boundary control of the heat equation, *Calculus of Variations and Control Theory* (D. L. Russell, Ed.) Academic Press, New York (1976) 305-319

[3] H. O. Fattorini, The maximum principle for nonlinear nonconvex systems in infinite dimensional spaces, Proceedings of 2nd. International Conference on Distributed Parameter Systems, Vorau Monastery, July 1984, *Springer Lecture Notes on Control and Information Sciences* 75 (1986) 162-178

[4] H. O. Fattorini, A unified theory of necessary conditions for nonlinear nonconvex control systems, *Applied Math. Optim.* 15 (1987) 141-185

[5] H. O. Fattorini, Optimal control of nonlinear systems: convergence of suboptimal controls, I, Proceedings of Special Session on Operator Methods for Optimal Control Problems, Annual AMS Meeting, New Orleans, 1986, *Lect. Notes in Pure and Applied Math.* 108, Marcel Dekker, New York (1987), 159-199

[6] H. O. Fattorini, Optimal control of nonlinear systems: convergence of suboptimal controls, II, Proceedings of IFIP Working Conference on Control Problems Described by Partial Differential Equations and Applications, Gainesville, 1986, *Springer Lecture Notes on Control and Information Sciences* 97 (1987) 230-246

[7] H. O. Fattorini, Some remarks on convergence of suboptimal controls, to appear in A. V. Balakrishnan's anniversary volume

[8] H. O. Fattorini, Convergence of suboptimal controls for point targets, Proceedings of Conference on Optimal Control of Partial Differential Equations, Oberwolfach, 1986, *Internat. Series Numer. Math.* 78, Birkhauser Verlag, Basel (1987) 91-107

[9] H. O. Fattorini, Convergence of suboptimal controls: the point target case, to appear

[10] H. O. Fattorini and H. Frankowska, Necessary conditions for infinite dimensional control problems, Proceedings of 8th. International Conference on Analysis and Optimization of Systems, Antibes-Juan Les Pins, 1988

[11] H. O. Fattorini and H. Frankowska, Necessary conditions for infinite dimensional nonlinear nonconvex control problems, to appear

[12] H. O. Fattorini and H. Frankowska, Explicit convergence estimates for suboptimal controls, to appear

[13] H. Frankowska, A general multiplier rule for infinite dimensional optimization problems, to appear

[14] H. Frankowska and Cz. Olech, R-convexity of the integral of set-valued functions, *Amer Jour. Math.*(1982) 156-165

[15] H. Frankowska and Cz. Olech, Boundary solutions of differential inclusions, *J. Diff. Equations* 44 (1982) 156-165

[16] A. Plis, Accesible sets in control theory, *Int. Conf. on Differential Equations*, Academic Press (1975) 646-650

[17] D. L. Russell, A unified boundary controllability theory for hyperbolic and parabolic partial differential equations, *Studies Appl. Math.* 52 (1973) 189-211

[18] D. L. Russell, Exact boundary controllability theorems for wave and heat processes in star-complemented regions, *Differential Games and Control Theory*, Marcel Dekker, New York, (1974) 67-98

Set-Valued Analysis and Some Control Problems

Halina Frankowska, Professor, CNRS-Centre National de la Recherche Scientifique, CEREMADE, Universite Paris-Dauphine, Paris, France

1 Introduction

Set-valued maps do arise in a natural way in control theory both as an issue to be studied and as a tool to be applied. In most cases once the problem gets the set-valued setting it requires to develop a new differential calculus of maps, to prove theorems analogous to those from mathematical analysis of (single-valued) functions, from ODE etc. This work is worth to be done not only for the reasons of mathematical curiosity. On one hand it leads to a unified approach to apparently different questions of control and optimization, on the other in several cases it allows to prove new results and to understand some different aspects of the problem.

We discuss here several applications of set-valued analysis and differential inclusions to questions of optimality, controllability and to the Lipschitz behavior of controls.

Consider the parametrized and the implicit control systems

$$x' = g(x, u(t)), \quad u(t) \in U \text{ is measurable} \tag{1}$$

$$f(x, x', u(t)) = 0, \quad u(t) \in U \text{ is measurable} \tag{2}$$

where the functions $g : \mathbf{R}^n \times U \to \mathbf{R}^n$, $f : \mathbf{R}^n \times \mathbf{R}^n \times U \to \mathbf{R}^m$ and the set U are given.

Systems (1), (2) are usually seen as families of ODE depending on selections $u(\cdot)$ from the set U. Solutions depend on the chosen control $u(\cdot)$. This suggests to apply tools and results from ODE and from the differential geometry.

In the early $60^{\underline{ies}}$ Ważewski [16] observed that (1) can be written in the form of differential inclusion

$$x' \in F(x) \tag{3}$$

where F is the set-valued map $x \to F(x) := g(x,U) = \bigcup_{u \in U} g(x,u)$. This allowed to him and Filippov [3] to study the issues of relaxation in control theory and to discover some qualitative properties of the solution set.

The implicit control system (2) also can be written in the form (3) by setting

$$F(x) \quad = \quad \{v \mid \exists\, u \in U,\ f(x,v,u) \quad = \quad 0\} \tag{4}$$

There exist several (Filippov's type) results insuring that under very mild assumptions on f and g, the solutions of the control system and the differential inclusion do coincide.

Differential inclusion (3) is a generalized differential equation already studied in the $30^{\underline{ies}}$ by H.Marchaud and S.C.Zaremba. Its solution is a family of trajectories and its properties depend on the graph of the map F. Many results from ODE can be extended to differential inclusions in a straightforward way. For instance Gronwall's lemma, variational equation, semigroup properties of trajectories, Nagumo's theorem concerning the existence of solutions in a closed domain found their analogous in this more general context. In this way numerous results on differential inclusions may be applied to control systems as well.

These are not the only examples where differential inclusions do arise. For instance closed loop control systems, control systems with constraints, systems with uncertainties may be also written in the differential inclusion formulation (3). Let us finally mention that some models arising in biological and social sciences may be given directly in the form (3) and, usually, one should not expect that (3) admits the parametrized form (1). All results existing in the literature on the parametrization of F assume that it has convex compact values and is continuous (Lipschitz continuous) to prove the existence of a continuous (respectively Lipschitz continuous) function g such that for all x, $F(x) = g(x,U)$. So even if in some cases it is possible to rewrite (3) as (1) in general the function g is nondifferentiable and is not simple to study.

Beside dynamics of control systems one meets set-valued maps in control when studies such issues as controllability or properties of the reachable map

$$t \mapsto R(t,\xi) \quad := \quad \{\, x(t) \mid x \text{ is a solution of } (1),\ x(0) \ = \ \xi \,\}$$

For instance the small time local controllability: $\forall\, t > 0,\ \xi \in Int\ R(t,\xi)$ can be seen as the openness of the reachable map $R(\cdot,\xi)$ in the sense that for every $t > 0$, $R(0,\xi) \in Int\ R(t,\xi)$ and one would seek an extension of classical open mapping principle to set-valued maps.

The set-valued approach to control theory problems do rise new mathematical issues: for instance one has to devise differential calculus of set-valued maps to approach "classical" questions of control theory: maximum principle, controllability, observability, high order necessary conditions for optimality etc.

The above questions in the context of differential inclusions were studied in [2], [4]-[8]. The aim of this paper is to show some applications of the obtained results to the usual and to the implicit control system (2).

The outline of the paper is as follows. Section 2 recalls definitions of set-valued derivatives and two inverse mapping theorems from [10]-[12]. In Section 3 we provide sufficient conditions for Lipschitz continuity of the map F defined by (4). In Sections 4, 5 results from [7], [4], [2] are applied to prove the maximum principle and a sufficient condition for local controllability for the implicit control system (2). In Section 6 we investigate the Lipschitz behavior of controls for (1). Finally in the last section we apply a set-valued inverse mapping theorem to the question of small time local controllability.

2 Some definitions and results of set-valued analysis

Let X be a Banach space. We recall first the notions of Kuratowski's *limsup* and *liminf*: Let T be a metric space and $A_\tau \subset X$, $\tau \in T$ be a family of subsets of X. The Kuratowski *limsup* and *liminf* of A_τ at τ_0 are closed sets given by

$$\begin{aligned} \limsup_{\tau\to\tau_0} A_\tau &= \{ v \in X \ : \ \liminf_{\tau\to\tau_0} \ \text{dist}\,(v, A_\tau) = 0 \} \\ \liminf_{\tau\to\tau_0} A_\tau &= \{ v \in X \ : \ \lim_{\tau\to\tau_0} \ \text{dist}\,(v, A_\tau) = 0 \} \end{aligned}$$

Consider a set-valued map $F : \mathbf{R}^n \to \mathbf{R}^n$, that is for all $x \in \mathbf{R}^n$, $F(x)$ is a (possibly empty) subset of $\mathbf{R}^n$. The domain and the graph of F are given by

$$Dom\ (F) = \{x \in \mathbf{R}^n \mid F(x) \neq \emptyset\}, \quad Graph\ (F) = \{(x, y) \mid x \in Dom\ (F),\ y \in F(x)\}$$

F is called locally Lipschitz at $\overline{x} \in Dom\ (F)$ if there exist $L > 0$, $\varepsilon > 0$ such that

$$\forall\, x,\ x' \in \mathbf{R}^n \text{ with } \|x - \overline{x}\| \le \varepsilon,\ \|x' - \overline{x}\| \le \varepsilon,\ \ F(x) \subset F(x') + L\|x - x'\|\, B$$

where B denotes the closed unit ball in $\mathbf{R}^n$.

DEFINITION 2.1 ([5]) *Let $F : \mathbf{R}^n \to \mathbf{R}^n$ be a locally Lipschitz at x set-valued map and let $y \in F(x)$. The derivative of F at (x, y) is the set-valued map $dF(x, y) : \mathbf{R}^n \to \mathbf{R}^n$ given by*

$$\forall\, a \in \mathbf{R}^n, \quad dF(x, y)a \ := \ \liminf_{h\to 0+} \frac{F(x + ha) - y}{h}$$

DEFINITION 2.2 ([13]) *A set-valued map $A : \mathbf{R}^n \to \mathbf{R}^n$ is called a closed convex process if its graph is a closed convex cone. The adjoint process of A is the closed convex process $A^\star$ defined by $p \in A^\star q \iff \forall\,(a, b) \in Graph\ (A),\ \ < p, a > \ \le \ < q, b >$.*

Consider a complete metric space $(\mathcal{U}, d)$ and a set-valued map $G : \mathcal{U} \to X$. For all $u \in \mathcal{U}$, $h > 0$ let $B_h(u)$ denote the closed ball in $\mathcal{U}$ of center u and radius h.

The inverse map of G is defined by

$$\forall\, x \in X, \ G^{-1}(x) = \{u \in U \mid x \in G(u)\}$$

DEFINITION 2.3 ([9] - [11]) *Let $(u, x) \in Graph\ (G)$ and $k > 0$.*

i) The contingent variation of G at (u, x) is the closed subset of X

$$G^{(1)}(u, x) \ := \ \limsup_{h\to 0+} \frac{G(B_h(u)) - x}{h}$$

ii) The k-th order variation of G at (u, x) is the closed subset of X

$$G^k(u, x) \ := \ \liminf_{\substack{(u', x') \to_G (u, x) \\ h \to 0+}} \frac{G(B_h(u')) - x'}{h^k}$$

where $\to_G$ denotes the convergence in Graph (G). When G is single-valued we write $G^{(1)}(u)$ and $G^k(u)$ instead of $G^{(1)}(u, G(u))$ and $G^k(u, G(u))$ respectively.

PROPOSITION 2.4 ([9] - [11]) *For every* $(u,x) \in$ *Graph* (G) *and* $k > 0$ *we have*
i) For all $K \geq k$, $0 \in G^k(u,x) \subset G^K(u,x)$
ii) For all $s > 0$, $\mathbf{R}_+ G^k(u,x) \subset G^{k+s}(u,x)$
iii) For all $\lambda_i \geq 0$, $v_i \in G^k(u,x)$, $i = 0, ..., m$ *with* $\sum_{i=0}^m \lambda_i = 1$, $\sum_{i=0}^m \lambda_i^k \, v_i \in G^k(u,x)$
iv) *For every* $v \in \text{co}\, G^k(u,x)$ *there exists* $\varepsilon > 0$ *such that* $\varepsilon v \in G^k(u,x)$
v) $\bigcup_{\lambda \geq 0} \lambda \text{co}\, G^k(u,x) = \bigcup_{\lambda \geq 0} \lambda G^k(u,x)$
Moreover if the dimension of X *is finite, then* $0 \in Int\, G^k(u,x) \Longleftrightarrow 0 \in Int\; co\, G^k(u,x)$ $\Longleftrightarrow \bigcup_{\lambda \geq 0} \lambda co\, G^k(u,x) = X \Longleftrightarrow \exists\, v_1, ..., v_p \in G^k(u,x)$ with $0 \in Int\; co\,\{v_1, ..., v_p\}$ *and* $\bigcup_{\lambda \geq 0, k > 0} \lambda \text{co}\, G^k(u,x) = X \Longleftrightarrow \exists\, k > 0$ *such that* $\bigcup_{\lambda \geq 0} \lambda co\, G^k(u,x) = X$.

THEOREM 2.5 (INVERSE MAPPING THEOREM [12]) *Assume that the norm of* X *is Gâteaux differentiable away from zero and let* $G : U \to X$ *be a continuous single-valued map. If for some* $\varepsilon > 0$, $\rho > 0$

$$\rho B \subset \bigcap_{d(u,\overline{u}) \leq \varepsilon} \overline{co}\, G^{(1)}(u)$$

then for every $u \in B_{\frac{\varepsilon}{2}}(\overline{u})$, $x \in \mathbf{X}$ *satisfying* $\|x - G(u)\| < min\left\{\frac{\varepsilon}{8}, \frac{\rho\varepsilon}{4}\right\}$

$$\text{dist}\,(u, G^{-1}(x)) \;\leq\; \frac{1}{\rho}\, \|G(u) - x\| \tag{5}$$

THEOREM 2.6 (HIGH ORDER INVERSE FUNCTION THEOREM [11]) *Let* $\overline{x} \in G(\overline{u})$. *If* $X = \mathbf{R}^n$ *and for some* $k \geq 1$ *the convex cone spanned by* $G^k(\overline{u}, \overline{x})$ *is equal to* X, *then there exists* $L > 0$ *such that for all* $(u, x, y) \in$ *Graph* $(G) \times X$ *near* $(\overline{u}, \overline{x}, \overline{x})$

$$\text{dist}\left(u,\; G^{-1}(y)\right) \;\leq\; L\, \|x - y\|^{\frac{1}{k}}$$

Remark — The proofs of the above theorems are based on the constructive Ljusternik's scheme. □

3 Implicit control systems and differential inclusions

Let U be a compact metric space and $f : \mathbf{R}^n \times \mathbf{R}^n \times U \to \mathbf{R}^m$ be a continuous function. Consider the control system (2). An absolutely continuous function $x : [0,T] \to \mathbf{R}^n$, $T > 0$ is called a solution of (2) if there exists a measurable selection $u(t) \in U$ such that $f(x(t), x'(t), u(t)) = 0$ a.e. in $[0,T]$.

Define the set-valued map $F : \mathbf{R}^n \to \mathbf{R}^n$ by (4) and consider the differential inclusion (3). Recall that an absolutely continuous function $x : [0,T] \to \mathbf{R}^n$, $T > 0$ is called a trajectory of (3) if $x'(t) \in F(x(t))$ a.e. in $[0,T]$. Clearly every solution of the control system (2) is a trajectory of the inclusion (3). The following lemma implies the converse statement.

LEMMA 3.1 *Every trajectory of (3) is also a solution of the control system (2).*

Proof— Fix a trajectory x of the differential inclusion (3) defined on the time interval $[0,T]$. We have to show existence of a measurable selection $u(t) \in U$ such that

$$f(\; x(t),\; x'(t),\; u(t)\;) \;=\; 0 \text{ a.e. in } [0,T] \tag{6}$$

Define the measurable set-valued map $t \to V(t) = (x(t), x'(t), U)$. Since $Kerf$ is closed, for every closed set $C \subset U$

$$\{t \mid V(t) \cap \; Ker\; f \cap \mathbf{R}^n \times \mathbf{R}^n \times C \neq \emptyset\} \text{ is Lebesgue measurable} \tag{7}$$

Setting $Z(t) = \{v \in U \mid (x(t), x'(t), v) \in Ker\; f\}$ we obtain that $Z(t) \neq \emptyset$ almost everywhere in $[0,T]$. From (7) follows that $Z(\cdot)$ is a measurable set-valued map with closed nonempty (almost everywhere) images and thereby it admits a measurable selection $u(t) \in Z(t)$. By definition of $Z(\cdot)$, $f(x(t), x'(t), u(t)) = 0$ a.e. in $[0,T]$. □

The introduced set-valued map F have a closed graph. If moreover for all $\overline{x} \in \mathbf{R}^n$

$$\exists\, \varepsilon > 0 \text{ such that } \liminf_{\|v\| \to \infty} \min_{x \in B_\varepsilon(\overline{x}), u \in U} \|f(x, v, u)\| \; > \; 0 \tag{8}$$

then F has compact images and is upper semicontinuous ([1, p.42]). Our aim is to investigate the lipschitz continuity.

THEOREM 3.2 *Assume that the derivative* $\frac{\partial f}{\partial(x,v)}(\cdot,\cdot,\cdot)$ *is continuous and for an open set* $\mathcal{N} \subset \mathbf{R}^n$ *and for every* $\overline{x} \in \mathcal{N}$ *the relation (8) holds true. Furthermore assume that*

$$\forall\, (x, v, u) \in Ker\; f \text{ with } x \in \mathcal{N}, \text{ the map } \frac{\partial f}{\partial v}(x, v, u) \text{ is surjective}$$

Then F *is locally Lipschitz on Dom* $F \cap \mathcal{N}$ *and for every* $(x, v, u) \in Ker\; f$ *with* $x \in \mathcal{N}$

$$Ker\; \frac{\partial f}{\partial(x,v)}(x, v, u) \; \subset \; Graph\; (dF(x, v))$$

Proof Fix $x \in \mathcal{N}$, $(x, v, u) \in Kerf$ and consider the function $\mathcal{N} \times \mathbf{R}^n \ni (y, w) \to \varphi(y, w) = (y, f(y, w, u))$. Since the derivative $\frac{\partial \varphi}{\partial(y,w)}(x, v)$ is surjective, there exists $\rho' > 0$ such that $\rho' B \subset \frac{\partial \varphi}{\partial(y,w)}(x, v)B$. This, the continuity of the derivative and (8) imply the existence of $\varepsilon > 0$, $\rho > 0$ independent of u such that $B_\varepsilon(x) \subset \mathcal{N}$ and

$$\forall\, y \in B_\varepsilon(x), \;\; \forall\, w \in F(x) + \varepsilon B, \;\; \rho B \subset \frac{\partial \varphi}{\partial(y, w)}(y, w)B \; \subset \; \varphi^{(1)}(y, w)$$

From Theorem 2.5 for all $(y, w) \in B_{\frac{\varepsilon}{2}}(x, F(x))$ and $z \in \mathbf{R}^{2n}$ with $\|z - \varphi(y, w)\| < min\left\{\frac{\varepsilon}{8}, \frac{\rho\varepsilon}{4}\right\}$ we have $dist\,((y, w), \varphi^{-1}(z)) \leq \frac{1}{\rho}\|\varphi(y, w) - z\|$. Since F is upper semicontinuous there exist $0 < \delta < \delta' < min\left\{\frac{\varepsilon}{8}, \frac{\rho\varepsilon}{4}\right\}$ such that for every $y \in B_\delta(x)$ we have $F(y) \subset F(x) + \delta' B$. In particular if $(y, w, u) \in Kerf$, then

$$\forall\, y' \in B_\delta(y), \;\; dist\,\Big((y, w), \varphi^{-1}(y', 0)\Big) \;\; \leq \;\; \frac{1}{\rho}\; \|(y, 0) - (y', 0)\| \;\; = \;\; \frac{1}{\rho}\; \|y - y'\|$$

$F(y')$ being closed we deduce from the last inequality that for some $w' \in F(y')$, $\|w - w'\| \leq \|(y, w) - (y', w')\| \leq \frac{1}{\rho}\|y - y'\|$ and thus F is Lipschitz continuous on $B_{\frac{\delta}{2}}(x)$. To prove the last statement fix $(a, b) \in Ker\; \frac{\partial f}{\partial(x,v)}(x, v, u)$. Then $f(x + ha, v + hb, u) = f(x, v, u) + h\frac{\partial f}{\partial(x,v)}(x, v, u)(a, b) + o(h) = o(h)$. This yields that $dist\,((x + ha, v + hb), \varphi^{-1}(x + ha, 0)) \leq \frac{1}{\rho}\|(x + ha, f(x + ha, v + hb, u)) - (x + ha, 0)\| = o(h)$. Hence for some b_h satisfying $v + hb_h \in F(x + ha)$ we have $\|v + hb - v - hb_h\| = o(h)$. Thus $b_h \to b$ as $h \to 0+$. □

THEOREM 3.3 *Under all the assumptions of Theorem 3.2 let* $\overline{x} \in \mathcal{N}$. *Then there exist* $\varepsilon > 0$ *and a constant* $L > 0$ *such that for all* $x \in B_\varepsilon(\overline{x})$ *and* (y, u) *satisfying* $f(x, y, u) = 0$ *the set-valued map* $A_{(x,y,u)}$ *defined by*

$$\forall\, a \in \mathbf{R}^n,\ A_{(x,y,u)}(a) = \{b \mid (a, b) \in Ker\ \frac{\partial f}{\partial(x, v)}(x, y, u)\} = -\frac{\partial f}{\partial v}(x, y, u)^{-1}\frac{\partial f}{\partial x}(x, y, u)a$$

is L-Lipschitz on $\mathbf{R}^n$ *closed convex process.*

Proof — Clearly for all (x, y, u) the map $A_{(x,y,u)}$ is a closed convex process. On the other hand the linear operator $(v, w) \to (v, \frac{\partial f}{\partial(x,v)}(x, y, u)(v, w))$ is surjective. Arguing exactly as in the proof of Theorem 3.2 we verify that for some $L, \varepsilon > 0$ and for all $(x, y, u) \in Ker\ f$ with $x \in B_\varepsilon(\overline{x})$, $A_{(x,y,u)}$ is L-Lipschitz on a neighborhood of zero with L independent from (x, y, u). But, since $A_{(x,y,u)}$ is a closed convex process, this yields it is L−Lipschitz on $\mathbf{R}^n$. □

THEOREM 3.4 *Let* $\overline{u} \in U$. *Define the set-valued map* $F_{\overline{u}} : \mathbf{R}^n \to \mathbf{R}^n$ *by*

$$F_{\overline{u}}(x) \;=\; \{\, v \mid f(x, v, \overline{u}) = 0 \,\} \tag{9}$$

Assume that for some $\overline{x} \in Dom\ (F_{\overline{u}})$, $f(\cdot, \cdot, \overline{u})$ *is continuously differentiable on a neighborhood of* $(\overline{x}, F_{\overline{u}}(\overline{x}))$ *and for all* $v \in F_{\overline{u}}(\overline{x})$ *the derivative* $\frac{\partial f}{\partial v}(\overline{x}, v, \overline{u})$ *is surjective. If*

$$\exists\, \varepsilon > 0 \text{ such that } \liminf_{\|v\| \to \infty} \min_{x \in B_\varepsilon(\overline{x})} \|f(x, v, \overline{u})\| \;>\; 0$$

then there exists a neighborhood $\mathcal{N}$ *of* $\overline{x}$ *such that*
a) $F_{\overline{u}}$ *has compact nonempty images and is locally Lipschitz on* $\mathcal{N}$
b) for all $x \in \mathcal{N}$ *and* $y \in F_{\overline{u}}(x)$ *the map* $dF_{\overline{u}}(x, y)$ *is Lipschitz closed convex process. Moreover*

$$Graph\ (dF_{\overline{u}}(x, y)) \;=\; Ker\ \frac{\partial f}{\partial(x, v)}(x, y, \overline{u}) \tag{10}$$

$$\forall\, \overline{v} \in F_{\overline{u}}(\overline{x}),\ \forall\, a \in \mathbf{R}^n,\ dF_{\overline{u}}(\overline{x}, \overline{v})a \;=\; \liminf_{(x,y) \to_{F_{\overline{u}}} (\overline{x}, \overline{v}),\ h \to 0+} \frac{F_{\overline{u}}(x + ha) - y}{h}$$

Proof — We apply Theorems 3.2 and 3.3 to the function $\hat{f}(x, v) = f(x, v, \overline{u})$ to prove a) and b). In order to prove (10) it is enough to show that for all $(x, y) \in Graph\ (F_{\overline{u}})$ we have $Graph\ (dF_{\overline{u}}(x, y)) \subset Ker\ \frac{\partial f}{\partial(x,v)}(x, y, \overline{u})$. Fix any $b \in dF_{\overline{u}}(x, y)a$ and let $b_h \to b$ be such that $y + hb_h \in F_{\overline{u}}(x + ha)$. Then $0 = f(x + ha, y + hb_h, \overline{u}) = f(x, y, \overline{u}) + h\frac{\partial f}{\partial(x,v)}(x, y, \overline{u})(a, b) + o(h) = h\frac{\partial f}{\partial(x,v)}(x, y, \overline{u})(a, b) + o(h)$. Dividing by h and taking the limit yields $\frac{\partial f}{\partial(x,v)}(x, y, \overline{u})(a, b) = 0$. Hence (10). On the other hand for every $(a, b) \in Ker\ \frac{\partial f}{\partial(x,v)}(\overline{x}, \overline{v}, \overline{u})$ and for all sequences $(x_i, v_i) \to (\overline{x}, \overline{v})$, satisfying $f(x_i, v_i, \overline{u}) = 0$ we have $f(x_i + ha, v_i + hb, \overline{u}) = f(x_i, v_i, \overline{u}) + h\frac{\partial f}{\partial(x,v)}(\overline{x}, \overline{v}, \overline{u})(a, b) + o(h) = o(h)$. In the same way as at the end of the proof of Theorem 3.2, we deduce that $b \in \liminf_{(x,y) \to_{F_{\overline{u}}} (\overline{x}, \overline{v}),\ h \to 0+} \frac{F_{\overline{u}}(x+ha)-y}{h}$.

4 Maximum principle

Let $\varphi : \mathbf{R}^n \times \mathbf{R}^n \to \mathbf{R}$ be a differentiable function and K be a subset of $\mathbf{R}^n$. In this section we study necessary conditions for optimality for the problem

$$\text{minimize } \varphi(x(0), x(1)) \tag{11}$$

over the solutions of the implicit control system

$$f(\ x,\ x',\ u(t)) \ = \ 0, \ \ u(t) \in U \text{ is measurable}, \ \ x(0) \in K \tag{12}$$

We assume that f satisfies the assumptions of Theorem 3.2 with $\mathcal{N} = \mathbf{R}^n$. By the results of Section 3, the system (2) may be replaced by (3) with F defined by (4). Recall that the contingent cone to K at $x \in K$ is defined by

$$T_K(x) \ = \ \limsup_{h \to 0+} \frac{K - x}{h}$$

THEOREM 4.1 *Let $(z, \overline{u})$ be a trajectory-control pair solving the problem (11)-(12). Then there exists an absolutely continuous function $p : [0,1] \to \mathbf{R}^n$ with $p' \in L^\infty(0,1)$ satisfying the adjoint equation*

$$p'(t) \ = \ \frac{\partial f}{\partial x}(z(t), z'(t), \overline{u}(t))^\star \frac{\partial f}{\partial v}(z(t), z'(t), \overline{u}(t))^{\star^{-1}} p(t) \ \text{ a.e. in } [0,1] \tag{13}$$

the maximum principle

$$< p(t), z'(t) > \ = \ \max \{ < p(t), \ v > \mid 0 \in \ f(z(t), v, U) \ \} \ \text{ a.e. in } [0,1] \tag{14}$$

and the transversality conditions

$$p(0) \in \frac{\partial \varphi}{\partial x_1}(z(0), z(1)) \ + \ T_K(z(0))^-, \ \ p(1) \ = \ -\frac{\partial \varphi}{\partial x_2}(z(0), z(1)) \tag{15}$$

Proof — By Theorem 3.2, using (8), we show that F is Lipschitz on a neighborhood of $z([0,T])$ and has compact images. For all $t \in [0,1]$ with $f(z(t), z'(t), \overline{u}(t)) = 0$ define the closed convex process

$$\forall\, a \in \mathbf{R}^n, \ \ A(t)a \ = \ -\frac{\partial f}{\partial v}(z(t), z'(t), \overline{u}(t))^{-1} \frac{\partial f}{\partial x}(z(t), z'(t), \overline{u}(t))a$$

We claim that there exists $L > 0$ such that for almost every $t \in [0,T]$, $A(t)$ is L-Lipschitz. Indeed applying Theorems 3.2, 3.3 we prove that for all $\overline{t} \in [0,T]$ there exist $\varepsilon > 0$, $L(\overline{t}) > 0$ such that for almost all $t \in B_\varepsilon(\overline{t}) \cap [0,T]$ the map $A(t)$ is $L(\overline{t})$-Lipschitz. The compactness of $[0,T]$ ends the proof of our claim. On the other hand for every $x \in \mathbf{R}^n$ the map $t \to A(t)x$ is measurable. For every $\xi \in K$ let

$$R_K(1, \xi) \ = \ \{ \ x(1) \mid x \ \text{ is a solution of } \ (2), \ x(0) = \xi \ \}$$

Since z is an optimal trajectory we easily verify that

$$\forall\ (a,b) \in T_{Graph\, R_K(1,\cdot)}\,(z(0), z(1)), \quad \varphi'(z(0), z(1))(a,b) \ \geq \ 0$$

Fix $\nu \in T_K(z(0))$. Then from the maximum principle for differential inclusions proved in [7, Theorem 4.9] follows the existence of absolutely continuous function $p \in W^{1,\infty}(0,1)$ satisfying the adjoint equation

$$p'(t) \in A(t)^\star(-p(t)) \text{ a.e. in } [0,1], \quad p(1) = -\frac{\partial \varphi}{\partial x_2}(z(0), z(1)) \tag{16}$$

the maximum principle (14) and the transversality condition $p(0) - \frac{\partial \varphi}{\partial x_1}(z(0), z(1)) \in (\mathbf{R}_+\nu)^-$. To prove (13) we compute $A(t)^\star$. Observe first that $r \in A(t)^\star(-q)$ if and only if

$$\begin{aligned}(r,q) \in (Graph\ A(t))^\perp &= Ker\ \tfrac{\partial f}{\partial(x,v)}(z(t), z'(t), \overline{u}(t))^\perp = Im\ \tfrac{\partial f}{\partial(x,v)}(z(t), z'(t), \overline{u}(t))^\star \\ &= Im\left(\tfrac{\partial f}{\partial x}(z(t), z'(t), \overline{u}(t))^\star \times \tfrac{\partial f}{\partial v}(z(t), z'(t), \overline{u}(t))^\star\right)\end{aligned}$$

Hence $r \in A(t)^\star(-q)$ if and only if for some $a \in \mathbf{R}^n$

$$r = \frac{\partial f}{\partial x}(z(t), z'(t), \overline{u}(t))^\star a, \quad q = \frac{\partial f}{\partial v}(z(t), z'(t), \overline{u}(t))^\star a$$

Since $\frac{\partial f}{\partial v}(z(t), z'(t), \overline{u}(t))$ is surjective, its adjoint is an injective linear operator and, hence, invertible on $Im\ \frac{\partial f}{\partial v}(z(t), z'(t), \overline{u}(t))^\star = Ker\ \frac{\partial f}{\partial v}(z(t), z'(t), \overline{u}(t))^\perp$. Thus

$$q \in Ker \frac{\partial f}{\partial v}(z(t), z'(t), \overline{u}(t))^\perp, \; r = \frac{\partial f}{\partial x}(z(t), z'(t), \overline{u}(t))^\star \frac{\partial f}{\partial v}(z(t), z'(t), \overline{u}(t))^{\star^{-1}} q$$

Therefore p is the unique solution of (16) and it verifies (13). Since $\nu \in T_K(z(0))$ is arbitrary we obtain (15).

5 Controllability at the point of equilibrium

Consider again the system (2), where $f : \mathbf{R}^n \times \mathbf{R}^n \times U \to \mathbf{R}^m$ is a continuous function and let $\xi \in \mathbf{R}^n$. The reachable set $R(T)$ of (2) from ξ at time T is defined by

$$R(T) = R(T,\xi) := \{\, x(T) \mid x \text{ is a solution of (2) on } [0,T],\ x(0) = \xi \,\}$$

In this section we provide a sufficient condition for

$$\exists\, T > 0 \text{ such that } \xi \in Int\ R(T) \tag{17}$$

applying the linearization techniques. We assume that ξ is an equilibrium, i.e., for some $\overline{u} \in U$, $f(\xi, 0, \overline{u}) = 0$ and we replace the system (2) by the differential inclusion

$$x' \in F_{\overline{u}}(x), \quad x(0) = \xi \tag{18}$$

with $F_{\overline{u}}$ defined by (9). Our theorem is based on the results of [4], [2].

THEOREM 5.1 *Assume that for some $\overline{u} \in U$, $f(\xi, 0, \overline{u}) = 0$ and that there exists $\varepsilon > 0$ such that $f(\cdot,\cdot,\overline{u})$ is continuously differentiable on $B_\varepsilon(\xi) \times \mathbf{R}^n$,*

$$\liminf_{\|v\|\to\infty} \min_{x \in B_\varepsilon(\xi)} \|f(x, v, \overline{u})\| > 0$$

and that for all $v \in F_{\overline{u}}(\xi)$ *the derivative* $\frac{\partial f}{\partial v}(\xi, v, \overline{u})$ *is surjective. Set*

$$L = \bigcup_{\lambda \geq 0} \lambda co \{v \mid f(\xi, v, \overline{u}) = 0\}, \quad A = \frac{\partial f}{\partial v}(\xi, 0, \overline{u}), \quad B = \frac{\partial f}{\partial x}(\xi, 0, \overline{u})$$

If for some $m \geq 1$

$$L + A^{-1}BL + \ldots + (A^{-1}B)^m L = \mathbf{R}^n = L - A^{-1}BL + \ldots + (-1)^m (A^{-1}B)^m L \quad (19)$$

then the system (2) is locally controllable at ξ *in the sense that (17) holds true. Furthermore if* L *is a subspace of* $\mathbf{R}^n$*, then (19) holds true if and only if*

$$L + A^{-1}BL + \ldots + (A^{-1}B)^{n-1}L = \mathbf{R}^n$$

In this case for all $t > 0$, $\xi \in \text{Int } R(t)$*, i.e. (2) is small time locally controllable at* ξ*.*

Proof — Observe that the map $a \to \mathcal{A}(a) := -A^{-1}Ba + L$ is a Lipschitz convex process and the map $a \to \overline{\mathcal{A}}(a) := \overline{\mathcal{A}(a)}$ is a Lipschitz closed convex process. Moreover, using an induction argument, we prove that

$$\begin{aligned} \overline{\mathcal{A}}^{m+1}(0) \supset \quad \mathcal{A}^{m+1}(0) &= L - A^{-1}BL + \ldots + (-1)^m (A^{-1}B)^m L \\ (-\overline{\mathcal{A}})^{m+1}(0) \supset (-\mathcal{A})^{m+1}(0) &= -L - A^{-1}BL - \ldots - (A^{-1}B)^m L \end{aligned}$$

Hence (19) yields $\overline{\mathcal{A}}^{m+1}(0) = \mathbf{R}^n = (-\overline{\mathcal{A}})^{m+1}(0)$. This and results of [2] imply the existence of $T > 0$ such that the reachable set $R^L(T)$ at time T of the inclusion

$$w' \in \overline{\mathcal{A}}(w), \quad w(0) = 0 \quad (20)$$

is equal to the whole space. By Theorem 3.4, $F_{\overline{u}}$ is Lipschitz on a neighborhood of ξ and has compact images. Then, from the last statement of Theorem 3.4 and the results of [4], controllability of the "linearized" inclusion (20) at time T imply that $\xi \in Int\ R(T)$. Assume next that L is a subspace of $\mathbf{R}^n$ and consider the subspaces

$$P_1 = L, \ \ldots, \ P_k = A^{-1}BP_{k-1} + P_{k-1}$$

If for some $k \geq 1, P_{k+1} = P_k$ then for all integer $s \geq 1$, $P_{k+s} = P_k$. Let k be the smallest integer having the above property. Since P_i are subspaces of the $n-$dimensional space, $k \leq n$. To prove the last statement observe that for all $t > 0$, $R^L(t)$ is a subspace of $\mathbf{R}^n$. Fix $v \in L \subset \overline{\mathcal{A}}(0)$. Using [1, p.120] it is not difficult to show that there exists a trajectory w of (20) satisfying $w'(0) = v$. Hence for every $t > 0$, $v \in T_{R^L(t)}(0)$ and therefore $L \subset T_{R^L(t)}(0) = R^L(t)$. Consider the subspace $P = \bigcap_{t>0} R^L(t)$. Then $L \subset P$. We show next that P is invariant by $A^{-1}B$. Indeed let $x \in P$, $v \in A^{-1}B(x)$ and fix $t > 0$. Using again [1, p.120] we prove that there exists a soluion w of the differential inclusion

$$w'(s) \in A^{-1}Bw(s), \quad s \geq t, \quad w(t) = x, \quad w'(t) = v \quad (21)$$

Since $\{R^L(s)\}$ is a nondecreasing family of subspaces we get that for every $s > t$, $v \in T_{R^L(s)}(x) = T_{R^L(s)}(0) = R^L(s)$. Since $s > t > 0$ are arbitrary we deduce that $v \in P$. This yields that $L + A^{-1}BL + \ldots + (A^{-1}B)^{n-1}L \subset P$ and, consequently, for every $t > 0$, $R^L(t) = \mathbf{R}^n$. But this and [4] together imply that for every $t > 0$, $\xi \in Int\ R(t)$.

6 Lipschitz behavior of controls

Let U be a topological space and $f : \mathbf{R}^n \times U \to \mathbf{R}^n$ be a continuous, differentiable in the first variable function. We assume that
a) f is locally Lipschitz in the first variable uniformly on U, i.e. for every $x \in \mathbf{R}^n$ there exist $L > 0$ and $\varepsilon > 0$ such that

$$\forall\, u \in U,\ \forall\, x', x'' \in B_\varepsilon(x),\ \ \|f(x', u) - f(x'', u)\| \le L\,\|x' - x''\|$$

b) for all $u \in U$ the derivative $\frac{\partial f}{\partial x}(\cdot, u)$ is continuous
c) for all $x \in \mathbf{R}^n$ the set $f(x, U)$ is bounded.

A (Lebesgue) measurable function $u : [0,T] \to U,\ T \ge 0$ is called an admissible control. Let $\mathcal{U}_T$ denote the set of all admissible controls defined on the time interval $[0,T]$. Consider a metric on $\mathcal{U}_T$:

$$\forall\, u,\ v\ \in \mathcal{U}_T,\quad d_T(u,v) = \mu(\ \{\ t \in [0,T] \mid\ u(t) \ne v(t)\ \}\ )$$

where μ denotes the Lebesgue measure. Then the space $(\mathcal{U}_T, d_T)$ is complete.

Let $\xi \in \mathbf{R}^n$ and consider the control system

$$x' \ =\ f(\ x,\ u(t)),\ \ u\ \in\ \mathcal{U}_T,\ \ x(0)\ =\ \xi \tag{22}$$

The reachable set of (22) at time $T \ge 0$ is given by

$$R(T)\ =\ \{\ x(T) \mid x \in W^{1,1}(0,T)\ \text{ is a solution of }\ (22)\ \}$$

where $W^{1,1}(0,T)$ is the (Sobolev) space of absolutely continuous functions on $[0,T]$. Fix a trajectory z of (22) defined on the time interval $[0,T]$ and let $\overline{u}$ be a corresponding control. In this section we provide a sufficient condition for

$$z(T)\ \in\ Int\ R(T) \tag{23}$$

and study the "regularity of the inverse". Consider the linear control system

$$w'(t) = \frac{\partial f}{\partial x}(z(t), \overline{u}(t))w(t) + v(t),\quad v(t) \in \overline{\text{co}}\ f(z(t), U) - f(z(t), \overline{u}(t)),\quad w(0) = 0 \tag{24}$$

and let $R^L(T)$ denote its reachable set at time T. Let $X_{\overline{u}}(t;s),\ 0 \le s \le t \le T$ be the fundamental solution of the equation

$$Z'(t)\ \ =\ \ \frac{\partial f}{\partial x}(z(t), \overline{u}(t))Z(t) \tag{25}$$

where $X_{\overline{u}}(s;s) = Id$. Then

$$R^L(T) = \{\int_0^T X_{\overline{u}}(T;s)v(s)ds \mid v(s) \in \overline{\text{co}}\ f(z(s), U) - f(z(s), \overline{u}(s))\} \tag{26}$$

THEOREM 6.1 *Assume that* $0 \in Int\ R^L(T)$. *Then* $z(T) \in Int\ R(T)$ *and for some* $\varepsilon > 0,\ L > 0$ *and for every control* $u \in \mathcal{U}_T$ *satisfying* $\mu(\{t \in [0,T] \|\ u(t) \ne \overline{u}(t)\}) \le \varepsilon$ *and all* $b \in B_\varepsilon(z(T))$, *there exists a trajectory-control pair* $(x_{\hat{u}}, \hat{u})$ *of (22) which verifies*

$$x_{\hat{u}}(T)\ =\ b;\ \ \mu(\{\ t \in [0,T] \mid\ \hat{u}(t)\ \ne\ u(t)\ \})\ \le\ L\,\|b - x_u(T)\| \tag{27}$$

In particular with every $b \in B_\varepsilon(z(T))$ *we can associate a control* $u \in \mathcal{U}_T$ *such that*

$$x_u(T)\ =\ b,\ \ \mu(\{\ t \in [0,T] \mid u(t) \ne \overline{u}(t)\ \})\ \le\ L\,\|b - z(T)\|$$

Remark — It is well known that whenever $0 \in Int\ R^L(T)$ we have $z(T) \in Int\ R(T)$. The proof of such result is usually based on a fixed point argument. The novelty of the above theorem lies in the estimation of controls (27) and the proof, which follows from the inverse mapping Theorem 2.5. Let us emphasize that the proof of Theorem 2.5 is based on a constructive argument. Thanks to this it is possible to estimate ε and L which depend on $max\ \{\rho > 0 \mid \rho B \subset R^L(T)\}$, on T and modulus of continuity of f and its partial derivative $\frac{\partial f}{\partial x}$ and to obtain $\hat{u}$ as the limit of a Cauchy sequence of controls constructed via Ljusternik's scheme. □

Proof — Replacing t by $\frac{t}{T}$ we may assume that $T = 1$. For all $u \in \mathcal{U}_1$, let x_u be the solution of (22) (when it exists on $[0,1]$) corresponding to the control u. From the Gronwall inequality follows that for some $\delta > 0$, the map $u \to x_u$ from $B_{2\delta}(\overline{u})$ to $W^{1,1}(0,1)$ is well defined and is Lipschitzian. For all $u \in B_\delta(\overline{u})$, let $X_u(\cdot;\cdot)$ denote the fundamental solution of the equation (25) with $(z,\overline{u})$ replaced by (x_u, u). Fix $u \in B_\delta(\overline{u})$, $v \in U$ and let t_0 be a Lebesgue point of the function $f(x_u(\cdot), u(\cdot))$. For all $h > 0$ consider the needle perturbations of controls

$$u_h(t) = \begin{cases} v & \text{if } t \in [t_0 - h, t_0] \\ u(t) & \text{otherwise} \end{cases}$$

Let x_h denote the solution of (22) corresponding to the control u_h. It is well known that

$$\lim_{h \to 0+} \frac{x_h(1) - x_u(1)}{h} = X_u(1;t_0)(f(x_u(t_0), v) - f(x(t_0), u(t_0))) \tag{28}$$

Set $V_u(t) = f(x_u(t), U) - f(x_u(t), u(t))$ and define the continuous map $G : B_{2\delta}(\overline{u}) \to \mathbf{R}^n$ by $G(u) = x_u(1)$. From (28) for every arbitrary but fixed $u \in B_\delta(\overline{u})$, for almost all $t_0 \in [0,1]$ and all $v \in V_u(t_0)$, $X_u(1;t_0)v \in G^{(1)}(u)$. Therefore $X_u(1;t_0)\ \overline{co}\ V_u(t_0) \subset \overline{co}\ G^{(1)}(u)$. From the mean value theorem

$$\int_0^1 X_u(1;t)\ \overline{co}\ V_u(t)dt \subset \overline{co}\ G^{(1)}(u) \tag{29}$$

Let $\rho > 0$ be such that $\rho B \subset \{\int_0^1 X_{\overline{u}}(1;t)v(t)dt \mid v(t) \in \overline{co}\ V_{\overline{u}}(t)\}$. The Gronwall inequality implies that $\{X_u(1;\cdot)\}$ converge uniformly to $X_{\overline{u}}(1;\cdot)$ when $u \to \overline{u}$ and

$$\lim_{u \to \overline{u}} \int_0^1 \mathcal{H}(\overline{co}\ V_{\overline{u}}(t),\ \overline{co}\ V_u(t))dt = 0$$

where $\mathcal{H}$ states for the Hausdorff distance. Since the left-hand side of (29) is convex and closed this yields that for some $\delta' > 0$ and all $u \in B_{\delta'}(\overline{u})$

$$\frac{\rho}{2} B \subset \{ \int_0^1 X_u(1;t)\ y(t)dt \mid y(t) \in \overline{co}\ V_u(t) \} \subset \overline{co}\ G^{(1)}(u)$$

Theorem 2.5 ends the proof. □

Remark — In [11] the above result was extended to the semilinear infinite dimensional control system

$$x' = Ax + f(x, u(t)),\quad u(t) \in U,\quad x(0) = \xi \tag{30}$$

where A is the infinitesimal generator of a strongly continuous semigroup $\{S(t)\}_{t \geq 0}$ on a Banach space E with the norm Gâteaux differentiable away from zero.

7 Small time local controllability

Consider a set-valued map $F : \mathbf{R}^n \to \mathbf{R}^n$ and the differential inclusion (3). Let $\xi \in \mathbf{R}^n$. For all $t \geq 0$ the reachable set of (3) from ξ at time t is given by

$$R(t) \;=\; R(t,\xi) \;:=\; \{\; x(t) \mid \; x \in W^{1,1}(0,t) \text{ is a trajectory of } (3),\; x(0) = \xi \;\}$$

The inclusion (3) is called small time locally controllable (s.t.l.c.) at ξ if for every $t > 0$, $\xi \in Int\ R(t)$. The following result was proved on the basis of High Order Inverse Function Theorem 2.6.

THEOREM 7.1 ([6]) *Let $F : \mathbf{R}^n \to \mathbf{R}^n$ be a set-valued map with compact, convex images and assume that ξ is a point of equilibrium, i.e. $0 \in F(\xi)$. Further assume that F is Lipschitz on a neighborhood of ξ. For all integer $k \geq 1$ set*

$$\mathcal{V}_k \;=\; \{\; v \in \mathbf{R}^n \mid \exists\, l(t) \in R(t) \text{ satisfying } l(t) = \xi + t^k v + o(t^k) \;\}$$

where $\lim_{t\to 0+} \frac{o(t^k)}{t^k} = 0$. Then $\mathcal{V}_k = R^k(0,\xi)$. Consequently if $\bigcup_{\lambda\geq 0, k\geq 1} \lambda co\mathcal{V}_k = \mathbf{R}^n$, then the differential inclusion (3) is s.t.l.c. at ξ.

Observe that in the definition of $\mathcal{V}_k$ we do not require any regularity of the selection l.

Of course to apply such general theorem it is necessary to know to compute variations $\mathcal{V}_k$. In the results below we show how some variations can be computed. We introduce the following notations: for all integer $k \geq 1$

$$V_k \;:=\; \{\; v \in \mathbf{R}^n \mid \forall\, t > 0,\; \xi \;+\; t^k v \in R(t) \;+\; O(t^{k+1})\;\},\quad Q \;=\; \bigcup_{\lambda\geq 0,\; k\geq 1} \lambda V_k \qquad (31)$$

where $\limsup_{t\to 0+} \frac{O(t^{k+1})}{t^{k+1}} < \infty$. If $0 \in F(\xi)$, then V_k is a starshaped set contained in $\mathcal{V}_k$.

PROPOSITION 7.2 *Under all assumptions of Theorem 7.1, the sequence $\{V_k\}_{k\geq 1}$ is nondecreasing and $(n+1)^{-k} co\ V_k \subset V_k$. Consequently Q is a convex subcone of $\bigcup_{\lambda\geq 0,k\geq 1} \lambda\ \mathcal{V}_k$.*

Proof— For all $v \in V_k$, $0 \leq t \leq 1$, $\xi + t^{k+1}v \in R(t^{\frac{k+1}{k}}) + O(t^{\frac{(k+1)^2}{k}}) \subset R(t) + O(t^{k+1})B$. Hence $V_k \subset V_{k+1}$ for every $k \geq 1$. Fix $v_i \in V_k$, $\lambda_i \geq 0$, $i = 0, ..., m$ such that $\sum_{i=0}^m \lambda_i = 1$. We first show that $\sum_{i=0}^m \lambda_i^k v_i \in V_k$. By the definition of V_k, $\xi + t^k \lambda_0^k v_0 \in R(\lambda_0 t) + O(t^{k+1})$. We proceed by the induction. Assume that we already proved that for some $0 \leq s \leq m-1$, $\xi + t^k \sum_{i=0}^s \lambda_i^k v_i \in R(t\sum_{i=0}^s \lambda_i) + O(t^{k+1})$. Let x_t be such trajectories of (3) starting at ξ that $x_t(t\sum_{i=0}^s \lambda_i) = \xi + t^k \sum_{i=0}^s \lambda_i^k v_i + O(t^{k+1})$ and $v_t \in \mathbf{R}^n$ be such that $\xi + t^k v_t \in R(t)$, $\|v_t - v_{s+1}\| = O(t)$. For all $t \geq 0$ define the absolutely continuous function $y_t : [t\lambda_{s+1}, t\sum_{i=0}^{s+1} \lambda_i] \to \mathbf{R}^n$ by $y_t(t\lambda_{s+1} + \tau) = t^k \lambda_{s+1}^k v_{t\lambda_{s+1}} + x_t(\tau)$. Then from the Lipschitz continuity of F there exists $L > 0$ such that for all small $t > 0$ and almost all $\tau \in [0, t\sum_{i=0}^s \lambda_i]$, $dist(y_t'(t\lambda_{s+1} + \tau), F(y_t(t\lambda_{s+1} + \tau))) = dist(x_t'(\tau), F(x_t(\tau) + t^k\lambda_{s+1}^k v_{t\lambda_{s+1}})) \leq Lt^k \|v_{t\lambda_{s+1}}\|$. From the Filippov theorem [1, p.120] we deduce that $y_t(t\sum_{i=0}^{s+1} \lambda_i) = \xi + t^k \sum_{i=0}^{s+1} \lambda_i^k v_i + O(t^{k+1}) \in R(t\sum_{i=0}^{s+1} \lambda_i) + O(t^{k+1})$. Applying this last inclusion with $s = m-1$ we obtain that $\sum_{i=0}^m \lambda_i^k v_i \in V_k$. Fix next $v \in co\ V_k$ and let $\mu_i \geq 0$, $v_i \in V_k$ be such that $\sum_{i=0}^n \mu_i = 1$ and $\sum_{i=0}^n \mu_i v_i = v$. Set $\lambda_i = \mu_i^{\frac{1}{k}}/(n+1)$ for $i = 0, ..., n$ and $\lambda_{n+1} = 1 - \sum_{i=0}^n \lambda_i$. By the choice of μ_i, λ_i we have $\lambda_{n+1} \geq 0$, $\sum_{i=0}^{n+1} \lambda_i = 1$. Then $v = \sum_{i=0}^n \mu_i v_i = (n+1)^k \left(\sum_{i=0}^n \lambda_i^k v_i + \lambda_{n+1}^k 0\right) \in (n+1)^k V_k$. This ends the proof.

THEOREM 7.3 (AN INVARIANCE PROPERTY OF Q) *Under assumptions of Theorem 7.1 let $v \in Q$ be such that $-v \in Q$ and let $f(x) \in F(x)$ be a selection of F twice continuously differentiable at ξ with $f(\xi) = 0$. Then $f'(\xi)v \in Q$.*

Proof — Since $\{V_k\}$ is a nondecreasing sequence of starshaped sets there exist $M > 0$, $k \geq 1$ and selections $l(t)$, $l_1(t) \in R(t)$ with

$$l(t) = \xi + t^k \frac{v}{M} + O(t^{k+1}), \quad l_1(t) = \xi - t^k \frac{v}{M} + O(t^{k+1})$$

Because Q is a cone replacing v by $\frac{v}{M}$ we may assume that $M = 1$. Fix $t > 0$ and consider the differential equation

$$x'(s) = f(x(s)), \quad s \geq t^2, \quad x(t^2) = l(t^2)$$

Then for all $\tau > 0$

$$x(t^2+\tau) = l(t^2) + \int_{t^2}^{t^2+\tau} f(x(s))ds = l(t^2) + f(l(t^2))\tau + \int_0^\tau (f(x(t^2+s)) - f(l(t^2)))ds =$$
$$l(t^2) + f'(\xi)(l(t^2) - \xi)\tau + O(\|l(t^2) - \xi\|^2)\tau + \int_0^\tau (f(x(t^2+s) - f(l(t^2)))ds$$

Let L be a Lipschitz constant of f on a neighborhood of zero. Thus for all small $t, \tau > 0$

$$\left\|x(t^2+\tau) - l(t^2)\right\| \leq \|f'(\xi)v\| \; t^{2k}\tau + O(t^{2k+2})\tau + L\int_0^\tau \left\|x(t^2+s) - l(t^2)\right\| ds$$

From the Gronwall inequality we deduce that for some $M > 0$, $\varepsilon > 0$ and all $t, s \in [0, \varepsilon]$, $\|x(t^2+s) - l(t^2)\| \leq Mt^{2k}s$. Therefore

$$\left\|x(t^2+\tau) - l(t^2) - f'(\xi)vt^{2k}\tau\right\| \leq O(t^{2k+2})\tau + \int_0^\tau LMt^{2k}sds = \tau O(t^{2k+2}) + \frac{LM}{2}t^{2k}\tau^2$$

and we proved that $x(t^2+t) = \xi + t^{2k}v + f'(\xi)vt^{2k+1} + O(t^{2k+2})$. For all $t > 0$ let $y_t : [0, t^2] \to \mathbf{R}^n$ be a solutions of (3) starting at ξ and satisfying $y_t(t^2) = l_1(t^2)$. For every $t > 0$ consider the absolutely continuous function

$$z_t(t^2 + t + s) = y_t(s) + x(t^2+t) - \xi, \quad s \in [0, t^2]$$

Then from the Lipschitz continuity of F there exists $K > 0$ such that for all small $t > 0$ and almost all $s \in [0, t^2]$

$$dist\,(z_t'(t^2+t+s), F(z_t(t^2+t+s))) = dist\,(y_t'(s), F(y_t(s) + x(t^2+t) - \xi)) \leq Kt^{2k}$$

By the Filippov theorem [1, p.120] for some $M_1 > 0$ and all small $t > 0$, $dist(R(2t^2+t), z_t(2t^2+t)) \leq M_1\, t^{2k+2}$. Since $z_t(2t^2+t) = \xi - t^{2k}v + t^{2k}v + f'(\xi)vt^{2k+1} + O(t^{2k+2}) = \xi + f'(\xi)vt^{2k+1} + O(t^{2k+2})$, we deduce that for all small $t > 0$, $\xi + (2t)^{2k+1}\left(\frac{1}{2}\right)^{2k+1} f'(\xi)v \in R(2t^2+t) + O(t^{2k+2}) \subset R(2t) + O(t^{2k+2})$. Using that Q is a cone we finally obtain that $f'(\xi)v \in Q$. □

Using [8] we can check easily that under the assumptions of Theorem 7.1, $F(\xi) \subset Q$. We provide next few examples of elements from Q for the control system

$$x' = \sum_{i=1}^k u_i(t)g_i(x), \quad u_i(t) \geq 0 \text{ is measurable}, \quad \sum_{i=1}^k u_i(t) = 1, \quad x(0) = \xi \tag{32}$$

where $g_i : \mathbf{R}^n \to \mathbf{R}^n$ are C^∞-functions. Sufficient conditions of s.t.l.c. of (32) were studied by many authors (see for example [15] and the bibliography contained therein). We do not devise here the calculus of variations for (32). Our aim is rather to illustrate how the convexity of the cone Q can be exploited to study variations of reachable sets.

LEMMA 7.4 *For every i, $g_i(\xi) \in Q$.*

Proof — Fix i and consider the solution x of the equation (32) with $u_i \equiv 1$. Then $x \in C^\infty$ and $x(t) = \xi + g_i(\xi)\ t + O(t^2) \in R(t)$. □

LEMMA 7.5 *Assume that for some $1 \le r \le k$, $0 \in rint\ co\ \{g_i(\xi) \mid 1 \le i \le r\}$ (relative interior). Then for every twice continuously differentiable at ξ selection*

$$f(x) \in \{ \sum_{s=1}^{k} u_s\, g_s(x) \mid u_s \ge 0,\ \sum_{s=1}^{k} u_s = 1 \}$$

and for all integers $m \ge 1$, $1 \le i, j \le r$ we have

$$f'(\xi)^m g_i(\xi) \in Q, \quad [g_i, g_j](\xi) \in Q, \quad f'(\xi)^m [g_i, g_j](\xi) \in Q$$

Proof — The first statement follows from Theorem 7.3 and Lemma 7.4. Fix $1 \le i, j \le r$ and let $\lambda_s > 0$ be such that $\sum_{s=1}^r \lambda_s g_s(\xi) = 0$, $\sum_{s=1}^r \lambda_s = 1$. Set $f = \sum_{s=1}^r \lambda_s g_s$, $c = 1 + \lambda_i + \lambda_j$. Then, by the Campbell-Hausdorff formula and the Gronwall lemma

$$\begin{gathered} exp\left(t\left(\textstyle\sum_{s=1, s\ne i,j}^r \lambda_s g_s + (\lambda_i + \lambda_j) f\right)\right) exp(\, t\lambda_i g_i\,)\ exp(\, t\lambda_j g_j\,)(\xi) = \\ exp\left(t\left(\textstyle\sum_{s=1, s\ne i,j}^r \lambda_s g_s + (\lambda_i + \lambda_j) f\right)\right) exp(t(\lambda_i g_i + \lambda_j g_j) + \tfrac{1}{2}t^2 \lambda_i \lambda_j [g_i, g_j] + \ldots)(\xi) = \\ exp(\, tcf + \tfrac{1}{2}\, t^2 \lambda_i \lambda_j\, [g_i, g_j] + \tfrac{c}{2}\, t^2\, [f, \lambda_i g_i + \lambda_j g_j]\ + \ldots\,)(\xi) = \\ \xi + \tfrac{1}{2}\, t^2 \lambda_i \lambda_j\, [g_i, g_j](\xi) + \tfrac{c}{2}\, t^2 [f, \lambda_i g_i + \lambda_j g_j](\xi) + O(t^3) \in R(2t) \end{gathered}$$

Hence

$$\lambda_i \lambda_j\, [g_i, g_j](\xi) + c\, [f, \lambda_i g_i + \lambda_j g_j](\xi) \in Q \tag{33}$$

From the first statement we obtain that

$$-\, c\, [f, \lambda_i g_i + \lambda_j g_j](\xi) = c\, f'(\xi)(\lambda_i g_i(\xi) + \lambda_j g_j(\xi)) \in Q \tag{34}$$

Convexity of the cone Q and (33), (34) yield that $\lambda_i \lambda_j [g_i, g_j](\xi) \in Q$. Since $[g_i, g_j](\xi) = -[g_j, g_i](\xi)$, Theorem 7.3 ends the proof. □

LEMMA 7.6 *In Lemma 7.5 assume that $r = 2$. Then $[g_1, [g_1, g_2]](\xi)$ & $[g_2, [g_2, g_1]](\xi)$ are contained in Q.*

Proof — Let $0 < \lambda < 1$ be such that $\lambda g_1(\xi) + (1 - \lambda) g_2(\xi) = 0$. Set $f = \lambda g_1 + (1 - \lambda) g_2$, $h_1 = \lambda g_1$, $h_2 = (1 - \lambda) g_2$. Then, by the Campbell-Hausdorff formulae and the Gronwall lemma,

$$\begin{gathered} exp(th_1) exp(th_2) exp(th_2) exp(th_1)(\xi) = exp(tf + \tfrac{1}{2}t^2[h_1, h_2] + \tfrac{1}{12}t^3[h_1 - h_2, [h_1, h_2]] + \ldots) \times \\ exp(tf - \tfrac{1}{2}t^2[h_1, h_2] + \tfrac{1}{12}t^3[h_1 - h_2, [h_1, h_2]] + \ldots)(\xi) = exp(\, 2tf + \tfrac{1}{6}t^3[h_1 - h_2, [h_1, h_2]] - \\ \tfrac{1}{2}t^3[f, [h_1, h_2]] + \ldots)(\xi) = \xi + \tfrac{1}{6}t^3[h_1 - h_2, [h_1, h_2]](\xi) - \tfrac{1}{2}t^3[f, [h_1, h_2]](\xi) + O(t^4) \in R(2t) \end{gathered}$$

Hence, using that Q is a cone, we deduce that

$$\frac{1}{6}\, [h_1 - h_2,\ [h_1, h_2]](\xi) - \frac{1}{2}\, [f,\ [h_1, h_2]](\xi) \in Q \tag{35}$$

By Lemma 7.5 the vector $a := (\frac{1}{6} + \frac{1}{2})[h_1 + h_2, [h_1, h_2]](\xi) = \frac{2}{3}\lambda(1 - \lambda) f'(\xi)[g_2, g_1](\xi) \in Q$. Adding a to the both sides of the inclusion (35) and using convexity of the cone Q, we get $[h_1, [h_1, h_2]](\xi) = \lambda^2(1 - \lambda)[g_1, [g_1, g_2]](\xi) \in Q$. Hence $[g_1, [g_1, g_2]](\xi) \in Q$. The proof of the second statement is analogous and is omitted.

References

[1] AUBIN J.-P. & CELLINA A. (1984) DIFFERENTIAL INCLUSIONS. Springer-Verlag (Grundlehren der Math. Wissenschaften), Vol.264, 1-342

[2] AUBIN J.-P., FRANKOWSKA H. & OLECH C. (1986) *Controllability of convex processes.* SIAM J. Control & Optimization, 24, 1192-1211

[3] FILIPPOV A.F. (1967) *Classical solutions of differential equations with multivalued right hand side.* SIAM J. Control & Optimization, 5, 609-621

[4] FRANKOWSKA H. (1987) *Local controllability and infinitesimal generators of semi-groups of set-valued maps.* SIAM J. Control & Optimization, 25, 412-432

[5] FRANKOWSKA H. (1987) *The maximum principle for an optimal solution to a differential inclusion with end point constraints.* SIAM J. Control & Optimization, 25, 145-157

[6] FRANKOWSKA H. (1989) *Local controllability of control systems with feedbacks.* J. Optimization Theory & Applications, (to appear)

[7] FRANKOWSKA H. (to appear) *Contingent cones to reachable sets of control systems.* SIAM J. Control & Optimization

[8] FRANKOWSKA H. (to appear) *Optimal trajectories associated to a solution of contingent Hamilton-Jacobi equation.* Applied Mathematics & Optimization

[9] FRANKOWSKA H. (1987) *An open mapping principle for set-valued maps.* J. Math. Analysis & Appl., 127, 172-180

[10] FRANKOWSKA H. (1987) *Théorèmes d'application ouverte et de fonction inverse.* CRAS, 305, 773-776

[11] FRANKOWSKA H. (to appear) *Some inverse mapping theorems.*

[12] FRANKOWSKA H. (to appear) *High order inverse function theorems.* Proceedings of the Conference on Nonlinear Analysis, Perpignan, June 1987

[13] ROCKAFELLAR R.T. (1970) CONVEX ANALYSIS. Princeton University Press, Princeton, NJ

[14] ROXIN E. (1965) *On the generalized dynamical systems defined by contingent equations*, J. Diff. Eqs., 1, 188-205

[15] SUSSMANN H. (1987) *A general theorem on local controllability.* SIAM J. Control & Optimization, 25, 158-194

[16] WAŻEWSKI T. (1964) *On an optimal control problem.* Proc. Conference on Differential Equations and their Applications, Prague, 229-242

Caratheodory Type Selections of Aumann Integrals

Andrzej Fryszkowski, Professor, Institute of Mathematics, Technical University of Warsaw, Warsaw, Poland

1. INTRODUCTION

Let T be a compact topological space with a σ-field L of measurable subsets of T given by a nonnegative, regular, finite Borel measure "dt". By $L^1(T,R^n)$ we denote the Banach space of all integrable functions u from T into an euclidean space R^n, with the norm $||u|| = \int_T |u(t)|dt$, where $|\cdot|$ stands for the euclidean norm.

A set $K \subset L^1(T,R^n)$ is called decomposable, if for all $u,v \in K$ and $A \in L$

$$u \cdot \chi_A + v \cdot (1 - \chi_A) \in K, \tag{1}$$

where x_A is the characteristic function of the set A. The family of nonempty, closed and decomposable subsets of $L^1(T,R^n)$ we denote by $dec(L^1)$.

For an arbitrary $K \in dec(L^1)$ the Aumann integral, denoted by $\int_T K$, is defined as the set $\{\int_T u(t)dt\colon u \in K\}$. The well-known result of Olech (1967) states that $\int_T K$ is convex and if $\int_T K$ is bounded, then it is also compact (see also Aumann (1965)).

Let X and S be, respectively, a polish space and a locally compact polish space with σ-fields $\Sigma \subset 2^X$ given by a nonnegative, rerular, finite, Borel measure and $\mathcal{B} \subset 2^S$ of Borel subsets. We shall consider a Caratheodory type mapping (C-mapping) K(x,s) from X × S into $dec(L^1)$. The latter means that

i) for every $x \in X$ K(x,.) is lower semicontinuous (l.s.c.) in s and

ii) K(.,.) is $\Sigma \otimes \mathcal{B}$-measurable.

One can easily check that in this case cl $\int_T K(x,s)$ is also a C-mapping from X × S into the family $clco(R^n)$ of nonempty, closed and convex subsets of R^n.

By Caratheodory selection (C-selection) of a C-mapping K(x,s) we mean a mapping k(x,s) from X × S into $L^1(T,R^n)$ which is Σ-measurable in first variable, continuous in the second and $k(x,s) \in K(x,s)$ for all $(x,s) \in X \times S$. Because every l.s.c. mapping K(s) from S into $dec(L^1)$ admits a continuous selection (Fryszkowski (1983a), Bressan and Colombo (1985)), then the existence of C-selections $k_0(x,s)$ for a C-mapping K(x,s) into $dec(L^1)$ can be concluded from the result of Artstein and Prikry (1987) (see also Castaing (1975), Fryszkowski (1977) and (1983b), Rybinski (1985) and Kim et al. (1987)). For a similar reason (and Michael Selection Thm) we have the existence of a C-selection l(x,s) for cl $\int_T K(x,s)$. Note, moreover, that a C-selection of cl $\int_T K(x,s)$ can be found in the "integral" form

$$l(x,s) = \int_T k_0(x,s)(t)dt, \tag{2}$$

which cannot be directly concluded from Artstein and Prikry (1987) arguments.

The purpose of this paper is to show that the above situation can be, in a sense, reversed. Namely, for every C-selection $l(x,s)$ of cl $\int_T K(x,s)$ and every $\varepsilon > 0$ there exists a C-selection $k(x,s)$ of $K(x,s)$ such that for all $x \in X$ and $s \in S$

$$|l(x,s) - \int_T k(x,s)(t)dt| < \varepsilon. \tag{3}$$

Note that not every C-selection $l(x,s)$ of cl $\int_T K(x,s)$ is of the form (2) what is true for dependent only on x mappings into $\mathrm{dec}(L^1)$ (see also Artstein (1988)).

The above theorem is a generalization of Gamkrelidze Lemma (1965) and the author's earlier work (1987), where case of l.s.c. mapping $K(s)$ from S into $\mathrm{dec}(L^1)$ was considered. We state and prove the main result in Section 3 and in Section 4 we formulate all results in terms of vector measures. Section 2 contains some necessary notations and facts.

2. NOTATIONS AND SOME SUPPLEMENTARY FACTS

Let X, Σ, S and $\mathcal{B}$ be as in the previous paragraph. We shall consider mappings $A(x)$, $B(s)$ and $C(x,s)$ from, respectively, X, S and $X \times S$ into L; $A(x), B(s), C(x,s) \in L$. Below we identify sets A with their characteristic functions χ_A and all regularity assumptions about the mapping $A(x)\colon X \to L$ are understood in terms of the mapping $\chi_{A(x)}\colon X \to L^1(T,R)$. Similarly it is for $B(s)\colon S \to L$ and $C(x,s)\colon X \times S \to L$.

Let Λ be a given set. A family $\{A_\alpha\}_{\alpha\in\Lambda}$ of measurable sets $A_\alpha \in \Sigma$ we shall call a decomposition of X if they are pairwise disjoint and $\cup A_\alpha = X$. If only a finite number of A_α is nonempty we have a finite decomposition of X. We will also consider decomposition and finite decomposition $\{B_\alpha\}$ of T.

Let for every $x \in X$ $\{A_\alpha(x)\}$ be a family of decompositions of T. If all mappings $A_\alpha\colon X \to L$ are Σ-measurable, we call $\{A_\alpha\}$ a Σ-measurable family of decompositions. Similarly we define a continuous and Caratheodory type (shortly C-decomposition) family of decompositions of T.

For a given multivalued mapping $U(t)$ from T into closed subsets of R^n denote by

$$K_U = \{u \in L^1(T,R^n) : u(t) \in U(t) \quad \text{a.e.} \quad \text{in } T\}. \tag{4}$$

Obviously $K_U \in \text{dec}(L^1)$ and in this case the integral $\int_T K_U$ we shall also denote by $\int_T U(t)dt$.

Below we shall consider continuous mappings $k_1(s), \ldots, k_m(s)$ from S into $L^1(T,R^n)$ and denote by

$$U_s(t) = \{k_1(s)(t), \ldots, k_m(s)(t)\}. \tag{5}$$

and

$$K_0(s) = K_{U_s}. \tag{6}$$

Clearly $K_0(s)$ is a l.s.c. mapping as well as $\int_T K_0(s)$. Moreover, from Aumann (1965) and Olech (1967) results follows that

$$\int_T K_0(s) = \int_T \text{clco } U_s(t)dt = \text{clco } \int_T U_s(t)dt. \tag{7}$$

The following result has been proved in [11]:

Proposition 1. Consider a l.s.c. mapping $K(s)$ *from* S *into* $\text{dec}(L^1)$ *and let* $l(s)$ *be a continuous selection of* $\text{cl} \int_T K(s)$. *Then for every* $\varepsilon > 0$ *there exists a continuous selection* $k(s)$ *of* $K(s)$ *such that for all* $s \in S$

$$|l(s) - \int_T k(s)(t)dt| < \varepsilon. \tag{8}$$

Moreover, if $K(s) = K_0(s)$ *is given by* (6) *then the selection* $k(s)$ *can be chosen in the form*

$$k(s) = \sum_{i=1}^{m} k_i(s) \cdot \chi_{A_i(s)}, \tag{9}$$

where $A_i(s)$ *is a finite, continuous family of decompositions of* T.

3. THE MAIN RESULT

We shall consider a C-mapping $K(x,s)$ from $X \times S$ into $\text{dec}(L^1)$. Clearly, $L(x,s) = \text{cl} \int_T K(x,s)$ is also a C-mapping from $X \times S$ into $\text{clco}(R^n)$ and every C-selection $k(x,s)$ of $K(x,s)$ "produces" a

C-selection $l(x,s)$ of $L(x,s)$ by the formula

$$l(x,s) = \int_T k(x,s)(t)dt. \tag{10}$$

This situation can be, in a sense, reversed.

Theorem 1. Consider a C-*mapping* $K(x,s)$ *from* $X \times S$ *into* $\mathrm{dec}(L^1)$. *Then for every* C-*selection* $l(x,s)$ *of* $\mathrm{cl} \int_T K(x,s)$ *and every* $\varepsilon > 0$ *there exists a* C-*selection* $k(x,s)$ *of* $K(x,s)$ *such that for all* $x \in X$ *and* $s \in S$

$$|l(x,s) - \int_T k(x,s)(t)dt| < \varepsilon. \tag{11}$$

Proof: We split the proof into two parts. First we prove the theorem for compact sets and next extend it to the general case. We shall assume that $l(x,s)$, $\varepsilon > 0$ and $\eta \in (0,\varepsilon)$ are fixed.

1. *S is compact.* In this case the space $\mathbf{C}$ of continuous mappings $\phi: S \to L^1(T,R^n)$ is a separable Banach space with the standard norm $||\phi|| = \sup\{||\phi(s)||: s \in S\}$. Let $k(x,s)$ be a fixed C-mapping from $X \times S$ into $L^1(T,R^n)$. For every $x \in X$ denote by

$$\mathcal{K}(x) = \{\phi \in \mathbf{C}: \phi(s) \in K(x,s) \text{ for all } s \in S\},$$
$$\kappa(x)(s) = k(x,s),$$
$$\ell(x)(s) - l(x,s) \text{ and}$$
$$(\int_T \kappa(x)(t)dt)(s) = \int_T k(x,s)(t)dt.$$

From the Proposition 1 follows that the sets $\mathcal{K}(x) \neq \emptyset$. One can check that the mappings $\mathcal{K}(x)$, $\kappa(x)$, $\ell(x)$ and $\int_T \ell(x)(t)dt$ are Σ-measurable as well as the mapping

$$\mathcal{P}(x) = \mathrm{cl}\{\phi \in \mathcal{K}(x): ||\ell(x) - \int_T \phi(x)(t)dt|| < \eta\}.$$

Now the Kuratowski and Ryll-Nardzewski Theorem (1975) ends the proof.

2. *S is locally compact.* Let for $m = 1,2, \ldots,$ S_m be such compact sets that

$$S_m \subset \mathrm{Int}\, S_{m+1} \quad \text{and} \quad \cup S_m = S.$$

From the first part we can conclude that there exist mappings $k_m(x,s)$

from $X \times S_m$ into $L^1(T,R^n)$ which are C-selections of $K(x,s)$ restricted to $X \times S_m$ and satisfying (11) on $X \times S_m$. We now define $k(x,s)$ to be equal to $k_m(x,s)$ if $(x,s) \in X \times S_m$ and $(x,s) \notin X \times S_i$ for $i < m$. Clearly $k(x,s)$ is a required selection.

As a corollary let us consider the following, special case of Theorem 1. Let $k_1(x,s), \ldots, k_m(x,s)$ and $p_1(x,s), \ldots, p_m(x,s)$ be C-mappings from $X \times S$ into, respectively, $L^1(T,R^n)$ and $L^1(T,R)$. We will assume that $p_1(x,s), \ldots, p_m(x,s)$ are nonnegative and such that for every $s \in S$

$$\sum_{i=1}^{m} p_i(x,s)(t) = 1 \quad \text{a.e.} \quad \text{in } T. \tag{12}$$

Define

$$U(x,s,t) = \{k_1(x,s)(t), \ldots, k_m(x,s)(t)\} \tag{13}$$

and let

$$K_0(x,s) = K_{U(x,s)}. \tag{14}$$

Since $k_i(x,s)$ and $p_i(x,s)$ are for all $i = 1,\ldots,m$ continuous, then $K_0(x,s)$ is a C-mapping and the mapping

$$l_0(x,s) = \int_T \left(\sum_{i=1}^{m} p_i(x,s)(t) \cdot k_i(x,s)(t) \right) dt \tag{15}$$

is a C-selection of $\text{cl} \int_T K_0(x,s)$. In this case the Theorem 1 leads to the following:

Theorem 2. For given C-*mappings* $k_1(x,s), \ldots, k_m(x,s)$ *and* $p_1(x,s), \ldots, p_m(x,s)$ *with* (12) *and for every* $\varepsilon > 0$ *there exists a* C-*selection* $k_0(x,s)$ *of the mapping* $K_0(x,s)$ *given by* (14) *such that for all* $x \in X$ *and* $s \in S$

$$\left| \int_T k_0(x,s)(t)dt - \int_T \left(\sum_{i=1}^{m} p_i(x,s)(t) \cdot k_i(x,s)(t) \right) dt \right| < \varepsilon .$$

The mapping $k_0(x,s)$ *can be chosen in the form*

$$k_0(x,s) = \sum_{i=1}^{m} k_i(x,s) \cdot \chi_{B_i(x,s)}, \tag{16}$$

for some C-decomposition $\{B_i(x,s)\}$ *of* T. *The last part follows from Proposition 1.*

4. CARATHEODORY SELECTIONS OF RANGE OF VECTOR MEASURES

Let M be an arbitrary set of measures μ on L taking values in $[0,\infty)$. By $\sup\{\mu: \mu \in M\}$ we shall denote the reference measure (or essential maximum) for M i.e. a measure $\mu_0: L \to [0,\infty]$ such that

i) $\mu \leq \mu_0$ for all $\mu \in M$

and

ii) if for a measure ν is $\mu \leq \nu$ for all $\mu \in M$, then $\mu_0 \leq \nu$.

(for details see Dunford and Schwartz (1956) and Oxtoby (1980)).

For an arbitrary measure $\mu: L \to R$ denote by

$$\mu^+ = \sup\{\mu,0\}, \quad \mu^- = \sup\{-\mu,0\} \quad \text{and} \quad |\mu| = \sup\{\mu^+,\mu^-\}.$$

We shall consider below the space $M = M(T,L,R^n)$ of all finite, nonatomic measures $\vec{\mu} = (\mu_1,\ldots,\mu_n)$ with the topology given by $||\vec{\mu}|| = \sup\{|\mu_1|,\ldots,|\mu_n|\}(T)$. Consider vector measures $\vec{\mu}(x,s) \in M$ and for every $x \in X$ and $s \in S$ denote by $R(x,s)$ the range of measure $\vec{\mu}(x,s)$; there is $R(x,s) = \{\vec{\mu}(x,s)(A): A \in L\}$. Because of the Liapunov Theorem the sets $R(x,s)$ are convex and compact. One can check that if $\vec{\mu}(x,s)$ is a C-mapping from $X \times S$ into M, then $R(x,s)$ is a C-mapping from $X \times S$ into $clco(R^n)$. We shall now assume that

$$\mu_0 = \sup\{|\vec{\mu}(x,s)|: x \in X,\ s \in S\} \tag{17}$$

is a finite measure and consider a C-mapping $A(x,s): X \times S \to L$. Clearly then the function

$$r(x,s) = \vec{\mu}(x,s)[A(x,s)] \tag{18}$$

is a C-selection of $R(x,s)$. Not every C-selection of $R(x,s)$ is of the form (14), however we have the following analogue of the Theorem 1:

Theorem 3. Let $r(x,s)$ *be a C-selection of* $R(x,s)$. *Then for every* $\varepsilon > 0$ *there exists a C-mapping* $A(x,s)$ *from* $X \times S$ *into* $L^1(T,R,\mu_0)$ *such that for all* $x \in X$ *and* $s \in S$

$$|\vec{\mu}(x,s)[A(x,s)] - r(x,s)| < \varepsilon. \tag{19}$$

Proof: Fix $r(x,s)$ and $\varepsilon > 0$. Consider the space $L^1(T,R,\mu_0)$ of μ_0-integrable real functions and let for every $x \in X$ and $s \in S$, $f(x,s) \in L^1(T,R,\mu_0)$ be such that the function for every $A \in L$

$$\vec{\mu}(x,s)(A) = \int_A f(x,s)(t)\,d\mu_0. \tag{20}$$

Since $\vec{\mu}(x,s)$ is a C-mapping then $f(x,s)$ is a C-mapping too. It is easy to see that $R(x,s) = \int_T U(x,s,t)\,d\mu_0$, where $U(x,s,t) = \{f(x,s)(t),0\}$. Now our claim is a consequence of Theorem 2.

REFERENCES

1. Z. Artstein and K. Prikry, Caratheodory Selections and the Scorza-Dragoni Property, *J. of Math. Anal. and Appl. 127:* 540-547 (1987).
2. Z. Artstein, Parametrized integration of multifunctions with application to control and optimization, in preprint: (1988).
3. J. P. Aubin and A. Cellina, Differential inclusion, Springer-Verlag, vol. 284, Berlin (1984).
4. R. J. Aumann, Integrals of set-valued functions, *J. of Math. Anal. and Appl. 127:* 1-12 (1965).
5. C. Castaing, Sur l'existence des sections separement measurable et separement continues d'une multi-applications. Seminaire d'Analyse Convexe, vol. 14, Montpellier (1975).
6. J. Diestel and J. J. Uhl, Vector Measures, Springer-Verlag, New York (1975).
7. N. Dunford and J. T. Schwartz, Linear Operators, J. Wiley, New York (1958).
8. A. Fryszkowski, Continuous selections for a class of nonconvex multivalued maps, *Studia Math. 76:* 163-174 (1983).
9. A. Fryszkowski, Caratheodory type selectors of set-valued maps of two variables, *Bull. Polon. Acad. Sci. 25:* 41-46 (1976).
10. A. Fryszkowski, Caratheodory type selections of multivalued mappings and their application to random differential inclusions, *Coll. Math., in prep.* (1983).
11. A. Fryszkowski, Continuous selections of Aumann integrals, *J. of Math. Anal. and Appl.:* to appear (1987).
12. R. V. Gamkrelidze, On some extremal problems in the theory of differential equations with applications to optimal control, *SIAM J. of Control 3:* 106-130, (1965).

13. T. Kim, K. Prikry and N. C. Yannelis, Caratheodory type selections and random fixed point theorems, *J. of Math. Anal. and Appl., 122:* 393-407 (1987).

14. K. Kuratowski and C. Ryll-Nardzewski, A general theorem of selectors, *Bull. Polon. Acad. of Sci., 6:* 397-401 (1975).

15. E. Michael, Continuous selection I, *Ann. of Math., 63:* 363-401 (1956).

16. C. Olech, Lexicographical order, range of integrals and "bang-bang" principle, *Math. Theory of Control*: 35-45, New York (1967).

17. J. Oxtoby, Measure Theory, 2nd Edition, Springer-Verlag, New York (1980).

18. L. Rybinski, On Caratheodory type selections, *Fund. Math., 125:* 187-193 (1985).

Efficient General Numerical Methods in Optimal Control Theory

John Gregory, Professor, Department of Mathematics, Southern Illinois University at Carbondale, Carbondale, Illinois

Cantian Lin, Department of Mathematics, Southern Illinois University, Carbondale, Illinois

ABSTRACT

New, accurate, efficient numerical methods are given for a general class of optimal control problems. These methods have an a priori error estimate of $O(h^2)$ and a Richardson error of $O(h^4)$ and include equality, inequality and transversality conditions. Of practical importance is that multiplier rules are used to reduce these problems to the calculus of variations setting in

---

This work was supported in part by Office of Naval Research Grant Number N00014-88-K-0081.

m-dependent variables. The resulting problem is efficiently solved by considering a tridiagonal system of nonlinear equations.

1. INTRODUCTION

Although the importance of optimal control problem theory is now universally accepted and outstanding theoretical results have been obtained little real headway has been made on the numerical solution of these problems. This is not surprising considering how difficult these problems are when compared, for example, with ordinary differential equations. However, the lack of general algorithms seems to have hindered the use of this theory.

The purpose of this paper is to describe a large class of accurate, efficient algorithms which can solve general problems in optimal control theory. It will be shown that these problems can be converted to equivalent problems in the calculus of variations and then solved with an a priori error of $O(h^2)$ or $O(h^4)$.

The remainder of this paper is as follows. In Section 2 we describe the numerical results for the calculus of variations problem and give a nontrivial example. In Section 3 we give a broad overview of the general problem we expect to completely solve and theoretical justification for the solution. In Section 4 we describe how to convert a basic problem to one of Section 2.

Similar conversion is done for inequalities in Section 5 and numerical transversality conditions in Section 6. In the spirit of this conference we focus in these two sections on illustrative examples. Finally, in Section 7 we present other results.

2. THE CALCULUS OF VARIATIONS PROBLEM

The problem is to find numerical solutions for extremal solutions of

$$I(x) = \int_a^b f(t,x,x')dt \tag{1}$$

where $x(t)$ is an m-vector. This will be done by finding

approximate numerical solutions of the first variational problem

$$I'(x,y) = \int_a^b (y^T f_x + y'^T f_{x'})dt = 0 \tag{2}$$

for numerical admissible variations $y(t)$. The setting and background is given in Hestenes [4; pp. 57-62]. In particular, we require the $m \times m$ matrix $f_{x'x'}$ to be invertible for each t in $[a,b]$, enough smoothness on f to yield a unique piecewise smooth solution and that (1) have no conjugate points in $[a,b]$.

Letting $\pi = (a = a_0 < a_1 < \cdots < a_N = b)$ be a partition of $[a,b]$ with $a_{k+1} - a_k = h = (b-a)/N$; $z_k(t)$ be the spline hat functions with $z_k(a_k) = 1$, $Z_k(t) = z_k(t)I_{m\times m}$;

$$x_h(t) = \sum_{k=0}^{N} Z_k(t)C_k \quad \text{and} \quad y_h(t) = \sum_{k=1}^{N-1} Z_k(t)D_k \tag{3}$$

be, respectively, the numerical solution to our problem and the numerical admissible variation; and utilizing the linearity of $y(t)$ in (2) we have the algorithm

$$\begin{aligned} &f_{x'}\left[a_{k-1}^*, \frac{x_k+x_{k-1}}{2}, \frac{x_k-x_{k-1}}{h}\right] + \frac{h}{2} f_x\left[a_{k-1}^*, \frac{x_k+x_{k-1}}{2}, \frac{x_k-x_{k-1}}{h}\right] \\ &\quad - f_{x'}\left[a_k^*, \frac{x_k+x_{k+1}}{2}, \frac{x_{k+1}-x_k}{h}\right] \\ &\quad + \frac{h}{2} f_x\left[a_k^*, \frac{x_k+x_{k+1}}{2}, \frac{x_{k+1}-x_k}{h}\right] = 0 \end{aligned} \tag{4}$$

for $k = 1,2,\ldots,N-1$. In the above $a_k^* = (a_k + a_{k+1})/2$ and $x_k = x_h(a_k)$ is the computed value of the solution $x(t)$ at a_k.

We note that (4) is a block tridiagonal system of equations which is easily solved in practice by Newton's method with the accuracy described in Theorem 2 below. For the two-point boundary value problem with $x(a) = x_a$ and $x(b) = x_b$, (4) is a system of $m(N-1)$ nonlinear equations in $m(N-1)$ unknowns. For the initial value problem with $x(a) = x_a$ and $x'(a) = x'_a$, we have a nonlinear equation in the m variables x_{k+1} for each $k = 1,2,\ldots,N-1$.

The first theorem involves long, but elementary calculations with the local truncation error, (see Henrici [3]).

THEOREM 1. Between corners of $x(t)$, the local truncation error is $h^3Q(a_k) + O(h^5)$, for $k = 1,2,\ldots,N-1$.

In practice, Theorem 1 tells us that we have chosen a good algorithm. The vector $Q(t)$ depends only on the solution $x(t)$ and its derivatives and f and its derivatives. Thus,

THEOREM 2. For $h > 0$ sufficiently small there exist $C > 0$ independent of h so that for any component e of the error $e_h(a_k) = x(a_k) - x_h(a_k)$ we have $|e| \leq Ch^2$. In addition, the Richardson solution $x_h^R(t)$ where $x_h^R(a_k) = [4x_{h/2}(a_k) - x_h(a_k)]/3$ has a maximum component, pointwise error satisfying $|e^R| \leq Ch^4$ where e^R is any component of $e_h^R(a_k) = x(a_k) - x_h^R(a_k)$.

The proof of this result is too long and difficult to be given here and will appear elsewhere.

To illustrate the accuracy and efficiency we give a nontrivial example.

Our problem is to find the extremal solution for $f(t,x,x')$ in (1) or (2) where $x(x_1,x_2,x_3)^T$ and

$$f(t,x,x') = \frac{1}{3}e^{-t}x_1'^3 + \frac{1}{2}x_2'^2 + \frac{1}{4}x_3'^4 + \frac{1}{2}e^{-2t}x_1^4 + 48e^{6t}x_3$$
$$- e^{-t}x_3x_1 - tx_3' - x_2 \sin t$$

subject to $x(0) = (1,0,1)^T$ and $x(1) = (e,\sin 1,e^2)^T$ on the interval $[a,b] = [0,1]$. The reader may verify that $x(t) = (e^t,\sin t,e^{2t})$ is the unique solution to this problem on the inverval $[0,1]$.

In Figure 1 we list as follows: Column 1 is the independent variable t, Columns 2-4 are the values of our solution $x(t)$ at these times, Columns 5-7 are the values of the errors $x_k-x(a_k)$ for $h = \frac{1}{8}$. Column 1 of the "second line" is the value of t, Columns 2-4 and 5-7 are the errors for $h = \frac{1}{16}$ and $h = \frac{1}{32}$ respectively. Note that for each time t and each component of e_h, the values are

t		$x(t)$		$e_h(t)$	$(h = \frac{1}{8})$	
.125	0.113D+01	0.125D+00	0.125D+01	-0.712D-03	-0.127D-04	-0.30
.250	0.128D+01	0.247D+00	0.160D+01	-0.124D-02	-0.241D-04	-0.45
.375	0.145D+01	0.366D+00	0.207D+01	-0.153D-02	-0.331D-04	-0.50
.500	0.165D+01	0.479D+00	0.267D+01	-0.157D-02	-0.383D-04	-0.49
.625	0.187D+01	0.585D+00	0.345D+01	-0.141D-02	-0.386D-04	-0.44
.750	0.212D+01	0.682D+00	0.445D+01	-0.107D-02	-0.329D-04	-0.34
.875	0.240D+01	0.768D+00	0.573D+01	-0.588D-03	-0.204D-04	-0.19

t	$e_h(t)$	$(h = \frac{1}{16})$		$e_h(t)$	$(h = \frac{1}{32})$	
.125	-0.182D-03	-0.317D-05	-0.746D-02	-0.458D-04	-0.793D-06	-0.18
.250	-0.317D-03	-0.603D-05	-0.113D-01	-0.795D-04	-0.151D-05	-0.28
.375	-0.389D-03	-0.826D-05	-0.128D-01	-0.977D-04	-0.206D-05	-0.32
.500	-0.401D-03	-0.956D-05	-0.127D-01	-0.101D-03	-0.239D-05	-0.31
.625	-0.359D-03	-0.964D-05	-0.113D-01	-0.900D-04	-0.241D-05	-0.28
.750	-0.272D-03	-0.823D-05	-0.874D-02	-0.683D-04	-0.206D-05	-0.22
.875	-0.150D-03	-0.509D-05	-0.503D-02	-0.376D-04	-0.127D-05	-0.12

t	$e_h^R(t)$	$(h = \frac{1}{8})$		$e_h^R(t)$	$(h = \frac{1}{16})$	
.125	-0.579D-05	0.434D-08	0.208D-03	-0.324D-06	0.271D-09	0.13
.250	-0.853D-05	0.825D-08	0.120D-04	-0.493D-06	0.515D-09	0.23
.375	-0.990D-05	0.113D-07	-0.161D-03	-0.585D-06	0.706D-09	-0.86
.500	-0.100D-04	0.131D-07	-0.255D-03	-0.600D-06	0.816D-09	-0.15
.625	-0.890D-05	0.132D-07	-0.276D-03	-0.539D-06	0.823D-09	-0.16
.750	-0.671D-05	0.113D-07	-0.236D-03	-0.410D-06	0.703D-09	-0.14
.875	-0.366D-05	0.696D-08	-0.144D-03	-0.225D-06	0.435D-09	-0.90

Figure 1

decreasing by a factor of four as h is halved, since $e_h = O(h^2)$. Column 1 of the "third line" is the time, while Columns 2-4 are the Richardson errors at $h = \frac{1}{8}$ and Columns 5-7 are the Richardson errors at $h = \frac{1}{16}$. By the Richardson value we mean $x_h^R(t) = [4x_{h/2}(t) - x_h(t)]/3$. We note that the expected error is $e_h^R(t) = O(h^4)$ and hence the values in Columns 5,6,7, respectively, should be approximately 1/16*th* the values in Columns 2,3,4.

Finally, we note that the initial guess for Newton's method was the linear values between x(0) and x(1). We considered our Newton method to have "converged" if $\|\Delta x\|_\infty < 10^{-8}$ which required (respectively 36 iterations when $h = \frac{1}{8}$, 72 iterations when $h = \frac{1}{16}$ and 119 iterations when $h = \frac{1}{32}$). We note that the Richardson values which are significantly better than the regular values are "free" in that they are simple linear combinations of existing values.

3. A BROAD OVERVIEW

The purpose of this brief section is to consider a general problem in optimal control theory and the justification for the numerical results. For this section we will change notation to correspond to Hestenes [4, pp. 346-351].

Hestenes shows that "A General Control Problem of Bolza" defined by the conditions

$$I_\rho(x) = g_\rho(b) + \int_{t^0}^{t^1} L_\rho(t,x(t),u(t))dt \qquad (\rho = 0,1,\ldots,p),$$

$$\varphi_\alpha(t,x,u) \leq 0 \quad (1 \leq \alpha \leq m'), \quad \varphi_\alpha(t,x,u) = 0 \quad (m' < \alpha \leq m),$$

$$\dot{x}^i = f^i(t,x,u), \quad \text{and}$$

$$t^s = T^s(b), \quad x^i(t^s) = X^{is}(b) \qquad (i = 1,\ldots,n;\ s = 0,1)$$

$$I_\gamma(x) \le 0 \quad (1 \le \gamma \le p'), \qquad I_\gamma(x) = 0 \quad (p' < \gamma \le p),$$

has a minimizing solution for $I_0(x)$ of the form

$$x_0\colon\ x_0(t),\ \ b_0\ ,\ \ u_0(t) \quad (t^0 \le t \le t^1)$$

if there exist multipliers

$$\lambda_0 \ge 0,\ \ \lambda_\gamma\ ,\ \ p_i(t),\ \ \mu_\alpha(t)$$

$$(\gamma = 1,\dots,p;\ i = 1,\dots,n;\ \alpha = 1,\dots,m),$$

not vanishing simultaneously, and functions

$$H(t,x,u,p,\mu) = p_i f^i - \lambda_\rho L_\rho - \mu_\alpha \varphi_\alpha,\ \ G(b) = \lambda_\rho g_\rho$$

$$(\rho = 0,1,\dots,p)$$

satisfying the usual, expected conditions (see [5; pp. 348-350]).

Finally, we claim that

THEOREM 3. The definitions of $x^{n+1}(t),\dots,x^{n+q+m}(t)$ given by

$$\dot{x}^i = u^{i-n},\quad x^i(a) = 0 \qquad (i = n+1,\dots,n+q)$$

$$\dot{x}^i = \mu_\alpha^{i-n-q},\quad x^i(a) = 0 \qquad (i = n+q+1,\dots,n+q+m)$$

allow us to convert the General Control Problem of Bolza to a problem of the form (1) and allows a numerical solution with the errors described in Theorem 2 above.

4. THE BASIC PROBLEM

The purpose of this section is to show that the basic optimal control problem can be converted to a problem in the calculus of variations and solved as in Section 2. For convenience, we revert to more familiar notation. We assume that all problems have unique minimum solutions.

Thus, we desire to minimize

$$J_1(x) = \int_a^b f(t,x,u)dt, \tag{5}$$

subject to the trajectory equation

$$x'(t) = g(t,x,u) \tag{6}$$

and the boundary conditions

$$x(a) = x_a, \; x(b) = x_b \qquad \text{or} \tag{7a}$$
$$x(a) = x_a, \; x(b) \text{ arbitrary.} \tag{7b}$$

We assume that x is an n-vector, u is an m-vector, and that f and g are smooth enough to obtain the results given below.

The optimal control problem (5)-(7) is converted to an equivalent calculus of variations problem by defining

$$x_1(t) = x(t), \tag{8a}$$

$$x_2(t) = \int_a^t u(s)ds, \qquad \text{and} \tag{8b}$$

$$x_3(t) = \int_a^t \lambda(s)ds \tag{8c}$$

where $\lambda(t)$ is the Lagrange multiplier associated with (6) given in $J_2(X)$ below.

Using (8) we define the 2n+m vector

$$X(t) = \left[x_1^T(t), x_2^T(t), x_3^T(t)\right]^T \tag{9}$$

and

$$\begin{aligned} J_2(X) &= \int_a^b \{f(t,x_1,x_2') + x_3'[x_1' - g(t,x_1,x_2')]\}dt \\ &= \int_a^b F(t,X,X')dt. \end{aligned} \tag{10}$$

Except for the natural boundary conditions for $x_2(b)$ and $x_3(b)$ considered below, algorithm (4) applied to (10) yield the results given in Theorem 1 and 2. We note that the above problem includes the "Linear Regulator Problems" (quadratic f, linear g) where (4) reduces to a system of linear equations which is easily solved by Gaussian elimination methods.

5. INEQUALITIES

We now consider the basic problem with inequality constraints of the form $\varphi(t,x,x',u) \leq 0$. For a variety of reasons we consider illustrative example problems and not general theory. In addition our examples are "simplistic in form". This is not done with concern for our methods but because we want problems where a closed form solution can be found for comparison purposes.

As our example, we wish to minimize

$$J(x) = \int_0^b \left[\frac{1}{2}u^2 - x^2\right] dt, \quad b > 0$$

subject to $x' = x+u$ and $|u| \leq \alpha \doteq \frac{\sqrt{6}}{2}$. One of the reasons we have chosen this problem is that we want an example where the optimal solution $u(t)$ takes on the values $\pm\alpha$ as well as values $|u| < \alpha$. In particular, it can be shown that the conditions $x(0) = 0$, $x'(0) = 1$ leads to the optimal control solution pictured in Figure 2 where $\bar{t} = \frac{7\pi}{12} \cong 1.8326$, $t^* = \frac{7\pi}{12} + \ell n\left[-\tan\frac{7\pi}{12}\right] \cong 3.1496$ and $t^{**} = t^* + \bar{t} + \frac{\pi}{12} \cong 5.2440$.

Before we begin with inequality problems we note that if $b \leq \bar{t}$ then the solution is the same as for the nonconstrained problem and that our basic problem on $[0,\bar{t}]$ has errors "identical" to the nontrivial example in Section 2 for the appropriate initial value problem or boundary value problem.

To handle inequality constraints we will consider three methods. The first is designated by the words "Kuhn-Tucker", the second is a two-sided slack variable approach. The third will feature a one-sided slack variable approach. The reader should easily see how to form more complicated problems using these examples.

In the Kuhn-Tucker case we define $x_1(t) = x(t)$, $x_2'(t) = u(t)$, $x_3'(t) = \lambda_1(t)$, $x_4'(t) = \lambda_2(t)$, $x_3'(t) = \lambda_3(t)$ where $x_2(0) = 0$ and for $i = 1,2,3$, $x_{2+i}(0) = 0$ and $\lambda_i(t)$ is the multiplier below. In particular, $|u| \leq \alpha$ implies $u-\alpha \leq 0$ and $-u-\alpha \leq 0$ and hence our problem becomes to minimize

$$J_2(X) = \int_0^b \left[\frac{1}{2}x_2'^2 - x_1^2 + x_3'(x_1' - x_1 - x_2') + x_4'(x_2' - \alpha) + x_5'(-x_2' - \alpha)\right] dt \quad (11)$$

subject to the Kuhn-Tucker conditions

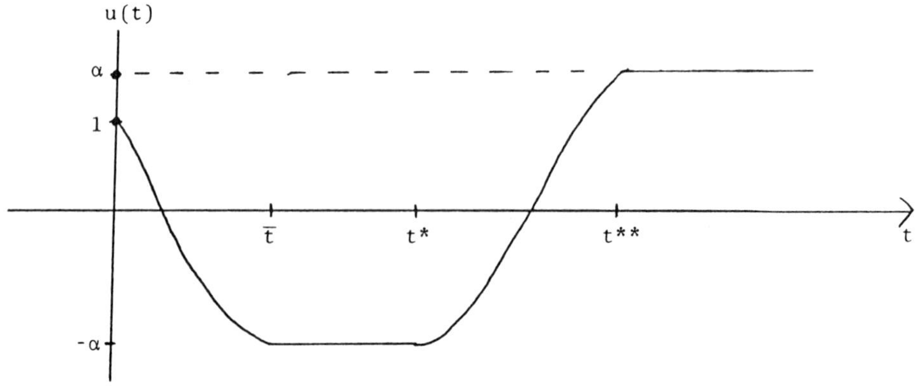

Figure 2

$$\begin{cases} \text{if } x_2'-\alpha < 0 & \text{then } x_4' = 0 \\ \text{if } -x_2'-\alpha < 0 & \text{then } x_5' = 0. \end{cases} \qquad (12)$$

Note that if the inequalities were equalities (both can not be) or if (11) was a calculus of variations problem, except for the right hand boundary conditions, we would use (4) with

$$f_{X'} = \begin{bmatrix} x_3' \\ x_2'-x_3'+x_4'-x_5' \\ x_1'-x_1-x_2' \\ x_2'-\alpha \\ -x_2'-\alpha \end{bmatrix} \quad \text{and} \quad f_X = \begin{bmatrix} -2x_1-x_3' \\ 0 \\ 0 \\ 0 \\ 0 \end{bmatrix}.$$

In the "free region" where $|u| < \alpha$ we can use the first three components of $f_{X'}$ and f_X in (13) along with the conditions $x_4' = 0$ and $x_5' = 0$. Similarly, in the region where $x_2' = -\alpha$ we use all but the fourth component of $f_{X'}, f_X$ in (13) along with the condition $x_4' = 0$ while in the region $x_2' = \alpha$ we use all but the fifth component of $f_{X'}, f_X$ in (13) along with the condition $x_5' = 0$.

If we know which region we are in we use the appropriate formulas as described in the last paragraph and obtain a numerical solution with errors as before. Of course, in general we will not have this information. However by combining this algorithm with one of the slack algorithms discribed below which are "singular" and do not quite give $O(h^2)$ errors, we can determine the boundaries of the regions and then use the Kuhn-Tucker method.

For a well defined initial value problem, the Kuhn-Tucker conditions work because of the local nature of the problem. Thus, if we know X_{k-1} and X_k we find X_{k+1} by solving (4). If the condition $|u| \leq \alpha$ is satisfied with this X_{k+1} we accept this value. If $u > \alpha$ or $u < -\alpha$ we compute X_{k+1} from the modified form of (13) described in the paragraph before the last one.

The second method is for two-sided inequalities which occur often. For example, $|u| \leq \alpha$ implies $\alpha^2-u^2 > 0$ so we define x_1, x_2, x_3 as above and x_4' by

$$x_4'^2+x_2'^2-\alpha^2 = 0 \quad \text{or} \tag{14a}$$

$$x_4' = \sqrt{\alpha^2-x_2'^2} \ . \tag{14b}$$

We still require a multiplier x_5' with $x_5(0) = 0$ so that our new integral to be minimized is

$$J_3(X) = \int_0^b \left[\frac{1}{2}x_2^2-x_1^2+x_3'(x_1'-x_1-x_2')+x_5'(\alpha^2-x_2'-x_4'^2)\right]dt. \tag{15}$$

This method has advantages and disadvantages over the Kuhn-Tucker method. The major advantage is that we now have only equalities. The major disadvantage is that an original "linear" problem is now nonlinear. We also note that even when a closed form solution of x_3' is known as in this case the right hand side of (14b) involves a singular integral. As remarked above, test cases give answers which are not as good as described in Section 2 but this method can be combined with the Kuhn-Tucker method.

Our final method involves one-sided inequalities. For convenience we assume only the constraint $u \geq -\alpha$ which is equivalent to $|u| \leq \alpha$ by assuming $0 \leq b \leq t^{**}$. In this case we introduce a slack variable x_4' and a multiplier x_5' to obtain

$$J_4(X) = \int_a^b \left[\frac{1}{2}x_2'^2-x_1^2+x_3'(x_1'-x_1-x_2')+x_5'(x_2'+\alpha-x_4'^2)\right]dt. \tag{16}$$

The comments are similar to those of the two-sided inequality case.

6. TRANSVERSALITY CONDITIONS

For convenience, we assume $(a,x(a))$ is given but if not, our methods for the terminal time $t = b$ can be easily adopted if a or $x(a)$ is not given. We note that there are two kinds of transversality conditions. In the first case, which is also the easiest, the terminal time is fixed but the x value, $x(b)$, is unknown. The more difficult case is when the terminal time is unknown.

For the case when $a, x(a)$ and $b = a_N$ are given we return to the derivation of algorithm (4). Choosing $z_N = 1+(t-a_N)/h$ on

$[a_{N-1},a_N]$ and zero otherwise, $Z_N(t) = z_N(t)I_{m\times m}$ and using (2) with $y = Z_N(t)$ we obtain the first two terms of (4) with $k = N$. Thus, with (4) and

$$f_{x'}\left[a^*_{N-1}, \frac{x_N+x_{N-1}}{2}, \frac{x_N-x_{N-1}}{h}\right] + \frac{h}{2}f_x\left[a^*_{N-1}, \frac{x_N+x_{N-1}}{2}, \frac{x_N-x_{N-1}}{h}\right] = 0 \quad (17)$$

we have a nonlinear system of dimension $m(N-1)+m = mN$ in the mN unknowns $x_1,x_2,\ldots,x_m$. We will see in the next paragraph that if components of $x(b)$ are not given the corresponding components of (16) hold as natural boundary conditions.

For optimal control problems with $b = a_N$ given we must use (17) even when $x_1(b) = x(b)$ is specified. This is because the values of $x_2(b),x_3(b),\ldots$ in Sections 4 and 5, although well defined, are not known. Thus, if $x_1(b) = x(b)$ is not specified we use "all" of (17) while if $x_1(b) = x(b)$ is specified we use the remaining component equations in (17).

It can be shown that the local truncation error for (17) is $O(h^2)$ and that a modification of our proof in Theorem 2 leads to a global error of $O(h^2)$. Computer examples confirm these results.

In the more difficult case when $t = b$ is not fixed we must derive numerical transversality conditions. For convenience, we assume $m = 1$ but the case with $m > 1$ follows immediately.

Thus, we assume that b is not known but that $(b,x(b))$ is on the graph of $(b,\varphi(b))$ where $y = \varphi(x)$ is a fixed given curve. Letting $y(t)$ be a variation in $x(t)$ we require

$$\int_a^{b+h} f(t,x+y,x'+y')dt = I(x+y) \geq I(x) = \int_a^b f(t,x,x')dt$$

which leads to (with $h > 0$)

$$hf + [\varphi(a_{N+1})-x_{N+1}][f_{x'} + \frac{h}{2}f_x] = 0 \quad (18)$$

where $f,f_{x'},f_x$ are evaluated at $(a^*_N,(x_N+x_{N+1})/2,(x_{N+1}-x_N)/h)$.

Similarly, we obtain a second "equivalent" condition

$$hf + [\varphi(a_{N-1})-x_{N-1}][-f_{x'} + \frac{h}{2}f_x] = 0 \quad (19)$$

where $f,f_{x'},f_x$ are evaluated at $(a^*_N,(x_N+x_{N-1})/2,(x_N-x_{N-1})/h)$.

It can be shown that (18) and (19) have local truncation error of $O(h^2)$. To avoid confusion we restate the result that a numerical "necessary condition" for a minimizing solution is that h,N and b = a+Nh hold so that (4) with k = 1,...,N-1 and (19) hold or (4) with k = 1,...,N and (18) hold.

7. OTHER PROBLEMS

We note that the methods given above can be applied to obtain further results.

Our first comment is to note that Section 4 includes "linear regulation problems" (see [6]) where (4) is a linear system of equations. Our second comment is to note that our methods immediately lead to numerical solutions for "abnormal problem" (see [1]) where extremal solutions vanish on nontrivial subintervals. These results generalize the ideas of conjugate points to conjugate intervals. We also note that our methods will solve minimum time problems by taking $f \equiv 1$ in (5) and letting $t = b$ be free.

Finally, we note that higher order algorithms similar to (4) can be constructed. For example,

$$-f_{x'}\left[a_k+\frac{h}{2},\ \frac{-2x_{k+2}+15x_{k+1}+12x_k-x_{k-1}}{24},\ \frac{-x_{k+2}+27x_{k+1}-27x_k+x_{k-1}}{24h}\right]$$

$$+ f_{x'}\left[a_k-\frac{h}{2},\ \frac{-x_{k+1}+12x_k+15x_{k-1}-2x_{k-2}}{24},\ \frac{-x_{k+1}+27x_k-27x_{k-1}+x_{k-2}}{24h}\right]$$

$$+ \frac{h}{2}f_x\left[a_k+\frac{h}{\sqrt{12}},\ \frac{(1-\sqrt{12})x_{k+2}+(-2+4\sqrt{12})x_{k+1}+(25-3\sqrt{12})x_k}{24},\right.$$

$$\left.\frac{-4x_{k+2}+(24+\sqrt{12})x_{k+1}+(-12-4\sqrt{12})x_k+(-8+2\sqrt{12})x_{k-1}}{24h}\right]$$

$$+ \frac{h}{2}f_x\left[a_k-\frac{h}{\sqrt{12}},\ \frac{(25-3\sqrt{12})x_k+(-2+4\sqrt{12})x_{k-1}+(1-\sqrt{12})x_{k-2}}{24},\right.$$

$$\frac{(8-2\sqrt{12})x_{k+1}+(-12-4\sqrt{12})x_k+(24+\sqrt{12})x_{k-1}+4x_{k+2}}{24h}\Bigg] = 0.$$

The associated local truncation error is $O(h^5)$ and we can prove that the two point boundary value problem has a global error $O(h^4)$ and a Richardson error of $O(h^6)$. This algorithm was surprisingly hard to obtain.

REFERENCES

1. Gregory, J. (1980). *Quadratic Form Theory and Differential Equations*, Academic Press, Mathematics in Science and Engineering Series, Volume 152.
2. Gregory, J., Zeman, M., and Badiey, M. (1985). A Finite Element Approximation for the Initial-Value Problem for Nonlinear Second Order Differential Equations, *Journal of Mathematical Analysis and Applications*, Volume 111, No. 1, 90-104.
3. Henrici, P. (1962). *Discrete Variable Methods in Ordinary Differential Equations*, John Wiley and Son.
4. Hestenes, M. R. (1966). *Calculus of Variations and Optimal Control Theory*, John Wiley and Son.
5. Keller, H. B. (1966). *Numerical Methods for Two-Point Boundary-Value Problems*, Blaisdell Publishing Co.
6. Kirk, D. E. (1970). *Optimal Control Theory*, Prentice Hall.
7. Leitmann, G. (1981). The Calculus of Variations and Optimal Control Theory, Plenum Press, *Mathematical Concepts and Methods in Science and Engineering Series*, Volume 24.
8. Lyubushin, A. A. (1979). Convergence Rate Estimate for Difference-Quadrature Approximations of Variational Problems, *U.S.S.R. Comput. Maths Math. Phys.*, Pergamon Press, Ltd., Vol. 18, 24-35.

Interconnection Between Ordinary and Delay Differential Equations

Istvan Győri, Professor, Department of Mathematics, The University of Rhode Island, Kingston, Rhode Island

1. INTRODUCTION

One of the most important differences between ordinary and delay differential equations is that in the ordinary case the dimension of the phase space is finite while in the delay case the phase space is infinite dimensional.

To avoid the difficulties arising from the gap between infinite and finite dimensional spaces, in 1965 Repin [10] initiated a method ensuring the arbitrarily close approximation of any sufficiently smooth solution of a delay differential equation by corresponding solutions of some associated ordinary differential equations. This technique termed the chain method. Janushevsky, in his book entitled "Control of Object with Delays" ([8]) , showed that this approximation method is very powerful tool to generalize several well known controability and observability results from ordinary differential systems to delay ones.

On the other hand this approximation technique derives powerful numerical methods (see, e.g. [2]) and it has interesting applications in the mathematical biology (see, e.g. [1]).

The main purpose of this paper is to give a very short introduction to the reader how to construct the approximating ordinary differential equations and to show some characteristic convergence theorems for some well understood scalar equations.

In Sec. 2 we give the construction of the basic chain method while in Sec. 3 we give an interesting modification of the basic chain method starting from the so called pipe-compartmental models which was initiated by the author in [4] and [5].

The main attraction of the modified chain method is that in that case the delay equation and its approximating ordinary equations have the same steady state solutions, as it is showed in Sec. 4. This shows promise to prove that the convergence of the approximation is uniform not only on compact intervals but on the whole time interval $[0,\infty)$. Indeed at the end of the paper we give a conjecture about the convergence on $[0,\infty)$ remarking that we can prove the conjecture in only a very restricted linear case. The importance of this result is that the convergence of the chain method on $[0,\infty)$ is not appear in the literature known to us.

We remark that the methods and statements in this paper can be generalized to delay differential systems with several discrete and/or distributed delays but because of space limitation we cannot present these results.

2. THE BASIC CHAIN METHOD

In this section we derive the basic chain method and we state a convergence theorem.

Consider the delay differential equation

$$\dot{x}(t)=f(t,x(t),x(t-\tau)), \quad t\geqslant 0, \tag{2.1}$$

where $f:R_+ xRxR\to R$, $(R_+=[0,\infty)$, $R=(-\infty,\infty))$, is a continuous function and $\tau>0$ is a constant. The initial conditions associated with (2.1) are the following type

$$x(s)=\phi(s) \quad , \quad -\tau\leqslant s\leqslant 0 \tag{2.2}$$

where $\phi\in C$ and $C=C([-\tau,0],R)$.

A solution of (2.1) and (2.2) will be denoted by $x(\phi)(t)$ and we will assume in the following that $x(\phi)(t)$ exists and is unique on $[-\tau,\infty)$.

We now choose $\phi\in C_0^1=\{\phi\in C^1:\dot{\phi}(0-)=f(0,\phi(0),\phi(-\tau))\}$ where C^1 denotes the set of continuously differentiable maps of $[-\tau,0]$ into R. Then the solution $x(t)=x(\phi)(t)$ of (2.1) and (2.2) is continuously differentable on $[-\tau,0)$

and $(0,\infty)$, moreover

$$\dot{x}(0-)=\dot{\phi}(0-)=f(0,\phi(0),\phi(-\tau))=\dot{x}(0+)$$

and clearly $x(t)$ is continuously differentiable on $[-\tau,\infty)$.

Let N be a positive integer. Then by Taylor's formula one has

$$\dot{x}(t-i\frac{\tau}{N})=-\frac{N}{\tau}x(t-i\frac{\tau}{N})+\frac{N}{\tau}x(t-(i-1)\frac{\tau}{N})+\varepsilon_i^{(N)}(t) , \qquad (2.3)$$

for all $i=1,\ldots,N$ and for all $t\geqslant 0$, where the error $\varepsilon_i^{(N)}(t)$ satisfies the following relation on any compact interval $[0,t_1]$:

$$\max_{1\leqslant i\leqslant N}\ \max_{0\leqslant t\leqslant t_1}|\varepsilon_i^{(N)}(t)|=$$

$$\max_{1\leqslant i\leqslant N}\ \max_{0\leqslant t\leqslant t_1}\left|-\frac{N}{\tau}x(t-i\frac{\tau}{N})+\frac{N}{\tau}x(t-(i-1)\frac{\tau}{N})-\dot{x}(t-i\frac{\tau}{N})\right|\to 0,$$

as $N\to+\infty$.

Set $x_i^{(N)}(t)=x(t-i\frac{\tau}{N})$ for all $i=0,\ldots,N$ and for all $t\geqslant 0$. Then (2.1) - (2.3) yield that $x^{(N)}(t)=(x_0^{(N)}(t),\ldots,x_N^{(N)}(t))^T$ is a solution of the ordinary differential equations

$$\dot{x}_0(t)=f(t,x_0(t),x_N(t)) , \qquad t\geqslant 0,$$

$$\dot{x}_i(t)=-\frac{N}{\tau}x_i(t)+\frac{N}{\tau}x_{i-1}(t)+\varepsilon_i^{(N)}(t) , \qquad 1\leqslant i\leqslant N , \ t\geqslant 0 ,$$

with the initial conditions

$$x_i(0)=\phi(-i\frac{\tau}{N}) , \qquad 0\leqslant i\leqslant N .$$

By dropping the error terms $\varepsilon_i^{(N)}(t)$ which are small for large N we obtain the so called chain approximation ordinary differential equations

$$\left.\begin{array}{l}\dot{y}_0(t)=f(t,y_0(t),y_N(t)) , \quad t\geqslant 0 , \\ \dot{y}_i(t)=-\frac{N}{\tau}y_i(t)+\frac{N}{\tau}y_{i-1}(t) , \quad t\geqslant 0 , \ 1\leqslant i\leqslant N ,\end{array}\right\} \qquad (2.4)$$

with the initial conditions

$$y_i(0)=\phi(-i\frac{\tau}{N}) \quad , \quad 0\leqslant i\leqslant N . \tag{2.5}$$

The solution of system (2.4) with initial conditions (2.5) is denoted by $y^{(N)}(\phi)(t)=(y_0^{(N)}(\phi)(t),\ldots,y_N^{(N)}(\phi)(t))^T$.

From some general approximation results proved by the author in [6], we state the following convergence theorem.

THEOREM 2.1 Assume that $\tau>0$ is a constant and $f:R_+ \times R\times R\to R$ is a continuous function and there is a number $m>0$ such that

$$|f(t,u_2,v_2)-f(t,u_1,v_1)|\leqslant m(|u_2-u_1|+|v_2-v_1|)$$

for all $(u_i,v_i)\in R\times R$, $i=1,2$.

Then

(a) for any $\phi\in C$ the solutions $x(\phi)(t)$ and $y^{(N)}(\phi)(t)=(y_0^{(N)}(\phi)(t),\ldots,y_N^{(N)}(\phi)(t))^T$ of (2.1) and (2.4), respectively, exists on $[0,\infty)$ and for all $t_1>0$

$$\sup_{0\leqslant t\leqslant t_1} |x(\phi)(t)-y_0^{(N)}(t)|\to 0 \quad , \quad \text{as } N\to+\infty .$$

(b) if the solution $x(\phi)(t)$ of (2.1) is twice continuously differentiable on $[-\tau,\infty)$ then for all $t_1>0$ there exists a constant $c(\phi,t_1)$ such that

$$|x(\phi)(t)-y_0^{(N)}(\phi)(t)|\leqslant c(\phi,t_1)\frac{\tau}{N} \quad , \quad 0\leqslant t\leqslant t_1 .$$

3. THE MODIFIED CHAIN METHOD

In this section we give a modification of the chain method to use a simple case of the compartmental systems with pipes. The general theory of linear compartmental systems with pipes were introduced and discussed by Gyori and Eller [3] and for the nonlinear case we refer to Gyori [4] and [5].

Consider the delay equation

$$\dot{x}(t)=-[g_0(x(t))+g(x(t))]+g(x(t-\tau)) \quad , \quad t\geqslant 0 , \tag{3.1}$$

where $g_0(x)$ and $g(x)$ are nonnegative continuous functions and $\tau>0$ is a constant.

Eq. (3.1) is related to a special compartmental system with one compartment and one pipe (see Fig. 3.1), where $x(t)$ denotes the amount of the material in the compartment at the moment $t, g_0(x(t))$ is the outflow rate from the system, moreover $g(x(t))$ is the speed of the material entering into the pipe from the compartment and τ denotes the transfer time in the pipe.

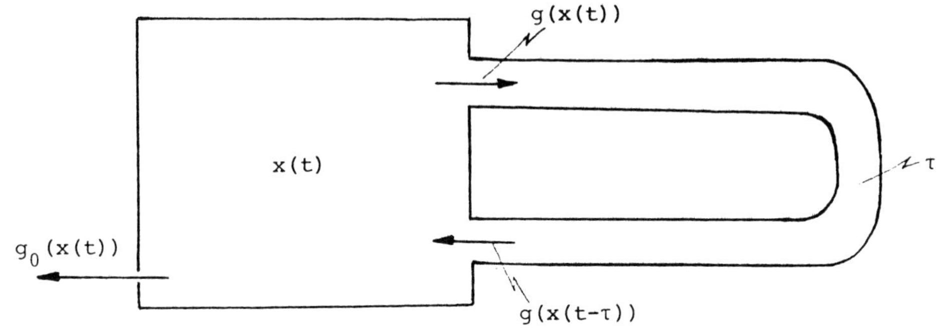

FIGURE 3.1 Pipe - compartmental system with one compartment and one pipe with discrete transit time τ.

The initial state of the system is defined by the initial condition

$$x(t)=\phi(t) \quad , \quad -\tau\leq t\leq 0, \ \phi\in C. \tag{3.2}$$

Equation (3.1) with initial condition (3.2) is equivalent to the following integral equation which represents the law of material conservation in the system:

$$x(t)+\int_{t-\tau}^{t} g(x(s))ds+\int_{0}^{t} g_0(x(s))ds=c[\phi] \quad , \quad t\geq 0 \quad , \tag{3.3}$$

where

$$c[\phi]=\phi(0)+\int_{-\tau}^{0} g(\phi(s))ds \ . \tag{3.4}$$

The integral $\int_{t-\tau}^{t} g(x(s))ds$ denotes the amount of the material in the pipe at the time $t\geq 0$.

Let N be a positive integer. Then for all $t\geq 0$ and $i\in\{1,\dots,N\}$ the amount of the material in the pipe over the time interval $[t-i\frac{\tau}{N}, t-(i-1)\frac{\tau}{N}]$ is given by

$$x_i^{(N)}(t)=\int_{t-i\frac{\tau}{N}}^{t-(i-1)\frac{\tau}{N}} g(x(s))\,ds \quad .$$

From this we find that $(x_1^{(N)}(t),\ldots,x_N^{(N)}(t))^T$ satisfies the system of differential equations

$$\dot{x}_1^{(N)}(t)=-\frac{N}{\tau}x_1^{(N)}(t)+g(x(t))+\mu_1^{(N)}(t) \quad ,$$

and

$$\dot{x}_i^{(N)}(t)=-\frac{N}{\tau}x_i^{(N)}(t)+\frac{N}{\tau}x_{i-1}^{(N)}(t)+\mu_i^{(N)}(t)-\mu_{i-1}^{(N)}(t) \quad , \quad 2<i<N,$$

where

$$\mu_j^{(N)}(t)=\frac{N}{\tau}\int_{t-j\frac{\tau}{N}}^{t-(j-1)\frac{\tau}{N}} g(x(s))\,ds-g(x(t-j\tfrac{\tau}{N})),$$

for all $j=1,\ldots,N$ and for all $t\geq 0$.

From the delay equation (3.1) it follows that $x(t)$ is a solution of the ordinary equation

$$\dot{x}_0^{(N)}(t)=-[g_0(x_0^{(N)}(t))+g(x_0^{(N)}(t))]+\frac{N}{\tau}x_N^{(N)}(t)-\mu_N^{(N)}(t)$$

for all $t\geq 0$.

The initial conditions for the functions $x_i^{(N)}(t)$ are given by

$$x_0^{(N)}(0)=\phi(0) \quad \text{and} \quad x_i^{(N)}(0)=\int_{-i\frac{\tau}{N}}^{-(i-1)\frac{\tau}{N}} g(\phi(s))\,ds \quad , \quad 1\leq i\leq N \ .$$

Since $g(x(t))$ is a continuous function we find that for all $t_1>0$

$$\max_{1\leq j\leq N}\ \max_{0\leq t\leq t_1} |\mu_j^{(N)}(t)|\to 0 \quad , \quad \text{as} \quad N\to+\infty \ .$$

Thus the approximating ordinary differential system for the delay equation (3.1) is the following

$$
\left.
\begin{aligned}
& z_0(t) = -[g_0(z_0(t)) + g(z_0(t))] + \frac{\tau}{N} z_N(t) \ , \\
& \dot{z}_1(t) = -\frac{N}{\tau} z_1(t) + g(z_0(t)) \\
\text{and} \quad & \dot{z}_i(t) = -\frac{N}{\tau} z_i(t) + \frac{N}{\tau} z_{i-1}(t) \quad , \quad 1 \leq i \leq N \ ,
\end{aligned}
\right\} \quad (3.5)
$$

with the initial conditions

$$
z_0(0) = \phi(0) \quad \text{and} \quad z_i(0) = \int_{-i\frac{\tau}{N}}^{-(i-1)\frac{\tau}{N}} g(\phi(s))\,ds \quad , \quad 1 \leq i \leq N \ . (3.6)
$$

The solution of (3.5) and (3.6) is denoted by $z^{(N)}(\phi)(t) = (z_0^{(N)}(\phi)(t), \ldots, z_N^{(N)}(\phi)(t))^T$.

The differential system (3.5) is related to a classical compartmental system given in Fig. 3.2 where the chain of the "small" compartments approximates the pipe in the original pipe - compartmental system showed in Fig. 3.1

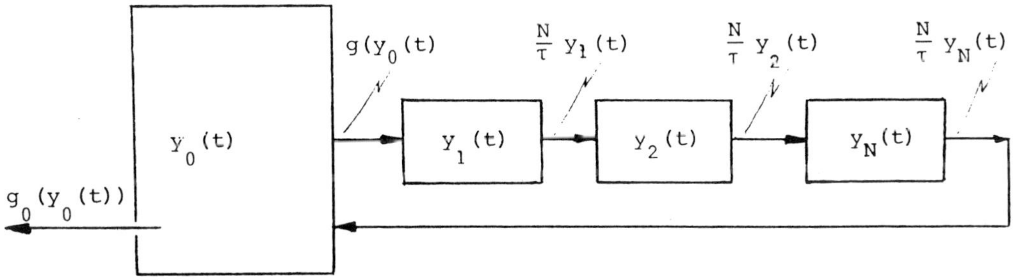

FIGURE 3.2 Approximate model for the pipe - compartmental system showed in Fig. 3.1

By Theorem 3.4 of the author ([6]) we may state the following approximation result.

THEOREM 3.1 Suppose that $\tau > 0$ is a real number and $g_0(x)$ and $g(x)$ are continuously differentiable nonnegative functions on R.

Then for any $\phi \in C$, the solution $x(\phi)(t)$ of Eq. (3.1) with initial condition (3.2) and the solution $z^{(N)}(\phi)(t)=(z_0^{(N)}(\phi)(t),\dots,z_N^{(N)}(\phi)(t))^T$ of system (3.5) with initial condition (3.6) exists on $[0,\infty)$ and for all $t_1>0$,

$$\max_{0\le t\le t_1} |x(\phi)(t)-z_0^{(N)}(\phi)(t)|\to 0 \quad , \quad \text{as} \quad N\to+\infty .$$

4. STEADY STATE SOLUTIONS AND APPROXIMATION ON $[0,\infty)$

We now investigate the steady state solutions of Eq. (3.1) and system (3.5), in each of the following two cases:

(a) $g_0(x)x>0$ for all $x\neq 0$. Then Eq. (3.1) has exactly one steady state solution $x=0$ and system (3.4) has also exactly one steady state solution $z^{(N)}=(0,\dots,0)^T$.

(b) $g_0(x)=0$ for all x. Then for all constant, $x(t)=c$, $(t\geq-\tau)$, is a steady state solution of Eq. (3.1). Therefore Eq. (3.1) does not give enough information to find a steady state corresponding to an initial function $\phi\in C$.

But since $g_0(x)=0$, $(x\in R)$, we have that Eq. (3.1) is equivalent to the following integral equation

$$x(t)+\int_{t-\tau}^{t} g(x(s))ds=c[\phi] \quad , \quad t\geq 0 \quad , \tag{4.1}$$

where

$$c[\phi]=\phi(0)+\int_{-\tau}^{0} g(\phi(s))ds . \tag{4.2}$$

Thus the unique steady state of Eq. (4.1) with respect to ϕ is the unique solution of the equation

$$d+\tau g(d)=c[\phi] \quad , \tag{4.3}$$

say $d[\phi]$.

We now consider the system (3.4) under the condition $g_0(x)=0$, $(x\in R)$. In that case $(z_0(t),\dots,z_N(t))^T$ is a steady state solution of system (3.5) if and only if

$$z_0(t)=c \quad \text{and} \quad z_1(t)=\dots=z_N(t)=\frac{\tau}{N}g(c) \quad , \quad t\geq 0 \quad , \tag{4.4}$$

where c is an arbitrary constant.

On the other hand since $g_0(x)=0$, $(x \in R)$, from (3.5) and (3.6) we find that

$$\sum_{i=0}^{N} z_i(t)=\sum_{i=0}^{N} z_i(0)=c[\phi] \quad , \quad t\geq 0 \; , \tag{4.5}$$

where $c[\phi]$ is defined by (4.2).

Combining (4.4) and (4.5) we have that $(z_0(t),\ldots,z_N(t))^T$ is a steady state solution of system (3.5) if and only if $z_0(t)=c$ and $z_i(t)=\frac{\tau}{N}\, g(c)$, $1\leq i\leq N$, $t\geq 0$, where

$$c+\sum_{i=1}^{N} \frac{\tau}{N}\, g(c)=c+\tau g(c)=c[\phi].$$

This means that c is a solution of Eq. (4.3), that is $c=d[\phi]$.

Consider the solution $x(\phi)(t)$ of Eq. (3.1) with initial condition (3.2) and the solution $z^{(N)}_{(\phi)}(t)=(z_0^{(N)}(\phi)(t),\ldots,z_N^{(N)}(\phi)(t))^T$ of system (3.5) with initial condition (3.6).

If $g_0(x)$ and $g(x)$ are continuously differentiable monotone nondecreasing functions then from [7] we know that $x(\phi)(\infty)=\lim_{t\to+\infty} x(\phi)(t)$ exists and from [9] we know that $z_i^{(N)}(\phi)(\infty)=\lim_{t\to+\infty} z_i^{(N)}(\phi)(t)$ also exists for all $i=0,1,\ldots,N$.

Moreover $x(\phi)(\infty)$ and $(z_0^{(N)}(\phi)(\infty),\ldots,z_N^{(N)}(\phi)(\infty))^T$ are steady state solutions of Eq. (3.1) and system (3.5), respectively.

Therefore in cases (a) and (b), we have that

$$x(\phi)(\infty)=d[\phi]=z_0^{(N)}(\phi)(\infty) \tag{4.6}$$

where $d[\phi]$ is the unique solution of (4.3).

CONJECTURE 4.1. Assume that $\tau>0$ is a constant and $g_0(x)$ and $g(x)$ are continuously differentiable monotone nondecreasing functions such that $g(x)x>0$ and either $g_0(x)=0$ or $g_0(x)x>0$ for all $x=0$.

Then for all $\phi\in C$ the solution $x(\phi)(t)$ of Eq.(3.1) with initial condition (3.2) and the solution $z^{(N)}(\phi)(t)=(z_0^{(N)}(\phi)(t),\ldots,z_N^{(N)}(\phi)(t))^T$ of system (3.5) with initial condition (3.6) satisfy the relation

$$\sup_{t\geq 0}|x(\phi)(t)-z_0^{(N)}(\phi)(t)|\to 0\ , \text{ as } N\to+\infty \qquad (4.7)$$

REMARK 4.1. We can prove the above conjecture in a very restricted case, namely if $g_0(x)=ax$ and $g(x)=bx$, where $a>0$ and $b>0$ are constants. In this case the proof is based on the fact that Eq. (3.1) reduces to the linear equation

$$\dot{x}(t)=-(a+b)x(t)+bx(t-\tau),$$

which is exponentially stable.

REMRK 4.2. We can also prove the above conjecture for exponentially stable linear delay differential systems.

The generalizations of Theorems 2.1 and 3.1 for systems with several and/or distributed time dependent delays can be found in [6].

REFERENCES

1. H. T. Banks, Delay systems in biological models: approximation techniques in "Nonlinear Systems and Applications" (ed. V. Lakshmikantham), Academic Press, New York, 21-28, (1977).

2. H. T. Banks and F. Kappel, Spline approximations for functional differential equations, J. Diff. Eq., 34, 496-522, (1979).

3. I. Gyori and J. Eller, Compartmental system with pipes, Math. Biosch. 53, 223-247, (1981).

4. I. Gyori, Delay differential and integro - differential equations in biological compartmental models, Systems Science 8, 76-83, (1983).

5. I. Gyori, Connections between compartmental systems with pipes and integro - differential equations, Mathematical Modelling, 7, 1215-1238, (1986).

6. I. Gyori, Two approximation techniques for functional differential equations, Comput. Math. Appl. (to appear).

7. J. R. Haddock and J. Terjeki, Liapunov - Razumikhin functions and invariance principle for functional differential equations, J. Diff. Eq. 48, 95-122, (198).

8. P. T. Janushevskij, Control of Object with Delays (in Russian), Nauka, Moscow, (1978).

9. H. Maeda and S. Kodama, Some results on nonlinear compartmental systems, IEEE-CAS-26, 3, 203-204, (1979).

10. Ju. M. Repin, On the approximate substitution of systems with delays by dynamic systems, (in Russian), Prikl. Math. Mech. 29, 226-235, (1965).

On a Tracking Problem

Sherry E. Hammel,* Naval Underwater Systems Center, Newport, Rhode Island

Pan-Tai Liu,** Professor, Department of Mathematics, University of Rhode Island, Kingston, Rhode Island

INTRODUCTION

A variety of state estimation problems arise in practice that involve tracking a moving vehicle from noisy observations. In particular, bearings-only tracking has continued to be a widely studied problem of both theoretical and practical interest [1-6]. This paper examines the effects of control inputs on solution accuracy for

*This research was supported by Chief of Naval Research Program Element 6276N, Project ZF66112001.

**The research of this author was partially supported by the Office of Naval Research under Grant No. N00014-88-K-0012.

bearings-only estimation. Since the analysis involves a moving observer, the terms "control inputs" and "observer motion" are synonymous here. The potential impact of observer motion on estimation accuracy has been previously addressed [7-9]. However, optimal observer paths have not been developed in a rigorous context.

Bearings-only tracking represents an important class of problems in which many encumbrances associated with an estimation process arise. Here, a moving observer passively monitors noisy angle-of-arrival data emitted from an acoustic source. Given the source state dynamics and functional dependence of the parameters on the observations, source position and velocity in two dimensions are estimated according to some criterion that reduces measurement and modelling errors. The nonlinear relation of the observations and states preclude application of conventional linear analysis. Also, since the data are extracted from a single sensor, only directional information can be obtained. Consequently, the process is observable only upon assuming the source maintains constant velocity while the observer executes at least one velocity change [10]. It is this prerequisite maneuver which distinguishes bearings-only tracking from more conventional techniques, such as classical triangulation ranging, and introduces added complexity to the problem.

Despite the aforementioned difficulties, numerous techniques have been devised for bearings-only tracking [11]. Among these include automatic estimation algorithms that process bearings sequentially (i.e. first order extended Kalman filter) or via batch type estimations based on Gauss-Newton procedures [12]. The effects of coordinate systems on solution behavior have also been addressed along with higher order estimators and modified measurement relations [13-16]. Since characteristics of the data are functionally dependent on the source/observer geometry, another significant factor affecting estimation accuracy is the role of the observer, and is the subject of this work. Emphasis is on realizing performance gains via observer inputs as opposed to employing better estimation algorithms.

THE CRITERION

Optimality is defined here using the Cramer-Rao lower bound (CRLB) as the index of optimality. The CRLB specifies the best performance expected under given conditions for any unbiased estimator of deterministic variables. The CRLB is achievable by linearizing the error covariance matrix (P) about the true states [17]. Generating observer paths will be based on minimizing the one sign uncertainty region formulated in the CRLB sense via

$$A_{1\sigma} = \pi \cdot \sqrt{|P|} \tag{1}$$

where $|\cdot|$ denotes determinant.

Inverting the covariance matrix yields the Fisher information matrix (FIM), which represents the information content in the observation model. Hence, the error region is also minimized by maximizing the |FIM|. This is equivalent to maximizing system observability which can be achieved through observer maneuvers since the determinant is a function of vehicle motion. When examining the sensitivity of the CRLB to observer motion, the theoretical aspects are discerned by using error free states in the FIM to assess system observability. The specific role of the observer in the overall estimation process is to create a source/observer geometry that maximizes system observability, thereby minimizing the region of uncertainty to enhance estimation accuracy.

An explicit analytical solution is difficult to obtain since the uncertainty region is a four-dimensional hyper-ellipsoid and leads to a quadruple integral cost function that is not amenable to classical control theory. It is shown here, however, that through decomposition of the FIM, the elements can be analytically evaluated under the assumption of constant observer motion over individual observation intervals. The determinant of the FIM is then computed and a numerical search procedure implemented to yield optimal observer motion consisting of a series of constant course trajectories. Optimal paths are characterized by the effective baseline-to-range ratio describing the initial scenario encountered.

PROBLEM FORMULATION

The bearings-only tracking problem is depicted in Fig. 1 where both source position and velocity are estimated. Let $\underline{x} = [r_x, r_y, v_x, v_y]^T$ define the system state vector whose components $\underline{r} = [r_x, r_y]^T$ and $\underline{v} = [v_x, v_y]^T$ represent position and velocity, respectively, relative to the source. Assuming source velocity is constant, the system state dynamics may be written in the form [13]

$$\dot{\underline{x}} = A\underline{x} + \underline{w} \tag{2}$$

where

$$A = \begin{bmatrix} 0 & 0 & 1 & 0 \\ 0 & 0 & 0 & 1 \\ 0 & 0 & 0 & 0 \\ 0 & 0 & 0 & 0 \end{bmatrix} \tag{3}$$

$$w = \begin{bmatrix} 0 \\ 0 \\ a_{0x} \\ a_{0y} \end{bmatrix} \tag{4}$$

and (a_{0x}, a_{0y}) are observer acceleration components. The observation model is

$$\beta = \tan^{-1} r_y/r_x + \eta \tag{5}$$

where η is assumed to be zero-mean Gaussian white noise with constant variance σ^2.

For later use, Eq. (2) is integrated to yield the state equation

$$\underline{x}(t) = \Phi(t,t^*)\ \underline{x}(t^*) + \underline{u}(t) \tag{6}$$

where $\Phi(t,t^*)$ is the state transition matrix referred to some arbitrary fixed time t^* given by

$$\Phi(t,t^*) = \begin{bmatrix} 1 & 0 & (t-t^*) & 0 \\ 0 & 1 & 0 & (t-t^*) \\ 0 & 0 & 1 & 0 \\ 0 & 0 & 0 & 1 \end{bmatrix} \tag{7}$$

and $\underline{u}(t)$ accounts for observer acceleration via

$$\underline{u}(t) = \begin{bmatrix} \int_{t^*}^{t} (t-\mu)\, a_{0x}(\mu)\, d\mu \\ \int_{t^*}^{t} (t-\mu)\, a_{0y}(\mu)\, d\mu \\ \int_{t^*}^{t} (a_{0x}(\mu))\, d\mu \\ \int_{t^*}^{t} (a_{0y}(\mu))\, d\mu \end{bmatrix} . \tag{8}$$

Earlier work has shown that although time translations involve a change in shape and reorientation, the volume of the hyper-ellipsoid is invariant with shifts in reference time for bearings-only tracking [7]. Since this study is directed toward minimizing the total region of uncertainty as opposed to individual components of error, the reference time, t*, is henceforth chosen to be the initial time, $t_0 = 0$, for simplicity without loss of generality.

The problem now is to determine an observer path via $\underline{u}$ that optimizes estimation accuracy for a given scenario. To this end, the FIM is examined for information content. The FIM for bearings-only tracking can be expressed as

$$\text{FIM} = \frac{1}{\sigma^2} \int_0^T \Phi^T(\tau,0)\, \frac{\sigma\beta}{\sigma \underline{x}}^T \frac{\sigma\beta}{\sigma \underline{x}}\, \Phi(\tau,0)\, d\tau$$

$$= \frac{1}{\sigma^2} \begin{bmatrix} \int_0^T \frac{\sin^2\beta}{r^2} d\tau & -\frac{1}{2}\int_0^T \frac{\sin 2\beta}{r^2} d\tau & \int_0^T \frac{\sin^2\beta}{r^2} d & -\frac{1}{2}\int_0^T \frac{\tau \sin 2\beta}{r^2} d\tau \\ -\frac{1}{2}\int_0^T \frac{\sin 2\beta}{r^2} d\tau & \int_0^T \frac{\cos^2\beta}{r^2} d\tau & -\frac{1}{2}\int_0^T \tau \frac{\sin 2\beta}{r^2} d\tau & \int_0^T \tau \frac{\cos^2\beta}{r^2} d\tau \\ \int_0^T \tau \frac{\sin^2\beta}{r^2} d\tau & -\frac{1}{2}\int_0^T \frac{\sin 2\beta}{r^2} d\tau & \int_0^T \tau \frac{2\sin^2\beta}{r^2} d & -\frac{1}{2}\int_0^T \tau \frac{2\sin 2\beta}{r^2} d\tau \\ -\frac{1}{2}\int_0^T \tau \frac{\sin 2\beta}{r^2} d\tau & \int_0^T \tau \frac{\cos^2\beta}{r^2} d\tau & -\frac{1}{2}\int_0^T \tau \frac{2\sin 2\beta}{r^2} d\tau & \int_0^T \tau \frac{2\cos^2\beta}{r^2} d\tau \end{bmatrix} \tag{9}$$

Note that the total observation interval, [0,T] is considered. Again, β and $r = \sqrt{r_x^2 + r_y^2}$ are the true bearing and range parameters. The initial bearing is zero for simplicity.

Observe that the problem has been formulated in a continuous context as opposed to the discrete version utilized in practice. This has been implemented for analytical purposes since the continuous form renders insight that may otherwise not be discernable. Embedded in the FIM is observer motion via the state equations. Hence, if the observer pursues a trajectory that maximizes the |FIM|, maximum system observability will be realized and the error ellipse will be minimized. As alluded to earlier, however, an explicit analytical solution using classical control theory is difficult to obtain since the determinant is a quadruple integral. In the next section, an alternative approach is considered to yield the optimal paths.

OPTIMAL PATHS

For general observer motion, the integrals of the FIM do not lend themselves to analysis. However, the individual terms of Eq. (9) can be evaluated upon assuming constant observer speed (S_0) throughout [0,T] and constant observer course, θ_i, over $[t_{i-1},t_i]$ for i = 1,2,...,N. To this end, Eq. (6) is rewritten in the form

$$\dot{r}_x = \dot{r}\cos\beta - r\dot{\beta}\sin\beta \tag{10a}$$

$$\dot{r}_y = \dot{r}\sin\beta - r\dot{\beta}\cos\beta \tag{10b}$$

where $\dot{\beta}$ and $\dot{r}$ denote bearing-rate and range-rate respectively. Differentiating Eqs. (10a) and (10b) assuming constant observer motion leads to the following result for each observation interval $[t_{i-1},t_i]$,

$$\frac{1}{r^2} = \frac{\dot{\beta}}{r_{i-1}^2\dot{\beta}_{i-1}} \tag{11}$$

and

$$\dot{\beta} = [S_0\sin(\theta_i - \beta) - S_c\sin(\theta_c - \beta)]/r \tag{12}$$

where S_c and θ_c are source speed and course respectively. Also, the bearing relation defined by Eq. (5) on $[t_{i-1},t_i]$ reduces to

$$\beta = \tan^{-1}\left(\frac{a + (t-t_i)b}{c + (t-t_i)d}\right) \tag{13}$$

where

$$a = r_{i-1}\sin\beta_{i-1}$$
$$b = S_0\sin\theta_i - S_c\sin\theta_c$$
$$c = r_{i-1}\cos\beta_{i-1}$$
$$d = S_0\cos\theta_i - S_c\cos\theta_c.$$

Upon rearranging Eq. (13), time can be expressed as a function of bearing over each observation interval as

$$t = t_i + \left[\frac{a\cos\beta - c\sin\beta}{d\sin\beta - b\cos\beta}\right]. \tag{14}$$

With Eqs. (11), (12), and (14) substituted into the FIM expression, each of the integrals in Eq. (9) can be explicitly evaluated over $[t_{i-1},t_i]$ by incorporating some trigonometric identities and the determinant computed over [0,T]. Minimization of the hyper-ellipsoid is finally achieved by implementing a numerical search method to generate N observer course changes that maximizes the |FIM|.

The particular nonlinear optimization scheme implemented here is based on quasi-Newton methods [18]. Inputs to the technique are observer speed, total observation time with $[t_i - t_{i-1}] = T/N$ and an estimate of the distance from the observer's initial position to the source's final position (r'). Note that the maximization involves relating information content over the entire observation interval [0,T]. Optimal paths for N = 20 and $S_0 = 45^{\circ}$ are depicted in Fig. 2. Observer paths for other source courses can be obtained via a mathematical transformation. It is noted that as N increases, the series of constant course trajectories that maximize |FIM| converge rather rapidly to a continuous solution. Increasing the number of observer course changes beyond 20 had negligible effects on the trajectories and corresponding values of the determinant. Indeed, for practical applications, 5 to 10 course changes may yield sufficiently large values of |FIM|.

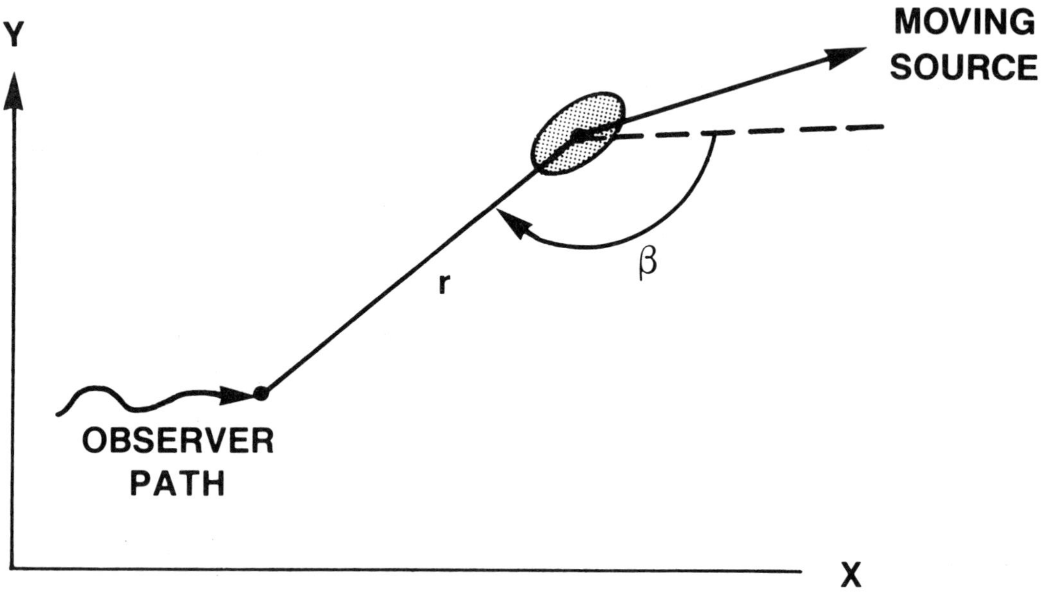

Fig. 1

In Fig. 2, the paths are associated with an $\alpha = S_0T/r'$ product. This product theoretically describes any scenario initially encountered. Observe that the optimal paths involve a trade-off between increasing bearing-rate and decreasing range, depending on the α product. For relatively small α, paths that minimize the error ellipse emphasize increasing bearing-rate, whereas for larger α, the paths entail more of a reduction in range. The singular case is at $\alpha = 1$, where the observer heads directly for the source and collision occurs at $t = T$. Although, in general, a zero bearing-rate course implies an unobservable solution, here the $|FIM|$ approaches infinity. Thus initial scenarios with an α product of 1 allow for an infinite amount of information to be obtained so that at final time T, the source is precisely located.

COMPARISON OF PATHS

It is interesting that the general trend for optimal observer paths exhibit zig-zag motion. Indeed, numerous studies of the past support that motion of this kind generally enhances estimation accuracy [6,7,13]. Previous analytical efforts have approximated the initial geometry as a long range scenario which essentially corresponds to a rather small α' product as defined here. Under such conditions, simplified ad hoc estimation techniques reveal that solutions can be enhanced for large bearing-rate geometries [6,7]. This is supported by the results presented here. However, Fig. 2 reveals that there is a large spectrum of possible source/observer geometries in which optimal observer motion involves a trade-off between increasing bearing-rate and decreasing range. The framework developed in this work leads to a procedure for delineating the appropriate trade-off.

A consideration in the bearings-only estimation process has been how much time to allocate each observer course. Although Fig. 2 was generated for N = 20, perhaps for practical applications the paths can be interpolated with three or so observer course changes, thereby answering the question of time allocation. Note, however, that the total observation interval, T, is considered in delineating

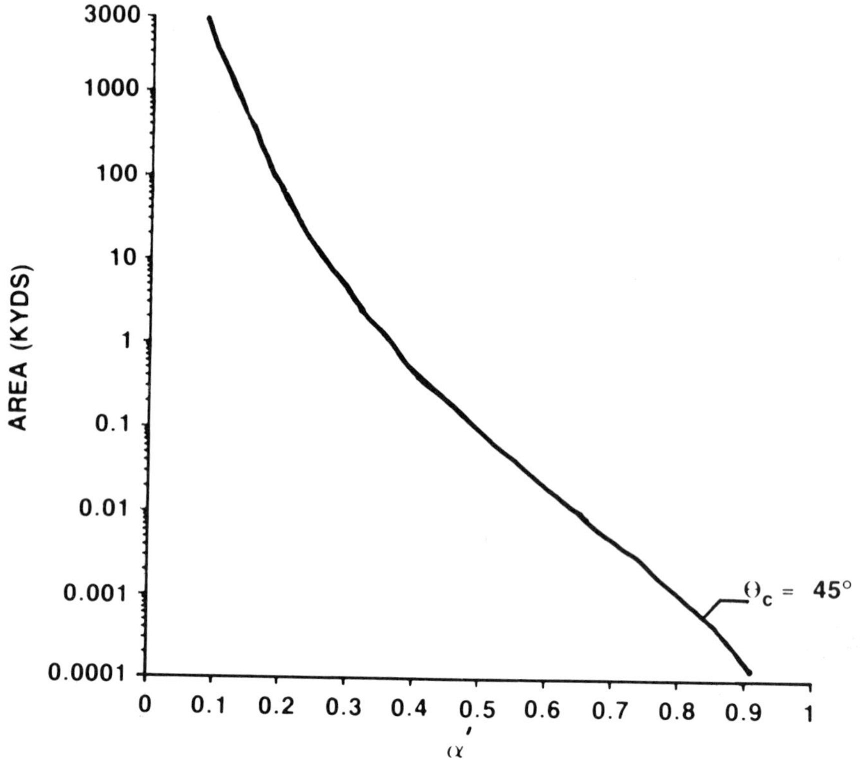
3000
1000
100
10
1
0.1
0.01
0.001
0.0001
AREA (KYDS)
0
0.1
0.2
0.3
0.4
0.5
0.6
0.7
0.8
0.9
1
α'
θc = 45°

Fig. 2

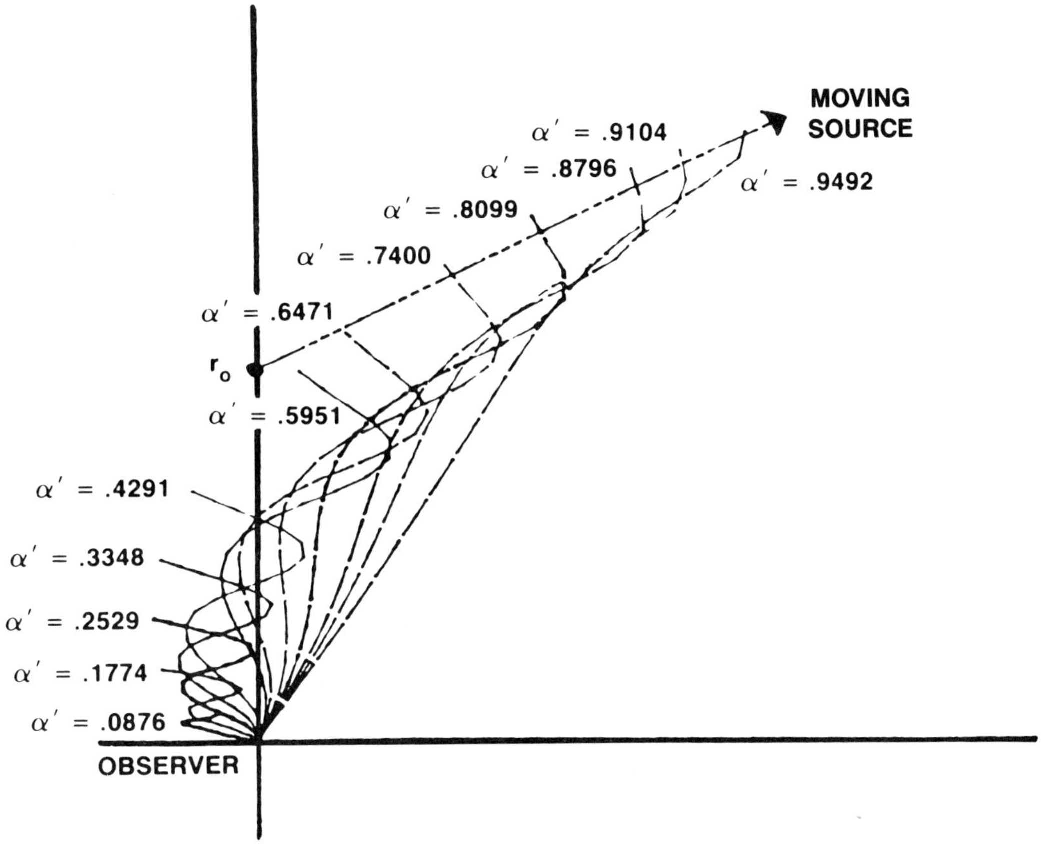
MOVING
SOURCE
α′ = .9104
α′ = .8796
α′ = .9492
α′ = .8099
α′ = .7400
α′ = .6471
r₀
α′ = .5951
α′ = .4291
α′ = .3348
α′ = .2529
α′ = .1774
α′ = .0876
OBSERVER

Fig. 3

appropriate observer motion.

This framework also provides a means for assessing the effects of observer motion on estimation accuracy. The areas of the hyper-ellipsoids corresponding to the optimal paths are plotted against their respective α products in Fig. 3 for each of the trajectories in Fig. 2. These areas represent the best performance expected with regard to observer motion for given S_0, T, r_0' and σ values. Determination of the effects of appropriating any particular path versus the optimal on estimation accuracy can be ascertained by computing the areas of the hyper-ellipsoids for the actual scenario initially encountered. Larger values of α generally result in higher system observability and smaller hyper-ellipsoids.

SUMMARY

Optimal observer paths that minimize the region of uncertainty were generated for the general bearings-only tracking problem. To this end, the FIM was decomposed under the assumption of constant observer motion and the individual elements evaluated explicitly. A numerical search procedure was then implemented to maximize the |FIM| to yield optimal observer paths consisting of a series of constant course trajectories. All the paths were characterized by their respective α product and contact course describing the initial scenario.

Optimal observer motion entails a trade-off between increasing bearing-rate and decreasing range. The areas of the hyper-ellipsoids associated with the trajectories represent the best performance expected with regard to observer motion. The effects of pursuing an alternate path versus the optimal on estimation accuracy can be ascertained upon computing the corresponding areas for the scenario initially encountered. These results reveal that the potential impact of observer motion on solution accuracy can be significant in the nonlinear estimation process.

REFERENCES

1. Foy, W.H. "Position location solutions by Taylor-series estimation." *IEEE Trans. Aerosp. and Electron Syst.*, Vol. 12, pp. 187-193, March 1976.

2. Tenney, R.R., R.S. Hebbert, and N.R. Sandell, Jr. "A tracking filter for maneuvering sources." *IEEE Trans. Automatic Control*, April 1977.

3. Poirot, J.L. and G.V. McWilliams. "Navigation by back triangulation." *IEEE Trans. Aerosp. and Electron. Syst.*, Vol. AES-12, pp. 270-274, March 1976.

4. Kolb, R.C. and F.H. Hollister. "Bearings-only target estimation." Proc. 1st Asilomar Conf. Circuits and Syst., 1967.

5. Lindgren, A.G. and K.F. Gong. "Position and velocity estimation via bearing observations." *IEEE Trans. Aerosp. Electron. Syst.*, Vol. AES-14, pp. 564-577, July 1978.

6. Murphy, D.J. "Noisy bearings-only target motion analysis." Ph.D. Dissertation, Department of Electrical Engineering, Northeastern University, 1969.

7. Nardone, S.C. "Fundamental properties and performance of bearings-only target localization and motion analysis." Ph.D. Dissertation, Department of Electrical Engineering, University of Rhode Island, Kingston, 1982.

8. Liu, P.T. and P.L. Bongiovanni. "On a passive vehicle tracking problem and max-minimization." *IEEE Trans. on Automatic Control*, Vol. AC-28, No. 2, pp. 233-235, February 1983.

9. Liu, P.T. and P.L. Bongiovanni. "A differential game approach to determine passive tracking maneuvers." Proceedings of the 4th MIT/ONR Workshop on C Systems, Report No. LISDS-R-1156, pp. 149-170, October 1981.

10. Nardone, S.C. and V.J. Aidala. "Observability criteria for bearings-only target motion analysis." *IEEE Trans. Aerosp. Electron. Syst.*, Vol. AES-17, pp. 162-166, March 1981.

11. R.W. Bass et al. "ASW target motion and measurement models," Computer Software Analysis, Inc., Tech. Rep. TR-72-024-01, Sept. 1972.

12. Jazwinski, A.H. *Stochastic Processes and Filtering Theory*. New York: Academic Press.

13. Aidala, V.J. "Kalman filter behavior in bearings-only tracking applications." *IEEE Trans. Aerosp. and Electron. Syst.*, AES-15, pp. 29-39, January 1979.

14. Aidala, V.J. and S.C. Nardone. "Biased estimation properties of the pseudo-linear tracking filter." *IEEE Trans. Aerosp. Electron. Syst.*, Vol. AES-18, No. 4, pp. 432-441, July 1982.

15. Aidala, V.J. and S.E. Hammel. "Utilization of modified polar coordinates for bearings-only tracking." *IEEE Trans. Automat. Contr.*, Vol. AC-28, No. 3, pp. 382-394, March 1983.

16. Hammel, S.E., V.J. Aidala, K.F. Gong, and A.G. Lindgren. "Recursive versus batch processing algorithms for bearings-only tracking." *IEEE Ocean '83*, August 1983.

17. Taylor, J.H. "The Cramer-Rao estimation error lower bound computation for deterministic nonlinear systems." *IEEE Trans. on Automatic Control*, Vol. AC-24, pp. 343-344, April 1979.

18. Fletcher, R. "FORTRAN subroutines for minimization by quasi-Newton methods." Report R7125, AERE, Howell, England, June 1972.

Nilpotent Approximations of Control Systems

Henry Hermes, Professor, Department of Mathematics, University of Colorado, Boulder, Colorado

Introduction. This paper surveys, develops, extends and illustrates, in a self-contained fashion, work of the last several years on nilpotent approximations of affine control systems. The goal is to constructively approximate the real analytic vector fields $X^0, \ldots, X^k$ which describe the n-dimensional, affine, control system

$$\dot{x} = X^0(x) + \sum_{i=1}^{k} u_i X^i(x), \quad x(0) = 0, \quad X^0(0) = 0 \tag{1}$$

by vector fields $Y^0, \ldots, Y^k$ for which either the lie algebra they generate, $L(Y^0, \ldots, Y^k)$, or an appropriate subalgebra, is nilpotent. The approximations considered were originally motivated by the problem of small time, local controllability (STLC) about the rest solution zero for system (1), i.e. were such that STLC for (1) is implied by STLC at zero of the approximating system

$$\dot{x} = Y^0(x) + \sum_{i=1}^{k} u_i Y^i(x), \quad x(0) = 0, \quad Y^0(0) = 0 \tag{2}$$

or, in special cases, (1) is STLC if and only if (2) is. Most proofs of high order sufficient conditions for STLC of systems such as (1) rely on nilpotent approximations.

The control values, throughout, need only satisfy $|u_i| \leq \epsilon_1$, $\epsilon_i > 0$, $i = 1, \ldots, k$. All vector fields are assumed real analytic. For any set, S, of vector fields on $\mathbb{R}^n$ (or locally on an n-manifold M^n), $L(S)$ will denote the lie algebra they generate; $S(0)$ the elements of S evaluated at 0; $[X,Y] = (\text{ad}X, Y)$ the commutator (or lie product) of vector fields X, Y and inductively $(\text{ad}^{l+1}X, Y) = [X, (\text{ad}^l X, Y)]$.

The first approximation considered is motivated by work of P. Crouch, [1], and was studied in [2]. We shall refer to this as the C-H approximation. For system (1) assign the weight 0 to X^0 and weight 1 to X^i for $i = 1, \ldots, k$. The weight assigned to a commutator will always be the sum of the weights of its constituent factors. For $j = 1,2, \cdots$ let

$$S_X^j \text{ be the set of commutators of } \{X^0, \ldots, X^k\} \text{ of weight } j.$$

Equivalently, $S_X^1 = \{(\text{ad}^\nu X^0, X^i)\colon\ i = 1, \ldots, k;\ \nu = 0,1, \cdots\}$ and S_X^j is the set of all products of j-tuples of elements of S_X^1. We assume throughout that

$$\dim L(S_X^1)(0) = n \tag{4}$$

This research was sponsored by NSF grant DMS-8721917.

which implies, [3], [4], that for all $t > 0$, the attainable set of system (1) at time t has nonempty interior in $\mathbb{R}^n$.

The notation X^π will denote a product of specific vector fields from the set $\{X^0, \ldots, X^k\}$ taken in a specific order. If we deal with the vector fields $Y^0, \ldots, Y^k$, then S_Y^j will denote the equivalent set to S_X^j and Y^π the equivalent commutator to X^π.

Theorem 1 deals with the C-H nilpotent approximation. Specifically, it yields a constructive way to generate approximating vector fields $Y^0, \ldots, Y^k$ such that $L(S_Y^1)$ is nilpotent and for each $j \geq 1$, $\dim \operatorname{span}(\bigcup_{i=1}^{j} S_Y^i(0)) = \dim \operatorname{span}(\bigcup_{i=1}^{j} S_X^i(0))$. This theorem is similar to, but an extension of, theorem 1 as stated in [2].

Stefani, [5], generates nilpotent approximations, including the C-H approximation and the "diagonal approximation" of theorem 4, here, starting from filtrations of $L = L(X^0, \ldots, X^k)$. Our approach differs only in that we assign a dilation δ_t on $\mathbb{R}^n$ (motivated by known sufficiency tests for (1) to be STLC at 0) which, in turn, induces a filtration. Indeed, for $x = (x_1, \ldots, x_n)$ local coordinates for $\mathbb{R}^n$, suppose $\delta_t x = (t^{r_1}x_1, \ldots, t^{r_n}x_n)$ where $r_1 \leq r_2 \leq \cdots \leq r_n$ are positive integers. Then, see §1, a vector field $X = 0$ if X is homogeneous of degree $l > r_n$ with respect to δ_t while if we define F_i as those vector fields which can be written as a sum of vector fields homogeneous of degree $\leq i$, then $\{F_i\}_{i \geq 0}$ gives a filtration of L.

Following Stefani, [5], we show both C-H and diagonal nilpotent approximations are (see corollary 2 to theorem 5) such that STLC of the approximating system implies STLC of (1). For another example of the use of theorem 1 in the STLC problem, recall

THEOREM 2 ([6], [7]). A sufficient condition that system (1) be STLC at 0 is that $\dim L(S_X^1)(0) = n$ and for all odd integers j,

$$\dim \operatorname{span}(\bigcup_{i=1}^{j} S_X^i(0)) = \dim \operatorname{span}(\bigcup_{i=1}^{j+1} S_X^i(0)) . \tag{5}$$

Call system (1) *odd* if (5) is satisfied. An immediate consequence of theorem 1 is

COROLLARY 1. An odd system is STLC at 0 if and only if its C-H nilpotent approximation is.

The local controllability test of theorem 2 results from examining the elements in $\bigcup_{j \geq 1} S_X^j(0)$ relative to a partial order based on the number of factors $X^1, \ldots, X^k$ in a commutator and independent of the number of factors X^0. For example, commutators in S_X^1 have one factor from $\{X^1, \ldots, X^k\}$, commutators in S_X^2 have two factors from $\{X^1, \ldots, X^k\}$, etc. The

C-H approximation "respects" this ordering. In [8], Sussmann gave a far reaching extension of theorem 2 based on examining commutators relative to different partial orderings. (See also [9].) For ease of notation and presentation we now restrict attention to the case where (1) is a single input system, i.e. $k = 1$. A commutator $W \in S_X^m$, which therefore has m factors X^1, is said to be of type (m, l) if it has l factors X^0. Note that there may be several elements of a given type (m, l) which are not related via the Jacobi identity. Define a partial order on $\bigcup_{m\geq 1} S_X^m$ as follows. If W is of type (m, l) we assign to it the weight $w(W) = m + l$ and define $W^1 < W^2$ if $w(W^1) < w(W^2)$.

Pictorially, see figure 1, this ordering is "relative to diagonals."

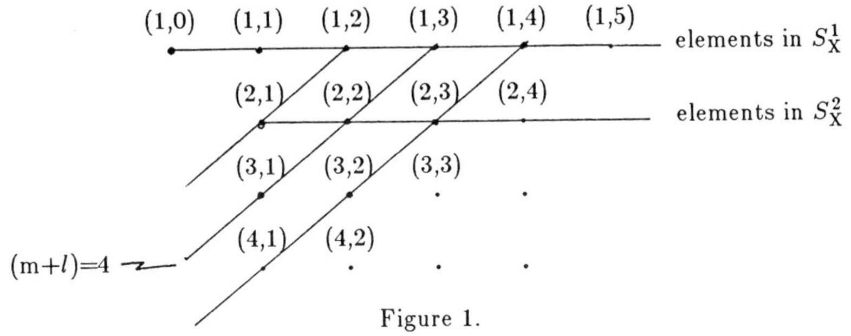

Figure 1.

A consequence of the main theorem in [8] may be stated as

THEOREM 3. The single input system (1) is STLC at 0 if $\dim L(S_X^1)(0) = n$ and every element W of type $(m,l) = $ (even, odd) satisfies $W(0) \in \text{span}\{V(0)\colon\ V < W\}$.

REMARK 1. Theorem 3, as stated above, is far weaker than theorem 7.3, [8], or algorithm 1, [9]. For example, here we exclude "neutralizing" a (2,3) element by a (1,4) element since these lie on the same diagonal. In the terminology of [9], we have made the *a priori* choice $\theta = 1$. (In [9], we usually required $\theta < 1$, with the difficulties of $\theta = 1$ illustrated in example 3.5, but this creates no problem since here we do not compare elements on the same diagonal.) Theorem 4 constructs an approximating system (which we shall call the *diagonal nilpotent approximation*) of the form (2) for the single input system (1) such that $L(Y^0,Y^1)$ is nilpotent and for each $j \geq 1$, $\text{span}\{X^\pi(0)\colon\ w(X^\pi) \leq j\} = \text{span}\{Y^\pi(0)\colon\ w(Y^\pi) \leq j\}$. This is an approximation tailored to the sufficient condition given for STLC in theorem 3. Stefani, [5], shows how to generate nilpotent approximating systems tailored to values $0 < \theta < 1$, i.e., to the more general form of Sussmann's theorem.

Section 2 derives, and extends, the Bressan rescaling results, [10], as an immediate consequence of the proof of the nilpotent approximation theorem. These were used by Bressan, [11], to examine the local structure of a time optimal feedback control for a single input, affine system on $\mathbb{R}^3$. The further use of nilpotent approximations in the feedback synthesis problem will be the topic of a forthcoming paper [12].

The last section, §3, of this paper consists of illustrative examples.

§1. Nilpotent Approximations. Choose coordinates $x = (x_1, \ldots, x_n)$ on $\mathbb{R}^n$. A *dilation*, relative to this coordinate choice, is a map $\delta_t\colon \mathbb{R}^n \to \mathbb{R}^n$ of the form $\delta_t x = (t^{r_1}x_1, \ldots, t^{r_n}x_n)$ where $1 \leq r_1 \leq \cdots \leq r_n$ are integers.

DEFINITION 1. A polynomial $h\colon \mathbb{R}^n \to \mathbb{R}^1$ is homogeneous of degree j with respect to δ_t if $t^j h(x) = h(\delta_t x)$. Let H_j denote the polynomials homogeneous of degree j and set $H_j = \{0\}$ if $j < 0$.

EXAMPLE 1.1. On $\mathbb{R}^4$ let $\delta_t x = (tx_1, tx_2, t^2x_3, t^3x_4)$. Then $h \in H_0 \Rightarrow h(x) = c$, $h \in H_1 \Rightarrow h(x) = c_1x_1 + c_2x_2$, $h \in H_2 \Rightarrow h(x) = c_1x_1^2 + c_2x_1x_2 + c_3x_2^2 + c_4x_3$, $h \in H_3 \Rightarrow h(x) = c_1x_1^3 + c_2x_1^2x_2 + c_3x_1x_2^2 + c_4x_2^3 + c_5x_1x_3 + c_6x_2x_3 + c_7x_4$, etc.

DEFINITION 2. A vector field X on $\mathbb{R}^n$ is homogeneous of degree m (with respect to δ_t) if $Xh \in H_{j-m}$ whenever $h \in H_j$, $j = 0, 1, \cdots$. Equivalently, in local coordinates, if $X(x) = \sum_{i=1}^{n} a_i(x)\, \partial/\partial x_i$, then X is homogeneous of degree m if $a_i \in H_{r_i - m}$, $i = 1, \ldots, n$.

Some useful properties are:

$$f \in H_j,\ g \in H_l \text{ and } h(x) = f(x)g(x) \Rightarrow h \in H_{j+l}\,. \tag{6}$$

$$h \in H_j \text{ implies } \frac{\partial h}{\partial x_i} \in H_{j-r_i}\,. \tag{7}$$

$$\begin{gathered} X \text{ homogeneous of degree } m,\ Y \text{ homogeneous} \\ \text{of degree } l \text{ and } h \in H_j \Rightarrow Xh \in H_{j-m}, \\ Yh \in H_{j-l} \text{ so } [X,Y]h = Y(Xh) - X(Yh) \in H_{j-(m+l)} \\ \text{i.e., } [X,Y] \text{ is homogeneous of degree } (m+l). \end{gathered} \tag{8}$$

PROPOSITION 1. Let L_i denote the vector fields homogeneous of degree $\geq i$ with respect to a given dilation $\delta_t x = (t^{r_1}x_1, \ldots, t^{r_n}x_n)$. Then each L_i is a lie algebra; L_1 is a nilpotent ideal in L_0; $L_m = \{0\}$ if $m > r_n$.

Indeed, $L_0 \supset L_1 \supset \cdots \supset L_{r_n} \supset L_{r_n+1} = \{0\}$ while (8) shows $[L_1, L_i] \subset L_{i+1}$ so the descending central series of L_1 terminates showing L_1 is nilpotent.

The C-H Nilpotent Approximation. For system (1) assume $\dim \text{span} S^1_X(0) = n_1$, $\dim \text{span}(S^1_X(0) \cup S^2_X(0)) = n_2$, etc. with N the smallest integer (which exists by (4)) such that $\dim \text{span}(\bigcup_{i=1}^{N} S^i_X(0)) = n$, i.e. $n_N = n$.

THEOREM 1 (Extension of Thm. 1, [2]). Consider the system (1) with $\dim L(S^1_X)(0) = n$. Let $X^{\pi_1}, \ldots, X^{\pi_{n_1}} \in S^1_X$ be linearly independent when evaluated at 0; adjoin $X^{\pi_{n_1+1}}, \ldots, X^{\pi_{n_2}} \in S^2_X$ such that $X^{\pi_1}(0), \ldots, X^{\pi_{n_2}}(0)$ are independent and continue in this way until one has $X^{\pi_1}, \ldots, X^{\pi_n}$ with $X^{\pi_1}(0), \ldots, X^{\pi_n}(0)$ independent. Then there exist (we construct) vector fields $Y^0, \ldots, Y^k$ on $\mathbb{R}^n$ which describe the C-H approximating system (2) and satisfy, in the chosen coordinates:

(a) $Y^0(0) = 0$

(b) $Y^{\pi_i}(0) = X^{\pi_i}(0), \; i = 1, \ldots, n$

(c) $\dim \text{span}(\bigcup_{i=1}^{j} S^i_Y(0)) = \dim \text{span}(\bigcup_{i=1}^{j} S^i_X(0)), \; j = 1, \ldots, N$

(d) $L(S^1_Y)$ is nilpotent.

(e) If $X^\pi(x) = \sum_{i=1}^{n} a_i(x)\, \partial/\partial x_i$ is a commutator of $\{X^0, \ldots, X^k\}$ of weight $j \geq 1$ and $Y^\pi(x) = \sum_{i=1}^{n} b_i(x)\, \partial/\partial x_i$ is the corresponding commutator of $\{Y^0, \ldots, Y^k\}$, then $a_i(0) = b_i(0), \; n_{j-1}+1 \leq i \leq n$.

The construction of $Y^0, \ldots, Y^k$ will be explicit and these vector fields will have polynomial coefficients relative to the preferred coordinates used in the proof of theorem 1.

Proof.

A. We first change to coordinates $x = (x_1, \ldots, x_n)$ such that relative to these, $X^{\pi_i}(0) = \partial/\partial x_i$, $i = 1, \ldots, n$. Constructively, if in previous coordinates $y = (y_1, \ldots, y_n)$ we had $X^{\pi_i}(y) = \sum_{j=1}^{n} a_{ij}(y)\, \partial/\partial y_j$, $i = 1, \ldots, n$ let $a^i(y)$ denote the column vector $(a_{i1}(y), \ldots, a_{in}(y))$. Form the matrix B having columns $a^1(0), \ldots, a^n(0)$. Since $X^{\pi_1}(0), \ldots, X^{\pi_n}(0)$ are independent, B is nonsingular. Make the linear coordinate change $x = B^{-1}y$ in which case $X^{\pi_i}(x) = \sum_{j=1}^{n} \alpha_{ij}(x)\, \partial/\partial x_j$ with $\alpha^i(x) = B^{-1}a^i(Bx)$ hence in these coordinates $X^{\pi_i}(0) = \partial/\partial x_i$.

B. Recall the notation $\dim \text{span}(\bigcup_{i=1}^{j} S^i_X(0)) = n_j$ with N the smallest integer such that $n_N = n$. From (4) and $X^0(0) = 0$ one sees not all $X^1(0), \ldots, X^k(0)$ can be zero hence

$n_1 \geq 1$.

Choose the dilation, relative to the above preferred coordinates, as

(i) $\delta_t x = (tx_1, \ldots, tx_{n_1}, t^2 x_{n_1+1}, \ldots, t^2 x_{n_2}, \ldots, t^N x_n)$ or for ease of reference in the proof,

(ii) $\delta_t x = (t^{r_1} x_1, \ldots, t^{r_{n_1}} x_{n_1}, t^{r_{n_1+1}} x_{n_1+1}, \ldots, t^{r_n} x_{n_N})$, i.e. the notation is such that

(iii) $r_i = j$ for $n_{j-1}+1 \leq i \leq n_j$.

PROPOSITION 2. If $Y(x) = \sum_{i=1}^{n} b_i(x)\, \partial/\partial x_i \in L_j$, i.e. Y is a sum of vector fields homogeneous of degree $\leq j$, then

$$b_i = 0 \text{ if } 1 \leq i \leq n_{j-1}\,. \tag{9}$$

Indeed, write $b_i = \sum_{l=j}^{\infty} b_{il}$ where $b_{il} \in H_{r_i - l}$, $i = 1, \ldots, n$. From (iii), $r_i - j < 0$ if $1 \leq i \leq n_{j-1}$ which means for these values of i, $b_{il} \in H_m$ with $m < 0$.

PROPOSITION 3. If $W(x) = \sum_{i=1}^{n} d_i(x)\, \partial/\partial x_i$ is a sum of vector fields homogeneous of degree $\leq j$, then $d_i(0) = 0$ for $n_{j+1} \leq i \leq n$.

Here, write $d_i = \sum_{l \leq j} d_{il}$ where $d_{il} \in H_{r_i - l}$. Then for $r_i > j$ these are all homogeneous of positive degree which means $d_i(0) = 0$ if $r_i > j$, i.e. if $n_j+1 \leq i \leq n$.

PROPOSITION 4. If $X^\pi(x) = \sum_{i=1}^{n} a_i(x)\, \partial/\partial x_i$ is a commutator of weight $j \geq 1$, then $a_i(0) = 0$ for $n_j+1 \leq i \leq n$.

This follows since X^π of weight $j \geq 1$ implies $X^\pi \in S_X^j$ which means $X^\pi(0) \in \text{span}\{X^{\pi_{n_{j-1}+1}}(0), \ldots, X^{\pi_{n_j}}(0)\}$ hence $a_i(0) = 0$ for $n_j+1 \leq i \leq n$.

C. Let $X^0(x) = \sum_{j=1}^{n} a_{0j}(x)\, \partial/\partial x_j$ be the original vector field of weight 0. The approximating vector field $Y^0(x) = \sum_{j=1}^{n} b_{0j}(x)\, \partial/\partial x_j$ is obtained by letting b_{0j} be the truncation of the expansion of a_{0j} in homogeneous polynomials of degree $\leq r_j$. Specifically, if we expand $a_{0j}(x) = \sum_{l=0}^{\infty} a_{0j}^l(x)$ where $a_{0j}^l \in H_j$, then

$$b_{01}(x) = \sum_{l=0}^{r_1} a_{01}^l(x), \ldots, b_{0n}(x) = \sum_{l=0}^{r_n} a_{0n}^l(x)\,. \tag{10}$$

If $X^i(x) = \sum_{j=1}^{n} a_{ij}(x)\, \partial/\partial x_j$, $i = 1, \ldots, k$ is of weight 1, we let $Y^i(x) = \sum_{j=1}^{n} b_{ij}(x)\, \partial/\partial x_j$

where now b_{ij} is the truncation, at degree (r_j-1), of a_{ij}, i.e. for $1 \le i \le k$,

$$b_{i1}(x) = \sum_{l=0}^{r_1-1} a_{i1}^{l}(x), \ldots, b_{in} = \sum_{l=0}^{r_n-1} a_{in}^{l}(x) . \tag{11}$$

Define the vector fields

$$Y^{0,0} = \sum_{j=1}^{n} a_{0j}^{r_j} \frac{\partial}{\partial x_j} \tag{12a}$$

and for $0 \le i \le k$

$$Y^{i,1} = \sum_{j=1}^{n} a_{ij}^{r_j-1} \frac{\partial}{\partial x_j}, \ldots, Y^{i,r_n} = \sum_{j=1}^{n} a_{ij}^{r_j-r_n} \frac{\partial}{\partial x_j} . \tag{12b}$$

Then $Y^{0,0}$ is homogeneous of degree 0 while $Y^{i,l}$ is homogeneous of degree l, $1 \le l \le r_n$. Recalling that a vector field homogeneous of degree $l > k$ is zero, for any $0 \le j \le k$ we can expand X^j in terms of vector fields $X^{j,l}$ homogeneous of degree l, i.e.

$$X^j = X^{j,r_n} + X^{j,r_n-1} + \cdots + X^{j,0} + X^{j,-1} + \cdots .$$

Now, from (12a,b) we relate the X^j and approximates Y^j as follows:

$$X^0 = Y^{0,r_n} + Y^{0,r_n-1} + \cdots + Y^{0,0} + X^{0,-1} + \cdots = Y^0 + X^{0,-1} + X^{0,-2} + \cdots \tag{13}$$

$$X^i = Y^{i,r_n} + Y^{i,r_n-1} + \cdots + Y^{i,1} + X^{i,0} + X^{i,-1} + \cdots = Y^i + X^{i,0} + X^{i,-1} + \cdots, \quad i \le i \le k.$$

CLAIM 1. From the choice of dilation and coordinates,

$$Y^{0,1} = \cdots = Y^{0,r_n} = 0, \quad Y^{i,2} = \cdots = Y^{i,r_n} = 0, \quad 1 \le i \le k$$

i.e.

$$Y^0 = Y^{0,0}, \quad Y^i = Y^{i,1}, \quad i = 1, \ldots, k .$$

Proof. The basic idea is that if claim 1 were false, one could construct a commutator $X^\pi(x) = \sum_{i=1}^{n} a_i(x)\, \partial/\partial x_i$ of weight j having some $a_i(0) \ne 0$ for $i \ge n_j+1$, thereby contradicting proposition 4. Suppose, for example, $Y^{0,1}(x) = \sum_{i=1}^{n} e_i(x)\, \partial/\partial x_i$, $e_i \in H_{r_i-1}$, is not zero. Then $e_1, \ldots, e_{n_1}$ are constants and $X^0(0) = 0$ implies $e_1 = \cdots = e_{n_1} = 0$. Thus $Y^{0,1} \ne 0$ implies some $e_j \in H_{r_j-1}$ is not zero for $j > n_1$. But then a commutator $C^\pi(x) = \sum_{i=1}^{n} c_i(x)\, \partial/\partial x_i$ of $Y^{0,1}$ and some X^{π_s}, $1 \le s \le n_1$, will have $c_i(0) \ne 0$ for some $i \ge n_1+1$. But such a commutator C^π is of weight 1, contradicting the choice of n_1, i.e. contradicting proposition 4. The argument that $Y^{i,2}, \ldots, Y^{i,r_n}$ must be zero for $0 \le i \le k$ is similar. □

From claim 1, we now may write

$$X^0 = Y^{0,0} + X^{0,-1} + X^{0,-2} + \cdots \tag{14}$$

$$X^i = Y^{i,1} + X^{i,0} + X^{i,-1} + \cdots$$

CLAIM 2. (Part (e) of Thm. 1.) If $X^\pi(x) = \sum_{i=1}^{n} a_i(x)\,\partial/\partial x_i$ is a commutator of $\{X^0, \ldots, X^k\}$ of weight $j \geq 1$ and $Y^\pi = \sum_{i=1}^{n} b_i(x)\,\partial/\partial x_i$ is the corresponding commutator of the approximating vector fields $\{Y^0, \ldots, Y^k\}$, then

$$a_i(0) = b_i(0), \quad n_{j-1}+1 \leq i \leq n\,. \tag{15}$$

Remark 2. From proposition 2, we see $b_i(0) = 0$ for $1 \leq i \leq n_{j-1}$, however we need *not* have $a_i(0) = 0$ for these values i, i.e. we do not get $X^\pi(0) = Y^\pi(0)$ for all commutators of length $j \leq N$.

Proof of Claim 2. Let L_0 be the subalgebra of smooth vector fields X on $\mathbb{R}^n$ which vanish at 0. Then from (14) and claim 1, we write

$$X^0 = Y^{0,0} + X^0_{-1} = Y^0 + X^0_{-1} \text{ where } X^0_{-1} = X^{0,-1} + X^{0,-2} + \cdots \tag{16}$$

$$Y^i = Y^{i,1} + X^i_0 = Y^i ++ X^i_0 \text{ where } X^i_0 = X^{i,0} + X^{i,-1} + \cdots,\ 1 \leq i \leq k\,.$$

Then $X^0(0) = 0 \Rightarrow Y^0(0) = 0$ (i.e. part (a) of theorem 1), i.e. $Y^0 \in L_0$ and also $X^0_{-1} \in L_0$, $X^i_0 \in L_0$, $1 \leq i \leq k$. Since L_0 is a subalgebra, it follows that

$$X^\pi = Y^\pi + W \tag{17}$$

where $W(x) = \sum_{i=1}^{n} d_i(x)\,\partial/\partial x_i$ is a sum of vector fields homogeneous of degree $\leq (j-1)$. By proposition 3 this means $d_i(0) = 0$, $n_{j-1}+1 \leq i \leq n$ and hence from (17), $a_i(0) = b_i(0)$, $n_{j-1}+1 \leq i \leq n$. □

D. Verification of conclusions (a), (b), (c), (c) of theorem 1. First, $Y^0(0) = 0$ was shown in the proof of claim 2. Next, consider the specifically chosen commutators X^{π_l}, $1 \leq l \leq n$, which in our preferred coordinates satisfy $X^{\pi_l}(0) = \sum_{i=1}^{n} a_{li}(0)\,\partial/\partial x_i = \partial/\partial x_l$. If X^{π_l} has weight j, then $n_{j-1}+1 \leq l \leq n_j$ so if $Y^{\pi_l}(x) = \sum_{i=1}^{n} b_{li}(x)\,\partial/\partial x_i$, then $Y^{\pi_l} \in L_j$ hence by proposition 2, $b_{li} = 0$ if $1 \leq i \leq n_{j-1}$. From claim 2, $b_{li}(0) = a_{li}(0)$ for $n_{j-1}+1 \leq i \leq n$, i.e. $Y^{\pi_l}(0) = X^{\pi_l}(0)$, $l = 1, \ldots, n$. This gives (b) of theorem 1. Next, from our choice of the X^{π_l} and (b), $\dim \operatorname{span}(\bigcup_{i=1}^{j} \mathcal{S}^i_X(0)) = \dim \operatorname{span}(X^{\pi_1}(0), \ldots, X^{\pi_{n_j}}(0))$

$= \dim \operatorname{span}(Y^{\pi_1}(0), \ldots, Y^{\pi_{n_j}}(0)) = \dim \operatorname{span}(\bigcup_{i=1}^{j} S_Y^i(0))$, giving (C). Finally, since elements in S_Y^1 are homogeneous of degree 1, (d) follows from proposition 1. □

Remark 3. If, in (11), we were to make the approximation $b_{ij}(x) = \sum_{l=1}^{r_j} a_{ij}^l(x)$, $1 \leq i \leq k$, $i \leq j \leq n$, the resulting system gives a "higher order" approximation but would have $L(S_Y^1)$ finite dimensional rather than nilpotent.

The Diagonal Nilpotent Approximation. Here, for ease of presentation, we consider the single input case, i.e. $k = 1$. The extension to the multi-input case is straightforward. Recall that now the weight of a commutator X^π of type (m,l) is $w(X^\pi) = m+l$.

There is one element, X^1, of weight 1 which is not zero at 0. Since $X^0(0) = 0$, (4) implies $X^1(0) \neq 0$. Let $X^{\pi_1} = X^1$ and $n_1 = 1$. Next (see figure 1) we consider elements of weight 2, i.e. if $[X^0, X^1](0)$ is independent of $X^1(0)$ we choose it as X^{π_2} and $n_2 = 2$. If not, $n_2 = 1$ and we examine elements of weight 3, etc. Inductively, let R^j denote the elements of weight j. If $\dim \operatorname{span}(\bigcup_{i=1}^{j-1} R^j(0)) = n_{j-1}$, $\dim \operatorname{span}(\bigcup_{i=1}^{j} R^i(0)) = n_j$ adjoin to the already chosen elements $X^{\pi_1}, \ldots, X^{\pi_{n_{j-1}}}$ commutators $X^{\pi_{n_{j-1}+1}}, \ldots, X^{\pi_{n_j}} \in R^j$ such that $X^{\pi_1}(0), \ldots, X^{\pi_{n_j}}(0)$ are independent. Continue, obtaining $X^{\pi_1}, \ldots, X^{\pi_n}$ with $X^{\pi_1}(0), \ldots, X^{\pi_{n_N}}(0)$ linearly independent, i.e. again $n_N = n$.

THEOREM 4. Consider the single input system (1), i.e. $k = 1$, with $\dim L(S_X^1)(0) = n$, and let $X^{\pi_1}, \ldots, X^{\pi_n}$ be chosen as above. Then there exist (we construct) vector fields Y^0, Y^1 on $\mathbb{R}^n$ which describe the diagonal nilpotent approximating system (2) and satisfy, in the chosen coordinates:

(a) $Y^0(0) = 0$

(b) $Y^{\pi_i}(0) = X^{\pi_i}(0)$, $i = 1, \ldots, n$

(c) $\dim \operatorname{span}\{Y^\pi(0)\colon w(Y^\pi) \leq j\} = \dim \operatorname{span}\{X^\pi(0)\colon w(X^\pi) \leq j\}$, $j \geq 1$.

(d) $L(Y^0, Y^1)$ is nilpotent.

(e) If $X^\pi(x) = \sum_{i=1}^{n} a_i(x)\, \partial/\partial x_i$ is a commutator of $\{X^0, X^1\}$ of weight $j \geq 1$ and $Y^\pi(x) = \sum_{i=1}^{n} b_i(x)\, \partial/\partial x_i$ is the corresponding commutator of $\{Y^0, Y^1\}$, then $a_i(0) = b_i(0)$, $n_{j-1}+1 \leq i \leq n$.

Proof. (Sketch; it is similar to the proof of theorem 1.)

First, again select coordinates $x=(x_1,\dots,x_n)$ for $\mathbb{R}^n$ such that $X^{\pi_i}=\partial/\partial x_i$, $i=1,\dots,n$. Choose the dilation, relative to the n_j, as

$$\delta_t x=(t^{r_1}x_1,\dots,t^{r_{n_{j-1}+1}}x_{n_{j-1}+1},\dots,t^{r_{n_j}}x_{n_j},\dots,t^{r_n}x_{n_N})$$

with

$$r_i=j \text{ if } n_{j-1}+1\le i\le n_j\,. \tag{18}$$

Now, for $i=0,1$, if $X^i(x)=\sum_{j=1}^{n}a_{ij}(x)\,\partial/\partial x_j$ the approximating vector field is $Y^i(x)=\sum_{j=1}^{n}b_{ij}(x)\,\partial/\partial x_j$ with b_{ij} the truncation, to degree $\le r_j-1$, of the expansion of a_{ij} in homogeneous polynomials (relative to δ_t) of degree $\le r_j-1$. Specifically, for $i=0,1$:

$$b_{i1}(x)=\sum_{l=0}^{r_1-1}a_{i1}^l(x),\dots,b_{in}(x)=\sum_{l=0}^{r_n-1}a_{in}^l(x) \tag{19}$$

where $a_{ij}^l\in H_l$. Next, for $i=0,1$ write

$$X^i=Y^{i,r_n}+\cdots+Y^{i,1}+X^{i,0}+X^{i,-1}+\cdots=Y^i+X^{i,0}+X^{i,-1}+\cdots \tag{20}$$

where $Y^{i,l}$, $X^{i,l}$ are homogeneous of degree l. As in the proof of claim 1, theorem 1, we conclude that $Y^{i,r_n}=\cdots=Y^{i,2}=0$, $i=0,1$, i.e. $Y^i=Y^{i,1}$, $i=0,1$. Now write, for $i=0,1$:

$$X^i=Y^i+X_0^i=Y^{i,1}+X_0^i \text{ where } X_0^i=X^{i,0}+X^{i,-1}+\cdots\,. \tag{21}$$

We next verify properties (a)-(e) of theorem 4. Since $X^0(0)=0$ and $X_0^i(0)=0$, $Y^0(0)=0$ from (21), i.e. (a). From (21) and (8), if $X^\pi(x)=\sum_{i=1}^{n}e_i(x)\,\partial/\partial x_i$ is a commutator of $\{X^0,X^1\}$ with $w(X^\pi)=j$, then

$$X^\pi=Y^\pi+W \tag{22}$$

where Y^π is homogeneous of degree j and W is the sum of vector fields homogeneous of degree $\le(j-1)$. Thus if $W(x)=\sum_{i=1}^{n}d_i(x)\,\partial/\partial x_i$, proposition 3 yields $d_i(0)=0$ for $n_{j-1}+1\le i\le n$. From this, and (22), we see that if $X^\pi(x)=\sum_{i=1}^{n}a_i(x)\,\partial/\partial x_i$ has weight j and $Y^\pi(x)=\sum_{i=1}^{n}b_i(x)\,\partial/\partial x_i$, then (conclusion (e))

$$a_i(0)=b_i(0),\quad n_{j-1}+1\le i\le n_j\,. \tag{23}$$

Also, proposition 2 yields $b_i(0)=0$, $1\le i\le n_{j-1}$ hence, since $X^{\pi_i}(0)=\partial/\partial x_i$, $i=1,\dots,n$ it follows that $Y^{\pi_i}(0)=X^{\pi_i}(0)$, $i=1,\dots,n$, i.e. (b). Thus (c) follows easily while (d) is a consequence of proposition 1 and that both Y^0, Y^1 are homogeneous of degree 1. □

§2. Bressan Scaling. Bressan, [10], introduces a nilpotent approximation (analogous to our diagonal approximation) and uses a rescaling to obtain a precise relationship between solutions of the approximating system and those of the original system. We next develop this result, for both the C-H and diagonal approximations, as a quick consequence of the proofs of theorems 1 and 4.

For the case of the C-H approximation, let $x = (x_1, \ldots, x_n)$ be the preferred coordinates as in part A of the proof of theorem 1 and again assume $X^i(x) = \sum_{j=1}^{n} a_{ij}(x)\, \partial/\partial x_j$, $i = 0, \ldots, k$. Then from (12a), (12b) and (14), with notation as in part C,

$$X^0(x) = \sum_{j=1}^{n} \Big(a_{0j}^{r_j}(x) + \sum_{l=r_j+1}^{\infty} a_{0j}^{l}(x)\Big) \frac{\partial}{\partial x_j} \tag{24a}$$

$$X^i(x) = \sum_{j=1}^{n} \Big(a_{ij}^{r_j-1}(x) + \sum_{l=r_j}^{\infty} a_{ij}^{l}(x)\Big) \frac{\partial}{\partial x_j} \tag{24b}$$

where, again, a_{ij}^{l} is homogeneous of degree l relative to δ_t.

THEOREM 5 (Bressan rescaling.)

(a) Let $(x_1, \ldots, x_n)$ denote the preferred coordinates for the C-H approximation; $\delta_t x = (t^{r_1}x_1, \ldots, t^{r_n}x_n)$ the dilation as in the proof of theorem 1 and $t \to x(t,u)$ the solution of (1) for admissible control u. Let $\epsilon > 0$, u be defined on $[0,\epsilon]$, and for $0 \le t \le 1$ define

$$u_i^{\epsilon}(t) = \epsilon^{-1}u_i(\epsilon t), \quad i = 1, \ldots, k; \quad x_j^{\epsilon}(t,u^{\epsilon}) = \epsilon^{-(r_j+1)}x_j(\epsilon t,u), \quad j = 1, \ldots, n. \tag{25}$$

Then $t \to x^{\epsilon}(t,u^{\epsilon})$ satisfies

$$\dot{x}(t) = X_{\epsilon}^0(x(t)) + \sum_{i=1}^{k} u_i^{\epsilon} X_{\epsilon}^i(x(t)), \quad x(0) = 0 \tag{26}$$

where

$$X_{\epsilon}^0(x) = \sum_{j=1}^{n} \Big(a_0^{r_j}(x) + \sum_{l=r_j+1}^{\infty} \epsilon^{l-r_j} a_{0j}^{l}(x)\Big) \frac{\partial}{\partial x_j} \tag{27a}$$

$$X_{\epsilon}^i(x) = \sum_{j=1}^{n} \Big(a_{ij}^{r_j-1}(x) + \sum_{l=r_j}^{\infty} \epsilon^{l-r_j+1} a_{ij}^{l}(x)\Big) \frac{\partial}{\partial x_j}, \quad i = i = 1, \ldots, k. \tag{27b}$$

and hence $X_{\epsilon}^0 \to Y^0$, $X_{\epsilon}^i \to Y^i$, $i = 1, \ldots, k$, as $\epsilon \to 0$ in the C^{∞} topology (on bounded sets) where $Y^0, \ldots, Y^k$ are the C-H approximating vector fields.

(b) Let $(x_1, \dots, x_n)$ be the preferred coordinates for the diagonal nilpotent approximation, δ_t the dilation, $t \to x(t,u)$ the solution of (1) for control u and $\epsilon > 0$. For u defined on $[0,\epsilon]$, define

$$u^\epsilon(t) = u(\epsilon t), \quad x_j^\epsilon(t,u^\epsilon) = \epsilon^{-r_j} x_j(\epsilon t,u), \quad j = 1, \dots, n \tag{28}$$

where $0 \le t \le 1$. Then $t \to x^\epsilon(t,u^\epsilon)$ satisfies Eq. (26) with now $X_\epsilon^i(x)$ given by (27b) for $i = 0,1$ and hence $X_\epsilon^0 \to Y^0$, $X_\epsilon^1 \to Y^1$ as $\epsilon \to 0$ in the C^∞ topology where Y^0, Y^1 are the "diagonal" approximating vector fields.

Proof. We give the proof of (a), that of (b) being similar. Differentiation of $x_j^\epsilon(t,u^\epsilon)$ and use of (24a), (24b) yields:

$$\dot{x}_j^\epsilon(t,u^\epsilon) = \epsilon^{-r_j}\dot{x}_j(\epsilon t,u) = \epsilon^{-r_j}(a_{0j}^{r_j}(\delta_\epsilon x^\epsilon) + \sum_{l=r_j+1}^{\infty} a_{0j}^l(\delta_\epsilon x^\epsilon))$$

$$+ \epsilon^{-r_j+1}\sum_{i=1}^{k} u_i^\epsilon(a_{ij}^{r_j-1}(\delta_\epsilon x^\epsilon) + \sum_{l=r_j}^{\infty} a_{ij}^l(\delta_\epsilon x^\epsilon))$$

$$= (a_{0j}^{r_j}(x^\epsilon) + \sum_{l=r_j+1}^{\infty} \epsilon^{l-r_j} a_{0j}^l(x^\epsilon)) + \sum_{i=1}^{k} u_i^\epsilon(a_{ij}^{r_j-1}(x^\epsilon) + \sum_{l=r_j}^{\infty} \epsilon^{l-r_j+1} a_{ij}^l(x^\epsilon))$$

This shows, componentwise, that $x^\epsilon(t,u^\epsilon)$ satisfies Eq. (26). The conclusion that $X_\epsilon^0 \to Y^0$, $X_\epsilon^i \to Y^i$, $i = 1, \dots, k$, is immediate (see (12a), (12b) and claim 1 of the proof of theorem 1). □

REMARK 4. The basic result of theorem 5 can be loosely stated in words as follows. Any property of the nilpotent approximating system which persists under arbitrarily small C^∞ perturbations of the describing vector fields $Y^0, \dots, Y^k$ is also a property of the original system.

REMARK 5. The control scaling, in (23), does not create difficulties as $\epsilon \to 0$. Indeed, one studies system (2.1) with $|u^\epsilon|$ small which implies properties for (1) with $u = \epsilon u^\epsilon$.

COROLLARY 2. (Stefani, [5, Thm. 2.2]. Application of Theorem 5 *a la* Remark 4.) If either the C-H or diagonal nilpotent approximation of system (1) is STLC at zero, then system (1) is STLC at zero.

Proof. Let $\Omega^\alpha[0,t]$ denote the set of Lebesgue measurable maps $\tau \to u(\tau)$ having $|u_i(\tau)| \le \alpha$, $\alpha > 0$,. $0 \le \tau < t$, $i = 1, \dots, k$, and with the relative topology of $L_2[0,t]$. Let $A_X^\alpha(t)$, $A_Y^\alpha(t)$, $A_\epsilon^\alpha(t)$ denote, respectively, the attainable sets at time t for system (1), the nilpotent approximating system (C-H or diagonal) and system (26), with admissible control set $\Omega^\alpha[0,t]$. STLC of the approximating system means that $0 \in \text{int } A_Y^\alpha(t)$ for all $t,\alpha > 0$ hence

$0 \in \text{int } A_Y^\alpha(1)$.

Choose an origin centered ball, B_ρ, of radius $\rho > 0$ contained interior to $A_Y^\alpha(1)$. For each $p \in B_\rho$ let Ω_p^α denote those controls $u \in \Omega^\alpha[0,1]$ having a corresponding solution to the nilpotent approximating system which reaches p at time 1. Then $\Omega_p^\alpha \neq \phi \ \forall p \in B_\rho$. For $\rho > 0$ and sufficiently small the map $p \rightarrow \Omega_p^\alpha$, $p \in B_\rho$, admits a continuous selection, [5, lemma 2.1]; denote this by $p \rightarrow u_p \in \Omega_p^\alpha$ and again let $t \rightarrow x^\epsilon(t,u_p)$ denote the solution of (26) for control u_p. The map $p \rightarrow x^\epsilon(1,u_p)$ is then a continuous map of B_ρ into $\mathbb{R}^n$ and one may choose $\epsilon > 0$ sufficiently small so that $|p - x^\epsilon(1,u_p)| < \rho/2$ for all p on the boundary of B_ρ. From the Brouwer fixed point theorem (specifically, see [13, cor. page 252] the map $p \rightarrow x^\epsilon(1,u_p)$, $p \in B_\rho$, then covers a $\rho/2$ ball about zero, i.e.

$$0 \in \text{int } A_\epsilon^\alpha(1), \ \forall \alpha > 0, \ \epsilon > 0 \text{ and sufficiently small.} \tag{29}$$

REMARK 6. The argument, to here, applies to perturbations of arbitrary systems, for example let

$$\Sigma_0\text{: } \dot{x}_1 = u, \ \dot{x}_2 = x_1^3 \qquad \Sigma_\epsilon\text{: } \dot{x}_1 = u, \ \dot{x}_2 = \epsilon x_1^2 + x_1^3 .$$

Denote the attainable sets of the above, for solutions initiating from zero, respectively by $A_0^\alpha(t)$, $A_\epsilon^\alpha(t)$. Clearly Σ_0 is STLC while, for $\epsilon > 0$, Σ_ϵ is not. The previous argument does show that $0 \in \text{int } A_\epsilon^\alpha(1)$ for $\epsilon > 0$ and sufficiently small, however one cannot conclude $0 \in \text{int } A_\epsilon^\alpha(t)$ for all $t > 0$. In our case, it is here where the dilation (nilpotent approximation) plays an essential role, as is next seen.

Returning to the proof of corollary 2, let $\delta_\epsilon A_\epsilon^\alpha(t) = \{\delta_\epsilon x\colon x \in A_\epsilon^\alpha(t)\}$. From (25), for the C-H approximation, or (28), for the diagonal approximation, we have (respectively) $A_X^{\epsilon\alpha}(\epsilon t) = \epsilon\delta_\epsilon A_\epsilon^\alpha(t)$, $A_X^\alpha(\epsilon t) = \delta_\epsilon A_\epsilon^\alpha(t)$. Thus from (29), $0 \in \text{int } A_X^{\epsilon\alpha}(\epsilon)$, or $0 \in \text{int } A_X^\alpha(\epsilon)$, for all $\alpha > 0$ and $\epsilon > 0$ sufficiently small (hence now for all $\epsilon > 0$) which shows STLC of system (1) at zero. □

§3. Examples.

For notational efficiency, vector fields will be written, in component form, as row vectors.

EXAMPLE 1. (A linear approximation of system (1) does not determine STLC; the C-H approximation does, but the diagonal approximation does not.) On $\mathbb{R}^4$, let $X^0(x) = (0, x_1, x_2, (x_1^4/24) + (x_2^3/12))$, $X^1 = (1, 0, 0, 0)$. Clearly a linear approximation of X^0 has zero last component and leads to a non-STLC system. For the C-H approximation choose $X^{\pi_1} = X^1$, $X^{\pi_2}(x) = [X^0, X^1](x) = (0, 1, 0, x_1^3/12)$, $X^{\pi_3}(x) = (\text{ad}^2X^0, X^1)(x) = (0, 0, 1, x_2^2/2)$ so $n_1 = 3$ and since no element in $S_X^2(0)$ is independent of $S_X^1(0)$, $n_2 = 3$ also. Next,

(i) $W^1 = [[(\mathrm{ad}^2 X^0, X^1), [X^0, X^1]], [X^0, X^1]] = (0, 0, 0, 1) \in S_X^3$

(ii) $W^2 = (\mathrm{ad}^4 X^1, X^0) = (0, 0, 0, 1) \in S_X^4$.

For the C-H approximation one must choose $X^{\pi_4} = W^1$. Coordinates are already such that $X^{\pi_i} = \partial/\partial x_i$ and $\delta_t x = (tx_1, tx_2, tx_3, t^3 x_4)$. The C-H nilpotent approximation has $Y^0(x) = (0, x_1, x_2, x_2^3/12)$, $Y^1 = X^1$ and is STLC, as is the original system, by theorem 2.

Since W^1 is of type (3,4) the element W^2, of type (4,1), precedes it via the ordering for the diagonal approximation. For this approximation, $X^{\pi_1}, X^{\pi_2}, X^{\pi_3}$ are as above while $X^{\pi_4} = W^2$. The dilation is $\delta_t x = (tx_1, t^2 x_2, t^3 x_3, t^4 x_4)$ giving $Y^0(x) = (0, x_1, x_2, x_1^4/4)$, $Y^1 = X^1$ and this system is not STLC.

EXAMPLE 2. (The diagonal approximation is STLC but the C-H approximation is not.) On $\mathbb{R}^3$ let $X^0(x) = (0, x_1, (x_2^2/2) + (x_1^3/6))$, $X^1 = (1,0,0)$. The relevant commutators are:

$$X^{\pi_1} = X^1, \quad X^{\pi_2}(x) = (\mathrm{ad}\, X^0, X^1)(x) = (0, 1, x_2^2/2),$$

$$W^1(x) = [(\mathrm{ad}^2 X^0, X^1), [X^0, X^1]] = (0,0,1) \in S_X^2 \text{ of type } (2,3),$$

$$W^2(x) = (\mathrm{ad}^3 X^1, X^0) = (0,0,1) \in S_X^3 \text{ of type } (3,1).$$

Both approximations have X^{π_1}, X^{π_2} as above. The C-H approximation has $X^{\pi_3} = W^1$ giving $\delta_t x = (tx_1, tx_2, t^2 x_3)$ and $Y^0(x) = (0, x_1, x_2^2/2)$, $Y^1 = (1,0,0)$, i.e. not STLC.

The diagonal approximation has $X^{\pi_3} = W^2$, dilation $\delta_5 x = (tx_1, t^2 x_2, t^4 x_3)$ and $Y^0(x) = (0, x_1, x_1^3/3)$, $Y^1 = X^1$. By theorem 2, this gives an STLC system.

EXAMPLE 3. (The original system is STLC but neither the C-H or diagonal approximation is.) On $\mathbb{R}^7$ let $X^0(x) = (0, x_1, x_2, x_3, x_4, x_5 + (x_1^4/24), (x_1^3/6) + (x_2^2/2))$, $X^1 = (1,0,0,0,0,0,0)$.

(a) For the C-H approximation we find $X^{\pi_i}(0) = \partial/\partial x_{i+1} = (\mathrm{ad}^i X^0, X^1)(0) \in S_X^1(0)$ for $i = 0, \ldots, 5$ while $X^{\pi_6}(0) = [(\mathrm{ad}^2 X^0, X^1), [X^0, X^1]](0) = \partial/\partial x_7$. This gives the dilation $\delta_t x = (tx_1, tx_2, tx_3, tx_4, tx_5, tx_6, t^2 x_7)$ and approximating vector fields $Y^0(x) = (0, x_1, x_2, x_3, x_4, x_5, x_2^2/2)$, $Y^1 = X^1$. The corresponding system is clearly not STLC.

(b) For the diagonal approximation, a change of coordinates is necessary. Computing: $X^{\pi_1} = X^1 = \partial/\partial x_1$ has $w(X^{\pi_1}) = 1$, $X^{\pi_2}(0) = (\mathrm{ad}\, X^0, X^1)(0) = \partial/\partial x_2$ with $w(X^{\pi_2}) = 2$, $X^{\pi_3}(0) = (\mathrm{ad}^2 X^0, X^1)(0) = \partial/\partial x_3$ with $w(X^{\pi_3}) = 3$, $X^{\pi_4}(0) = (\mathrm{ad}^3 X^0, X^1)(0) = \partial/\partial x_4$ with $w(X^{\pi_4}) = 4$ but now $(\mathrm{ad}^3 X^1, X^0)(0) = \partial/\partial x_7$ also has weight 4 while $(\mathrm{ad}^4 X^0, X^1) = \partial/\partial x_5$ and $(\mathrm{ad}^4 X^1, X^0)(0) = \partial/\partial x_6$ are of weight 5. Changing coordinates as in part A of the proof of theorem 1, and calling the new coordinates $y = (y_1, \ldots, y_7)$, in these coordinates $X^0(y) = (0, y_1, y_2, y_3, (y_1^3/6)+(y_2^2/2), y_4, y_6+(y_1^4/24))$, $X^1 = \partial/\partial y_1$ and with $X^{\pi_1}, \ldots, X^{\pi_4}$ as above, $X^{\pi_5} = (ad^3X^1, X^0)$, $X^{\pi_6} = (ad^4X^0, X^1)$, $X^{\pi_7} = (ad^4X^1, X^0)$ we have $X^{\pi_i}(0) = \partial/\partial y_i$, $i = 1, \ldots, 7$. The dilation is $\delta_t y = (ty_1, t^2y_2, t^3y_3, t^4y_4, t^4y_t, t^5y_6, t^5y_7)$ and the diagonal nilpotent approximating vector fields are $Y^0(y) = (0, y_1, y_2, y_3, y_1^3/6, y_4, y_1^4/24)$, $Y^1 = \partial/\partial y_1$. The corresponding system is clearly not STLC at zero.

Finally, the original system is STLC at zero as can be obtained by Sussmann's Theorem 7.3, [8], or its restatement, algorithm 1, [9], with $1/2 < \theta < 3/4$.

EXAMPLE 4. Consider the linear system $\dot{x} = Ax + ub$ on $\mathbb{R}^n$. Then $\mathrm{span}S_X^1(0) = \mathrm{span}\{b, Ab, \ldots, A^{n-1}b\}$ and if this has dimension n, $\delta_t x = (tx_1, \ldots, tx_n)$. The coordinates change part A, theorem 1, yields the system $\dot{y}_1 = u$, $\dot{y}_2 = y_1, \ldots, \dot{y}_n = y_{n-1}$ which is the nilpotent approximation and also the original system in canonical form.

References

1. Crouch, P.E., Solvable approximations to control systems, *SIAM J. Control & Opt.* **22** (1984), 40-54.
2. Hermes, H., Nilpotent approximations of control systems and distributions, *SIAM J. Control & Opt.* **24** (1986), 731-736.
3. Sussmann, H. and Jurdjevic, V.; Controllability of nonlinear systems, *J. Diff. Eq.* **12** (1972), 95-116.
4. Hermes, H., Local controllability and sufficient conditions in singular problems, *J. Diff. Eq.* **20** (1976), 213-232.
5. Stefani, G., Polynomial approximations to control systems & local controllability, *Proceedings 24th CDC*, I, (1985), 33-38.
6. Hermes, H., Control systems which generate decomposable lie algebras, *J. Diff. Eqs.* **44** (1982), 166-187.
7. Sussmann, H., Lie brackets and local controllability, a sufficient condition for scalar input systems, *SIAM J. Control & Opt.* **21** (1983), 686-713.
8. Sussmann, H., A general theorem on local controllability, *SIAM J. Control & Opt.* **25** (1987), 158-194.
9. Hermes, H. and Kawski, M., Local controllability of a single input affine system, *Nonlinear Anal. & Appl.*, Lecture Notes in Pure and Appl. Math #109, Lakshmikantham, ed., Marcel Dekker, Inc., New York (1987), 235-248.
10. Bressan, A., Local Asymptotic Approximations of Nonlinear Control Systems, *Int. J. Control* **41** (1985), 1331-1336.
11. Bressan, A., The Generic Local Time-Optimal Stabilizing Controls in Dimension 3, *SIAM J. Control & Opt.* **24** (1986), 177-190.
12. Hermes, H., Local Feedback Structures and Nilpotent Approximations, preprint.
13. Lee, E.B. and Markus, L.; Foundations of Optimal Control Theory, John Wiley & Sons, Inc., N.Y. (1967).

ERRATUM TO

Nilpotent Approximations of Control Systems

It has been pointed out to me by Prof. M. Kawski that an arbitrary linear coordinate change as made in part A of the constructive proof of theorem 1 will not always give the desired result, specifically claim 1. A correct coordinate change can be defined as the inverse of the map

$$y = \varphi(x) = \left(\exp\left(\sum_{i=1}^{n} x_i X^{\pi_i}\right)\right)(0)$$

i.e., take the x–coordinates as coordinates of the first kind relative to the vector fields $X^{\pi_1}, \ldots, X^{\pi_n}$. It readily follows that relative to the x–coordinates $X^{\pi_i}(0) = \partial/\partial x_i$, $i = 1, \ldots, n$. The result stated as claim 1 is nontrivial, but is a consequence of theorem 5, or specifically lemma 8.2, in the paper "Hypoelliptic Differential Operators and Nilpotent Groups," L. Rothschild and E. Stein, Acta Math. **81** (1976), 247–320.

Nonlinear Discrete-Time Systems

Accessibility Conditions

Bronislaw Jakubczyk[1], Professor, Institute of Mathematics, Polish Academy of Sciences, Warsaw, Poland

Eduardo D. Sontag[2], Professor, Department of Mathematics, Rutgers University, New Brunswick, New Jersey

1 INTRODUCTION

In this paper we study the problem of positive-time accessibility for control systems defined by difference equations. Lie-theoretic conditions are given for this property to hold, and it is related to transitivity (positive and negative time control).

Specifically, we are interested in nonlinear discrete time systems of the form

$$x(t+1) = f(x(t), u(t)), \qquad x(t) \in X, u(t) \in U, \tag{1}$$

where X is $\mathbb{R}^n$ or more generally a connected differentiable manifold of dimension n and U is a subset of $\mathbb{R}^m$ (or more generally a manifold itself). We assume that X is of class C^k, $k = \infty$ (the *smooth* case) or, when dealing with *analytic* systems, $k = \omega$. The function

$$f : X \times U \to X$$

is also assumed to be C^k, in a sense to be made more precise later. The time t takes values in the set of nonnegative integers $\mathbb{Z}_+$ (forward case), the set of nonpositive integers $\mathbb{Z}_-$ (backward case), or in the set of all integers $\mathbb{Z}$ (forward-backward case).

It is recognized that a better understanding of discrete time control systems is essential in the modeling of physical systems under digital control. However, discrete time systems have been studied much less than their continuous counterparts, and their properties may diverge considerably from those of the latter. The paper [15] and the monograph [16] considered various aspects of discrete systems defined by polynomial evolution equations. However, the general theory remained until recently much weaker than that possible in

[1] Work was done while a Visiting Professor at Rutgers, The State University of New Jersey.

[2] Research supported in part by US Air Force Grants 85-0247 and 88-0235

the more classical continuous time case, for which a large body of knowledge (see for example [1], [4], [5], [23]) is now available. One difficulty is that noninvertibility of the transition maps $f(\cdot, u)$ means that *semi*groups appear where groups would appear in the continuous case, so less algebraic structure is available. Another problem is that no analogue is available for difference equations of the infinitesimal information obtained by taking derivatives with respect to time. One way to try to avoid the first of these difficulties is to restrict attention to *invertible* systems, for which by assumption the above transition maps are indeed invertible for each fixed control value u. For the second problem, the lack of infinitesimal information, one may substitute derivations *with respect to control values*, provided that one restricts attention to situations in which enough differentiable structure exists in the control set U.

Fortunately, both of the above restrictions are very reasonable. Although invertibility is a priori a strong assumption in the context of general discrete time systems, it is the case that for systems that result from *sampling*, this assumption is always satisfied. Recall that sampling is the process under which the state of a continuous time system is measured at discrete instants, and control actions are taken also at discrete instants. Under such a process, the obtained transition maps as observed at the sampling times turns out to give rise to an invertible model. (Analogously, in classical dynamical systems one studies time-one diffeomorphisms and Poincaré maps associated to differential equations.) The second restriction, having enough directions in the control set, is satisfied in any situation in which controls can take unrestricted values, or even if they are subject to magnitude or various other types of algebraic constraints. Invertible discrete time systems are often also obtained in numerical schemes for discretizing continuous time models.

The paper [6] introduced the idea of studying invertible discrete nonlinear control systems, and developed a realization theory which parallels much of the continuous situation. A major related development was the discovery by M.Fliess and D.Normand-Cyrot ([2], [14]) that a large number of concepts from the continuous time theory admit generalizations to invertible discrete time systems, provided that appropriate objects –vector fields, distributions, and Lie algebras,– be defined; further work along these lines was carried out in [8], [20], [12], [17], and related papers.

In [10] a systematic and rigorous study of the structure needed for studying controllability problems for discrete time invertible systems is presented. This paper summarizes and discusses some of the results given there, in the special case where controls are assumed to be scalar valued and unrestricted, and where $X = \mathbb{R}^n$ rather than a general manifold. This will result in a simpler exposition than that in [10], which should be consulted for all proofs.

2 BASIC DEFINITIONS

We start by introducing basic notation and definitions. As stated previously, time takes integer values, $t \in \mathbb{Z}$ and we consider only real-valued, unrestricted controls, and state spaces which are Euclidean.

We introduce the following notations for the effect of shift operators:

$$x^+(t) = x(t+1) \text{ and } x^-(t) = x(t-1).$$

In this way we can write equation (1) in the more compact form

$$x^+ = f(x,u), \qquad x(t) \in \mathbb{R}^n, u(t) \in \mathbb{R}.$$

The system is *of class* C^k if the function $f : \mathbb{R}^n \times \mathbb{R} \to \mathbb{R}^n$ is of class C^k. By *smooth* we mean of class C^∞.

Associated to each such system there is a family of maps

$$f_u = f(\cdot, u) : \mathbb{R}^n \longrightarrow \mathbb{R}^n, \qquad u \in \mathbb{R}.$$

DEFINITION 2.1 The system (1) is *invertible* if for each $u \in \mathbb{R}$ the map f_u is a global diffeomorphism of $\mathbb{R}^n$. □

To any invertible system one can associate an *inverse* or *reversed-time* system with equations

$$x^- = f^-(x,u), \tag{2}$$

where $f^-(x,u) = f_u^{-1}(x)$. By the implicit mapping theorem, this is again of class C^k, and its inverse is the original system.

Unless otherwise stated, every system appearing in this paper will be assumed to be invertible.

The maps f_u and their inverses f_u^{-1} can be considered as "one step forward maps" (respectively, "one step backward maps"). If we apply a sequence of controls $u_1, \dots u_k$ then we obtain the composition of these maps denoted by

$$f_{u_k \cdots u_1} = f_{u_k} \circ \cdots \circ f_{u_1}. \tag{3}$$

Allowing backward as well as forward steps we obtain a larger family of maps

$$f_{u_k \cdots u_1}^{\epsilon_k \cdots \epsilon_1} = f_{u_k}^{\epsilon_k} \circ \cdots \circ f_{u_1}^{\epsilon_1}, \tag{4}$$

where each of $\epsilon_1, \dots, \epsilon_k$ takes a value ± 1.

We shall denote by $A_k^+(x)$ the set of points attainable from x in k forward steps, and by $A^+(x)$ the set of points attainable from x in any nonnegative number of forward steps. Replacing forward steps by backward steps we obtain other sets, $A_k^-(x)$ and $A^-(x)$, which consist of points controllable to x in k steps, and controllable to x in any nonnegative number of steps, respectively. Finally, the set of points attainable from x in any number of positive and negative steps is called the *orbit* of x, and is denoted by $A(x)$.

DEFINITION 2.2 The system (1) is *forward (backward) accessible from* x if its attainable set $A^+(x)$ (respectively, $A^-(x)$) has a nonempty interior. It is called *transitive from* x (*or forward-backward accessible from* x) if its orbit $A(x)$ has a nonempty interior (and so it is necessarily open).

Finally, the system is *forward (backward) accessible* if it is forward (backward) accessible from any $x \in \mathbb{R}^n$, and it is called *transitive* if it is transitive from any $x \in \mathbb{R}^n$. □

There is a straightforward criterion for accessibility of the discrete time system, based on the rank of the following map. For each fixed state x and integer k define

$$\psi_{k,x}(\mathbf{u}) := f_{u_k \cdots u_1}(x),$$

where $\mathbf{u} = (u_1, \dots, u_k)$ takes values in the k-th Cartesian product $\mathbb{R}^k$. Notice that the attainable set $A_k^+(x)$ is by definition equal to the image of this map. By Sard's theorem and the implicit function theorem one has the following:

PROPOSITION 2.3 Consider a smooth system (1). For any fixed x and k, the interior of the attainable set $A_k^+(x)$ is nonempty if and only if

$$\sup\{\text{rank}\,\frac{\partial}{\partial \mathbf{u}}\psi_{k,x}(u), \mathbf{u} \in \mathbb{R}^k\} = n$$

and thus

$$\sup\{\text{rank}\,\frac{\partial}{\partial \mathbf{u}}\psi_{k,x}(u), \mathbf{u} \in \mathbb{R}^k, k \geq 1\} = n$$

is necssary and sufficient for forward accessibility of system (1) from x. ∎

Remark 2.4 The orbit $A(x)$ is the (countable) union of the images of the maps (4), so we can use an analogous argument to give a criterion for transitivity from x, using the maps (4) rather then (3) in order to define a family of maps playing the role of the $\psi_{k,x}$'s. ∎

The above proposition and remark might appear to give satisfactory criteria for forward accessibility and transitivity. Unfortunately, this is not the case. Although for simple systems they may be used to decide whether a given system is forward accessible or not, for more complicated systems explicitly computing the functions $\psi_{k,x}$ may be highly nontrivial, since composition is hard to deal with computationally. As an example, consider for instance the problem of obtaining a general formula for the n-th composition of the quadratic function $g(x) = ax^2 + bx + c$ with itself or that of computing the function $\psi_{k,x}$ if $f(x, u) = g(x) + xu$. The problem becomes even more serious in the case of deciding the transitivity of the system, as this requires also finding the inverse maps f_u^{-1} needed for computing the composed maps (4). Even for classes like that of bilinear systems, proposition 2.3 doesn't seem to provide much useful information.

Also from a purely theoretical point of view proposition 2.3 is of little interest. This is because it gives too limited an insight into the geometry of our systems and it provides an even more limited tool for their study. The maps appearing in the criteria do not have much algebraic and geometric structure.

The main aim of the next section is to introduce a sort of "infinitesimal description" of the discrete time system. This is done by introducing certain vector fields associated to it. By doing so we immediately get a powerful tool and a rich algebraic and geometric structure based on the Lie product of vector fields. In particular, the accessibility properties of the system can be studied using natural Lie algebras of vector fields associated to the system. The idea of introducing vector fields corresponding to infinitesimal perturbations of control values is a natural generalization of the concept of actions of Lie groups, and it was originally proposed in the context of nonlinear control in [2]. These vector fields also find natural applications in the study of controllability properties and the feedback linearizability of sampled systems ([17], [9]).

2.1 Vector Fields Associated to the System

We associate the following four families of vector fields to our discrete time system (1), one vector field for each $u \in \mathbb{R}$:

$$X_u^+(x) = \left.\frac{\partial}{\partial v}\right|_{v=0} f_u^{-1} \circ f_{u+v}(x),$$

$$X_u^-(x) = \left.\frac{\partial}{\partial v}\right|_{v=0} f_u \circ f_{u+v}^{-1}(x),$$

$$Y_u^+(x) = \left.\frac{\partial}{\partial v}\right|_{v=0} f_{u+v}^{-1} \circ f_u(x),$$

$$Y_u^-(x) = \left.\frac{\partial}{\partial v}\right|_{v=0} f_{u+v} \circ f_u^{-1}(x).$$

The interrelations between these vector fields are explained in the next proposition.

The special case in which the function f happens to correspond to the flow of a (complete) vector field Z, that is $f(x,u) = \exp(uZ)$, is important when discussing continuous time systems within our framework. In that case all of the above vector fields are in fact independent of u, and they provide the same information about the system. This is because by the semigroup property of flows it holds that $f_{u+v} = f_u \circ f_v = f_v \circ f_u$, so that $X_u^+ = -X_u^- = Z = -Y_u^+ = Y_u^-$. These equalities help understand why the continuous time theory is considerably simpler than the discrete one.

Note that applying these definitions to the inverse system (2) instead of system (1) gives the same vector fields except that the pluses are changed for minuses and vice versa.

Given a vector field Y and a control value u, we can define another vector field from Y by applying a change of coordinates given by the diffeomorphism f_u,

$$(\mathrm{Ad}_u Y)(x) = (\mathrm{d}f_u(x))^{-1} Y(f_u(x)).$$

Here $\mathrm{d}f_u$ stands for the differential of f_u with respect to x. Using the diffeomorphisms (3), we may also define

$$(\mathrm{Ad}_{u_k \cdots u_1} Y)(x) = (\mathrm{d}f_{u_k \cdots u_1}(x))^{-1} Y(f_{u_k \cdots u_1}(x)),$$

and, applying the even more general family of diffeomorphisms (4) ,

$$(\mathrm{Ad}_{u_k \cdots u_1}^{\epsilon_k \cdots \epsilon_1} Y)(x) = (\mathrm{d}f_{u_k \cdots u_1}^{\epsilon_k \cdots \epsilon_1}(x))^{-1} Y(f_{u_k \cdots u_1}^{\epsilon_k \cdots \epsilon_1}(x)). \tag{5}$$

Clearly, the operators "Ad" so defined are linear operators acting on vector fields Y, and we have that

$$\mathrm{Ad}_{u_k \cdots u_1}^{\epsilon_k \cdots \epsilon_1} Y = \mathrm{Ad}_{u_1}^{\epsilon_1} \cdots \mathrm{Ad}_{u_k}^{\epsilon_k} Y. \tag{6}$$

(Note the reversal of indices.) We will use the abbreviated notation $\mathrm{Ad}_0^k Y$ for $\mathrm{Ad}_{0 \ldots 0} Y$ with $u = 0$ repeated k-times, if $k > 0$, and for $\mathrm{Ad}_{0 \ldots 0}^{-1 \cdots -1} Y$, if $k < 0$. Additionally, $\mathrm{Ad}_0^0 Y = Y$. With this notation we have that

$$(\mathrm{Ad}_0^k X_u^+)(x) = \left.\frac{\partial}{\partial v}\right|_{v=0} f_0^{-k} \circ f_u^{-1} \circ f_{u+v} \circ f_0^k(x)$$

and, more generally,

$$(\mathrm{Ad}_{u_k \cdots u_1} X_{u_0}^+)(x) = \left.\frac{\partial}{\partial v}\right|_{v=0} f_{u_k \cdots u_1}^{-1} \circ f_{u_0}^{-1} \circ f_{u_0+v} \circ f_{u_k \cdots u_1}(x).$$

Since our system is assumed to be invertible, we could apply all definitions to the inverse system (2) instead of (1). Then all the pluses in the superscripts change for minuses and Ad_u changes for Ad_u^{-1}, and vice versa. Therefore, we will have the following fact.

Reversion principle: *Any general property of systems of the type (1) which can be expressed in terms of the above defined vector fields is preserved if we change the pluses in the superscripts for the minuses and each* Ad_u *for* Ad_u^{-1}*, and vice versa.*

Remark 2.5 Some of the above defined vector fields can be equivalently defined as follows:

$$X_u^+(x) = (\mathrm{d}f_u(x))^{-1}\frac{\partial}{\partial u}f_u(x),$$

$$(\mathrm{Ad}_{u_k\cdots u_1} X_{u_0}^+)(x) = (\mathrm{d}f_{u_k\cdots u_1}(x))^{-1}X_{u_0}^+(f_{u_k\cdots u_1}(x)).$$

Since the inverses f_u^{-1} do not appear, the right-hand sides now make sense for *locally* invertible systems. Those of our results which can be stated exclusively in terms of the above vector fields will also hold for locally invertible systems. Furthermore, criteria stated in their terms can be checked without computing the inverse of any diffeomorphism; only matrix inversions are required. ∎

The basic interrelations between the vector fields $X_u^+, X_u^-, Y_u^+, Y_u^-$ are given by the following proposition.

PROPOSITION 2.6 The following equalities:

(a) $X_u^+ = -Y_u^+\,,\quad X_u^- = -Y_u^-.$

(b) $X_u^+ = -\mathrm{Ad}_u X_u^-\,,\quad Y_u^+ = -\mathrm{Ad}_u Y_u^-.$

hold for each $u \in \mathbb{R}$. ∎

In the next proposition and in the rest of the paper we shall use the following notational convention. Given a family of vector fields $\{Y_\alpha | \alpha \in A\}$, we denote by $\mathrm{Lie}\{Y_\alpha | \alpha \in A\}$ the Lie algebra generated by this family of vector fields and by $\mathrm{Lie}\{Y_\alpha | \alpha \in A\}(x)$ the subspace of the tangent space at x generated by the vector fields in this Lie algebra.

PROPOSITION 2.7 For analytic systems,

$$\mathrm{Ad}_0^k X_u^+(x) \in \mathrm{Lie}\{\mathrm{Ad}_0^{k+1} X_u^- \mid u \in \mathbb{R}\}(x)$$

and

$$\mathrm{Ad}_0^{-k} X_u^-(x) \in \mathrm{Lie}\{\mathrm{Ad}_0^{-k-1} X_u^+ \mid u \in \mathbb{R}\}(x)$$

for each $x \in \mathbb{R}^n$, each $u \in \mathbb{R}$, and each integer k. ∎

Note that it is not claimed that, for instance, X_u^+ is in the Lie algebra generated by the vector fields $\mathrm{Ad}_0 X_u^-$. The statement pertains only to the equality of the associated distributions, that is, of the tangent spaces at each point.

3 ACCESSIBILITY CRITERIA

To state our criteria we shall need the following families of vector fields:

$$\Gamma^+ = \{\mathrm{Ad}_{u_k\cdots u_1} X_{u_0}^+ | k \geq 0, u_0, \ldots, u_k \in \mathbb{R}\},$$

$$\Gamma^- = \{\mathrm{Ad}_{u_k\ \cdots\ u_1}^{-1\cdots -1} X_{u_0}^- | k \geq 0, u_0, \ldots, u_k \in \mathbb{R}\},$$

$$\Gamma = \{\mathrm{Ad}_{u_k\cdots u_1}^{\epsilon_k\cdots\epsilon_1} X_{u_0}^\sigma | k \geq 0, u_0, \ldots, u_k \in \mathbb{R}, \epsilon_1, \ldots, \epsilon_k = \pm 1, \sigma = \pm\}.$$

As previously, for a family of vector fields Δ, we denote by $\text{Lie}\{\Delta\}$ the Lie algebra of vector fields generated by Δ, by $\Delta(x)$ the linear space spanned by the vectors at x given by the vector fields in Δ, and by $\text{Lie}\{\Delta\}(x)$ the linear space of tangent vectors at x given by the vector fields in the Lie algebra.

The following theorem gives criteria for accessibility of smooth systems. It will be one of the main results of this paper.

THEOREM 1 *The following properties hold for any smooth system (1).*

(a) The system is forward accessible if and only if any of the following two equivalent conditions hold:

$$\dim \Gamma^+(x) = n \ \forall x \in \mathbb{R}^n, \quad \text{or} \quad \dim \text{Lie}\{\Gamma^+\}(x) = n \ \forall x \in \mathbb{R}^n.$$

(b) The system is backward accessible if and only if any of the following two equivalent conditions hold:

$$\dim \Gamma^-(x) = n \ \forall x \in \mathbb{R}^n, \quad \text{or} \quad \dim \text{Lie}\{\Gamma^-\}(x) = n \ \forall x \in \mathbb{R}^n.$$

(c) The system is transitive if and only if any of the following two equivalent conditions hold:

$$\dim \Gamma(x) = n \ \forall x \in \mathbb{R}^n, \quad \text{or} \quad \dim \text{Lie}\{\Gamma\}(x) = n \ \forall x \in \mathbb{R}^n. \qquad \blacksquare$$

To state a stronger version of our result, valid for analytic systems, we need the following Lie algebras of vector fields:

$$L^+ = \text{Lie}\{\text{Ad}_0^k X_u^+ \mid k \geq 0, u \in \mathbb{R}\},$$

$$L^- = \text{Lie}\{\text{Ad}_0^k X_u^- \mid k \leq 0, u \in \mathbb{R}\},$$

$$L = \text{Lie}\{\ \text{Ad}_0^k X_u^\sigma \mid k \in \mathbb{Z}, u \in \mathbb{R}, \sigma \in \{+,-\}\}.$$

The following inclusions are evident:

$$L^+ \subset \text{Lie}\,\Gamma^+, \quad L^- \subset \text{Lie}\,\Gamma^-, \quad L \subset \text{Lie}\,\Gamma.$$

In terms of this data, we now state another one of our main results.

THEOREM 2 *The following properties hold for any analytic system (1):*

(a) The system is forward accessible if and only if

$$\dim L^+(x) = n \ \text{ for any } x \in \mathbb{R}^n.$$

(b) The system is backward accessible if and only if

$$\dim L^-(x) = n \ \text{ for any } x \in \mathbb{R}^n.$$

(c) The system is transitive if and only if

$$\dim L(x) = n \ \text{ for any } x \in \mathbb{R}^n. \qquad \blacksquare$$

Remark 3.1 As a consequence of proposition 2.7, if we were to take in the definition of the Lie algebra L only $\sigma = +$, or alternatively only $\sigma = -$, a smaller set of vector fields may result, but the conclusions in the theorem would hold equally well. ∎

There is a pointwise version of the above results. An *equilibrium point* $x_0 \in \mathbb{R}^n$ is one such that $f(x_0, 0) = 0$.

THEOREM 3 *The following properties hold:*

(a) A smooth system (1) is transitive from x if and only if $\dim \Gamma(x) = n$ *(equivalently,* $\dim \text{Lie}\{\Gamma\}(x) = n$*). An analytic system (1) is transitive from x if and only* $\dim L(x) = n$.

(b) An analytic system (1) is forward (respectively, backward) accessible from an equilibrium point x_0 if and only if $\dim L^+(x_0) = n$ *(respectively,* $\dim L^-(x_0) = n$*).* ∎

The second part of theorem 3 will be strengthened as a consequence of the following.

PROPOSITION 3.2 If the system is analytic and x_0 is an equilibrium point then

$$L^+(x_0) = L^-(x_0) = L(x_0).$$ ∎

The above theorem and proposition immediately imply the following corollary.

COROLLARY 3.3 Assume that the system is analytic and x_0 is an equilibrium point. Then forward accessibility from x_0, backward accessibility from x_0, and transitivity from x_0 are all equivalent properties. ∎

The above results are consequences of the following somewhat stronger sufficiency and necessity results.

Define the following families of vector fields:

$$X_u^{+,i} = \frac{\partial^i}{\partial u^i} X_u^+, \qquad X_u^{-,i} = \frac{\partial^i}{\partial u^i} X_u^-.$$

THEOREM 4 *The following statements hold for any smooth system (1).*

(a) If

$$\dim \text{Lie}\{\Gamma^+\}(x) = n \;\; \textit{for all } x \in \mathbb{R}^n, \tag{7}$$

then the system is forward accessible.

(b) If x_0 is an equilibrium point and if

$$\dim \text{Lie}\{\text{Ad}_0^k X_0^{+,i} \mid k \geq 0,\; i \geq 0\}(x_0) = n, \tag{8}$$

then the system is forward accessible from x_0.

(c) The same statements hold for backward accessibility if we replace Γ^+ for Γ^- and $X_u^{+,i}$ for $X_u^{-,i}$. ∎

This is a converse to theorem 4:

THEOREM 5 *(a) If system (1) is of class C^1 and forward accessible from x,*

$$\dim \Gamma^+(x) = n.$$

(b) If system (1) is analytic and forward accessible from x,

$$\dim L^+(x) = n.$$

(c) Analogous results hold for backward accessibility with Γ^+, L^+ replaced by Γ^-, L^-. ∎

4 NONACCESSIBLE SYSTEMS

In this section we will briefly discuss nonaccessible and, more generally, nontransitive systems. The following "orbit theorem" is crucial in understanding such systems. The theorem has a long history starting with results of Chow, Nagano, ([13]), Sussmann ([22]) and Stefan ([21]) in the continuous time case. In the discrete time case analogous results to those in continuous time where provided in [6], [20], [8], and [17], the latter containing also a proof of a more abstract result dealing with a general notion of action on manifolds. These papers should be consulted for details of the proof, which we omit.

THEOREM 6 *Any orbit $A(x)$ of the smooth system (1) is an immersed submanifold of $\mathbb{R}^n$ with at most countably many connected components, whose the tangent space is given by*

$$T_y A(y) = \Gamma(y)$$

at each $y \in A(x)$. In the analytic case we have that

$$T_y A(y) = L(y)$$

holds too. ∎

As the attainable set from x lies in the orbit from x, there is no chance for forward or backward accessibility from x if there is no transitivity from x (that is, the orbit is not of full dimension). In this case it is reasonable to ask whether the attainable set has a nonempty interior in the orbit. In the case of analytic continuous time systems the answer is always positive, as proved by Sussmann and Jurdjevic ([24]). The following theorem generalizes this result to discrete time systems.

THEOREM 7 *If x_0 is an equilibrium point of an analytic system (1), then each of the attainable sets $A^+(x_0)$ and $A^-(x_0)$ has a nonempty interior in the orbit $A(x_0)$.*

Proof. If we restrict our system to the orbit then the problem reduces to proving that the system is forward (backward) accessible from x_0, if it is transitive from x_0. But this follows immediately from Theorem 3. ∎

Remark 4.1 The above provides an analogue of what is sometimes called the *positive form of Chow's lemma* for continuous time systems. In fact, the proof is related to that in the continuous time case. However, there is an interesting subtlety that appears here. Contrary to the continuous situation, it is *not* true now that the assumption that x_0 is an equilibrium state can be relaxed. In the paper [17], remark 9.15, an example is given of an analytic system with $n = 1$ and a state x such that $A(x) = \mathbb{R}^n$ but the system is not forward accessible from this x. In fact, the system in question arises from the sampling of a continuous time system. This means that the notion of transitivity is in the discrete time case probably too weak with respect to nonequilibrium states. ∎

The following families of vector fields will help us to understand better the geometry of the attainable sets $A^+(x)$ and $A^-(x)$ and , in particular, to estimate their dimensions. Define

$$\Delta_k^+ = \{\mathrm{Ad}_0^i X_u^+ \mid 0 \le i \le k-1, u \in \mathbb{R}\}, \quad L_k^+ = \mathrm{Lie}\,\Delta_k^+,$$

and

$$\Delta_k^- = \{\mathrm{Ad}_0^{-i} X_u^- \mid 0 \le i \le k-1, u \in \mathbb{R}\}, \quad L_k^- = \mathrm{Lie}\,\Delta_k^-.$$

For any family of vector fields Δ, let $\mathrm{Orb}_\Delta(x)$ denote the orbit of this family passing through x. This orbit has a natural structure of immersed second countable submanifold ([22], [21]). Further, the orbit of $\mathrm{Lie}\,\Delta$ coincides with the orbit of Δ.

PROPOSITION 4.2 For any smooth system,

$$A_k^+(x) \subset \mathrm{Orb}_{\Delta_k^-}(y), \text{for any } y \in A_k^+(x),$$

and

$$A_k^-(x) \subset \mathrm{Orb}_{\Delta_k^+}(y), \text{for any } y \in A_k^-(x).$$ ∎

From the above proposition we immediately conclude the following necessary conditions for accessibility.

COROLLARY 4.3 If an analytic system is forward accessible from x then

$$\dim L^-(y) = n \text{ for any } y \in A^+(x).$$

Similarly, if it is backward accessible from x, then

$$\dim L^+(y) = n \text{ for any } y \in A^-(x).$$ ∎

5 AN EXAMPLE

Consider the following invertible polynomial system with $n = 3$.

$$\begin{aligned} x^+ &= x(z^2+1)^2 \\ y^+ &= y(z^2+1)^3 \\ z^+ &= z+u, \end{aligned} \tag{9}$$

where we are using the superscript $+$ to denote time shift, and we denote coordinates as (x, y, z). Calculating, one obtains that $X_u^+ = -2z(z^2+1)^{-1}Z - X_u^-$ and $X_u^- = (0, 0, -1)'$, where Z is the vector field

$$\begin{pmatrix} 2x \\ 3y \\ 0 \end{pmatrix},$$

for each $u \in \mathbb{R}$. Since the basic vector fields X_u^+ and X_u^- turn out to be independent of u in this example, we drop the subscripts u from now on. Further,

$$\mathrm{Ad}_0 X^+ = \begin{pmatrix} -8xz(z^2+1)^{-1} \\ -12yz(z^2+1)^{-1} \\ 1 \end{pmatrix} = 2X^+ + X^-, \tag{10}$$

from which it follows that

$$\mathrm{span}\,\{X^+, \mathrm{Ad}_0 X^+\} = \mathrm{span}\,\{X^+, X^-\}.$$

The identity $\mathrm{Ad}_0 X^- = -X^+$ (c.f. proposition 2.6(b)) implies that

$$\mathrm{Ad}_0^2 X^+ = 2\mathrm{Ad}_0 X^+ - X^- \in \mathrm{span}\,\{X^+, X^-\},$$

so the linear span of the set of all generators of L^+, $\{\mathrm{Ad}_0^k X^+, k \geq 0\}$, coincides with the span of X^+ and X^-. Similarly, applying Ad_0^{-1} to both sides of (10),

$$\mathrm{Ad}_0^{-1} X^- = X^+ - 2\mathrm{Ad}_0^{-1} X^+ = X^+ + 2X^-,$$

so the span of the $\{\mathrm{Ad}_0^k X^-, k \leq 0\}$, the generators of L^-, is again the same. Finally,

$$[X^+, X^-] = 2(1-z^2)(z^2+1)^{-2} Z,$$

from which it follows that $\{X^-, X^+, [X^+, X^-]\}$ and $\{X^-, Z\}$ span the same C^∞ submodule of vector fields. The latter set is involutive, and we conclude that, for this example,

$$L^+ = L^- = L.$$

Thus the orbits have dimension 2 through each point except at those points with $x = y = 0$, where Z vanishes, and there the dimension is 1. The tangent spaces are given by the vectors $\partial/\partial z$ and $2x\partial/\partial x + 3y\partial/\partial y$. The forward and backward accessible sets contain open subsets of each orbit, by the equality of these Lie algebras.

Of course, in this very simple example one can analyze the system directly. The initial states (x_0, y_0, z_0) with $x_0 = y_0 = 0$ are such that the only possible directions of movement are those in which z changes, as is clear from the equations (9), consistently with the above conclusion about tangent spaces. The points where exactly one of x_0 or y_0 is nonzero are also easy to analyze. Take now a point with both x_0 and y_0 nonzero. Consider the set C consisting of all points (x, y, z) with

$$y_0^2 x^3 = x_0^3 y^2.$$

This is the cross product of a cusp with a line. The forward accessible set consists of all (x, y, z) in C with $\mathrm{sign}\, y = \mathrm{sign}\, y_0$ for which $|x| \geq |x_0|$ and $|y| \geq |y_0|$. The backward accessible has both these inequalities reversed, and the orbit consists of the branch of C

with just sign $y = \text{sign}\, y_0$. Note how each such set C, an algebraic variety, can be stratified into three submanifolds, which turn out to be its singular set (the orbit of $(0,0,0)$), the orbit of (x_0, y_0, z_0), and the orbit of the 'conjugate' point $(x_0, -y_0, z_0)$.

Thus in this example both the forward-accessible set and the orbit from each point are open subsets of an irreducible algebraic variety. More generally, similar behavior may be expected when dealing with invertible polynomial systems and equilibrium initial states. We conjecture that the orbit is an open subset of the *quasi-reachable* set in the sense of [15] and [16]. This is an algebraic variety, and it can be computed explicitly, via Jacobians of the n-step transition map. Note that polynomial invertible systems may exhibit highly nonlinear behavior, such as in the case $x^+ = x^3 + x + u$, where the inverse of the transition mapping is not even rational. We plan to study such systems in greater depth in the future.

6 REFERENCES

1. Brockett, R., "Nonlinear systems and differential geometry," *Proc. IEEE* **64**(1976): 61-72.

2. Fliess, M. and D. Normand-Cyrot, "A group-theoretic approach to discrete-time nonlinear controllability," *Proc. IEEE Conf. Dec. Control*, 1981.

3. Goodman, R., "Lifting vector fields to nilpotent Lie groups," *J. Math. Pures et Appl.* **57**(1978): 77-86.

4. Hermann, R. and A. J. Krener, "Nonlinear controllability and observability," *IEEE Trans. Autom. Ctr.* **22**:728-740.

5. Isidori, A., *Nonlinear Control Systems: An Introduction*, Springer, Berlin, 1985.

6. Jakubczyk, B., "Invertible realizations of nonlinear discrete time systems," *Proc. Princeton Conf. Inf.Sc. and Sys.* (1980):235-239.

7. Jakubczyk, B., "Feedback linearization of discrete-time systems," *Systems and Control Letters* **9** (1987): 441-416.

8. Jakubczyk, B., and D.Normand-Cyrot, "Orbites de pseudo-groupes de diffeomorphismes et commandabilité des systemes non linearires en temps discret," *C.R.Acad. Sc. Paris*, **298-I**(1984): 257-260.

9. Jakubczyk, B. and E. D. Sontag, "The effect of sampling on feedback linearization," *Proc. IEEE Conf. Decision and Control, Los Angeles, Dec. 1987*, pp. 1374-1379.

10. Jakubczyk, B. and E. D. Sontag, "Controllability of nonlinear discrete time systems: a Lie-algebraic approach," to appear in *SIAM J. Control and Optimization.* Also Report 88-09, *SYCON - Rutgers Center for Systems and Control.*

11. Kalman, R.E., Y.C.Ho, and K. S. Narendra, "Controllability of linear dynamical systems," *Contr. Diff. Eqs.* **1**(1963):189-213.

12. Monaco, S., D.Normand-Cyrot, and S. Stornelli, "On the linearizing feedback in nonlinear sampled-data control schemes," in *Proc. IEEE Conf. Dec. and Control*, Athens, Dec. 1986.

13. Nagano, T., "Linear differential systems with singularities and an application to transitive Lie algebras," *J. Math Soc. Japan* **18** (1966): 398-404.

14. Normand-Cyrot, Dorothée, *Théorie et Pratique des Systèmes Non Linéaires en Temps Discret*, Thèse de Docteur d'Etat, Université Paris-Sud, 1983.

15. Sontag, E.D. and Y. Rouchaleau, "On discrete-time polynomial systems," *J. Nonlinear Analysis* **1**(1976):55-64.

16. Sontag, E.D., *Polynomial Response Maps*, Springer, Berlin-NY, 1979.

17. Sontag, E.D., "Orbit theorems and sampling," in *Algebraic And Geometric Methods in Nonlinear Control Theory*, M. Fliess and M. Hazewinkel (Eds.), Reidel, Dordrecht, 1986, pp. 441-486.

18. Sontag, E.D., "An eigenvalue condition for sampled weak controllability of bilinear systems," *Systems and Control Letters* **7** (1986): 313-316.

19. Sontag, E.D., "A Chow property for sampled bilinear systems," to appear in the Proceedings of the *Conference on Mathematical Theory of Networks and Systems*, Tempe, June 1987.

20. Sontag, E.D. and H. J. Sussmann, "Accessibility under sampling," *Proc. IEEE Conf. Dec. and Control*, Orlando, Dec. 1982.

21. Stefan, P., "Attainable sets are manifolds," preprint, Univ. of Wales, 1973.

22. Sussmann, H.J., "Orbits of families of vector fields and integrability of distributions," *Trans.AMS* **180**(1973):171-188.

23. Sussmann, H.J., "Lie brackets, real analiticity, and geometric control," in *Differential Geometric Control theory* (R. W. Brockett, R. S. Millman, and H. J. Sussmann, eds.), Birkhauser, Boston, 1983.

24. Sussmann, H. J. and V. Jurdjevic, "Controllability of nonlinear systems," *J. Diff. Eqs.* **12**(1972):95-116.

Invariants and Optimality

Velimir Jurdjevic, Professor, Department of Mathematics, University of Toronto, Toronto, Ontario, Canada

I. Introduction

Much of the progress in the theory of non-liner systems is due to the geometric ideas originated by S. Lie, F. Klein and E. Cartan ([BMS]). The principal point of view is to study a differential system through its invariants and the associated group of symmetries. For a control system, this point of view yields elegant results through the systematic study of the invariants of its reachable sets. It is the aim of this paper to bring into focus several such results trough a common geometric perspective.

The fundamental theoretic tool is a theorem of Nagano ([N]) concerning the orbits of families of analytic vector fields and their integrability properties. The geometric interpretation of a control system as a family of vector fields, which makes connections with Nagano's theorem evident, has its roots in a much earlier paper of R. Hermann ([H]). Hermann understood the controllability rank condition for linear systems through a theorem of Chow ([CH]), which in turn was based on a theorem of Caratheodory in his treatment of the second law of thermodynamics (Math. Ann. 67 (1909)).

The fact that control systems are in general not time reversible accounts for the "semi-orbit" structure of the reachable sets. A natural theoretic device for their study is the notion of the Lie Saturate of the system ([JK1], and its weak form which appeared earlier in Ann. Fourier 3, 4 (1981), 151-179). Although initially conceived as a tool for proving controllability, the Lie Saturate is also useful in studies on optimality.

The basic ideas will be illustrated through the optimization of a quadratic integral functional over the trajectories of a linear control system. When such a problem is singular i.e., the string Legendre-Clebsch condition is not satisfied, there is a group of transformations G

Lecture given at the conference on 30 years of Modern Optimal Control held in honor of S. Lefschetz and J.P. LaSalle at the University of Rhode Island in June 1988.

which acts on the optimal trajectories of the system. This group is generated by certain vector fields in the Lie saturate of the system, called the jump vector fields. They correspond to what is also known as the "cheap controls" [KKR].

The Lie saturate interpretation of this group of symmetries elucidates the basic nature of the optimal synthesis: for any points a and b in the state space and any $T > 0$ there exist unique points $\bar{a}$ and $\bar{b}$ such that the optimal way to go from a to b in T units of time is to jump instantly to $\bar{a}$ via an element of the group, G follow an optimal trajectory to $\bar{b}$ and then to jump instantly to b by an element of G. Such synthesis is called the Generalized Turnpike Synthesis ([JK2]).

II. Analytic control systems

We will restrict our attention to systems which are real analytic in the state variable. That is we shall assume that the state space X is an n-dimensional real analytic manifold and that the control system on X is given by the following differential system:

$$\frac{dx}{dt} = f(x, u) \tag{1}$$

where u is the control assumed to belong to R^m.

It will be assumed that for each $u \in R^m$ $F(\cdot, u)$ is an analytic vector field on X. It is also necessary to assume some smoothing assumptions jointly in x and u, for instance that $F: X \times R^m \to TX$ is C^1. In all of the specific situations that we consider in this paper $F: X \times U \to TM$ is analytic.

Certainly linear control systems are in the above class, or more generally, systems in R^n described by affine vector fields:

$$\frac{dx}{dt} = (A_0x + a_0) + \sum_{i=1}^{m} (A_ix + a_i)u_i \tag{2}$$

with $A_0, \ldots, A_m$ being $n \times n$ matrices and $a_0, \ldots, a_m$ vectors in R^n.

Since affine vector fields generate a finite dimensional Lie algebra equation, (2) may be regarded as a projection of a right invariant control system on a Lie group G of the form

$$\frac{dx}{dt} = X_0(x) + \sum_{i=1}^{m} u_i X_i(x) \tag{3}$$

with $x(t) \in G$. ([J])

For all of these systems the starting point of study is the Hermann-Nagano theorem concerning the orbits of families of analytic vector fields. We will need several notations required to state this result. We will denote by $\{\exp t\, V\}$ the one parameter group of diffeomorphisms induced by a vector field V. For notational simplicity, we shall assume that all vector fields under consideration are complete. If F is a family of vector fields then $G(F)$ stands for the group of diffeomorphisms on X generated by $\cup\{\exp t\, V\colon t \in R\}$ as V varies over F. That is any element of $G(F)$ is of the form

$$\exp t_p V_p \circ \exp t_{p-1} V_{p-1} \circ \ldots \circ \exp t_1 V_1 \tag{4}$$

for some positive integer p, $(t_1, \ldots, t_p) \in R^p$ and $\{V_1, \ldots, V_p\} \subset F$.

The orbit of F through a point x in X is the action of $G(F)$ on x. That is, the orbit through x is the set $G(F)(x) = \{g(x)\ g \in G(F)\}$. The Hermann-Nagano theorem says that each orbit of an analytic family F is a submanifold of X with the tangent space at each of its points y equal to the evaluation of the Lie algebra generated by F at y. This result implies in particular that the Lie algebra generated by the elements of F is of constant rank over all points of an orbit of F.

We will use Lie (F) to denote the Lie algebra generated by F. Each differential system (1) defines an analytic family F given by $\{F(\cdot, u),\ u \in R^m\}$. Let $S(F)$ be the semigroup of diffeomorphisms of X generated by $\cup\{\exp t\, V\colon t \geq 0\}$, $V \in F$. The action of $S(F)$ through x is equal to the set of points of X that can be reached by the trajectories of (1) from x generated by the piecewise-constant controls with finitely many discontinuities ([L]).

More generally $A_F(x, t)$ will denote the set of all points y of X of the form

$$y = \exp t_p V_p \circ \ldots \circ \exp t_1 V_1(x) \quad t_1 \geq 0, \ldots, t_p \geq 0, \quad \sum_{i=1}^{p} t_i = t \tag{5}$$

and $\{V_1, \ldots, V_p\} \subset F$.

$A_F(x, t)$ is the set of all points reachable from x in t units of time by the composition of integral curves of F. We will also use $A_F(x, \leq t)$ to denote $\bigcup_{s \leq t} A_F(x, s)$. F is said to be *strongly controllable* if $A_F(x, \leq r) = X$ for each $x \in X$ and each $t > 0$. F is *controllable* if $A_F(x, < \infty) = X$ for each $x \in X$.

For linear systems the appropriate family of vector fields is $F = \{A(\cdot)+Bu\colon\ u \in R^m\}$. The Lie algebra generated by F is the vector space spanned by the linear field A and all the constant vector fields $\{A^kBu\colon\ u \in R^m,\ k \geq 0\}$. The Hermann-Nagano theorem implies that the rank of $\{A^kBu\colon\ k \geq 0\ \ u \in R^m\}$ must be equal to R^n if the system is to be controllable. Additional ideas are required to show that the system is strongly controllable when $\dim \{A^kBu\colon\ k \geq 0,\ u \in R^m\} = n$. An elegant way of recovering such a result is through the Lie saturable and through an explicit construction of the controllability spaces.

III. Invariants of the Reachable Sets and the Lie Saturate

The class of piecewise constant controls, although convenient for a geometric perspective through the equivalent family of vector fields, is in many situations not very natural. The boundary points of its reachable sets, which are of fundamental interest in optimal control theory, are sometimes reached only via controls with infinitely many discontinuities having points of accumulation as in the Fuller example. Additionally reachable sets need not be closed even by admitting all absolutely continuous paths of the system.

It is therefore natural to work with the closure of the reachable sets provided that not much information is lost about the original system. Such an approach seems particularly fruitful when applied to systems of analytic vector fields. For such systems,

$$\text{Int cl } A_F(x, \leq t) \subset \text{Int } A_F(x, \leq t + \varepsilon) \quad \text{for any} \quad \varepsilon > 0 \tag{6}$$

any $x \in X$ and any $t > 0$. Both the interior and the closure are taken in the submanifold topology of the orbit through x. The above inclusion depends fundamentally on the analyticity of F. In general one may not take $\varepsilon = 0$ as the following example shows.

Example 1. Let X be the unit circle S^1 and V be the rotation field on S^1 i.e., $V(x_1, x_2) = (-x_2, x_1)$. Take $F = \{f(u)V : u \in R\}$ where f is any monotone increasing function satisfying $\lim_{u \to -\infty} f(u) = -\frac{1}{2}$ and $\lim_{u \to \infty} f(u) = \frac{1}{2}$. Then $A_F(x, \le 1) = S^1 - \{-x\}$. Hence, Int cl $A_F(x, \le 1) = S^1$ which is not contained in Int $A_F(x, \le 1)$.

The Lie saturate is defined through with the notion of equivalence introduced in [JK1]. Families F and G of vector fields are said to be *equivalent* if cl $A_F(x, \le t) =$ cl $A_G(x, \le t)$ for each $x \in X$ and each t, $t > 0$. They are said to be *weakly equivalent* if cl $A_F(x, < \infty) =$ cl $A_G(x, < \infty)$ for each $x \in X$. On this level of generality equivalent families could have disparate reachable sets. The most striking example is obtained on a torus T^2 with a densely winding vector field V. For then $F = \{\lambda V : \lambda \in R\}$ is equivalent to the set G of *all* vector fields on T^2. $A_F(x, \le t)$ is a graph of a curve while $A_G(x, \le t) = T^2$ for each $x \in T^2$. To avoid such harsh situations we restrict equivalent families to satisfy certain algebraic conditions in terms of their Lie algebras. That is we additionally require that both $F \subset \text{Lie}(G)$ and $G \subset \text{Lie}(F)$. Then $\text{Lie}(F) = \text{Lie}(G)$ and F and G have the same orbits. The *Lie saturate of F* is the union of all subsets of Lie(F) equivalent to F. The *weak Lie saturate* is defined correspondingly through the weak equivalence. It is an easy consequence of the definitions (and inclusion (6)) that F is strongly controllable if and only if the Lie saturate of F is equal to Lie (F).

The following properties of the Lie saturate are basic:

(7a) LS(F) is a closed convex set. (We assume that the space of vector fields on X is topologized any topology stronger than C^1 topology.

(7b) If $V \in LS(F)$ then $\lambda V \in LS(F)$ for all λ with $0 \le \lambda \le 1$.

If $\Phi : X \to X$ is an analytic mapping such that Φ^{-1} is also analytic then $\{\Phi \circ \exp t V \circ \Phi^{-1}\}$ is a one parameter group of analytic transformations. We will use $\phi_\# V$ to denote its infinitesimal generator $\phi_\# V = d\phi \circ V \circ \phi^1$. The following evident property is also very basic:

(7c) Suppose that Φ is such that

$$\Phi \text{ cl } A_F(\Phi^{-1}(x), \le t) \subset \text{cl } A_F(x, \le t)$$

for each $x \in X$ and $t > 0$. Then $\phi_{\#} LS(F) \subset LS(F)$. We termed such a transformation Φ a normalizer of F. It is clear that the set of all normalizers is a group which leaves the Lie saturate of F invariant.

(7d) If S is a vector space in $LS(F)$ then Lie(S) is in $LS(F)$.

It is a simple matter to show strong controllability of linear systems through the above formalism:

The space $\{Bu\colon u \in R^m\}$ is in the Lie Saturate since $Bu = \lim_{\lambda \to \infty} \frac{1}{\lambda}(A(\cdot)+B\lambda u)$ (follows from (7b)). Therefore, each translation $\Phi(x) = x+Bu$ is in the normalizer of the system. Hence $(\Phi_{\#}A)(x) = A(x-Bu)$ is also in the Lie Saturate. Then $ABu = \lim_{\lambda \to \infty} \frac{1}{\lambda}(A(\cdot)+B\lambda u)$ and therefore $\{ABu\colon u \in R^m\}$ is contained in the Lie Saturate. The successive applications of these arguments show that each of the controllability spaces $\{A^kBu\colon u \in R^m\}$ is in the Lie Saturate. This statement, plus the Hermann-Nagano theorem, easily imply that $A(x, \leq T) = C$ for each $x \in C$ and each $T > 0$ where C is the controllability space $\bigcup_{k \geq 0} \{A^kBu\colon u \in R^m\}$.

The conceptual clarity of the preceding argument leads to natural extensions of the linear controllability results to wider classes of systems such as for instance those considered in [JK1].

IV. Optimality

If F is an arbitrary family of vector fields, we shall be interested in the trajectories of F which are in the boundary of the reachable sets. It is natural to enlarge the class of admissible trajectories of F to include the absolutely continuous paths $x(t)$ whose derivative $\frac{dx}{dt}$ belongs to $F(x(t))$ for almost all t in any compact time interval $[0, T]$.

Our discussion of optimality will be in terms of the Maximum Principle and the associated Hamiltonian formalism. For that reason it will be convenient to introduce additional concepts. We will use T^*X to denote the cotangent bundle of X. T_x^*X is the cotangent space at x, and $<v, p> = p(v)$ for $(v, p) \in T_xX \times T_x^*X$ is the canonical pairing. A vector field V on X has a natural lifting to a vector field in T^*X defined in terms of the Hamiltonian function H_V of V defined by $H_V(x, p) = <V(x), p>$ for each $(x, p) \in T_x^*X$. We will use $\vec{H}_V$ to denote the Hamiltonian vector field associated to H_V. That is the integral curves $(x(t), p(t))$ of $\vec{H}_V$ satisfy $\frac{dx}{dt} = \frac{\partial H}{\partial p}$ and $\frac{dp}{dt} = -\frac{\partial H}{\partial x}$. The Maximum Principle for F can be stated in the following way:

Suppose that x(t) is a trajectory of F on [0, T] such that $x(T) \in \partial A_F(x(0), T)$. Let V be any measurable selection from F defined by $\frac{dx}{dt} = V(t, x(t))$ a.e. on [0, T]. Then x(t) is the projection of an integral curve (x, p) of $\vec{H}_V$, with p not identically zero on [0, T] such that

$$<W(x(t)), p(t)> \leq H_V(x(t), p(t)) \tag{7}$$

a.e. in [0, T] for any vector field W in F. That is the Hamiltonian of V is maximal along its integral curve $(x(t), p(t))$ among all other Hamiltonians of the family F.

The Maximum Principle has a nice generalization in terms of the Lie saturate of F.

Suppose that the interior of $A_F(x(0), T)$ is equal to the interior of $A_{LS(F)}(x(0), T)$. Then the equality in (7) can be extended to all elements of $LS(F)$ to read:

$$<Wx(t), p(t)> \leq H_V(x(t), p(t)) \tag{8}$$

for each $W \in LS(F)$ and almost all t in [0, T].

Since the Lie saturate contains Lie brackets of arbitrarily high orders, it follows that the preceding generalization is a form of a High Order Maximum Principle as discussed by A. Krener ([Kr]). The geometric perspective allows for an effective treatment of optimization of linear systems with a singular quadratic integral cost. The remaining part of the paper will be devoted to such a problem.

The basic problem is to minimize $\int_0^T c(x, u)dt$ over the trajectories of a linear system $\frac{dx}{dt} = Ax+Bu$ in R^n. The cost functional c(x, u) is of the following form:

$$c(x, u) = \frac{1}{2}\langle u, Pu\rangle + \langle u, Qx\rangle + \frac{1}{2}\langle x, Rx\rangle \tag{9}$$

for some appropriate linear operators P, Q and R of appropriate dimensions. It will be assumed that P and R are symmetric. The minimitation is relative to fixed end-point conditions; that is, it is assumed that a, b in R^n and $T > 0$ are given and that the problem is to find the trajectory pair (x, u) of the system such that $x(0) = a$, $x(T) = b$ and $\int_0^T c(x, u)dt \leq \int_0^T c(y, v)dt$ for any other trajectory (y, v) with $y(0) = a$ and $y(T) = b$.

We will use v(a, b, T) to denote the optimal cost $\int_0^T c(x, u)dt$.

The subsequent analysis is based on the following assumptions:

(A1) $\frac{dx}{dt} = Ax+Bu$ is controllable. Therefore $v(a, b, T) < \infty$ for any a, b in R^n and $T > 0$.

(A2) It will be assumed that $V(a, b, T) > -\infty$ for any a, b in R^n and $T > 0$. This assumption implies that $\int_0^T c(x, u)dt \geq 0$ for any trajectory (x, u) with $x(0) = x(T) = 0$. Hence the zero trajectory is optimal for $a = b = 0$ and $T > 0$.

(A3) It will be assumed that the zero trajectory is the only optimal trajectory for $a = b = 0$ and any $T > 0$.

A very natural way to study optimality of such systems is through the reachable sets of the extended system

$$\frac{dx^0}{dt} = c(x, u), \quad \frac{dx}{dt} = Ax + Bu \tag{10}$$

defined for (x^0, x) in the extended state space $X = R^{n+1}$. Equations (10) define an equivalent family of vector fields F on X with $F = \{(c(\cdot, u), A(\cdot) + Bu) \; u \in R^m\}$. The problem of optimality is equivalent to studying the trajectories of F which remain in the boundary of its reachable sets. The precise connection is as follows.

Assume that (x, u) is an optimal trajectory relative to $a = x(0)$ and $x(T) = b$. Then the extended trajectory $(x^0(t), x(t))$ belongs to the boundary of $A_F(\tilde{a}, T)$ for $t = T$. Here $\tilde{a}$ denotes the initial extended state $(0, a)$. In addition $x^0(T) = v(a, b, T)$. Conversely if $x^0(T) = v(x(0), x(T), T)$, $(\tilde{x}(0) = (0, x(0)))$, then $(x^0(T), x(T)) \in \partial A_F(\tilde{x}(0), T)$. This assertion follows from the following simple argument.

Since $\int_0^T c(x, u)dt \geq 0$ for any (x, u) with $x(0) = x(T) = 0$, it follows from our uniqueness assumption (A3) that there is a such trajectory with $\int_0^T c(x, u)dt > 0$. Therefore $\sup \int_0^T c(x, u)dt = \infty$ over the trajectories (x, u) with $x(0) = a$ and $x(T) = b$, for any a, b in R^n and any $T > 0$. Hence, the half-line $\{(x^0(T) + \lambda, x(T)) : \lambda \geq 0\}$ is contained in $A_F(\tilde{x}(0), T)$, for any extended trajectory $(x^0(t), x(t))$.

More generally the following statement is true:

$$(x^0, x) \in \partial A_F(\tilde{a}, T) \quad \text{if and only if} \quad x^0 = v(a, x, T) . \tag{11}$$

It then follows that the boundary of $A_F(\tilde{a}, T)$ consists of the points reached by the optimal trajectories, and those which are not reachable, but are limits of the trajectories of the system. These points account for the "jump" directions and the "generalized turnpike" phenomena. The space of jump fields is contained in the Lie saturate of F and it, together with the original system, accounts for all the points on the boundary of $A_F(\tilde{a}, T)$.

It is easy to show that each element V of Lie(F) must be of the form (f_0, f), with $f_0(x)$ a quadratic function of x, and $f(x)$ an affine vector field in R^n with its linear part a scalar multiple of $A(x)$. Hence Lie (F) is a finite dimensional Lie algebra, and therefore, closed subset of the set of all smooth vector fields on X.

We define the space of all jump fields J as the set of all the vector fields $V = (f_0, f)$ in Lie (F) which satisfy:

(i) $\lambda J \in LS(\mathcal{F})$ for each $\lambda \in R$ and

(ii) f_0 is a linear function and f is a constant vector field.

Any two points on the orbit of $\mathcal{J}$ can be reached by a trajectory of $\mathcal{J}$ in arbitrarily short interval of time since $\mathcal{J}$ is closed under scalar multiplications.

Since $LS(\mathcal{F})$ is a closed convex set it follows that $\mathcal{J}$ is a vector space. It follows from (7d) that $\mathrm{Lie}(\mathcal{J}) \subset LS(\mathcal{F})$. The Lie bracket of any element of $\mathcal{J}$ is of the form $(c, 0)$ for some constant number c. It is easy to show that $c = 0$ based on the fact that $v(a, a, T) \leq 0$ for any $a \in R^n$ and any $T > 0$. Thus $\mathcal{J}$ is a *commutative Lie algebra.*

We will denote by D the projection of $\mathcal{J}$ on R^n. That is $d \in D$ if and only if there exists a linear function f_0 such that $(f_0, d) \in \mathcal{J}$. D will be called the *space* of *jump directions*. It follows from commutativity of $\mathcal{J}$ that there exists a symmetric linear operator $M\colon D \to (R^n)'$ such that $(f_0, d) \in \mathcal{J}$ if and only if $f_0(x) = Mx$. ($(R^n)'$ denotes the dual of R^n.)

For any $V \in \mathcal{J}$ with $V(x) = (\langle x, Md\rangle, d)$ its exponential map $\exp V$ is given by:

$$\exp V(a^0, a) = \left(a^0 + \langle a, Md\rangle + \frac{1}{2}\langle d, Md\rangle, a + d\right) \tag{12}$$

It is assumed that $V = (\langle Mx, d\rangle, d)$. We will use G to denote the groups of exponentials generated by the elements of $\mathcal{J}$. G stabilizes the sets of accessibility; that is,

$$G(\mathrm{cl}\, A_{\mathcal{F}}(\tilde{a}, \leq T)) \subset \mathrm{cl}\, A_{\mathcal{F}}(\tilde{a}, \leq T) \text{ and } \mathrm{cl}\, A_{\mathcal{F}}(G(\tilde{a}), \leq T) \subset \mathrm{cl}\, A_{\mathcal{F}}(\tilde{a}, \leq T) \tag{13}$$

for any $\tilde{a} = (a^0, a)$ in R^{n+1} and any $T > 0$.

The basic property of G which allows for a constructive generation of the elements of $\mathcal{J}$ is contained in the following symmetry statement:

For any optimal trajectory (x, u) (relative to $a = x(0)$ and $b = x(T)$ the curve $y(t) = gx(t)$ $0 \leq t \leq T$ satisfies

$$y(T) \in \partial A_{\mathcal{F}}(g(a), T) \quad \text{for any } g \in G\,. \tag{14}$$

We shall exploit this symmetry principle through the Maximum Principle.

Let (x, u) be an optimal trajectory for $x(0) = a$ $x(T) = b$. Then the extended trajectory $((x^0, x), u)$ must satisfy the Maximum Principle: that is there exists an absolutely continuous curve (p^0, p) defined on $[0, T]$ not identically zero with values in X' such that

$$\frac{dp^0}{dt} = -\frac{\partial H}{\partial x^0} \quad \frac{dp}{dt} = -\frac{\partial H}{\partial x} \tag{15a}$$

where $H(x^0, x, p^0, p, u) = p^0 c(x, u) + <p, Ax + Bu>$ and,

$$p^0 c(x(t), v) + <p(t), Ax + Bv> \leq H(x^0(t), x(t), p^0(t), p(t), u(t)) \tag{15b}$$

for each $v \in R^m$ and almost all t in $[0, T]$.

Since H does not depend explicitly on x^0, it follows that p^0 is a non positive number which can be normalized to be either equal to -1 or equal to zero. $p^0 = 0$ is not possible in this case since we have assumed controllability.

Therefore the system Hamiltonian H reduces to the usual form $H(x, p, u) = -c(x, u) + <p, Ax + Bu>$. The fact that the zero trajectory is optimal for $a = b = 0$ and any $T > 0$ easily implies that $P \geq 0$. Hence the system Hamiltonian H is concave in u for each x and p. Hence (15b) is equivalent to $\frac{\partial H}{\partial u}(x, p, u) = 0$. A triple of curves $(x(t), p(t), u(t))$ will be called extremal on an interval $[0, T]$ if:

$$\frac{dx}{dt} = \frac{\partial H}{\partial p}, \quad \frac{dp}{dt} = -\frac{\partial H}{\partial x} \quad \text{and} \quad \frac{\partial H}{\partial u}(x, p, u) = 0\,. \tag{16}$$

It can be easily proved, under our assumptions, that for each extremal triple (x, p, u) the pair (x, u) is optimal for $x(0) = a$ $x(T) = b$. Hence the study of optimality is reduced to the study of extremals given by (16). More explicitly these equations are

$$\frac{dx}{dt} = Ax + Bu, \qquad \frac{dp}{dt} = Rx - A'p + Q'u \tag{17}$$

$$\text{and } -Pu - Qx + B'p = 0$$

The problem is *regular* if $P > 0$. In such a case the preceding equation is solvable for u in terms of x and p. The optimal feedback law $u(x, p) = P^{-1}(B'p - Qx)$ defines a single quadratic Hamiltonian $H_0(x, p) = H(x, p, u(x)p)$. Equations (16) represent the Hamiltonian flow of H_0. In such a case H_0 defines a *field of extremals* and therefore for each a, b and T there exists a unique extremal (x, p) with $x(0) = a$ and $x(T) = b$. ([KOG]) The corresponding cost $x^0(T)$ is optimal and hence $A_F(\tilde{a}, T)$ is closed. In such a case, the optimal synthesis if *regular.*

When $\ker P \neq 0$ the problem is called singular. Then $-P - Qx + B'p = 0$ is not solvable for u in terms of x and p. Its complete resolution, the key for the generalized optimal synthesis, is intimately linked with the space of jump directions.

For any optimal triple (x(t), p(t), u(t)) the above equation satisfies

$$<v, B'p(t)\ Qx(t)> \ = \ 0 \qquad \text{for any } v \in \ker P\,. \tag{18}$$

We will use $\mathcal{L}_0$ to denote the space of all linear forms $<v, B'p-Qx>$ $v \in \ker P$. Equation (18) is equivalent to saying that optimal trajectories are contained in the zero set of the elements of $\mathcal{L}_0$. $-Pu-Qx+B'p = 0$ can be solved for u in terms of x and p modulo the elements of $\mathcal{L}^0$.

Let E be any subspace of R^m complementary to $\ker P$, and let $u = e \oplus v$ with $e \in E$ and $v \in \ker P$. Upon identifying E with $R^m/\ker P$ it follows that $-Pu - Qx + B'p = 0$ has a unique solution $e(x, p)$ modulo an element of $\mathcal{L}_0$. Let $H_0(x, p) = H(x, p, e(x, p))$. The quadratic Hamiltonian H_0 depends on the choice of the complementary space E. For if E′ is another such space then

$$H_0(x, p) \ = \ H_0'(x, p) + <v(x, p), B'p - Qx> \tag{19}$$

where v(x, p) is a linear mapping from $R^n \times (R^n)'$ into $\ker P$. Hence along optimal curves $H_0 = H_0'$.

Equations (17) can be recast in terms of the preceding notation in the following way:

For each function f let $\vec{f}$ denote the Hamiltonian field of f. Let $\vec{\mathcal{L}}_0$ be the distribution spanned by the elements $\vec{f}$ with $f \in \mathcal{L}_0$. Then a curve $\sigma(t) = (x(t),p(t))$ is optimal if and only if

$$\frac{d\sigma}{dt} \in \vec{H}_0 + \vec{\mathcal{L}}_0 \qquad \text{and} \qquad \mathcal{L}_0(\sigma(t)) \ = \ 0\,. \tag{20}$$

Equation (20), when viewed as a constrained Hamiltonian system, become a special case of the ones studied by Dirac in his attempt to provide mathematical foundation for quantum mechanics ([D]). The elements of L_0 correspond to the constraints that Dirac called the constraints of the first kind.

The complete resolution of (20), which results in a single Hamiltonian quadratic system on a symplectic subspace Ω in $R^n \times (R^n)'$, is fundamentally intertwined with the space of jump fields.

The elements of L_0 are the Hamiltonians of jump vector fields. The argument is simple and goes as follows:

The vector field V_v in F, defined by an element v in ker P, is of the form

$$V_v(x) = (\langle v, Qx\rangle + \frac{1}{2}\langle x, Rx\rangle, \quad Ax + Bv).$$

It follows by (7a) and (7c) that $\lim_{\lambda \to \infty} \frac{1}{\lambda} V_{\lambda v}$ is in the Lie saturate of F. Since this limit is equal to $(\langle v, Qx\rangle, Bv)$, $x \in R^n$, it follows that it is a jump field. Its Hamiltonian is equal to $\langle v, B'p - Qx\rangle$ for $(x, p) \in R^n \times (R^n)'$.

Let L_∞ denote the vector space of the Hamiltonians corresponding to the elements of J. It follows that L_∞ can be constructed explicitly in terms of an increasing sequence of subspaces

$$L_0 \subset L_1 \subset \ldots \subset L_k \subset \ldots \qquad L_\infty = \bigcup_{k \geq 0} L_k . \tag{21}$$

The construction is expressed most simply in terms of the Poisson bracket. Recall that if f and g are functions of x and p, their Poisson bracket $\{f, g\}$ is a function defined by $\{f, g\}(x, p) = \langle \frac{\partial f}{\partial p}, \frac{\partial g}{\partial x}\rangle - \langle \frac{\partial f}{\partial x}, \frac{\partial g}{\partial p}\rangle$. The connection between Poisson brackets and Lie brackets is the following:

Let $\vec{f}$ and $\vec{g}$ be the Hamiltonian fields corresponding to the functions f and g. Then,

$$[\vec{f}, \vec{g}] = \overrightarrow{\{ f, g \}} .$$

We will use ad f to denote the linear operation $\mathrm{ad}\, f(g) = \{f, g\}$ for any function g.

The spaces $\{L_k\}$ are defined recursively the following formula:

$$L_{k+1} = L_k + \{\mathrm{ad}H_0(h):\ h \in L_k \text{ and } \mathrm{ad}h(\mathrm{ad}H_0(h)) = 0\} \tag{22}$$

Alternatively, L_0 can be characterized in terms of a symmetric bilinear form β on the space of all linear functions of x and p given by $\beta(f, g) = \mathrm{ad}\, f\, \mathrm{ad}\, g(H_0)$. L_0 is the smallest vector subspace S of the space of all linear functions which contains L_0 and which satisfies $\mathrm{ad}H_0(S \cap \ker \beta) \subset S$.

Although the Hamiltonian H_0 is not an intrinsic function, since it is defined through an arbitrary complementary space E, it follows from (19) that the spaces $\{L_k\}$ are intrinsic. Analogously the bilinear form β is intrinsicly defined on the space L_∞.

The connection of L_∞ with equation (20) can be most naturally seen through the symmetry statement in (14), which imposes additional constraints on the space of optimal paths of the system. Additional details along with all the proofs will appear in [JK2].

REFERENCES

[BMS] Differential Geometric Control Theory, edited by R. Brockett, R. Milman and H. Sussmann, Progress in Mathematics, Vol 27 (Birkäser, Boston 1983).

[CH] W.L. Chow, Uber systeme von linearen partiellen differential gleichungen erster ordnung, Math. Ann. Vol 117 (1940), 98-105.

[D] P.A.M. Dirac, Generalized Hamiltonian dynamics, Can. J. Math., 2 (1950), 129-148.

[H] R. Hermann, On the accessibility problem in control theory, International Symposium on non linear differential equations and non linear mechanics, Academic Press, New York (1961), 325-332.

[HS] M.L.I. Hautus and L.M. Silverman, Systems structure and singular control, Linear Algebra and Its Applications, 50 (1983), 369-402.

[J] V. Jurdjevic, Affine control systems, **Systems and Control Encyclopedia**, Vol 8, Pergamon Press.

[JK1] V. Jurdjevic and I. Kupka, Polynomial control systems, Math. Ann. 272 (1985), 361-368.

[JK2] V. Jurdjevic and I. Kupka, Linear systems with singular quadratic integral cost, to appear.

[KKR] P. Kokotović, H. Khalil, J. O'Reilly, **Singular Perturbation Methods in Control: Analysis and Design**, Academic Press (1986).

[Kog] J. Kogan, Bifurcation of extremals in optimal control, Lecture notes in mathematics, Springer-Verlag (1986), no. 1216.

[Kr] A. Krener, The High Order Maximum Principle and its applications to singular extremals, SIAM Journal on Control and Optimization, Vol 151, No. 2 (1977), 256-292.

[L] C. Lobry, Contrôlabilité des systèmes non-linéaire, SIAM J. Control 11 (1973), 670-676.

[N] Linear Differential systems with singularities and an application to transitive Lie algebras, J. Math. Society, Japan 18 (1966), 398-404.

Optimal Control Problems Arising in Large Deviation Theory

Efthimios Kappos, Professor, Department of Electrical and Computer Engineering, Northeastern University, Boston, Massachusetts

1 Introduction

The purpose of this paper is to present large deviation theory for diffusion processes as an important instance where a nonlinear optimal control setting arises naturally in physical problems. This control problem has many aspects that are novel and the techniques used to treat it are also new. We also aim to show that, by making assumptions on the dynamics that are justified on physical grounds, solutions can be obtained to control problems that are generally too difficult to solve. Throughout, we use a Hamiltonian setting for the nonlinear optimal control problem and approach it from a qualitative, topological point of view. It is a feature of the problem to be described that the more modern approaches to control (using for example generalized analysis or Lie techniques) are not needed. Thus, we find ourselves going back to *old methods* in order to solve *new problems*. What makes possible the success of our approach is the input from the theory of dynamical systems. It tells us of the staggering complexity of possible dynamical behavior, yet it also gives ways of classifying and systematically examining the *global* features of dynamical systems.

There are several application areas of this research, some of which we include in this paper. A natural small–noise context makes it necesary to study the large deviation behavior of nonlinear devices and circuits (thermal noise) and large, interconnected power systems subject to random load disturbances.

1.1 Background and Motivation

In the design of optimal control systems, cost functionals should be consistent with control objectives. The Linear Quadratic (LQ) Regulator, with state equation

$$\dot{x} = Ax + Bu \tag{1}$$

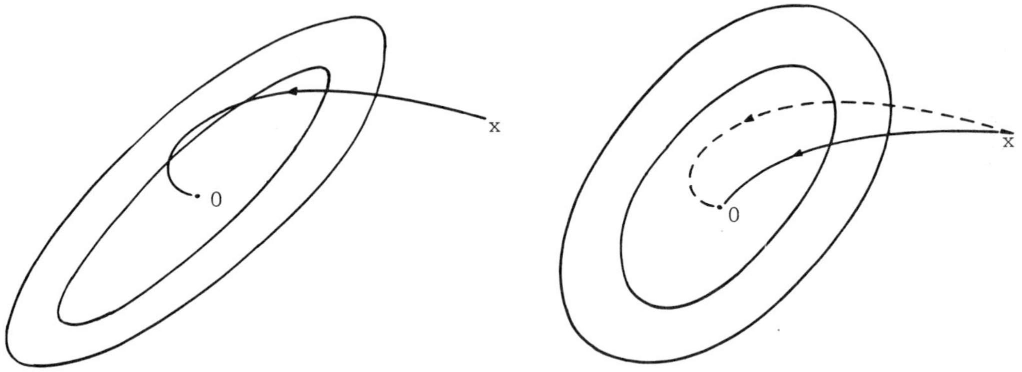

Figure 1: (a) Level sets of the state cost term $\frac{1}{2}x^TQx$ and a trajectory of the state dynamics $\dot{x} = Ax$ (the origin is a stable equilibrium for the case shown).

(b) Level sets of the optimal cost function $\frac{1}{2}x^TPx$ and the trajectory for the optimal feedback system $\dot{x} = (A - BK)x$.

is a good example. The quadratic cost functional with the form

$$J = \frac{1}{2}\int(x^TQx + u^TRu)dt \quad R > 0\ ,\ Q \geq 0 \tag{2}$$

has a first term penalizing deviations from the zero state and a second the control action. This functional translates the control objective of regulating the system state to the origin with minimum control effort (since we require this for all initial conditions, we are essentially requiring the origin to be asymptotically stable for the controlled system dynamics). It is then no surprise that the optimal regulator solution is a state feedback $u^*(x) = -Kx$ such that the resulting state equation

$$\dot{x} = (A - BK)x \tag{3}$$

is stable; furthermore, the optimal cost as a function of initial condition is a Lyapunov function for the feedback system (3) (under the assumptions that (A, B) is stabilizable and $(Q^{1/2}, A)$ is detectable, see [9]). In Figure 1(a), we show the level sets of the quadratic term $\frac{1}{2}x^TQx$ (which need not be a Lyapunov function for the dynamics $\dot{x} = Ax$) and in Figure 1(b) we show the level sets of the optimal cost function, which **is** a Lyapunov function for the dynamics (3).

Note that if the origin were a stable equilibrium for the state dynamics, the detectability condition would be trivially satisfied no matter what Q is. In other words, the state–weighting term is necessary only for performance purposes (speeding up the return to zero).

In other physical contexts, as we explain below, the control objective is to steer the system state **away** from a stable equilibrium point, with minimal control effort. A *'reasonable'* quadratic cost functional would then be obtained by replacing the nonnegative definite state–weighting matrix Q by a nonpositive definite one, to reward deviations from

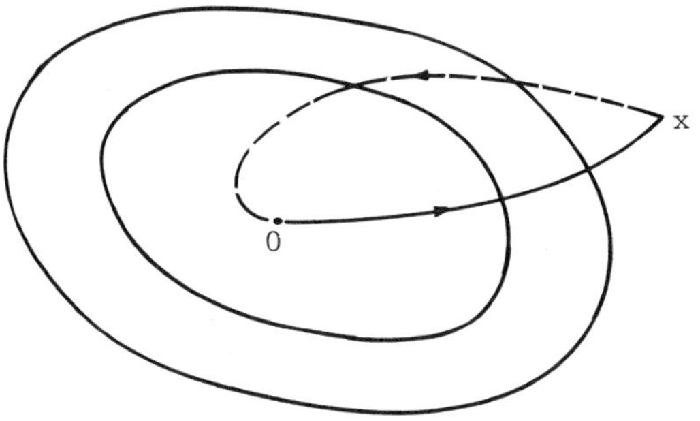

Figure 2: Level sets of the optimal cost function, which is a Lyapunov function for both the uncontrolled dynamics (dotted line) and the optimally controlled ones (solid line).

the zero state. In fact, since the equilibrium is assumed stable, we can consider a cost functional with only a quadratic cost–of–control. Furthermore, we will want to consider nonlinear systems and attractors more general than equilibrium points. It turns out that Lyapunov functions play a basic role, just as in the linear case before. Under a controllability condition (and some other technical assumptions), the optimal control is found to be in feedback form and the optimal cost functional can be defined as a function of the terminal point and is a Lyapunov function for **both** the uncontrolled and the optimally controlled dynamics (see Figure 2.)

1.2 Physical Basis for Nonlinear Optimal Control Problem

We now turn to a description of the circumstances under which the above nonlinear optimal control problem arises naturally. It is our claim that this control problem is fundamental in a manner similar to that in which classical mechanics is the natural approximation to more advanced physical theories such as quantum mechanics or statistical mechanics. In quantum mechanics, the classical mechanical particles furnish an approximation to the quantum mechanical description, in the limit as $h \downarrow 0$. In stochastic systems, paths of an associated optimal control system provide an approximation to the large deviation probabilistic problem, in the limit of small noise.

Indeed, consider stochastic systems that are described by stochastic differential equations of the form

$$dx_t^\epsilon = f(x_t^\epsilon)dt + \epsilon g(x_t^\epsilon)dw_t. \tag{4}$$

(The class of stochastic systems for which a large deviation theory can be developed is broader than the class of diffusion processes we present, but the control setting is most clear in this case and so we restrict ourselves to it in this paper.) As $\epsilon \downarrow 0$, with overwhelming probability the system trajectories follow closely the 'deterministic' trajectories of the system

$$\dot{x}_t = f(x_t) \tag{5}$$

over finite time–intervals. Paths other than those of (5) are extremely rare—the diffusion

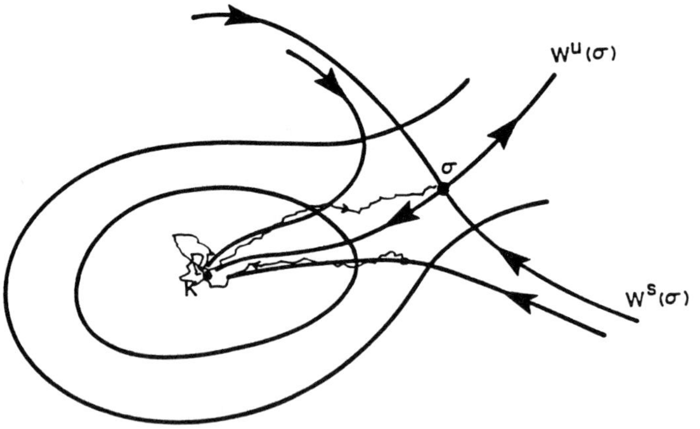

Figure 3: Possible large deviation behavior of path starting from a point in the region of attraction of the stable equilibrium x^0.

has exponentially small probability of following them. The exponent is of the form $\frac{-J}{\epsilon^2}$, where J is measuring the control energy needed to steer the state of the system

$$\dot{x}_t = f(x_t) + g(x_t)u \tag{6}$$

along the selected path. At this point it is necessary to assume something about the stability structure of the dynamics (5). This in fact amounts to separating a transient and a steady–state part of the dynamics. If, for example, f has an asymptotically stable equilibrium point x^0 and we take an initial condition for the diffusion in the region of attraction of x^0, the trajectory of the diffusion will first move towards x^0 and will stay close to it for a long time, since over finite time–intervals the diffusion stays close to the trajectories of (5). Occasionally, it will make large excursions away from x^0 that may lead to transitions from one region of attraction to another (see Figure 3).

Since all such paths are rare, we need a way of differentiating between them. This is done by the theory of *large deviations for diffusions* (see [3], [2]). First, let us notice that we need only consider paths that are obtained as controlled paths of the system (6). Now the probability of following a path that starts near x^0 and moves away from that point is measured by the cost–of–control functional

$$J(u) = \frac{1}{2} \int |u|^2 dt. \tag{7}$$

(the precise bounds are given, for example in [2] and [3]). The functional J is an **action functional** for this physical problem, just like the classical action. The important difference is that, in the probabilistic setting, it leads to a selections of the path with the **minimum cost** as the overwhelmingly likely path, asymptotically as $\epsilon \downarrow 0$, and not all its stationary points, as in the stationary phase method. This is a generalization of the **Laplace method** which we now explain in the simplest possible context.

Suppose we are given k independent events E_i, $i = 1, \ldots, k$ whose probabilities are exponentially small and given by

$$P(E_i) = e^{-\frac{J_i}{\epsilon^2}} \quad , \quad J_i > 0. \tag{8}$$

Clearly, $\lim_{\epsilon\downarrow 0} P(E_i) = 0$, $\forall i$. Assume the J_i are ordered as follows:

$$J_k \geq J_{k-1} \geq \cdots \geq J_2 > J_1, \tag{9}$$

and let the event E stand for $E_1 \cup \cdots \cup E_k$. Then we have the result that

$$\begin{aligned} &\lim_{\epsilon\downarrow 0} P(E_1/E) = 1 \\ &\lim_{\epsilon\downarrow 0} P(E_i/E) = 0 \quad , \quad i > 1. \end{aligned} \tag{10}$$

If the minimum is achieved by more than one event, a similar result holds. The proof follows easily by examining the ratio expressing the conditional probabilities above

$$P(E_i/E) = \frac{P(E_i)}{P(E)} = \frac{e^{\frac{-J_i}{\epsilon^2}}}{\sum_{j=1}^{k} e^{\frac{-J_j}{\epsilon^2}}}. \tag{11}$$

In the large deviation context, the E_i are the events that the diffusion follows a path of (6) that is not a path of f and the J_i are the values of the action functional.

In the next sections, we present the nonlinear optimal control results of the theory of large deviations for diffusions, as developed in [3], [2] and [4]. In section 2, a class of dynamical systems that has stability properties appropriate for the theory is defined and motivated by examples. Section 3 gives the optimal control and large deviation results and the examples are revisited in section 4. We close with a section summarizing the paper and drawing some conclusions. We keep the presentation brief and we aim to explain rather than to prove.

2 Dissipative Dynamics and Lyapunov Functions

It is well–known that nonlinear resistors can be used to model active electronic devices. Real devices, however, cannot generate energy indefinitely and, *eventually*, i.e. for large values of the circuit variables, will dissipate energy. This eventual passivity condition is reasonable on physical grounds, if only to effectively limit the range of operation of these devices, so that they are well–designed and do not exceed tolerance limits. A similar behavior should be expected of many other models of physical systems. In other words, real systems are dissipative, at least in the 'eventual' sense mentioned above. To make this notion precise mathematically is not trivial. In this paper, we specify a class of dynamical systems that has this dissipativeness property.

DEFINITION: *A vector field f in $\mathbf{R}^n$ will be called* **dissipative Morse–Smale** *if*
(i) f is a complete, smooth vector field that yields the global flow $\phi(t,x)$ for all $x \in \mathbf{R}^n$ and $t \in \mathbf{R}$.
(ii) It has a finite number of asymptotically stable attractors

$$\mathcal{A}(f) = \{K_1, \ldots, K_p\}$$

that are connected, compact submanifolds of $\mathbf{R}^n$ of dimension less than n.

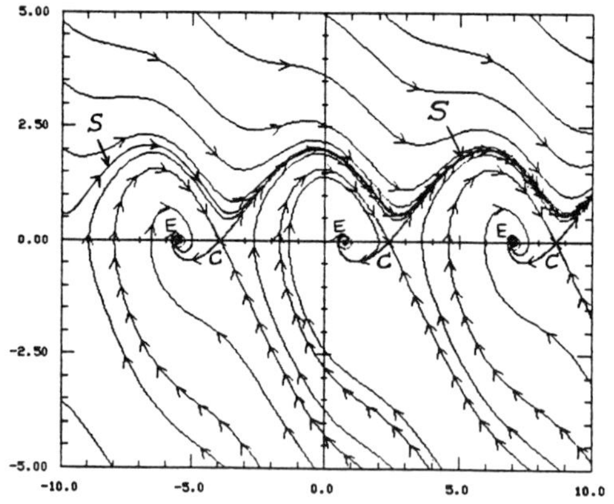

Figure 4: Phase portrait of the Josephson junction model (12) for $G = 0.5$ and $I = 0.7$.

(iii) There is a finite number of hyperbolic saddle critical elements (of index greater than zero)

$$\mathcal{C}(f) = \{\sigma_1, \ldots, \sigma_q\}$$

(iv) The limit set behavior is simple. With '∞' denoting the far–away points,

$$L_\alpha(f) = \mathcal{C}(f) \cup \{\infty\} \quad , \quad L_\omega(f) = \mathcal{C}(f) \cup \mathcal{A}(f) \text{and}$$

$$\Omega(f) = L_\alpha \cup L_\omega = \mathcal{A}(f) \cup \mathcal{C}(f) \cup \{\infty\}$$

and furthermore, the globally defined stable and unstable manifolds of the critical elements intersect transversely.

The sets $L_\alpha(f)$, $L_\omega(f)$ and $\Omega(f)$ are the alpha– and omega–limit sets and the non–wandering set of f (for details and complete definitions, see [3] and [6].) This class will be denoted $\mathcal{D}(\mathbf{R}^n)$ and is, of course, closely related to the class of Morse–Smale vector fields defined in a compact manifold. In fact, we chose to define this class in $\mathbf{R}^n$ and not on manifolds for simplicity. It agrees with intuitive notions of dissipativeness, since all trajectories either approach an attractor or a critical element, in positive time, and are either attracted to a critical element or run off to infinity, in negative time. In Figure 4, we plot the two–dimensional phase portrait of the Josephson junction model

$$\begin{aligned} \dot{x} &= y \\ \dot{y} &= -Gy - \sin x + I, \end{aligned} \tag{12}$$

in the range of the bifurcation parameters G and I where a stable equilibrium E coexists with a stable running orbit S, which is actually a limit cycle in the natural state manifold $S^1 \times \mathbf{R}$.

The notion of dissipativeness is intimately tied to the notion of Lyapunov functions. A Lyapunov function in engineering usage is a function defined near an equilibrium point that is nonincreasing on trajectories of the vector field. Its existence implies—and is implied by— the stability of the equilibrium point. For general nonlinear dynamical systems, the concept of a Lyapunov function is global. A Lyapunov function is a function defined in $\mathbf{R}^n \setminus \Omega(f)$ that is smooth and stictly decreasing along the paths of f. Several types of such functions exist for dissipative vector fields. They can be classified according to the ordering they induce on the set of critical elements and attractors of f (see [6]). The **orbit diagram** (or Smale diagram) of a vector field in $\mathcal{D}(\mathbf{R}^n)$ is of help in this classification. This diagram is a graph with levels according to index, nodes the attractors and critical elements and directed arrows signifying the existence of a trajectory with α–limit point the node of origin and ω–limit point the node of destination. Figure 5(a) gives two simple examples of orbit diagrams, while Figure 5(b) explains how to classify Lyapunov functions using the orbit diagram.

3 Optimal Control Problem and Large Deviation Results

In this section, we present the optimal control problem that arises in the small–noise limit of equation (4) when the drift vector field f is in the class $\mathcal{D}(\mathbf{R}^n)$. Solving this control problem amounts to finding the large deviation asymptotics, as was explained before. The results are a generalization of the ones in [2] in that the global picture is more precise (because of the assumption on f) and the diffusion matrix g is not required to be nonsingular for all x.

The connection between the diffusion process and the control system (6) was mentioned in the introduction. When the drift of the diffusion is in $\mathcal{D}(\mathbf{R}^n)$, for small ϵ we are likely to find the process near $\mathcal{A}(f)$ (since the critical elements have attracting sets of Lebesgue measure zero). For the diffusion to exit the region of attraction R_K of the attractor K, it must be possible to control the system (6) in directions away from K. This leads to a definition of controllability with respect to Lyapunov functions (see [8])

DEFINITION: *The control system (6) is called* **Lyapunov controllable** *if there is a Lyapunov function V for the vector field f such that the row vector of 1-forms $dV(g) \equiv (dV(g_i))_{i=1}^m$ is non-vanishing in R_K.*

Next we state the optimal control problem that is associated to the small–noise diffusion (4)

Optimal Control Problem: *Find the control u to minimize the cost functional $J_x(u) = \frac{1}{2}\int |u|^2 dt$ subject to the initial condition $x_0 \in K$ and the terminal condition $x_T = x \in R_K$, a point in the region of attraction of K.*

We actually have to be a little more careful: we allow the initial point to be on a closed ϵ–neighborhood of K in R_K and then let $\epsilon \downarrow 0$. The following result is from [3]:

THEOREM 1: *For the optimal control problem above and for every attractor K of f, suppose the control dynamics (6) is Lyapunov controllable for representatives of all types of Lyapunov functions in R_K and that for any $x \in R_K$ the optimal control is non-vanishing and the optimal trajectory has no equilibrium points.*

Then, there is a Lyapunov function V_K^* and a smooth feedback control u_K^*, both defined in R_K, such that

$$u_K^*(x) = g^T(x)\nabla V_K^*(x)$$

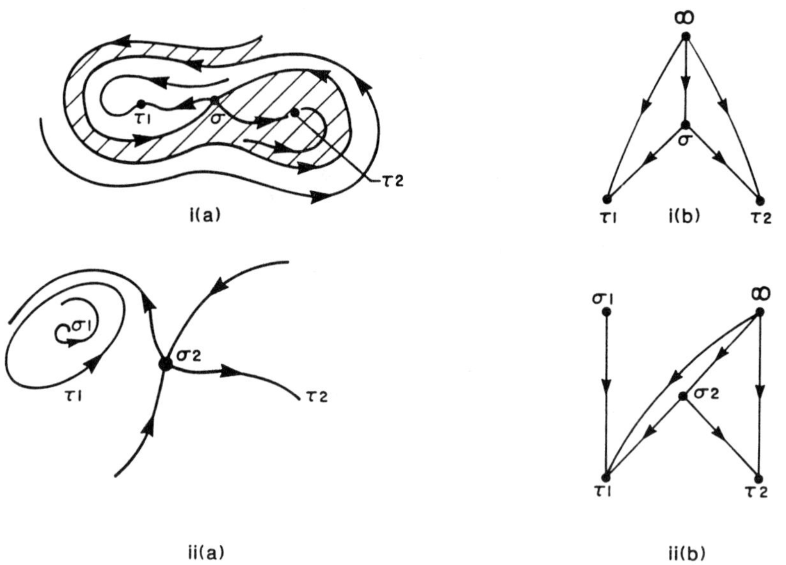

Figure 5(a): Phase portraits of two flows in $\mathcal{D}(\mathbf{R}^n)$ and their orbit diagrams.

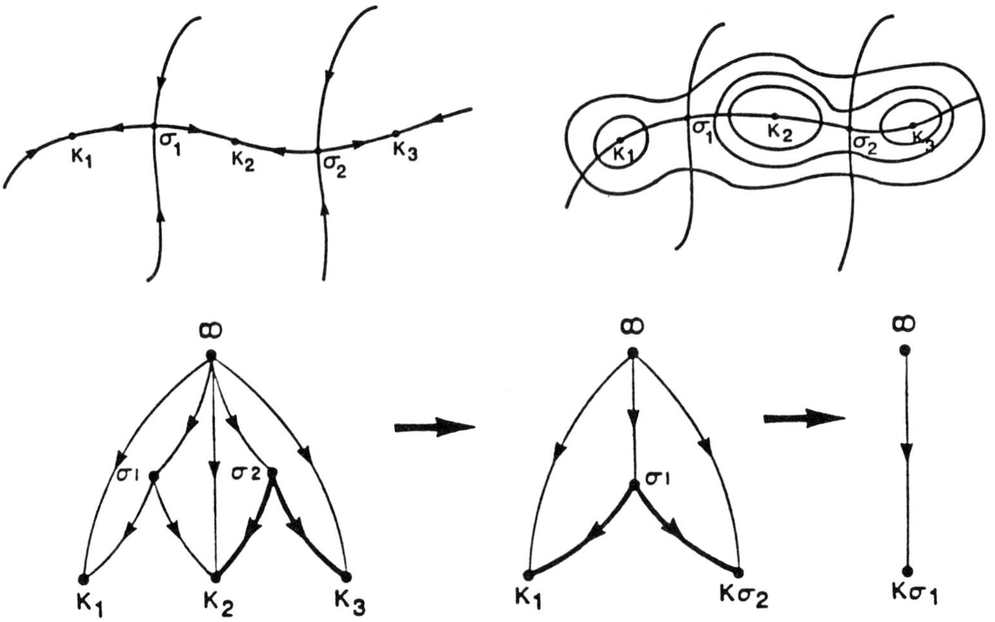

Figure 5(b): Correspondence between the Lyapunov function type and the grouping of nodes in the orbit diagram.

and the optimal trajectory to the point $x \in R_K$ is a trajectory of the system

$$\dot{x} = f(x) + g(x)g^T(x)\nabla V_K^*(x).$$

The Lyapunov function is also the optimal cost functional

$$V^*(x) = \inf_u J_x(u)$$

for the optimization problem. The optimal cost for exiting R_K is equal to $V_K^*(s^*)$, where s^* is the saddle critical element on $\partial \bar{R}_K$ minimizing V_K^* there.

(We only consider exits from R_K into another region of attraction.) This leads to the large deviation result

THEOREM 2: *For the diffusion process (4), suppose that the conditions of theorem 1 are satisfied for the associated control system (6). For every attractor K of f, denote V_K^* the Lyapunov function that solves the optimal control problem in R_K and s_K^* the saddle critical element minimizing V_K^* on $\partial \bar{R}_K$.*

Then, for any $\delta > 0$ and $x \in R_K$, if the first exit time from R_K is

$$\tau_K^\epsilon \equiv \inf\{t\,;\, x_t^\epsilon \notin \bar{R}_K\},$$

$$\lim_{\epsilon \downarrow 0} P\{\, |x_{\tau_K^\epsilon}^\epsilon - s_K^*| < \delta\} = 1$$

and the asymptotic exit time satisfies

$$\lim_{\epsilon \downarrow 0} \epsilon^2 \ln E_x \tau_K^\epsilon = V_K^*(s_K^*).$$

Remarks:

1) Note that we did not have to specify a domain for the exit problem, as is done usually, since we have available the region of attraction and also Lyapunov surfaces that are transverse to the flow and contained in it.

2) In the case that the saddle s_K^* is a limit cycle, the distance defined in the theorem is of course the minimum of the distances of points in the limit cycle.

3) Finally, note that the results are independent of the starting point x in the region of attraction. This reflects the fact that, whatever the starting point, we first hit, with overwhelming probability, a neighborhood of the attractor K. It remains true in the singular case because, the controllability problem is divided in two parts: going towards and away from the attractor.

4 Examples

We give briefly two applications of the theory of the previous sections: the Josephson junction under thermal noise and a voltage stability problem for large, interconnected power systems.

The Josephson junction model given in section 2 has two attractors: the stable equilibrium E and the limit cycle S. The introduction of small thermal noise gives the state equations

$$\begin{aligned} \dot{x} &= y \\ \dot{y} &= -Gy - \sin x + I + \epsilon G n, \end{aligned} \tag{13}$$

where n is 'white–noise'. Note that g is singular and the usual large deviation theory is not applicable. Small noise can cause transitions from E to S, and we would like to find out which is more stable (in the sense that the invariant measure would concentrate on it, see [2]).

For exit from R_E, we can use the energy function

$$T(x,y) = \frac{1}{2}y^2 + U(x), \tag{14}$$

where $U(x) = -\cos x - Ix$. The function T solves the Hamilton–Jacobi equation for the associated optimal control problem and yields the optimal feedback

$$u^*(x,y) = 2Gy. \tag{15}$$

The large deviation exit is from the saddle equilibrium C. For exit from S, however, T is of no use, since it is not constant on S and the optimal control problem dictates that the optimal cost function should be constant on the attractors. The theory of section 3 is necessary. It can be shown (see [7]) that the Lyapunov controllability condition is satisfied and hence we have

THEOREM: *There is a Lyapunov function V defined in a neighborhood of R_S such that the optimal control problem is solved by the feedback control*

$$u^*(x,y) = 2Gp_y(x,y), \tag{16}$$

with $p_y = \frac{\partial V}{\partial y}$.

Optimal exit from R_S is from the saddle C.

The relative stability of the attractors E and S is measured by the optimal cost for exit from R_E and R_S respectively. For the case shown in Figure 4, it turns out that it takes less energy to exit from R_S and hence E is the more stable attractor (a fact that is by no means obvious from the phase portrait).

The power system model we will consider is taken from [1] (see also [5]). The state is composed of the vector of *generator angular velocities* ω, the vector of *relative phase angles* α and the vector of *bus voltage magnitudes* E:

$$x = \begin{bmatrix} \omega \\ \alpha \\ E \end{bmatrix}.$$

The state equation takes the form of the stochastic differential equation

$$dx_t = A\nabla V(x_t)dt + \epsilon(-A - A^T)^{1/2} dw_t, \tag{17}$$

where the matrix A is given by:

$$A = \begin{bmatrix} 0 & -M^{-1}T_1^T & 0 \\ T_1 M^{-1} & -T_2 D_\ell^{-1} T_2^T & 0 \\ 0 & 0 & -D_v^{-1} \end{bmatrix} \tag{18}$$

and V is the function:

$$V(x) \equiv \frac{1}{2}\omega^T M \omega - \frac{1}{2}\sum_{i=0}^{n}\sum_{k=0}^{n} B_{ik}E_iE_k\cos(\alpha_i - \alpha_k) +$$

$$+\frac{1}{2}\sum_{i=0}^{n}\sum_{k=0}^{n} B_{ik}E_i^0E_k^0\cos(\alpha_i^0 - \alpha_k^0) +$$

$$+\sum_{k=0}^{n}\int_{E_k^0}^{E_k} \frac{Q_{D,k}(x)}{x}dx - P^T(\alpha - \alpha^0). \tag{19}$$

Note that here: $f(x) = A\nabla V(x)$ and $g = (-A - A^T)^{1/2}$ is a constant matrix (independent of x). By construction, the matrix A is non-singular and negative definite and it is the sum of a diagonal and a skew-symmetric matrix (for details, see [1]). This implies that the equilibrium points of the *deterministic dynamics*:

$$\dot{x} = A\nabla V(x) \tag{20}$$

are precisely the points where $\nabla V(x) = 0$. To apply large deviation theory, we assume that ϵ is vanishingly small (exactly how small it is, is a matter of experiment and in any case the theory will only give asymptotic bounds). This is our way of modelling small load disturbances. Note that the diffusion matrix g is singular; in particular, the ω part of the state vector is unaffected by noise. Hence the results of Wentzell and Freidlin [2] are not applicable. The *associated control problem* has the dynamics:

$$\dot{x} = A\nabla V(x) + (-A - A^T)^{1/2}u, \tag{21}$$

In order to use the results of the previous section, we must check that the vector field satisfies the dissipative properties (i)-(iv) and that the diffusion is Lyapunov controllable for some Lyapunov function. In fact, it is enough to restrict attention when necessary to an open set containing the region of attraction of the stable operating point. That the vector field is complete is an easy conclusion from the state equations, since the sinusoidal terms are bounded and the rest grows no faster than linearly. The existence of the asymptotically stable equilibrium and the fact that the saddle equilibria on its boundary are hyperbolic are established easily from the form of the function V. The other conditions are more difficult to check, but it can be assumed, by appealing to genericity, that there will at least be parameter regions where this will be true.

As for the existence of a Lyapunov function making the diffusion controllable, this is immediate using the function V itself and checking first that it is a Lyapunov function:

$$f^T(x)\nabla V(x) = \nabla V^T(x)A^T\nabla V(x) < 0, \tag{22}$$

since A is negative definite and then that

$$\|g(x)\nabla V(x)\|^2 = \nabla^T V(x) g(x) g^T(x) \nabla V(x) \neq 0 \tag{23}$$

in our case:

$$\nabla^T V(x)(A + A^T)\nabla V(x) = 2\nabla^T V(x) A \nabla V(x) > 0 \tag{24}$$

The remarkable fact is that V *is also the optimal cost function of the associated control problem*, since it satisfies the Hamilton–Jacobi equation for the optimization problem. The optimal control paths that are obtained from V satisfy:

$$\begin{aligned} \dot{x} &= A\nabla V(x) + (-A - A^T)\nabla V(x) = \\ &= -A^T \nabla V(x). \end{aligned} \tag{25}$$

In the closure of the region of attraction of the operating point, the equilibria of this system are the same as those of the deterministic dynamics (20). However, the stability of each is 'reversed': the attractor becomes a repellor and a k-saddle an $(m - k)$ saddle (where m is the dimension of the state vector). We have thus arrived at the following result:

THEOREM: *The power system dynamics has a dissipative drift and is Lyapunov controllable with respect to the Lyapunov function V. Let Σ be the set of saddle equilibria in the subset of the boundary $\partial \overline{R_K}$ and s^* the $(n - 1)$ saddle in Σ minimizing V in this set.*

Then V is also the optimal control function and solves the optimal control problem in R_K by the feedback control:

$$u^*(x) = (-A - A^T)^{1/2} \nabla V(x).$$

The large deviation exit from R_K is from a neighborhood of s^* with probability one and the expected exit time satisfies:

$$\lim_{\epsilon \downarrow 0} \epsilon^2 E_x(\tau^\epsilon) = V(s^*).$$

5 Summary and Conclusions

An important nonlinear optimal control problem arises naturally in small–noise problems for stochastic processes. This control problem has the aim of controlling with least control energy away from attractors. Physical considerations, and input from dynamical systems theory make it reasonable to make a dissipativeness assumption on the dynamics. This in turn makes possible the solution of the nonlinear optimization problem, at least qualitatively. Global Lyapunov functions are an extremely useful tool in the analysis of nonlinear dynamical systems, even when control action is considered. Finally, system theory can benefit from a topological approach and from the insights of the mathematical theory of dynamical systems.

REFERENCES

[1] C.L. DeMarco, A. Bergen: *'A security measure for random load disturbances in nonlinear power system models'*, IEEE Trans. on Circ. and Systems, **34**–12, 1987, pp.1546–1557.
[2] M. Freidlin, A. Wentzell: *Random Perturbations of Dynamical Systems*, Springer 1984.
[3] E. Kappos: *'Large deviations for dissipative dynamics: an optimal control approach'*, Univ. of California, Berkeley, ERL Memo, M86/86, 1986.
[4] E. Kappos: *'Large deviations for singular diffusions with dissipative drift'*, *invited paper*, Princeton Conference on Information Systems and Control, 1988.
[5] E. Kappos: *'Large deviation theory applied to power system stability'*, IEE Conference 'Control '88', Oxford, 1988.
[6] E. Kappos, S. Sastry: *Global Lyapunov function theory*, University of California, Berkeley, ERL Memo M86/81, 1986.
[7] E. Kappos, S. Sastry: *'Large deviation exits from a running orbit in the Josephson junction'*, IEEE Symposium on Circuits and Systems, Helsinki, 1988.
[8] E. Kappos, S. Sastry: *'Lyapunov controllability and global optimal control'*, 25th Control and Decision Conference, Athens, 1986.
[9] H. Kwakernaak, R. Sivan: *Linear Optimal Control Systems*, Wiley 1972.

Periodic and Aperiodic Orbits in the Hamiltonian Formulation of a Model Magnetic System

Charles Kaufman, Professor, Physics Department, The University of Rhode Island, Kingston, Rhode Island

N. Srivastava, Physics Department, The University of Rhode Island, Kingston, Rhode Island

G. Müller, Physics Department, The University of Rhode Island, Kingston, Rhode Island

I. INTRODUCTION

Classical dynamics and quantum dynamics have influenced each other since the idea of a quantum mechanics originated. Classical dynamics came first, so its influence on quantum theory almost goes without saying. Quantum mechanics grew out of classical mechanics. The converse influence is often referred to in the abstract, but rarely in detail. One finds statements[1] roughly to the effect that the classical theory was developed more fully in order to use it to further elucidate the corresponding quantum dynamics. But specific examples of classical calculations, which were suggested by quantum results or ideas, are not common. One of these rare examples[2] is "rotators" or "classical spins," and that is the subject considered here.

This study makes contact with optimization theory at two places. First, the spin problem is initially expressed as an application of Hamilton-

---

This work was supported by the U. S. National Science Foundation Grant No. DMR-86-03036 and by a Grant-in-Aid of Research from Sigma Xi, The Scientific Research Society.

ian dynamics; that is, it is simply an explicit particular example of the principle of least action. In the course of solution, we uncover two qualitatively different types of behavior, viz "regular" and "chaotic," whose occurrence depends on the value of a (control) parameter. The chaotic solutions, moreover, are not equally chaotic; there is a more-or-less smooth progression into and back out of chaos as the parameter changes. The second contact with control theory then is a question: can the "intensity" of the chaos be quantified, and if so, is there a value of the parameter for which the system is maximally chaotic?

The paper has four sections. Section II is a review of classical dynamics, including a description of numerical techniques for distinguishing regular from chaotic motion. Section III describes how the quantum mechanical form of a classical dynamics problem is produced. Section IV discusses the exchange-type interactions relevant to classical spins and presents results of numerical integration for one specific such model.

II. CLASSICAL DYNAMICS

A problem in classical dynamics can be expressed in the Newtonian, Lagrangian, or Hamiltonian formalism.[1,3] Problems differ from each other by (or are specified by) choice of force, Lagrangian, or Hamiltonian, respectively. The formalism of the first two cases produces n, 2nd order ordinary differential equations, and, for the third, 2n, 1st order ordinary differential equations. n is the number of "degrees of freedom" of the problem. In each case, 2n initial values are needed. The problem is "integrable" if explicit functions $q_i(q_{io}, \dot{q}_{io}, t)$ can be found which give the coordinates, at any time, in terms of the initial values and of t; if the functions are not known, but can be expressed as definite integrals, the problem has been merely "reduced to quadrature," but is still integrable. In the case of the Hamiltonian formulation, a necessary and sufficient condition for integrability is the existence of n independent functions I_i which are "in involution" (defined below) with each other.

If the system is integrable, then all initial conditions lead to solutions which are periodic or quasiperiodic (a sum of periodic, but incommensurable, terms); there are n, possibly degenerate, frequencies. If the system is non-integrable, then at least some initial conditions lead to aperiodic solutions; these aperiodic solutions are the famous chaotic orbits of the recent literature. The so-called KAM theorem[4] states (roughly) that if an integrable Hamiltonian system is perturbed by a small non-integrable term then a finite fraction of the initial condition space leads to periodic or quasiperiodic solutions; i.e., regularity persists in the face of a small departure from integrability. The quasiperiodic trajectories lie on surfaces called KAM tori, by generalization from the $n = 2$ case: there the phase space has 4 dimensions; if there is a conserved energy, as there usually is, the motion lies in a 3-d subspace; and if doubly, but incommensurably, periodic, covers a 2-d toroidal surface .

Whether a particular initial configuration leads to a regular or to a chaotic orbit cannot be decided easily. One possibility is to measure the divergence of two nearby trajectories: it is linear for regular ones, exponential for chaotic ones. We use here the method of the Poincaré surface of section, or Poincaré cut. The system is numerically integrated, and the intersections of the trajectory with a specified surface in the phase space (the cut plane) is graphed. The graphs are finite sets of points for periodic trajectories, smooth curves for quasiperiodic ones, and swarms of points for chaotic ones.

This technique is illustrated in figures 1 and 2. Figure 1 shows two trajectories and figure 2 the intersections of these with the plane $\theta_2 = \pi/2$, for the system

$$\dot{\theta}_1 = J \sin\theta_2 \sin(\phi_1 - \phi_2) + (A_x - A_y)\cos\phi_1 \sin\phi_1 \sin\Theta_1,$$

$$\begin{aligned}\dot{\phi}_1 = {} & (J/\sin\theta_1)\sin\theta_2 \cos\theta_1 \cos(\phi_1 - \phi_2) \\ & - \cos\theta_1 (A_x \cos^2\phi_1 + A_y \sin^2\phi_1) ,\end{aligned} \tag{1}$$

and $1 \rightleftarrows 2$.

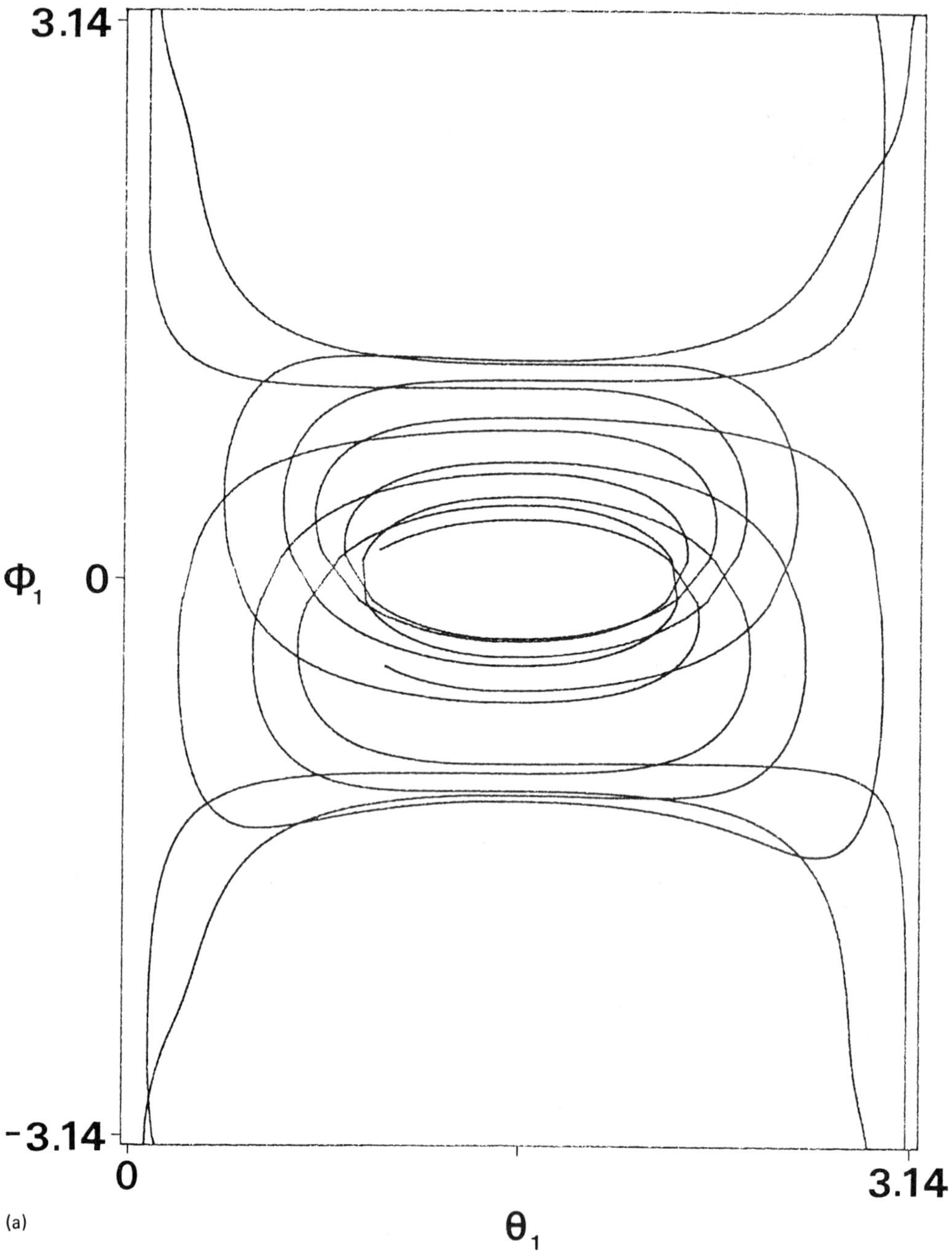

(a)

FIGURE 1

(a) Trajectory for the system given by Eq. (1) for initial data $(\theta_1 \theta_2 \phi_1 \phi_2)$ = (1.00, 2.00, 3.00, 1.26), projected onto the (θ_1, ϕ_1) plane, for $0 \leq t \leq 50$.

(b) As in (a), but with $(\theta_1 \theta_2 \phi_1 \phi_2)$ = (1.00, 2.00, 3.00, 1.29).

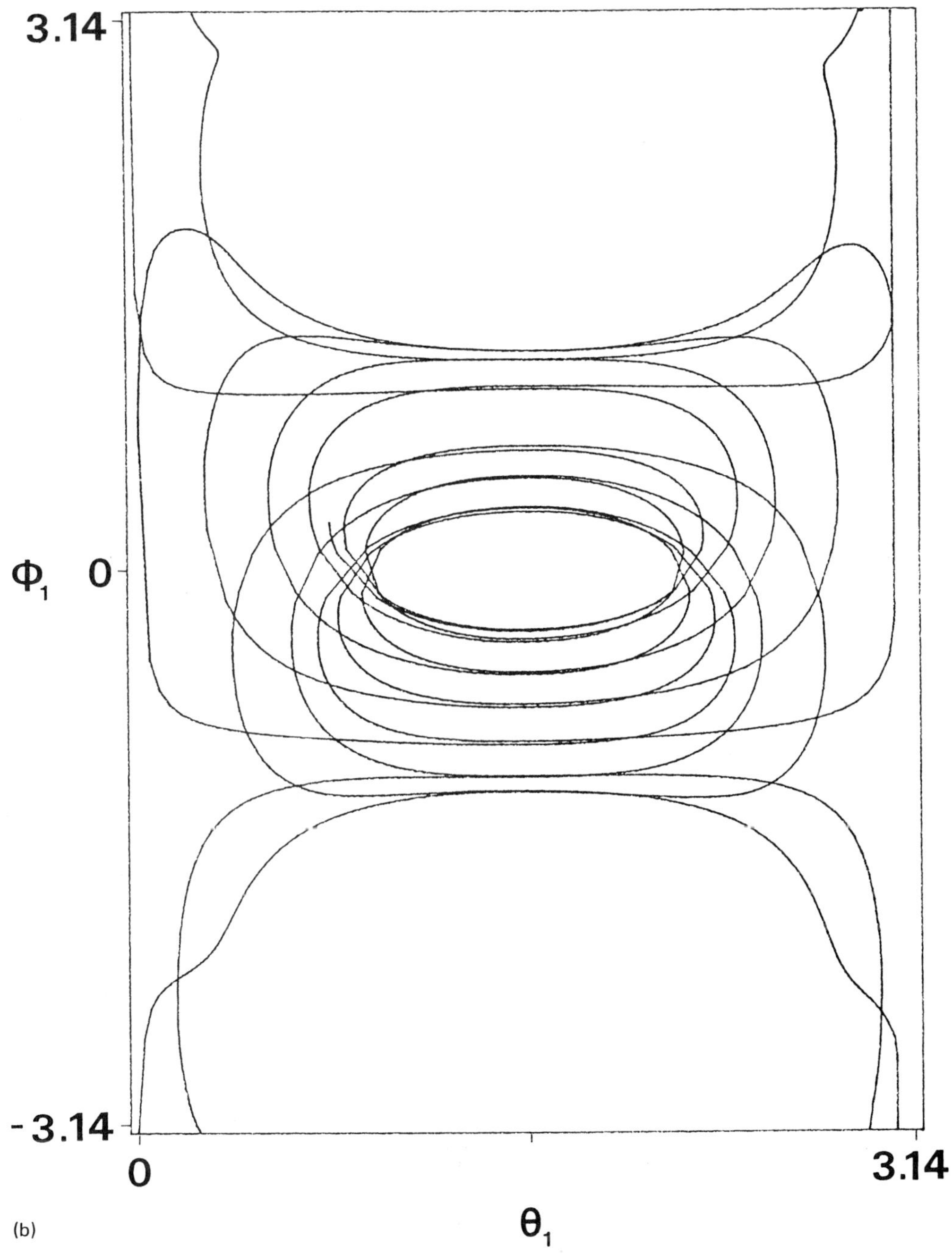

(b)

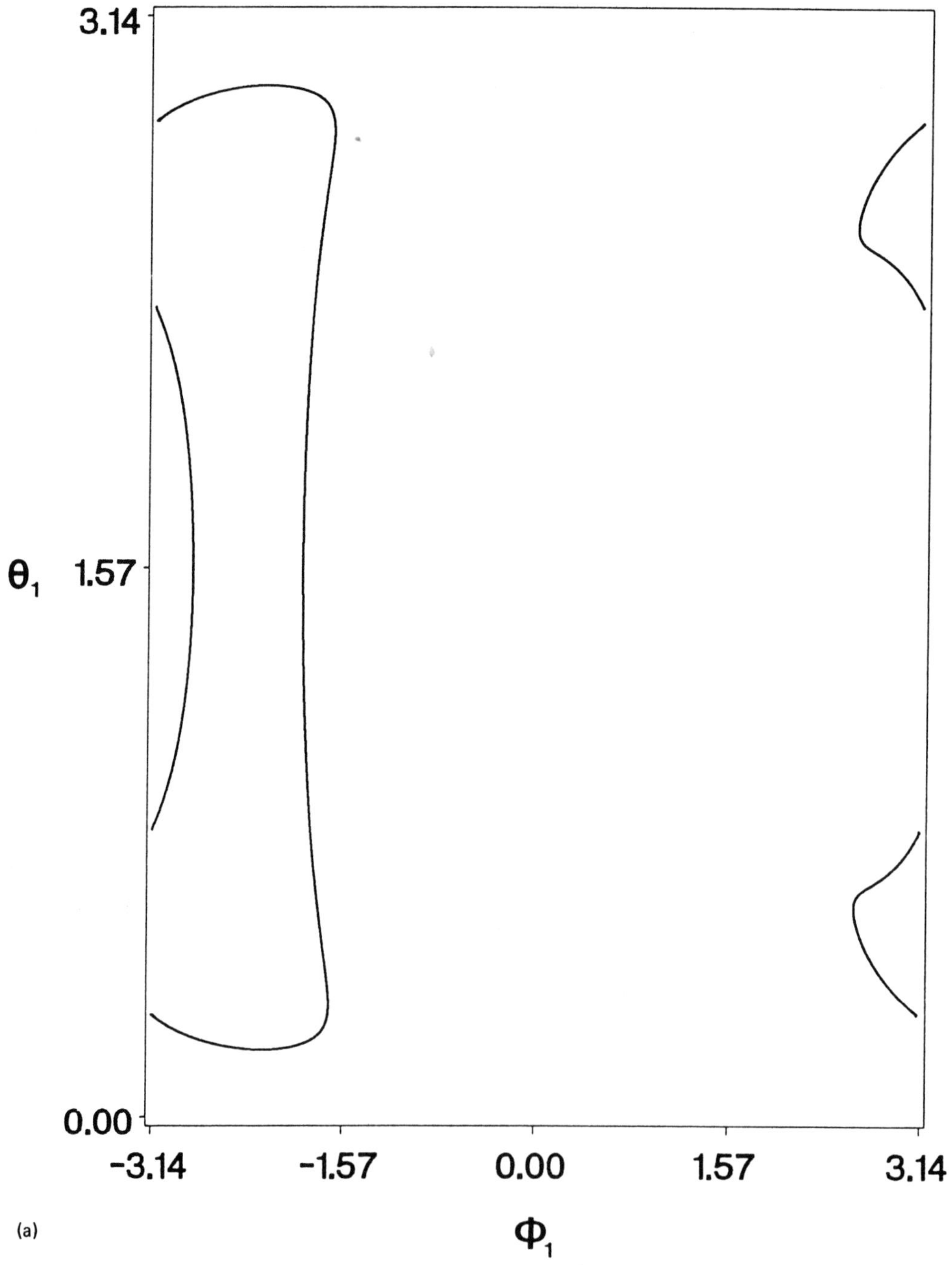

FIGURE 2

(a) Intersection of the trajectory of Fig. 1a with the plane $\theta_2 = \pi/2$, $0 \leq t \leq 18000$.

(b) Intersection of the trajectory of Fig. 1b with the plane $\theta_2 = \pi/2$, $0 \leq t \leq 18000$.

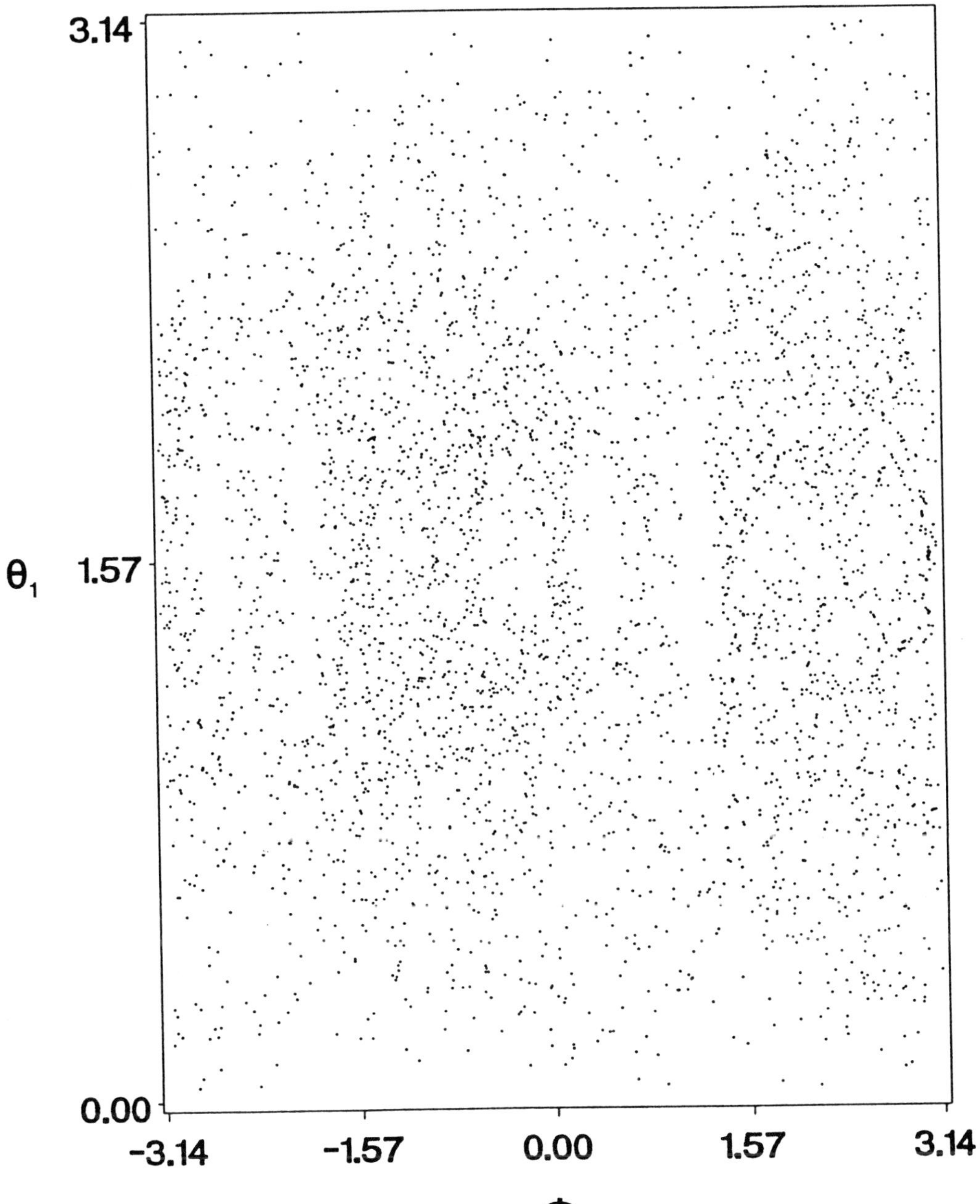

(b)

The initial conditions are $\theta_{10}=1.00$, $\theta_{20}=2.00$, $\phi_{10}=3.00$, and $\phi_{20}=1.26$ for (a), $= 1.29$ for (b). The parameters J, A_x, A_y are 1,2,-1, respectively. Study of the trajectories[5] (Fig. 1) gives no clue to the nature of the orbits: the pictures are effectively indistinguishable. Study of the intersections (Fig. 2) is hardly needed, nor is the reason for the use of the word chaos!

Given a system described by a set of generalized coordinates q_i, $i = 1, \ldots n$, the Lagrangian L is [kinetic energy - potential energy], expressed as a function of the q_i, their time derivatives, and time. The momentum canonically conjugate to q_i, called p_i, is given by

$$p_i = \frac{\partial L}{\partial \dot{q}_i}(q_i, \dot{q}_i, t) \tag{2}$$

The Hamiltonian function is

$$H = \sum_{i=1}^{n} p_i \dot{q}_i - L \tag{3}$$

expressed in terms of p_i, q_i and t.

The equations of motion for the system, derivable from the principle of least action, are then

$$\dot{q}_i = \frac{\partial H}{\partial p_i}, \quad \dot{p}_i = -\frac{\partial H}{\partial q_i} \tag{4}$$

The time derivative of any function of q_i, p_i, t is given by

$$\frac{dF}{dt} = \frac{\partial F}{\partial t} + \{F, H\}, \tag{5}$$

where

$$\{A, B\} \equiv \frac{\partial A}{\partial q_i}\frac{\partial B}{\partial p_i} - \frac{\partial A}{\partial p_i}\frac{\partial B}{\partial q_i}, \tag{6}$$

the "Poisson Bracket" of A and B. Two functions whose Poisson Bracket is zero are said to be "in involution" with each other.

III. FROM CLASSICAL TO QUANTUM AND BACK

In quantum mechanics, dynamical variables are represented by (possibly) non-commuting operators. The prescription[6] for forming a quantum problem that corresponds to a specific classical one is the replacement of Poisson brackets by commutators:

$$\frac{dA_c}{dt} = \{A, H_c\} + \frac{\partial A_c}{\partial t} \tag{7}$$

becomes

$$\frac{dA_q}{dt} = i\,[A_q, H_q] + \frac{\partial A_q}{\partial t}, \tag{8}$$

where $[A, B] \equiv AB - BA$ is the commutator of A and B. Here we deal with a problem in which the correspondence is used in the other direction; the classical problem is suggested by a quantum one; we know the commutator, and contrive classical variables whose Poisson brackets have the same algebraic structure. The Poisson bracket need not be explicitly dealt with; dq/dt can be written directly if the commutator is known in terms of the variables of the problem.

In the applications discussed here, the quantum variables are spin operators. Each one is a 3-vector and can be considered to represent an elementary magnet. If there are several spins, each is represented by its own vector (S_i^x, S_i^y, S_i^z). They satisfy the commutation rule

$$[S_i^x, S_j^y] = i\delta_{ij}\, S_i^z \quad (x, y, z \text{ cyclic}) \tag{9}$$

A specific problem corresponds to a specific choice of Hamiltonian

$H(\vec{S}_1, \vec{S}_2, \vec{S}_3 \ldots)$. For example, the simplest interaction between two spins is

$$H = -J\vec{S}_1 \cdot \vec{S}_2 \quad ,$$

one spin in an external field $B\hat{e}_z$ has

$$H = -BS^z;$$

etc.

A simple calculation shows that

$$\frac{d}{dt}[S_i^{x\,2} + S_i^{y\,2} + S_i^{z\,2}] = 0$$

for general H, and therefore each spin is a fixed length vector with only two independent components. For a particular H, then, the time development of the spin is given by

$$\dot{S}_i^x = i[S_i^x, H], \; \dot{S}_i^y = i[S_i^y, H], \; \dot{S}_i^z = i[S_i^z, H] \qquad (10)$$

(for simplicity we presume no explicit time dependence).

Now each commutator can be reduced to primitive commutators $[S_i^{\alpha}, S_j^{\beta}]$ and the replacement of each of these using equations (9), together with reinterpretation of the variables as classical ones, produces the classical problem from the quantum one.

This method of effecting the change from quantum to classical does not make explicit use of the canonical coordinates and momenta p and q. But if these are chosen to be cos θ and ϕ, respectively, and if the components of the spin vectors are given in terms of θ and ϕ by

$$S^x = \sin\theta \;\; \cos\phi$$

$$S^y = \sin\theta \, \sin\phi \qquad (11)$$

$$S^z = \cos\theta ,$$

then the identical equations of motion follow from Hamiltonian's equations (4).

IV. CHOICE OF INTERACTION

The energy of one spin $\vec{S}_1$, interacting in a spherically symmetric way with a second, of spin $\vec{S}_2$, is $\vec{S}_1 \cdot \vec{S}_2$. A more general form for this interaction is

$$\sum_{i,j=1}^{n} J_{ij}\vec{S}_i\cdot\vec{S}_j + \sum_{i=1}^{n} \vec{h}\cdot\vec{S}_i .$$

Specific choices of J_{ij} and $\vec{h}$ are known by particular names. For example, $H = -\vec{S}_1\cdot\vec{S}_2$ is the two-spin Heisenberg model, $H = -S_1^z S_2^z$ the two spin Ising model, $H = -S_1^x S_2^x - S_1^y S_2^y$ the two-spin X-Y model, etc. Models defined by such Hamiltonians are known as lattice models because there is no kinetic energy in them ; the primitive objects act as if fixed in position, nor do they have energy associated with the motion of changing orientation. And because we deal only with a small number of objects here, we speak of spin clusters rather than lattices. No explicit lattice structure is involved in our examples, however; the distance between spins plays no role. We have considered the dynamics for various J_{ij}, $\vec{h}$ in previous work[2,7]; here we present the results for the two-spin X-Y model with single-site anisotropy, defined by

$$H = -S_1^x S_2^x - S_1^y S_2^y + \frac{\alpha}{2}\,[S_1^{x^2} + S_2^{x^2} - S_1^{y^2} - S_2^{y^2}], \tag{12}$$

for $0 < \alpha < 1$.

The equations of motion which follow from this H are

$$\dot{S}_1^x = S_1^z S_2^y + \alpha S_1^y S_1^z$$

$$\dot{S}_1^y = S_1^z S_2^z + \alpha S_1^x S_1^z \qquad (13)$$

$$\dot{S}_1^z = S_1^x S_2^y - S_1^y S_2^x - 2\alpha S_1^x S_1^y$$

and $1 \rightleftarrows 2$.

The problem is integrable for $\alpha = 0$ and for $\alpha = 1$. H provides one of the two necessary invariants; the second invariant is

$$S_1^z + S_2^z$$

for the case $\alpha = 0$; for $\alpha = 1$,

$$- S_1^x S_2^x + S_1^y S_2^y + S_2^z S_2^z$$

has zero time derivative and commutes with H.

From $\alpha \neq 0$ or 1, the problem must be numerically integrated. Figures (3a) through (3f) show the results of numerical integration of these equations for $\alpha = 0$, 0.1, 0.5, 0.7, 0.85, and 1.0, each for a variety of initial conditions, for all of which H has the constant value -.09957501. Plotted are the intersections of the trajectories, with the plane $\theta_2 = \pi/2$, for which $\frac{d\theta_2}{dt} < 0$, projected onto the (ϕ_1, θ_1) plane. The progression from integrability, into chaos, and back out to integrability, as α varies from 0 to 1, is evident. For $\alpha \simeq 0$, and $\alpha \simeq 1$, the equations are "nearly" integrable, and almost all initial conditions shown lead to regular orbits, or tori. For $\alpha = 0.7$, a large fraction of the initial condition space apparently leads to chaotic orbits. For intermediate α, there is an intermediate fraction of the space covered by chaotic orbits.

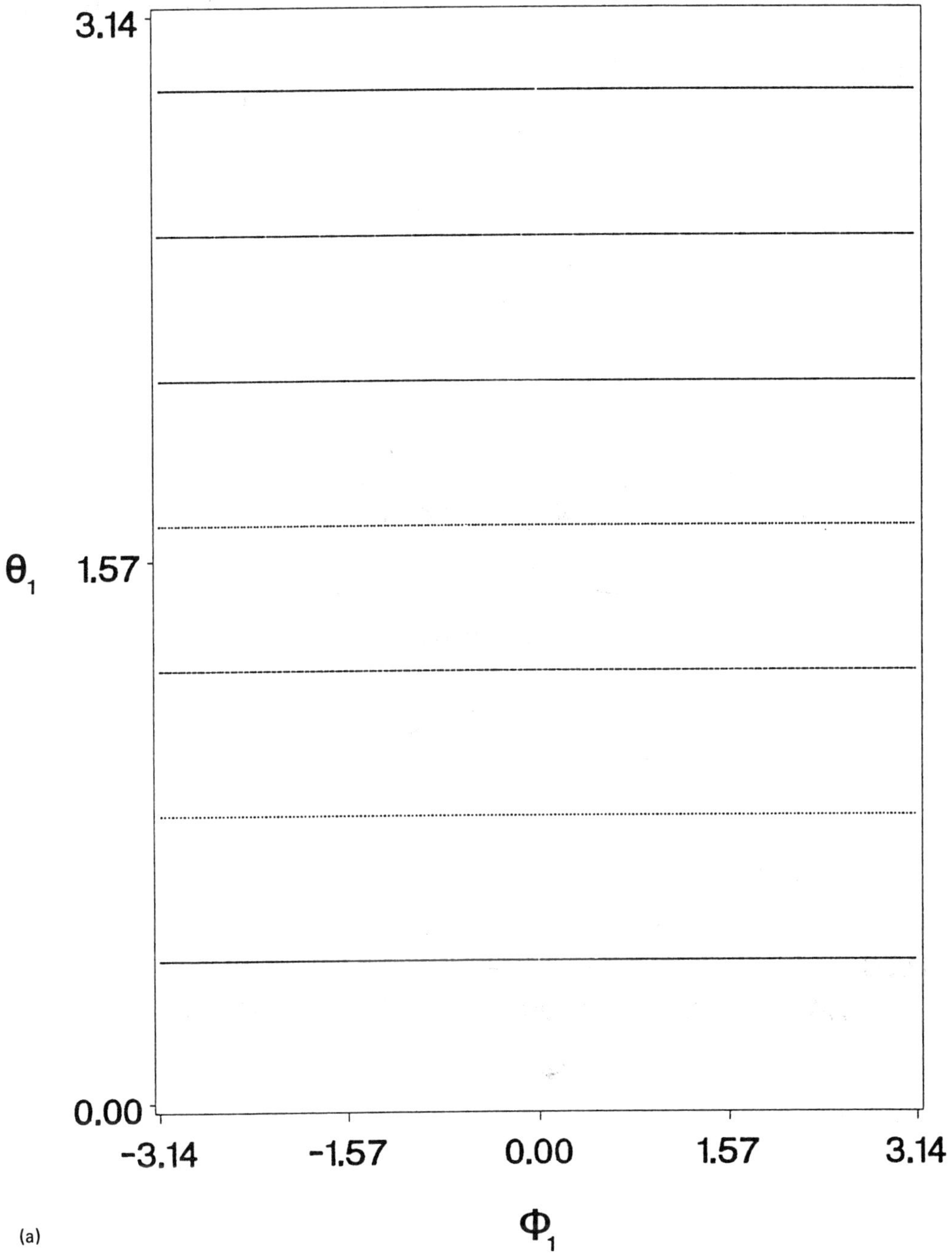

(a)

FIGURE 3

Projection onto the $(\theta_1\phi_1)$ plane, of the intersection of the trajectory of the system given by Eq. (13) with the plane $\theta_2 = \pi/2$. Only intersections for which $(d\theta_2/dt) < 0$ are plotted.

(a) $\alpha = 0.0$ (b) $\alpha = 0.1$ (c) $\alpha = 0.5$ (d) $\alpha = 0.7$
(e) $\alpha = 0.85$ (f) $\alpha = 1.0$

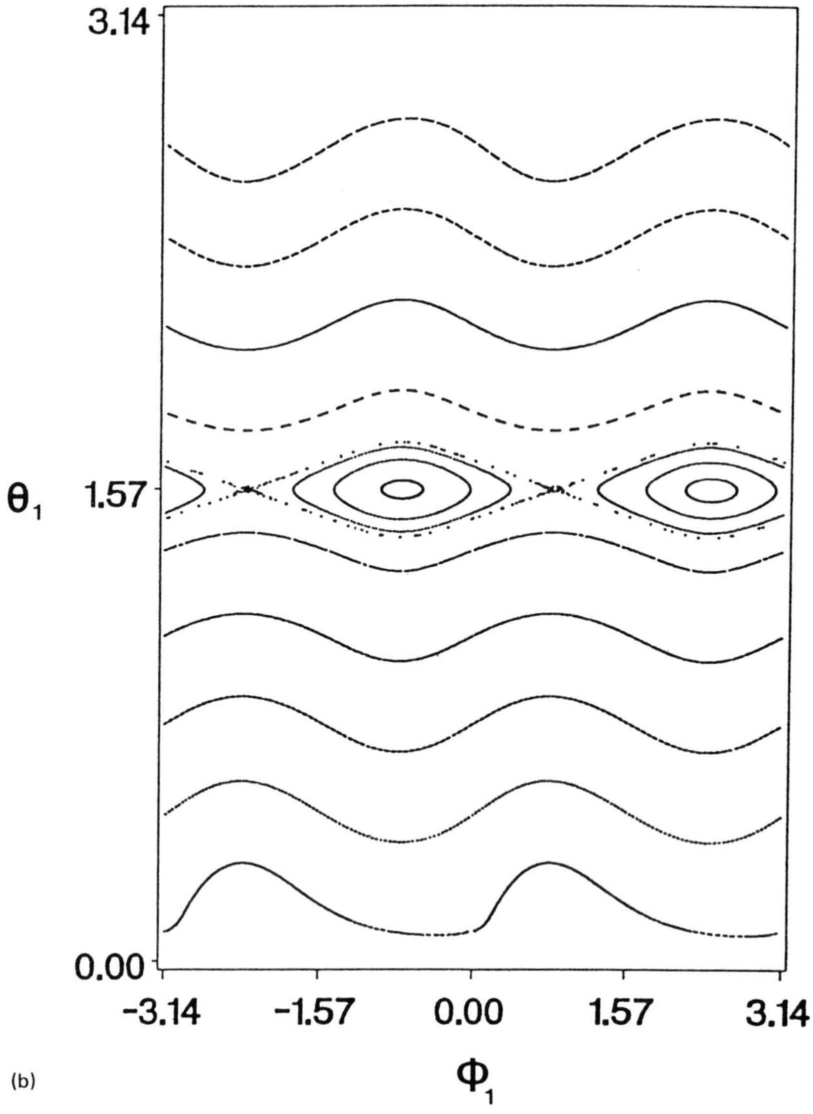

(b)

Figure 3 (continued)

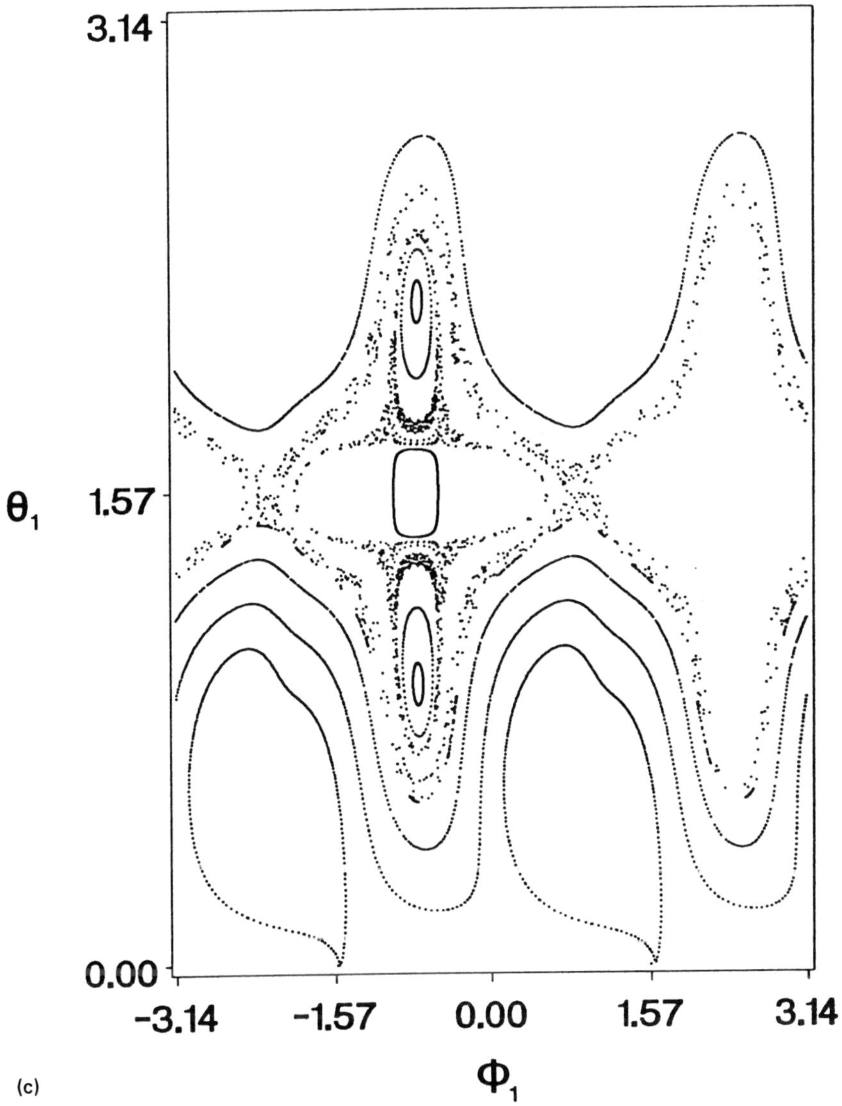

(c)

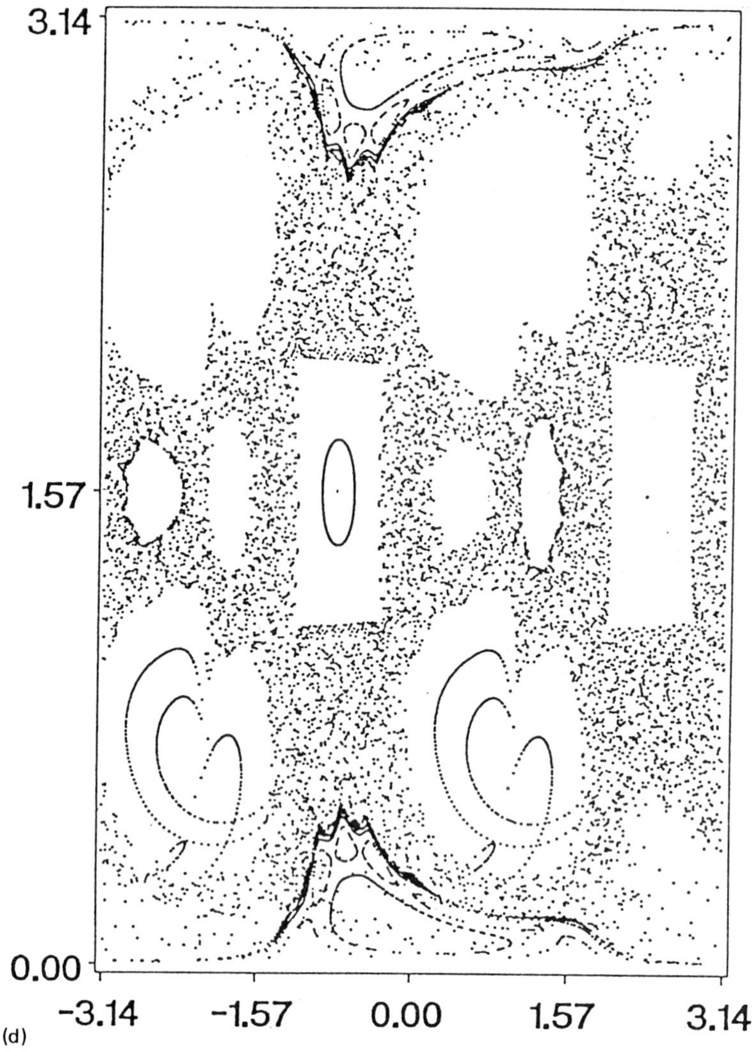

(d)

Figure 3 (continued)

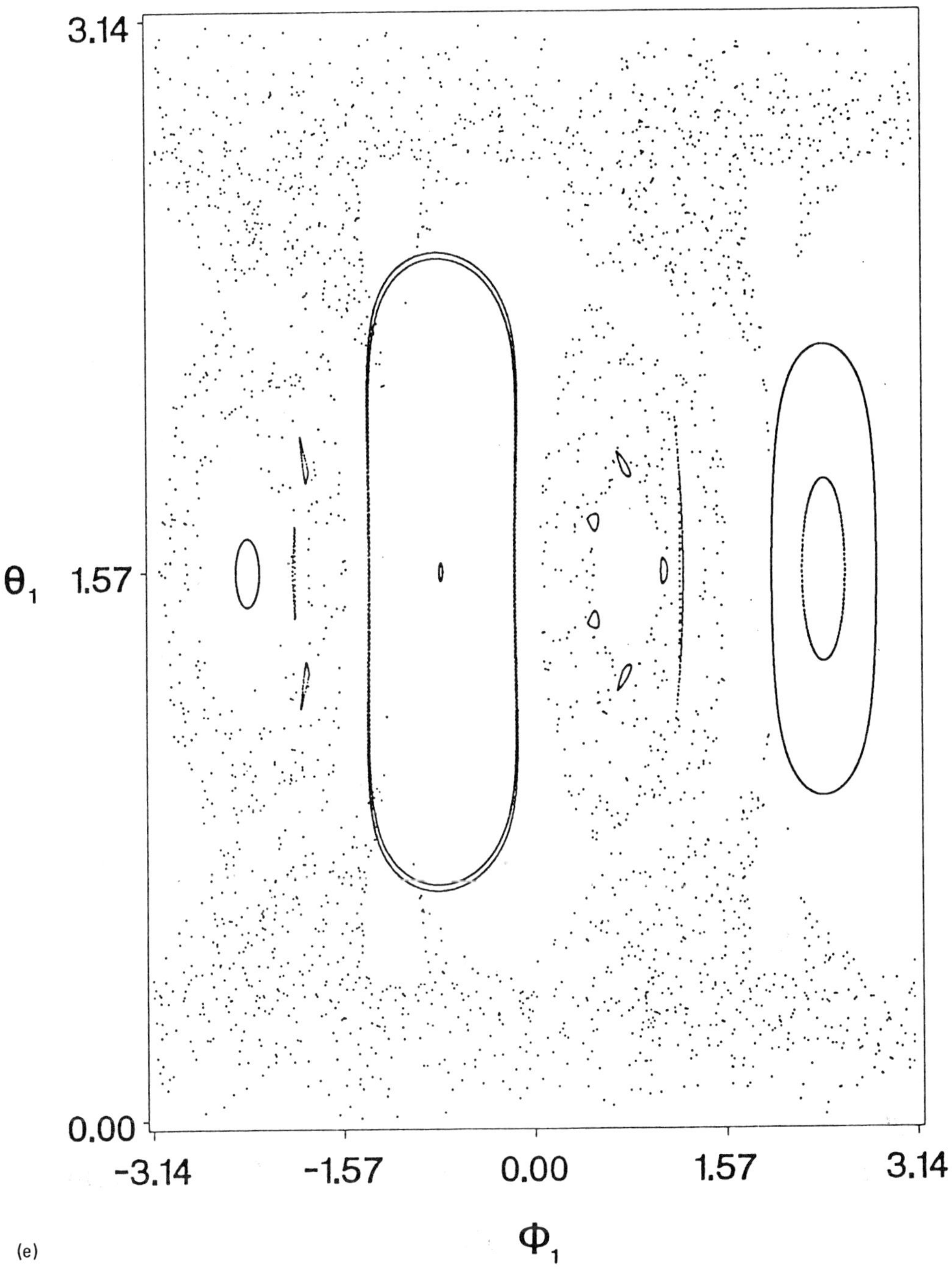

(e)

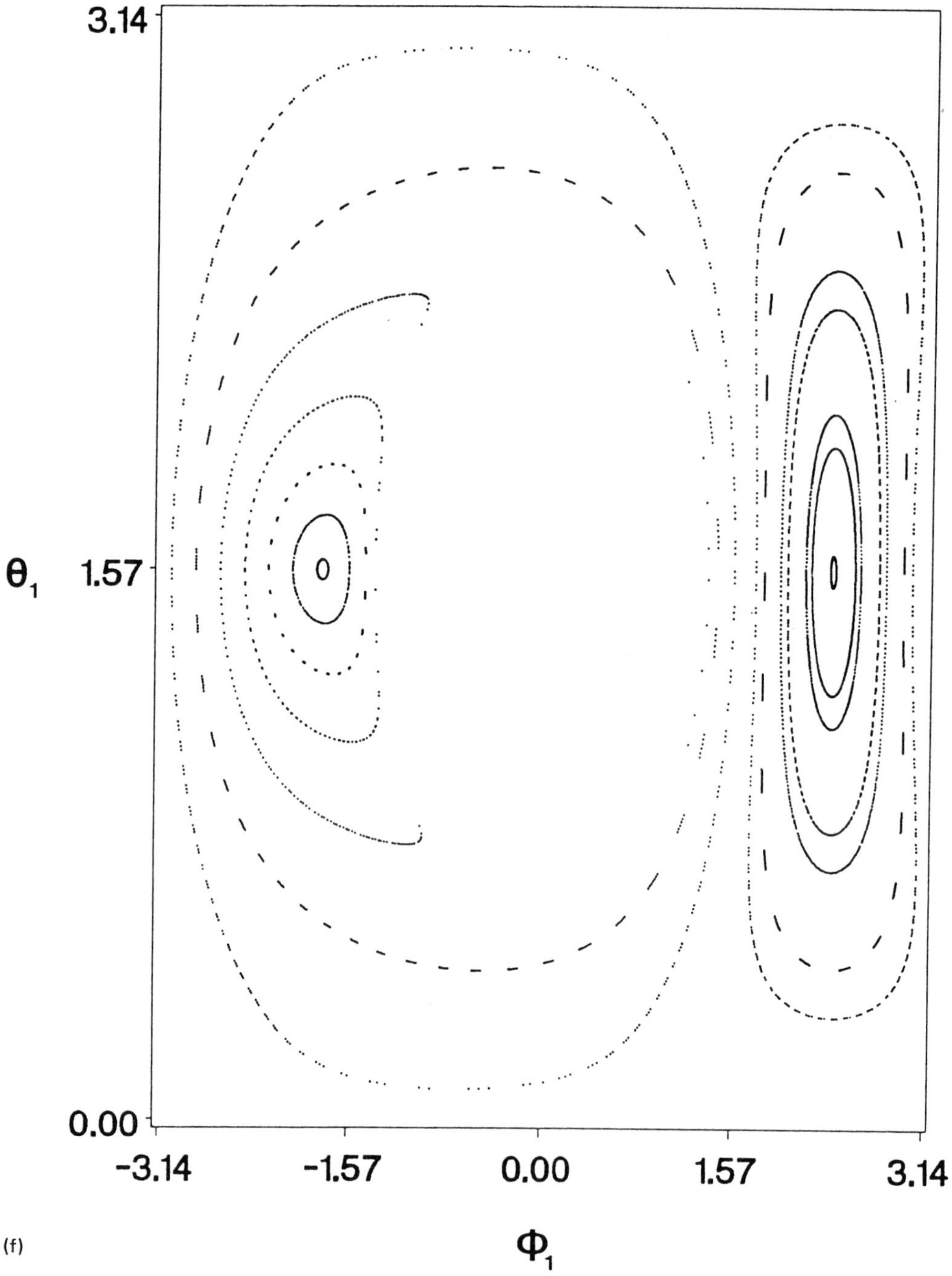

(f)

Figure 3 (continue)

These numerical statements suggest several questions. First, is there a convenient way to define the fraction (measure) of the initial condition space which leads to regular orbits? That is, is the measure easily extractable from the numerical information? Then, if so, is it expressible as a function of α? Finally, if that function exists, we can study the controlability of the system as α is changed. Specifically, we could find the value of α for which this model is most chaotic.

REFERENCES

1. H. Goldstein, *Classical Mechanics* Addison Wesley, Cambridge (1953).
2. E. Magyari, H. Thomas, R. Weber, C. Kaufman and G. Müller, Z. Phys. **B65**, 363 (1987).
3. W. Yourgrau and S. Mandelstam, *Variational Principles in Dynamics and Quantum Theory*, 3rd Ed. W. B. Saunders Company, Philadelphia (1968).
4. V. I. Arnold, *Mathematical Methods of Classical Mechanics*, Springer, New York (1978).
5. The integration times for Figures 1 and 2 are 50 and 18000 units, respectively. If the trajectories in Figure 1 were shown for a longer time, almost all of the graphs would be covered, and no trajectory would be discernable.
6. P. A. M. Dirac, *The Principles of Quantum Mechanics*, 4th ed. Oxford U. Press (1958).
7. N. Srivastava, C. Kaufman, G. Müller, R. Weber and H. Thomas, Z. Phys. **B70**, 251 (1988).

Existence of Minimizers and "Complete Figure" in Linear-Quadratic Problems of Optimal Control

Jacob Kogan, Professor, Department of Mathematics, University of Maryland, Baltimore County Campus, Baltimore, Maryland

1. INTRODUCTION

A field of extremals is a basic device for deriving sufficient conditions for optimality in the classical calculus of variations. For the simplest variational problem extremals of the field fill out an open domain of the space and define a vector of "optimal velocity" at each point of the domain. The above mentioned property of the field also holds for many Lagrange problems. However for many variational problems (especially when the Hamiltonian function H is not analytic) this property no longer holds, and the region covered by extremals does not have a uniform dimension (see e.g. [CA]).

In this paper we consider a linear–quadratic Bolza problem of optimal control which is defined by functions measurable in time. We first assume the existence of minimizers, study their structure, and show that the region R covered by minimizers in general does not have a uniform dimension. Our principal result is the following: *Let the Bolza problem be defined in a time interval* $[T_-, T_+]$. *Let* $R_t = \{x : \text{there exists a minimizer } x(\cdot) \text{ with } x(t) = x\}$. *Then there exists a final partition* $T_- \leq r_1 < r_2 < \ldots < r_k = T_+$ *of the time interval such that* $\dim R_t = d_i$ *whenever* $r_i < t \le r_{i+1}$, *and* $d_1 < d_2 < \ldots < d_{k-1} = n$. *For each* $i = 2, \ldots, k-1$ *there exists a minimizer* $x_i(\cdot)$ *such that*

$$x_i(\tau) = 0 \quad \text{on} \quad [T_-, r_i] \quad \text{and} \quad x_i(\tau) \neq 0 \quad \text{on} \quad (r_i, T_+].$$

We investigate a connection between the points $\{r_i\}$ and the points where the attainable set of the linear system blows up. Finally we drop the existence assumption and show that if extremals have the above described structure, than each extremal is a minimizer.

2. FRAMEWORK

Consider a linear–quadratic control problem in R^n

$$\dot{x}(\tau) = A(\tau)x(\tau) + B(\tau)u(\tau) \tag{1}$$

$$J(x(\cdot), u(\cdot), t, T_+) = C(x(\cdot), u(\cdot), t, T_+) + \frac{1}{2}x^*(T_+)\Gamma x(T_+) \tag{2}$$

where * indicates the transpose, and $C(x(\cdot), u(\cdot), t, T_+)$ is given by the formula

$$C(x(\cdot), u(\cdot), t, T_+) = \frac{1}{2}\int_t^{T_+} \begin{bmatrix} x(\tau) \\ u(\tau) \end{bmatrix}^* \begin{bmatrix} R_{11}(\tau) & R_{12}(\tau) \\ R_{21}(\tau) & R_{22}(\tau) \end{bmatrix} \begin{bmatrix} x(\tau) \\ u(\tau) \end{bmatrix} d\tau. \tag{3}$$

Entries of all the matrices in (1), (2), (3) are measurable and bounded functions on compact subintervals of the real line, $R_{22}(\tau) > \delta I$ for some positive δ, and all real τ.

DEFINITION 1. Let $x(\cdot)$ be an absolutely continuous function from $[t, T_+]$ to R^n and $u(\cdot)$ be a measurable and bounded function from $[t, T_+]$ to R^m. The pair $(x(\cdot), u(\cdot))$ is admissible if it satisfies (1).

Let (t, x) be an element of R^{n+1}. We define the value function $V_J(t, x)$ associated with the cost functional (2) by the formula

$$V_J(t, x) = \inf\{J(x(\cdot), u(\cdot), t, T_+) : (x(\cdot), u(\cdot)) \text{ is admissible}, x(t) = x\}. \tag{4}$$

DEFINITION 2. We shall say that (t, x) is a minimal point if $V_J(t, x) > -\infty$.

DEFINITION 3. For each $t \le T_+$ denote by $M(t)$ the set of minimal points at time t. Namely

$$M(t) = \{x : x \in R^n, V_J(t, x) > -\infty\}.$$

We shall say that an admissible pair $(x(\cdot), u(\cdot))$ is a minimal pair, and the trajectory $x(\cdot)$ is a minimizer on $[t, T_+]$ if

$$V_J(t, x(t)) = J(x(\cdot), u(\cdot), t, T_+). \tag{5}$$

Our main goal is to investigate the structure of minimizers and the dynamics of $M(t)$ in the linear-quadratic problem (1), (2).

3. CANONICAL FORM

Since for any matrix Q one has $x^*Qx = x^*\{\frac{1}{2}Q^* + \frac{1}{2}Q\}x$ we can assume that $R_{22}^*(\tau) = R_{22}(\tau)$ and $\Gamma = \Gamma^*$. Furthermore, without any loss of generality (due to an appropriate substitution) one can also assume that

$$A(\tau) \equiv 0,\ R_{12}(\tau) \equiv 0,\ R_{21}(\tau) \equiv 0,\ R_{11}(\tau) = R_{11}^*(\tau),\ R_{22}(\tau) \equiv I, \text{ and } \Gamma = \Gamma^*.$$

Indeed, consider the following Bolza problem.

$$\dot{y}(\tau) = \tilde{B}(\tau)w(\tau) \tag{6a}$$

$$\tilde{J}(y(\cdot), w(\cdot), t, T_+) = \frac{1}{2}\int_t^{T_+} \{y^*(\tau)\tilde{R}(\tau)y(\tau) + |w(\tau)|^2\}d\tau + \frac{1}{2}y^*(T_+)\Gamma y(T_+), \tag{7a}$$

where

$$\frac{d}{d\tau}\Phi(\tau, T_+) = [A(\tau) - \frac{1}{2}B(\tau)R_{22}^{-1}(\tau)\{R_{12}^*(\tau) + R_{21}(\tau)\}]\Phi(\tau, T_+) \text{ with } \Phi(T_+, T_+) = I,$$

$$\tilde{B}(\tau) = \Phi(T_+, t)B(\tau)R_{22}^{-\frac{1}{2}}(\tau),$$

$$\tilde{R}(\tau) = \Phi^*(t, T_+)[R_{11}(\tau) - \frac{1}{4}\{R_{12}^*(\tau) + R_{21}(\tau)\}^* R_{22}^{-1}(\tau)\{R_{12}^*(\tau) + R_{21}(\tau)\}]\Phi(\tau, T_+).$$

A straightforward computation shows that $(x(\cdot), u(\cdot))$ is an admissible pair for (1), (2) if and only if $(y(\cdot), w(\cdot))$ (where

$$y(\tau) = \Phi(T_+, \tau)x(\tau) \text{ and } w(\tau) = R_{22}^{\frac{1}{2}}(\tau)u(\tau) + \frac{1}{2}R_{22}^{-\frac{1}{2}}(\tau)\{R_{12}^*(\tau) + R_{21}(\tau)\}x(\tau)\ )$$

is an admissible pair for (6a), (7a). Since $J(x(\cdot), u(\cdot), t, T_+) = \tilde{J}(y(\cdot), w(\cdot), t, T_+)$ and $\Phi(T_+, \tau)$ is nonsingular there exists a one-to-one correspondence between minimizers of (1), (2) and those of (6a), (7a). Then the problem (1), (2), (3) will henceforth be given by

$$\dot{x}(\tau) = B(\tau)u(\tau) \tag{6}$$

$$J(x(\cdot), u(\cdot), t, T_+) = C(x(\cdot), u(\cdot), t, T_+) + \frac{1}{2}x^*(T_+)\Gamma x(T_+) \tag{7}$$

$$C(x(\cdot), u(\cdot), t, T_+) = \frac{1}{2}\int_t^{T_+} x^*(\tau)R(\tau)x(\tau) + |u(\tau)|^2 dt. \tag{8}$$

Here for simplicity sake we denote the matrix $R_{11}(\tau)$ just by $R(\tau)$.

4. NECESSARY CONDITIONS

It is clear that the existence of a minimal point (t,x) immediately implies that $(t,0)$ is also a minimal point. We now turn attention to the trajectories $z(\cdot)$ with $z(t) = z(T_+) = 0$.

DEFINITION 4. We shall say that the cost C:

1. is semipositive definite if for every admissible pair $(z(\cdot), v(\cdot))$ with $z(t) = z(T_+) = 0$ the cost $C(z(\cdot), v(\cdot), t, T_+) \geq 0$, and there exists a nontrivial admissible pair with zero boundary conditions and zero cost. We denote the semipositive definite cost by $c(t, T_+) \geq 0$.

2. is positive definite if $C(z(\cdot), v(\cdot), t, T_+) \geq 0$ for each admissible pair $(z(\cdot), v(\cdot))$ with $z(t) = z(T_+) = 0$, and the condition $C(z(\cdot), v(\cdot), t, T_+) = 0$ implies that $(z(\cdot), v(\cdot)) = (0,0)$ in $[t, T_+]$. We denote the positive definite cost by $c(t, T_+) > 0$.

3. is indefinite if there exists an admissible pair $(z(\cdot), v(\cdot))$ with $z(t) = z(T_+) = 0$ such that $C(z(\cdot), v(\cdot), t, T_+) < 0$. We denote the indefinite cost by $c(t, T_+) < 0$.

An elementary necessary condition for existence of a minimal point should include the following:

HYPOTHESIS 1. The cost $c(t, T_+)$ is semipositive definite.

Our first goal is to show that the existence of a minimal point (t,x) implies that $c(t, T_+)$ is necessarily positive definite.

LEMMA 1. Suppose that $c(t, T_+)$ is only semipositive definite. Then $V_J(t, 0) = -\infty$.

Proof. Since $c(t, T_+) \geq 0$ one can find a nontrivial admissible pair $(z(\cdot), v(\cdot))$ with

$$z(t) = z(T_+) = 0 \qquad \text{and} \qquad C(z(\cdot), v(\cdot), t, T_+) = 0.$$

Let T be the minimal point in $[t, T_+]$ such that beyond T the trajectory $z(\cdot)$ remains at the origin, i.e.

$$T = \inf\{\tau : \tau \in [t, T_+],\ z(s) = 0 \quad \text{for each} \quad s \in [\tau, T_+]\}.$$

Then $t < T \leq T_+$, $z(T) = 0$, $C(z(\cdot), v(\cdot), t, T) = 0$ (the proof of this equality is elementary, see e.g., [JK]), and in each left neighborhood of T there exists a scalar s such that $z(s) \neq 0$. Let $\{s_i\}$ be a sequence of scalars such that

$$s_i > T,\ \lim_{i \to \infty} s_i = T \text{ and } z(s_i) \neq 0.$$

Define a sequence of admissible pairs $(z_i(\cdot), v_i(\cdot))$ by the relations

$$(z_i(\tau), v_i(\tau)) = \frac{1}{|z(s_i)|}(z(\tau), v(\tau)), \quad \tau \in [t, T_+].$$

Then

$$z_i(t) = 0, \quad |z_i(s_i)| = 1, \quad z_i(T) = 0 \quad \text{and} \quad C(z_i(\cdot), v_i(\cdot), t, T) = 0.$$

Since we go with "a high speed" $C(z_i(\cdot), v_i(\cdot), s_i, T) \to \infty$ as $i \to \infty$, and $\lim_{i\to\infty} C(z_i(\cdot), v_i(\cdot), t, s_i) = -\infty$ (see [K2]). Finally define a sequence of admissible pairs $(\overline{z}_i(\cdot), \overline{v}_i(\cdot))$ by the formula:

$$\begin{aligned} (\overline{z}_i(\tau), \overline{v}_i(\tau)) &= (z_i(\tau), v_i(\tau)) & \tau \in [t, s_i] \\ (\overline{z}_i(\tau), \overline{v}_i(\tau)) &= (z_i(s_i), 0) & \tau \in (s_i, T_+]. \end{aligned}$$

Then

$$\lim_{i\to\infty} J(\overline{z}_i(\cdot), \overline{v}_i(\cdot), t, T_+) = \lim_{i\to\infty} C(z_i(\cdot), v_i(\cdot), t, s_i) + \frac{1}{2}\int_{s_i}^{T_+} z_i^*(s_i)R(\tau)z_i(s_i)d\tau +$$

$$+\frac{1}{2}z_i^*(s_i)\Gamma z_i(s_i) \le \lim_{i\to\infty} C(z_i(\cdot), v_i(\cdot), t, s_i) + \frac{1}{2}\int_t^{T_+} |R(\tau)|d\tau + \frac{1}{2}|\Gamma| = -\infty.$$

Hence $V_J(t, 0) = -\infty$. This completes the proof.

It is clear now that necessary conditions for existence of minimizers should include the following:

HYPOTHESIS 2. The cost $c(t, T_+) > 0$ (is positive definite).

Hypothesis 2 allows one to reduce the original minimization problem to a minimization problem in a finite dimensional Euclidean space (see Lemma 2). In order to proceed we recall the definition of the attainable set.

DEFINITION 5. The attainable set $AT(b; a, x_a)$ at time b for the control system (6) with initial state x_a and initial time t_a is the set of all points $x_b \in R^n$ such that for some trajectory $\varphi(\cdot)$ satisfying the condition $\varphi(a) = x_a$ the relation $\varphi(b) = x_b$ holds.

We define the value function $V_C(a, x_a, b, x_b)$ associated with the cost functional (8) as follows: Let (a, x_a, b, x_b) be an element of R^{2n+2} we set:

$$V_C(a, x_a, b, x_b) = \inf\{C(x(\cdot), u(\cdot), a, b) : (x(\cdot), u(\cdot)) \text{ is admissible}, x(a) = x_a, \ x(b) = x_b\}.$$

We say that a triple $(x(\cdot), u(\cdot), \eta(\cdot))$ is an extremal triple if $(x(\cdot), u(\cdot))$ is an admissible pair and the triple $(x(\cdot), u(\cdot), \eta(\cdot))$ satisfies the Pontryagin Maximum Principle. The proof of the next lemma is given in [K2].

LEMMA 2. The cost $c(t, T_+) > 0$ (is positive definite) if and only if the following condition holds: For each a, $b \in [t, T_+]$ such that $x_b \in AT(b; a, x_a)$ there exists a unique extremal pair $(x(\cdot), u(\cdot))$ with $x(a) = x_a$, $x(b) = x_b$. If $c(t, T_+) > 0$, then $C(x(\cdot), u(\cdot), a, b) = V_C(a, x_a, b, x_b)$.

Since the cost J is quadratic and the corresponding Hamiltonian system is linear there exist four matrices $W_{ij}(t,T_+)$, $i,j = 1,2$ such that the cost of an extremal triple $(x(\cdot),u(\cdot),\eta(\cdot))$ is given by

$$J(x(\cdot),u(\cdot),t,T_+) = \begin{bmatrix} x(t) \\ p(t) \end{bmatrix}^* \begin{bmatrix} W_{11}(t,T_+) & W_{12}(t,T_+) \\ W_{21}(t,T_+) & W_{22}(t,T_+) \end{bmatrix} \begin{bmatrix} x(t) \\ p(t) \end{bmatrix} \tag{9}$$

where $p(t) = \eta^*(t)$. We assume that $W_{22}^*(t,T_+) = W_{22}(t,T_+)$ for convenience.

This formula provides the main computational tool for evaluation of minimizers as follows: Let x be a given initial state. If p minimizes the quadratic expression

$$\begin{bmatrix} x \\ p \end{bmatrix}^* \begin{bmatrix} W_{11}(t,T_+) & W_{12}(t,T_+) \\ W_{21}(t,T_+) & W_{22}(t,T_+) \end{bmatrix} \begin{bmatrix} x \\ p \end{bmatrix},$$

and $(x(\cdot),u(\cdot),\eta(\cdot))$ is an extremal triple with $x(t) = x$ and $\eta^*(t) = p$; then

$$V_J(t,x) = J(x(\cdot),u(\cdot),t,T_+).$$

5. STRUCTURE OF MINIMIZERS

An extremal trajectory usually satisfies a first order necessary condition. The necessary condition depends on the problem under consideration. Since in this section we consider a Bolza problem it is convenient to introduce the following:

DEFINITION 6. An extremal triple $(x(\cdot),u(\cdot),\eta(\cdot))$ is called a Bolza extremal triple if it satisfies the transversality condition, that is

$$\eta(T_+) + x(T_+)\Gamma = 0. \tag{10}$$

We call the extremal pair $(x(\cdot),u(\cdot))$ a Bolza extremal pair, and the extremal trajectory $x(\cdot)$ will be called a Bolza extremal trajectory.

We first describe the set of minimal points at time t.

LEMMA 3. For each t the set $M(t)$ is a vector subspace of R^n. If $M(t) \neq \emptyset$, then the following two conditions are equivalent:

1. $x(\cdot)$ is a minimizer on $[t,T_+]$,
2. $x(\cdot)$ is a Bolza extremal on $[t,T_+]$.

Proof. Let $(x(\cdot), u(\cdot), \eta(\cdot))$ be an extremal triple with $x(t) = x$. Then, by (9)

$$J(x(\cdot), u(\cdot), t, T_+) = x^*(t)W_{11}(t, T_+)x(t) + x^*(t)[W_{12}(t, T_+) + W_{21}(t, T_+)]p(t) +$$
$$+ p^*(t)W_{22}(t, T_+)p(t).$$

So for fixed $x(t) = x$ the cost is a quadratic expression with respect to $p = p(t)$. The dominated term in the expression is $p^*W_{22}(t, T_+)p$. We consider now the following three cases:

1. $W_{22}(t, T_+)$ is indefinite. Then $M(t) = \emptyset$ (and indeed is a vector subspace of R^n).

2. $W_{22}(t, T_+) > 0$ (is positive definite). Then $M(t) = R^n$.

3. Finally let $W_{22}(t, T_+) \geq 0$ (be semipositive definite).

Denote by $O(t, T_+)$ the null subspace of $W_{22}(t, T_+)$, i.e.

$$O(t, T_+) = \{q : q \in R^n,\ q^*W_{22}(t, T_+)q = 0\}.$$

Then (t, x) is a minimal point if and only if

$$x^*[W_{12}(t, T_+) + W_{21}(t, T_+)]q = 0 \quad \text{for each} \quad q \in O(t, T_+).$$

This linear relation implies that $M(t)$ is a vector subspace of R^n. Moreover the proof shows that for each $x \in M(t)$ there exists a minimizer $x(\cdot)$ with $x(t) = x$.

We now pass to the second part of the lemma. Let $(x(\cdot), u(\cdot), \eta(\cdot))$ be a Bolza extremal triple defined on $[t, T_+]$. If $M(t) \neq \emptyset$, then $x(\cdot)$ is a minimizer on $[t, T_+]$. Indeed, let $(y(\cdot), w(\cdot))$ be an admissible pair with $x(t) = y(t)$. Then a straightforward computation shows that

$$J(y(\cdot), w(\cdot), t, T_+) - J(x(\cdot), u(\cdot), t, T_+) =$$
$$= \frac{1}{2}(y(T_+) - x(T_+))^*\Gamma(y(T_+) - x(T_+)) + \tag{11}$$
$$+ \frac{1}{2}\int_t^{T_+} [(y(\tau) - x(\tau))^* R(\tau)(y(\tau) - x(\tau)) + |w(\tau) - u(\tau)|^2] d\tau. \tag{12}$$

Since $0 \in M(t)$, and $y(\cdot) - x(\cdot)$ is an admissible trajectory initiating at the origin at time t the expression (11), (12) should be nonnegative. This shows that $x(\cdot)$ is a minimizer on $[t, T_+]$. Since every minimizer is a Bolza extremal our proof is completed.

REMARK 1. Notice that

1. $V_J(t, x) > -\infty$ whenever $x \in M(t)$,

2. $V_J(t,x) = -\infty$ whenever $x \in R^n - M(t)$.

REMARK 2. Let $t < s < T_+$. Then $M(t) \subseteq M(s)$.

As an immediate consequence we have:

COROLLARY 1. There exists a finite set $\{r_i\}_{i=0}^k$

$$-\infty = r_0 \leq r_1 < \ldots < r_k = T_+ \tag{13}$$

such that

1. $M(t) = \emptyset$ when $t \in (r_0, r_1)$,

2. $M(t) = M(s)$ whenever $t, s \in (r_{i-1}, r_i)$, $i = 1, \ldots, k$,

3. $M(t)$ is a proper subspace of $M(s)$ whenever $t < r_i < s$ for some $i = 1, \ldots, k-1$.

In the following example we compute $M(t)$ for each $t \in (-\infty, T_+]$.

EXAMPLE 1. Consider an optimal control problem:

$$\dot{x}(\tau) = b(\tau)u(\tau)$$

$$J(x(\cdot), u(\cdot), t, T_+) = \frac{1}{2}\int_t^{T_+} u^2(\tau)d\tau - \frac{1}{2}x^2(T_+)$$

where x, u are scalars, $b(\tau)$ is a scalar function defined by the relations

$$b(\tau) = \begin{cases} 0 & \text{if } \tau \in [0,1] \\ b(\tau) = 1 & \text{otherwise.} \end{cases}$$

In this example the feedback $\bar{u}(\tau, x, \eta) = b(\tau)\eta$, and the Hamiltonian system is

$$\begin{cases} \dot{x}(\tau) = b^2(\tau)\eta(\tau) \\ \dot{\eta}(\tau) = 0 \end{cases}$$

Let $T_+ = 2$. We consider the following three cases.

Case 1: $1 < t < 2$. The terminal point $x(2) = x + (2-t)p$ (where $x = x(t)$ and $p = \eta^*(t)$), and the cost (see (9)) is given by

$$J(x(\cdot), u(\cdot), t, 2) = \frac{1}{2}\left\{-x^2 - 2x(2-t)p + (2-t)(t-1)p^2\right\}.$$

Since $(2-t)(t-1) > 0$ whenever $1 < t < 2$ it follows (from Lemma 3, case 2) that $M(t) = R^1$.

Case 2: $0 \leq t \leq 1$. In this case $x(2) = x + p$ and

$$J(x(\cdot), u(\cdot), t, 2) = \frac{1}{2}\{-x^2 - 2xp\}.$$

This implies (see Lemma 3, case 3), that $M(t) = \{0\}$.

Case 3: $-\infty < t < 0$. Since $x(2) = x + p(1-t)$ the cost is

$$J(x(\cdot), u(\cdot), t, 2) = \frac{1}{2}\{-x^2 - 2x(1-t)p + (1-t)tp^2\}.$$

Since $(1-t)t < 0$ there exist no minimal points when $t < 0$, and $M(t) = \emptyset$. So in this example $r_0 = -\infty$, $r_1 = 0$, $r_2 = 1$, $r_3 = 2$.

The next theorems describe the form of the intersection of extremals and show that if extremals "split out" they never meet again. The proofs are similar to this of Lemma 1 and can be found in [K2].

THEOREM 1. Let $x_1(\cdot)$ and $x_2(\cdot)$ be two minimizers defined on $[t, T_+]$. If for some $s \in [t, T_+]$ the relation $x_1(s) \neq x_2(s)$ holds, then $x_1(\tau) \neq x_2(\tau)$ for each $\tau \in [s, T_+]$.

THEOREM 2. The maximal number of linearly independent minimizers (as functions on $[t, T_+]$) initiating from 0 at time t is $n - k$ if and only if $\dim M(t) = k$.

THEOREM 3. For each $i = 2, \ldots, k-1$ there exists a minimizer $x_i(\cdot)$ such that

$$x_i(\tau) = 0 \quad \text{on} \quad [T_-, r_i] \quad \text{and} \quad x_i(\tau) \neq 0 \quad \text{on} \quad (r_i, T_+].$$

A point r is a branching point of an extremal $x(\cdot)$ defined on the interval $[T_-, T_+]$ if there exists an extremal $y(\cdot)$ such that:

$$x(\tau) = y(\tau) \quad \text{for each} \quad \tau \in [T_-, r] \quad \text{and} \quad x(\tau) \neq y(\tau) \quad \text{for each} \quad \tau \in (r, T_+].$$

In linear-quadratic problems branching points are points where the attainable set "blows up"(see [K1]). Theorem 4 tells us that the set $\{r_2, \ldots, r_{k-1}\}$ is a subset of the set of branching points of the extremal $x(\tau) \equiv 0$ on $[T_-, T_+]$. The choice of this subset is governed by the matrix Γ, the choice of the point r_1 is governed by the problem, i.e., the matrices $B(t)$, $R(t)$ and Γ. In the following examples we illustrate the last statement.

EXAMPLE 2. Consider the following system:

$$\dot{x}_1(\tau) = b_1(\tau)u_1(\tau), \; \dot{x}_2(\tau) = b_2(\tau)u_2(\tau)$$

where $b_1(\tau)$, $b_2(\tau)$ are scalar valued functions given by

$$\begin{array}{llll} b_1(\tau) = 0 & \tau \in [0,2] & \text{and} \quad b_1(\tau) = 1 & \text{otherwise,} \\ b_2(\tau) = 0 & \tau \in [0,1] & \text{and} \quad b_2(\tau) = 1 & \text{otherwise.} \end{array}$$

The cost is given by the formula

$$J(x(\cdot), u(\cdot), t, T_+) = \frac{1}{2}\int_t^{T_+} [u_1^2(\tau) + u_2^2(\tau)]d\tau - \frac{1}{2}x_1^2 - \frac{1}{4}x_2^2$$

Let $T_+ = 3$. Then one can show that

$$r_0 = -\infty,\ r_1 = T_- = 0,\ r_2 = 1,\ r_3 = 2,\ r_4 = T_+ = 3, \text{ and}$$

$$\begin{array}{llll} M(t) & = R^2 & \text{for} & t \in (r_3, r_4] \\ M(t) & = R^1 & \text{for} & t \in (r_2, r_3] \\ M(t) & = \{0\} & \text{for} & t \in [r_1, r_2] \\ M(t) & = \emptyset & \text{for} & t < r_1. \end{array}$$

Notice, that the trivial extremal trajectory $x(\cdot) \equiv 0$ has two branching points $r = 1$ and $r = 2$ in $[T_-, T_+]$.

Let now the cost be given by

$$J(x(\cdot), u(\cdot), t, T_+) = \frac{1}{2} \int_t^{T_+} [u_1^2(\tau) + u_2^2(\tau)] d\tau - \frac{1}{2} x_1^2.$$

If $T_+ = 3$, then $r_0 = -\infty$, $r_1 = T_- = 0$, $r_2 = 2$, $r_3 = T_+ = 3$ and

$$\begin{array}{llll} M(t) & = R^2 & \text{for} & t \in (r_2, r_3] \\ M(t) & = R^1 & \text{for} & t \in [r_1, r_2] \\ M(t) & = \emptyset & \text{for} & t < r_1. \end{array}$$

So in this case $\{r_2\}$ is a proper subset of the set of branching points.

Notice that if $J(x(\cdot), u(\cdot), t, T_+) = \frac{1}{2} \int_t^{T_+} [u_1^2(\tau) + u_2^2(\tau)] d\tau$ and $T_+ = 3$, then $r_0 = r_1 = T_- = -\infty$, $r_2 = T_+ = 3$, and $M(\tau) = R^2$ for each $t \in (-\infty, 3]$.

6. FIELD OF EXTREMALS

The existence of minimizers implies a certain structure of Bolza extremals (see Theorem 3). We wish now to show that if Bolza extremals of (1), (2) have this structure then the cost functional (2) is semi-positive definite (the formal definition will be provided at the end of the section), and each Bolza extremal is a minimizer.

Our main assumption in this subsection is the following

HYPOTHESIS 3. For each Bolza extremal $x(\cdot)$ defined on $[T_-, T_+]$ and such that $x(s) = 0$ for some $s \in [T_-, T_+]$, the equality $x(\tau) = 0$ holds for each $\tau \in [T_-, s]$.

Our main goal is to show that under Hypothesis 3 each Bolza extremal is a minimizer, $V_J(T_-, 0) = 0$ and, in particular, $c(T_-, T_+) > 0$.

We first remind the reader the formal definitions of positive and semipositive cost functional (2) (see e.g. [CO], [M]).

DEFINITION 7. The quadratic functional (7) is semipositive definite on $[T_-, T_+]$ if for each admissible pair $(x(\cdot), u(\cdot))$ with $x(T_-) = 0$ the cost $J(x(\cdot), u(\cdot), T_-, T_+) \geq 0$, and the there exists a nontrivial admissible pair $(y(\cdot), w(\cdot))$ with $y(T_-) = 0$ and $J(y(\cdot), w(\cdot), T_-, T_+) = 0$.

DEFINITION 8. The quadratic functional (7) is positive definite on $[T_-, T_+]$ if for each nontrivial admissible pair $(x(\cdot), u(\cdot))$ with $x(T_-) = 0$ the cost $J(x(\cdot), u(\cdot), T_-, T_+) > 0$.

We now state two major results characterizing positive and semipositive second variation. (The fool proofs are given in [K2]).

THEOREM 4. The following statements are equivalent

1. the quadratic functional (7) is semipositive definite on $[T_-, T_+]$,
2. Hypothesis 3.1 is satisfied on $[T_-, T_+]$,
3. $M(T_-)$ is a proper nonempty subspace of R^n.

THEOREM 5. The following statements are equivalent

1. the quadratic functional (7) is positive definite on $[T_-, T_+]$,
2. there exists no points of intersection of Bolza extremals in $[T_-, T_+]$,
3. $M(t) = R^n$ for each $t \in [T_-, T_+]$.

REFERENCES

[CA] Caratheodory C., Calculus of Variations and Partial Differential Equations of the First Order. Part 2. Holden–Day, 1965.

[CO] Coppel W.A., Linear–quadratic optimal control. Proceedings of the Royal Society of Edinburgh, 73A, 18 1974/75, pp. 271-289.

[JK] Jurdjevic V., Kogan J., Optimality of extremals for linear systems with quadratic costs. Journal of Mathematical Analysis and Applications, (to appear).

[K1] Kogan J., Structure of extremals in optimal control problems. SIAM Journal on Control, Vol 25:4, 1987, pp. 951-966.

[K2] Kogan J., Structure of minimizers in linear–quadratic Bolza problems of optimal control. Journal of Optimization Theory and Applications, (to appear).

[M] Molinari B.P., Nonnegativity of a quadratic functional. SIAM Journal on Control, Vol. 13, No. 4, 1975, pp. 792-806.

On Near-Optimum Large-Scale Regulators

G. S. Ladde, Professor, Department of Mathematics, The University of Texas at Arlington, Arlington, Texas

O. Sirisaengtaksin, Professor, Department of Applied Mathematical Sciences, University of Houston-Downtown, Houston, Texas

ABSTRACT

In this paper we consider an n-time scale linear singularly perturbed regulator problem. By implementing a reduction scheme, the original system can be decomposed into slow and fast subsystem regulators. These subsystems provide approximations for both optimal control law and optimal performance of the original regulator system.

1. INTRODUCTION

The singular perturbation approach to linear regulator problems of systems having slow and fast modes has been applied successfully in [1,3-6]. The near optimal control law of the original system is shown to be the composite of control laws for the slow and fast subsystem regulators, which are obtained by a reduction process. Moreover, this composite control also yields the near optimum performance.

In this paper, we consider a regulator problem for n-time scale linear singularly perturbed system. We first formulate the slow and fast n subsystem regulators by employing a reduction procedure [1] successively. Then study the convergence of the optimal control law for the original control problem to the composite of n optimal control laws for the slow and fast subsystem regulators, in seperate time scales. This can be achieved by implementing a diagonalization technique in [2]. Furthermore, we obtain the near-optimum performance by using the composite of n optimal control laws for the slow and fast subsystem regulators.

2. PROBLEM FORMULATION

Consider a singularly perturbed control system of the form

$$\varepsilon_i \dot{x}_i = \sum_{j=1}^{n} A_{ij}(t) x_j + B_i(t) u(t), \; x_i(t_0) = x_i^0 \tag{2.1}$$

where $x_i \in \Re^{n_i}$, $i = J(1,n)$, $J(a,b) = (a, a+1, \ldots, b)$; $a, b \in Z^+ \cup \{0\}$ and $u \in \Re^m$; the dimension of the entire system (2.1) is $N = \sum_{i=1}^{n} n_i$; all the coefficient matrix functions are continuous in $\Re_+$ and have appropriate dimensions; parameters ε_i's are positive real numbers where $i \in J(1, n-1)$. We consider an observation system

$$y = \sum_{i=1}^{n} C_i(t) x_i \tag{2.2}$$

where $y \in \Re^l$, matrix functions C_i's are continuous on $\Re_+$ and have appropriate dimensions.

Our problem is to find a control law which minimizes the performance index J defined by

$$J(u) = x^T \Gamma x|_{t=t_f} + \int_{t_0}^{t_f} (y^T y + u^T R u) dt \tag{2.3}$$

where, for $i, j \in J(1,n)$

$$\Gamma = [\varepsilon_{ij} \Gamma_{ij}]_{n \times n}, \; \Gamma_{ji} = \Gamma_{ij}^T$$

$$\varepsilon_{ij} = \begin{cases} \varepsilon_{ij} & \text{for } i = j \\ \varepsilon_i \varepsilon_j & \text{for } i = j \end{cases}$$

and R is a square $m \times m$ matrix.

Our main objectives are (i) to decompose the regulator problem (2.1)-(2.3) into n-seperate lower order control problems, and (ii) to investigate the limiting behavior of the optimal control law for the original control problem to the optimal control laws for the lower order control problems; and the near optimality of the performance index.

3. REDUCED ORDER CONTROL PROBLEMS

In this section, reduced order control problems relative to (2.1)-(2.3) are formulated in different time scales by implementing a successive reduction scheme. The optimal control law for each reduced order control problem can be obtained in its own time scale.

For $m \in J(0, n-2)$, the ε_{n-m-1}-overall reduced system of (2.1)-(2.3) reducecs to

$$\left\{ \begin{array}{rcl} \dot{x}_{n-m-1}^m & = & A_{11}(m) x_{n-m-1}^m + A_{12}(m) x_{n-m}(m) + B_1(m) u_{n-m}^m, \\ & & x_{n-m-1}^m(t_0) = x_{n-m-1}^{m0}, \\ \varepsilon_{n-m} \dot{x}_{n-m}(m) & = & A_{21}(m) x_{n-m-1}^m + A_{22}(m) x_{n-m}(m) + B_2(m) u_{n-m}^m, \\ & & x_{n-m}(m)(t_0) = x_{n-m}^0(m), \\ y^m & = & C_1(m) x_{n-m-1}^m + C_2(m) x_{n-m}(m) + D_{m-1} u_{n-m}^m \\ J_{n-m}^m & = & \{x_{n-m-1}^{m^T} \Gamma(n-m-1) x_{n-m-1}^m + 2 x_{n-m-1}^{m^T} \Gamma_{n-m}^{n-m-1} x_{n-m}(m) \\ & & + \varepsilon_{n-m} x_{n-m}^T(m) \Gamma_{n-m,n-m} x_{n-m}(m)\}_{t=t_f} + \int_{t_0}^{t_f} \{y^{m^T} y^m \\ & & + u_{n-m}^{m^T} R(m-2) u_{n-m}^m\} dt \end{array} \right. \tag{3.1}$$

where for $m \in J(0,n-1)$, $r \in J(1,n)$

$$x_r^m = [x_1^T(m)\ x_2^T(m) \cdots x_r^T(m)]^T,$$

$$x_r^{m0} = [x_1^{0^T}(m)\ x_2^{0^T}(m) \cdots x_r^{0^T}(m)]^T,$$

$$[x_1^T(0)\ x_2^T(0) \cdots x_r^T(0)]^T = [x_1^T\ x_2^T \cdots x_r^T]^T,$$

$$[x_1^{0^T}(0)\ x_2^{0^T}(0) \cdots x_r^{0^T}(0)]^T = [x_1^{0^T}\ x_2^{0^T} \cdots x_r^{0^T}]^T,$$

for $m \in J(0,n-2)$

$$A^m = \begin{bmatrix} A_{11}(m) & A_{12}(m) \\ \frac{1}{\varepsilon_{n-m}} A_{21}(m) & \frac{1}{\varepsilon_{n-m}} A_{22}(m) \end{bmatrix},$$

$$B^m = \begin{bmatrix} B_1(m) \\ \frac{1}{\varepsilon_{n-m}} B_2(m) \end{bmatrix}, \quad C^m = \begin{bmatrix} C_1(m) C_2(m) \end{bmatrix},$$

$$\begin{aligned} A_{11}(m) &= [\tfrac{1}{\varepsilon_i} \hat{A}_{ij}^m]_{(n-m-1)\times(n-m-1)} \\ A_{12}(m) &= [\tfrac{1}{\varepsilon_1} \hat{A}_{1\,n-m}^{m^T}\ \tfrac{1}{\varepsilon_2} \hat{A}_{2\,n-m}^{m^T} \cdots \tfrac{1}{\varepsilon_3} \hat{A}_{n-m-1\,n-m}^{m^T}]^T \\ A_{21}(m) &= [\hat{A}_{n-m\,1}^m\ \hat{A}_{n-m\,2}^m \cdots \hat{A}_{n-m\,n-m-1}^m] \\ A_{22}(m) &= \hat{A}_{n-m\,n-m}^m \end{aligned}$$

$$B_1(m) = [\frac{1}{\varepsilon_i} \hat{B}_i^m]_{n-m-1\times 1},\ B_2(m) = \hat{B}_{n-m}^m,\ C_1(m) = [\hat{C}_j^m]_{1\times n-m-1},\ C_2(m) = \hat{C}_{n-m}^m$$

$$\hat{A}_{ij}^m = \begin{cases} A_{ij} & \text{for } m = 0 \text{ and } i,j \in J(1,n) \\ \hat{A}_{ij}^{m-1} - \hat{A}_{i\,n-m+1}^{m-1}[\hat{A}_{n-m+1\,n-m+1}^{m-1}]^{-1} \hat{A}_{n-m+1\,j}^{m-1} & \text{for } m \in J(1,n-1) \\ & \text{and } i,j \in J(1,n-m) \end{cases}$$

$$\hat{B}_i^m = \begin{cases} B_i & \text{for } m = 0 \text{ and } i \in J(1,n) \\ \hat{B}_i^{m-1} - \hat{B}_{i\,n-m+1}^{m-1}[\hat{A}_{n-m+1\,n-m+1}^{m-1}]^{-1} \hat{B}_{n-m+1}^{m-1} & \text{for } m c\ J(1,n-1) \\ & \text{and } i \in J(1,n-m) \end{cases}$$

$$\hat{C}_j^m = \begin{cases} C_j & \text{for } m = 0 \text{ and } j \in J(1,n) \\ \hat{C}_j^{m-1} - \hat{C}_{n-m+1}^{m-1}[\hat{A}_{n-m-1\times n-m-1}^{n-m}]^{-1} \hat{A}_{n-m+1,j}^{m-1} & \text{for } m \in J(1,n-1) \\ & \text{and } j \in J(1,n-m) \end{cases}$$

for $m = n-1$

$$A^{n-1} = A_{11}(n-1), B^{n-1} = B_1(n-1), C^{n-1} = C_1(n-1),$$

$$\Gamma(r) = [\varepsilon_{ij}\Gamma_{ij}]_{r\times r},\ r \in J(0,n-2)$$

$$\Gamma_r^p = [\varepsilon_{1r}\Gamma_{1r}^T\ \varepsilon_{2r}\Gamma_{2r}^T \cdots \varepsilon_{pr}\Gamma_{pr}^T],\ p \in J(1,r-1),\ r \in J(2,n),\ p < r,$$

$$R(m-2) = R + \sum_{k=o}^{m-2} D_K^T D_k \ \ with \ \ R(-2) = R(-1) = R,\ m \in J(0,n-2).$$

Assumption (A3.1) : $A_{22}(m)$ is nonsingular for $m \in J(0,n-2)$.

To obtain the ε_{n-m}-overall reduced order control problem for (3.1), we formally set $\varepsilon_{n-m} = 0$, this system is given by

$$
(3.2) \quad \begin{cases} \dot{x}^{m+1}_{n-m-1} &= A^{m+1}x^{m+1}_{n-m-1} + B^{m+1}u^{m}_{n-ms}, \; x^{m+1}_{n-m-1}(t_0) = x^{m+1\,0}_{n-m-1} \\ \bar{x}_{n-m}(m) &= -A^{-1}_{22}(m)A_{21}(m)x^{m+1}_{n-m-1} - A^{-1}_{22}(m)B_2(m)u^{m}_{n-ms} \\ y^{m+1} &= C^{m+1}x^{m+1}_{n-m-1} + D_m u^{m}_{n-ms} \\ J^{m+1}_{n-m-1} &= x^{m+1^T}_{n-m-1}\Gamma(n-m-1)x^{m+1}_{n-m-1}|_{t=t_f} \\ & \quad + \int_{t_0}^{t_f}(y^{m+1^T}y^{m+1} + u^{m^T}_{n-ms}R(m-1)u^{m}_{n-ms})dt \end{cases}
$$

The optimal control law for the ε_{n-m}-overall reduced order control problem (3.2) can be described by

$$
(3.3) \qquad u^{m}_{n-ms} = -F^T(n-m-1)x^{m+1}_{n-m-1}
$$

where $F(n-m-1) = [S(n-m-1)B^{m+1} + C^{m^T}D_m]R^{-1}(m)$, $S(n-m-1)$ satisfies

$$
(3.4) \quad \begin{aligned} -\dot{S}(n-m-1) &= A^{m+1^T}S(n-m-1) + S(n-m-1)A^{m+1} \\ & \quad + C^{m+1^T}C^{m+1} - F(n-m-1)R(m)F^T(n-m-1), \\ & \quad S(n-m-1)(t_f) = \Gamma(n-m-1). \end{aligned}
$$

By setting $x_{n-m}(m+1) = x_{n-m}(m) - \bar{x}_{n-m}(m)$, $u^{m}_{n-mf} = u^{m}_{n-m} - u^{m}_{n-ms}$ and $y_{n-mf} = y^m - y^{m+1}$, the ε_{n-m}-fast reduced order control problem is obtained as follows

$$
(3.5) \quad \begin{cases} \varepsilon_{n-m}\dot{x}_{n-m}(m+1) &= A_{22}(m)x_{n-m}(m+1) + B_2 + (m)u^{m}_{n-mf}, \\ & \quad x_{n-m}(m+1)(t_0) = x^{0}_{n-m}(m+1), \\ y_{n-mf} &= C_2(m)x_{n-m}(m+1) + D_{m-1}u^{m}_{n-mf}, \\ J_{n-m}(m) &= \varepsilon_{n-m}x^{T}_{n-m}(m+1)\Gamma_{n-m,n-m}x_{n-m}(m+1)|_{t=t_f} \\ & \quad + \int_{t_0}^{t_f}(y^{T}_{n-mf}y_{n-mf} + u^{m^T}_{n-mf}R(m-2)u^{m}_{n-mf})dt \end{cases}
$$

The optimal control law for the ε_{n-m}-fast reduced order control problem (3.5) is given by

$$
(3.6) \qquad u^{m}_{n-mf} = -F^{T}_{n-m}x_{n-m}(m+1)
$$

where $F_{n-m} = [S_{n-m}B_2(m) + C^{T}_{2}(m)D_{m-1}]R^{-1}(m-1)$, $S(n-m)$ satisfies

$$
\begin{aligned} -\varepsilon_{n-m}\dot{S}_{n-m} &= A^{T}_{22}(m)S_{n-m} + S_{n-m}A_{22}(m) + C^{T}_{2}(m)C_2(m) \\ & \quad - F_{n-m}R(m-1)F^{T}_{n-m}, \; S_{n-m}(t_f) = \Gamma_{n-m,n-m} \end{aligned}
$$

and its reduced system is

$$
(3.7) \quad 0 = A^{T}_{22}(m)\bar{S}_{n-m} + \bar{S}_{n-m}A_{22}(m) + C^{T}_{2}(m)C_2(m) - \bar{F}_{n-m}R(m-1)\bar{F}^{T}_{n-m},
$$

$$
\bar{F}_{n-m} = [\bar{S}_{n-m}B_2(m) + C^{T}_{2}(m)D_{m-1}]R^{-1}(m-1).
$$

The optimal control law for control problem (3.1) is given by

$$
(3.8) \qquad u^{m}_{n-m} = -[F^{c^T}(n-m-1)x^{m}_{n-m-1} + F^{c^T}_{n-m}x_{n-m}(m)]
$$

where

$$
F^c(n-m-1) = [S^c(n-m-1)B_1(m) + S^{c^T}_{n-m}(n-m-1)B_2(m) + C^{T}_{1}(m)D_{m-1}]R^{-1}(m-1)
$$

$$F^c_{n-m} = [\varepsilon_{n-m} S^{c^T}_{n-m}(n-m-1)B_1(m) + S^c_{n-m}B_2(m) + C_2^T(m)D_{m-1}]^{-1}(m-1)$$

$$(3.9)\quad \begin{cases} -\dot{S}^c(n-m-1) = & A_{11}^T(m)S^c(n-m-1) + S^c(n-m-1)A_{11}(m) \\ & +C_1^T(m)C_1(m) + A_{21}^T(m)S^{c^T}_{n-m}(n-m-1) \\ & +S^c_{n-m}(n-m-1)A_{21}(m) \\ & -F^c(n-m-1)R(m-1)F^{c^T}(n-m-1), \\ & S^c(n-m-1)(t_f) = \Gamma(n-m-1) \\ -\varepsilon_{n-m}\dot{S}^c_{n-m}(n-m-1) = & \varepsilon_{n-m}A_{11}^T(m)\dot{S}^c_{n-m}(n-m-1) + A_{21}^T(m)S^c_{n-m} \\ & +S^c(n-m-1)A_{12}(m) + C_1^T(m)C_2(m) \\ & +S^c_{n-m}(n-m-1)A_{22}(m) \\ & -F^c(n-m-1)R(m-1)F^{c^T}_{n-m}, \\ & S^c_{n-m}(n-m-1)(t_f) = \Gamma^{n-m-1}_{n-m} \\ -\varepsilon_{n-m}\dot{S}^c_{n-m} = & \varepsilon_{n-m}A_{12}^T(m)S^c_{n-m}(n-m-1) + A_{22}^T S^c_{n-m} \\ & +\varepsilon_{n-m}S^c_{n-m}(n-m-1)A_{12}(m) + S^c_{n-m}A_{22}(m) \\ & +C_2^{c^T}(m)C_2(m) - F^c_{n-m}R(m-1)F^{c^T}_{n-m}, \\ & S^c_{n-m}(t_f) = \Gamma_{n-m,n-m} \end{cases}$$

$$(3.10)\quad \begin{cases} -\dot{\bar{S}}^c(n-m-1) = & A_{11}^T(m)\bar{S}^c(n-m-1) + \bar{S}^c(n-m-1)A_{11}(m) \\ & +C_1^T(m)C_1(m) + A_{21}^T(m)\bar{S}^{c^T}_{n-m}(n-m-1) \\ & +\bar{S}^c_{n-m}(n-m-1)A_{21}(m) \\ & -\bar{F}^c(n-m-1)R(m-1)\bar{F}^{c^T}(n-m-1), \\ & \bar{S}^c(n-m-1)(t_f) = \Gamma(n-m-1) \\ 0 = & A_{21}^T(m)\bar{S}^c_{n-m} + \bar{S}^c_{n-m-1}A_{12}(m) \\ & +\bar{S}^c_{n-m}(n-m-1)A_{22}(m) + C_1^T(m)C_2(m) \\ & -\bar{F}^c(n-m-1)R(m-1)\bar{F}^{c^T}_{n-m} \\ 0 = & A_{22}^T(m)\bar{S}^c_{n-m} + \bar{S}^c_{n-m}A_{22}(m) + C_2^{c^T}(m)C_2(m) \\ & -\bar{F}^c_{n-m}R(m-1)\bar{F}^{c^T}_{n-m} \end{cases}$$

where

$$\bar{F}^c(n-m-1) = [\bar{S}^c(n-m-1)B_1(m) + \bar{S}^{c^T}(n-m-1)B_2(m) + C_1^T(m)D_{m-1}]R^{-1}(m-1)$$

$$\bar{F}^c_{n-m} = [\bar{S}^c_{n-m}B_2(m) + C_2^T(m)D_{m-1}]R^{-1}(m-1)$$

Note that this successive reduction scheme terminates after (n-1) steps. And at the end of (n-1) steps, the control problem (2.1)-(2.3) reduces into n-reduced order control problems in their own time scales. This enables us to obtain the optimal control law for each reduced control problem. The combination of these optimal control laws will be used as the near optimal optimal control law for control problem (2.1)-(2.3).

4. MAIN RESULTS

In this section, the limiting behavior of the optimal control law for control problem (2.1)-(2.3) and the near optimal performance are analyzed. The relationship between the optimal control law for control problem (2.1)-(2.3) and the suboptimal control laws in Section 3 is established. This enables us to achieve the optimal control law for th control problem (2.1)-(2.3) by solving n-reduced order control problems in seperate time scales. Furthermore, the near optimum performance is obtained by using these suboptimal control laws in Section 3.

Before we present the main results, the following result is needed, which provides a relationship between each complete control problem and reduced order control problems corresponding to the complete control problem.

THEOREM 4.1. Let the assumption (A3.1) hold. Furthermore $\tilde{A}_{22}(m) = A_{22}(m) - B_2(m)F^T_{n-m}$ is stable for each $m \in J(0, n-2)$ and the controls $u^m_{n-m} = -F^T(n-m-1)x^{m+1}_{n-m-1}$, $u^m_{n-mf} = -\bar{F}^T_{n-m}x_{n-m}(m+1)$ and

$$
(4.1)\qquad \begin{aligned} u^m_{n-mc} &= [(I - \bar{F}^T_{n-m}A^{-1}_{22}(m)B_2(m))F^T(n-m-1) \\ &\quad + \bar{F}^T_{n-m}A^{-1}_{22}(m)A_{21}(m)]x^m_{n-m-1} - \bar{F}^T_{n-m}x_{n-m}(m) \end{aligned}
$$

are applied to control problems (3.2), (3.4) and (3.1) respectively, where $\bar{F}_{n-m}$ is defined by (3.7). Then

$$
\begin{aligned}
x^m_{n-m-1} &= x^{m+1}_{n-m-1} + O\left(\tfrac{\varepsilon_{n-m}}{\varepsilon_{n-m-1}}\right) \\
x_{n-m}(m) &= -A^{-1}_{22}(m)[A_{21}(m) - B_2(m)F^T(n-m-1)]x^{m+1}_{n-m-1} \\
&\quad + x_{n-m}(m) + O\left(\tfrac{\varepsilon_{n-m}}{\varepsilon_{n-m-1}}\right) \\
u^m_{n-mc} &= u^m_{n-ms} + u^m_{n-mf} + O\left(\tfrac{\varepsilon_{n-m}}{\varepsilon_{n-m-1}}\right) \\
y^m &= y^{m+1} + y_{n-mf} + O\left(\tfrac{\varepsilon_{n-m}}{\varepsilon_{n-m-1}}\right)
\end{aligned}
$$

hold for all $t_0 \le t \le t_f$.

Proof: The feedback system (3.1) with u^m_{n-mc} reduces to

$$
\begin{bmatrix} \dot{x}^m_{n-m-1} \\ \varepsilon_{n-m}\dot{x}_{n-m}(m) \end{bmatrix} =
$$

$$
\begin{bmatrix}
A_{11}(m) - B_1(m)[I - \bar{F}^T_m A^{-1}_{22}(m)B_2(m)]F^T(n-m-1) & A_{12}(m) - B_1(m)\bar{F}^T_{n-m} \\
-B_1(m)\bar{F}^T_{n-m}A^{-1}_{22}(m)A_{21}(m) & \\
[A_{22}(m) - B_2(m)\bar{F}^T_{n-m}]A^{-1}_{22}(m) & A_{22}(m) - B_2(m)\bar{F}^T_{n-m} \\
\times[A_{21}(m) - B_2(m)F^T(n-m-1)] &
\end{bmatrix}
$$

$$
\times \begin{bmatrix} x^m_{n-m-1} \\ x_{n-m}(m) \end{bmatrix}.
$$

Apply a transformation T_m in [2] to the above feedback system, where T_m is described by

$$
T_m = \begin{bmatrix} I^{n-m-1} + \varepsilon_{n-m}M_mL_m & \varepsilon_{n-m}M_m \\ -L_m & I_{n-m} \end{bmatrix},
$$

$$
I^r = I_{\sum_{i=1}^r n_i \times \sum_{i=1}^r n_i}, \; I_r = I_{n_r \times n_r}, \; r \in J(1,n)
$$

$$
\begin{aligned}
L_m &= -\tilde{A}^{-1}_{22}(m)[A_{21}(m) - B_2(m)F^T(n-m-1)] + O\left(\tfrac{\varepsilon_{n-m}}{\varepsilon_{n-m-1}}\right) \\
M_m &= [A_{12}(m) - B_1(m)\bar{F}^T_{n-m}]\tilde{A}^{-1}_{22}(m) + O\left(\tfrac{\varepsilon_{n-m}}{\varepsilon_{n-m-1}}\right)
\end{aligned}.
$$

From the results in [2], we have

$$\begin{aligned} x^m_{n-m-1} &= x^{m+1}_{n-m-1} + O(\tfrac{\varepsilon_{n-m}}{\varepsilon_{n-m-1}}) \\ x_{n-m}(m) &= -A^{-1}_{22}(m)[A_{21}(m) - B_2(m)F^T(n-m-1)]x^{m+1}_{n-m-1} \cdot \\ &\quad + x_{n-m}(m+1) + O(\tfrac{\varepsilon_{n-m}}{\varepsilon_{n-m-1}}) \end{aligned}$$

Since $u^m_{n-mc} = -F^T(n-m-1)x^m_{n-m-1} - \bar{F}^T_{n-m}x_{n-m}(m)$, this implies that

$$u^m_{n-mc} = u^m_{n-ms} + u^m_{n-mf} + O(\frac{\varepsilon_{n-m}}{\varepsilon_{n-m-1}}).$$

To show the last result, it is enough to show that $y^m - y^{m+1} - y_{n-mf} = O(\frac{\varepsilon_{n-m}}{\varepsilon_{n-m-1}})$. From the above results and definitions of y^m, y^{m+1} and y_{n-mf}, we have

$$\begin{aligned} y^m - y^{m+1} - y_{n-mf} &= C^m x^{m+1}_{n-m-1} + C_2(m)A^{-1}_{22}(m)B_2(m)F^T(n-m-1)x^{m+1}_{n-m-1} \\ &\quad + C_2(m)x_{n-m}(m+1) + D^{m-1}u^m_{n-ms} + D^{m-1}u^m_{n-mf} \\ &\quad - C^m x^{m+1}_{n-m-1} - D^m u^m_{n-ms} + C_2(m)x_{n-m}(m+1) \\ &\quad - D^{m-1}u^m_{n-mf} + O(\tfrac{\varepsilon_{n-m}}{\varepsilon_{n-m-1}}) \\ &= C_2(m)A^{-1}_{22}(m)B_2(m)F^T(n-m-1)x^{m+1}_{n-m-1} \\ &\quad + C_2(m)A^{-1}_{22}(m)B_2(m)u^m_{n-ms} + O(\tfrac{\varepsilon_{n-m}}{\varepsilon_{n-m-1}}). \end{aligned}$$

From the definition of u^m_{n-ms}, the right hand side of the above expression is $O(\frac{\varepsilon_{n-m}}{\varepsilon_{n-m-1}})$. This completes the prove of the Theorem 4.1.

THEOREM 4.2. Let the assumptions of Theorem 4.1 hold. Furthermore assume that for $m \in J(0, n-2)$

(1) $(A^{m+1}, B^{m+1}, C^{m+1})$ and $(A_{22}(m), B_2(m), C_2(m))$ are stabilizable and detectable.

(2) $R(m)$ is postive definite.

Then u^m_{n-m} is the near optimal control for control problem (3.1) can be approximated by

$$u^m_{n-m} = u^m_{n-mc} + O(\frac{\varepsilon_{n-m}}{\varepsilon_{n-m-1}})$$

where u^m_{n-m} and u^m_{n-mc} are defined in (3.8) and (4.1), respectively.
Proof: From invertability of $\tilde{A}_{22}(m)$, definitions of $R(m)$ and $\bar{F}_{n-m}$ together with some manipulations, we obtain and identity

$$[I - \bar{F}^T_{n-m}A^{-1}_{22}(m)B_2(m)]R^{-1}(m) = R^{-1}(m)[I + B^T_2(m)\tilde{A}^{-1}_{22}(m)^T\bar{F}_{n-m}].$$

And by applying the above identity and the definitions of $F(n-m-1)$, u^m_{n-m}, in (4.1) can be written as follows:

$$\begin{aligned} u^m_{n-mc} &= -R^{-1}(m-1)\{[B^T_1(m)S(n-m-1) + B^T_2(m)S^m_{n-m-1}(n-m-1) \\ &\quad + D^T_{m-1}C_1(m)]x^m_{n-m-1}B^T_2(m)\bar{S}_{n-m} + D^T_{m-1}C_2(m)]x_{n-m}(m)\} \end{aligned} \tag{4.2}$$

where

$$\begin{aligned} S^T_{n-m}(n-m-1) &= \{\bar{S}^c(n-m-1)[B_1(m)\bar{F}^T_{n-m} - A_{12}(m)] - [A^T_{21}(m)\bar{S}^c_{n-m} \\ &\quad + C^T_1(m)C_2(m) - C^T_1(m)D_{m-1}\bar{F}^T_{n-m}]\}\tilde{A}^{-1}_{22}(m). \end{aligned}$$

It is known that the solution of the Riccati systm (3.9) can be approximated by the solution of the reduced system (3.10), i.e.

$$\begin{aligned} S^c(n-m-1) &= \bar{S}^c(n-m-1) + O(\tfrac{\varepsilon_{n-m}}{\varepsilon_{n-m-1}}) \\ S^c_{n-m}(n-m-1) &= \bar{S}^c_{n-m}(n-m-1) + O(\tfrac{\varepsilon_{n-m}}{\varepsilon_{n-m-1}}) \\ S^c_{n-m} &= \bar{S}^c_{n-m} + O(\tfrac{\varepsilon_{n-m}}{\varepsilon_{n-m-1}}) \end{aligned}$$

Obviously $\bar{S}^c_{n-m} \equiv \bar{S}_{n-m}$, where $\bar{S}^c_{n-m}$ and $\bar{S}_{n-m}$ are defined in (3.10) and (3.7). And by substituting $\bar{S}^c_{n-m}(n-m-1)$ in (3.10), one can shoow that $\bar{S}^c(n-m-1) \equiv \bar{S}(n-m-1)$, where $\bar{S}(n-m-1)$ is defined in (3.4). Hence $u^m_{n-m} = u^m_{n-mc} + O(\frac{\varepsilon_{n-m}}{\varepsilon_{n-m-1}})$.

COLLOLARY 4.3. Let the assumptions in Theorem 4.2 hold. Then the optimal control law u^0_n for control problem (2.1)-(2.3) can be approximated by

$$u^0_n = u^0_{nc} + \sum_{k=2}^{n} O(\frac{\varepsilon_k}{\varepsilon_{k-1}}) = u^{n-2}_{1s} + \sum_{k=2}^{n} u^{n-k}_{kf} + \sum_{k=2}^{n} O(\frac{\varepsilon_k}{\varepsilon_{k-1}})$$

where u^{n-2}_{1s}, u^{n-i}_{if} is the optimal control law for ε_2-overall reduced control problem and ε_i-fast reduced control problems.
Proof: Proof follows from Theorem 4.1 and 4.2.

THEOREM 4.4. Let the assumptions in Theorem 4.2 hold. Then the optimal performance index J^m_{n-m} in (3.1) can be approximated by

$$J^m_{n-mc} = J^m_{n-m} + O(\frac{\varepsilon^2_{n-m}}{\varepsilon^2_{n-m-1}})$$

where J^m_{n-mc} is the value of the performance index J^m_{n-m} with u^m_{n-mc}.
Proof: u^m_{n-mc} in (4.2) from Theorem 4.2 can be rewritten as

$$\begin{aligned} u^m_{n-mc} = -R^{-1}(m-1)\Bigg\{B^{m^T}\begin{bmatrix} S(n-m-1) & 0 \\ \varepsilon_{n-m}S^T_{n-m}(n-m-1) & \bar{S}_{n-m}\end{bmatrix} \\ + \begin{bmatrix} D^T_{m-1}C_1(m) \\ D^T_{m-1}C_2(m)\end{bmatrix}^T\Bigg\} x^m_{n-m} \\ = -R^{-1}(m-1)[B^{m^T}S^{mc} + C^{m^T}D_{m-1}]x^m_{n-m}. \end{aligned}$$

Similarly u^m_{n-m} in (3.8) can be rewritten as

$$u^m_{n-m} = -R^{-1}(m-1)[B^{m^T}M^m + C^{m^T}D_{m-1}] = -F^m x^m_{n-m}$$

where

$$S^m = \begin{bmatrix} S^c(n-m-1) & S^c_{n-m}(n-m-1) \\ \varepsilon_{n-m}S^{c^T}_{n-m}(n-m-1) & S_{n-m}\end{bmatrix},$$

$$-\dot{S}^m = A^{m^T}S^m + S^mA^m - F^mR^{-1}(m-1)F^m + C^{m^T}C^m \tag{4.3}$$

The value of the performance index J^m_{n-m} with u^m_{n-m} and u^m_{n-mc} can be described by

$$J^m_{n-m} = x^{m0^T}_{n-m}S^m x^{m0}_{n-m} \; and \; J^{mc}_{n-m} = x^{m0^T}_{n-m}P^m x^{m0}_{n-m}$$

where S^m satisfies (3.9) and P^m satisfies

$$-\dot{P}^m = [A^m - B^m F^{mc^T}]^T P^m + P^m [A^m - B^m F^{mc^T}] + C^{m^T} C^m \tag{4.4}$$

where $F^{mc} = [S^{mc} B^{m^T} + D_{m-1} C^{m^T}] R^{-1}(m-1)$.
Let $W^m = P^m - S^m$, from (4.3) and (4.4), we have

$$\begin{aligned} \dot{W}^m &= (A^m - B^m F^{mc^T})^T W^m + W^m (A^m - B^m F^{mc^T}) \\ &\quad + (F^m - F^{mc})^T R^{-1}(m-1)(F^m - F^{mc}). \end{aligned} \tag{4.5}$$

From Theorem 4.2, as $\varepsilon_{n-m} \to 0$, we have $F^m - F^{mc} = O(\frac{\varepsilon_{n-m}}{\varepsilon_{n-m-1}})$. Therefore the last term of (4.5) is of $O(\frac{\varepsilon^2_{n-m}}{\varepsilon^2_{n-m-1}})$. Since $\tilde{A}_{22}(m)$ and $A^{m+1} - B^{m+1} F^T(n-m-1)$ are stable, we can conlude that W^m is of $O(\frac{\varepsilon^2_{n-m}}{\varepsilon^2_{n-m-1}})$. This implies $P^m = S^m + O(\frac{\varepsilon^2_{n-m}}{\varepsilon^2_{n-m-1}})$. That is $J^m_{n-mc} = J^m_{n-m} + O(\frac{\varepsilon^2_{n-m}}{\varepsilon^2_{n-m-1}})$.

COLLOLARY 4.5. Let the assumptions of Theorem 4.4 hold. Then the optimal performance index J in (2.3) is approximated by

$$J^0_{nc} = J^0_n + \sum_{k=2}^{n} O(\frac{\varepsilon^2_{n-k}}{\varepsilon^2_{n-k-1}})$$

where J^0_n and J^0_{nc} are the values of the performance index J with the u^0_n and u^0_{nc}, respectively.
Proof: The proof follows from Theorem 4.4.

REFERENCES

[1] J.H. Chow and P.V. Kokotovic, A decomposition of near-optimum regulators for systems with slow and fast modes, *IEEE Tran. Automat. Contr.* AC-21(1976), pp.701-705.

[2] M. Kathirkamanayagan and G.S. Ladde, Diagonalization and stability of large-scale singularly perturbed linear systems, *J. of Math. Anal. and Appl.*, (in press).

[3] P.V. Kokotovic, H.K. Khalil and O'Reilly, *Singular Perturbation Methods in Control: Analysis and Design*, Academic Press, New York, 1986.

[4] G.S. Ladde and O. Sirisaengtaksin, On near-optimum regulator for large-scale systems, *the VIIth International Conference on Nonlinear Analysis and Applications, edited by V. Lakshmikantham*, Marcel Dekker, Inc., New York, 1987, pp.213-238.

[5] R.E. O'Mally, Jr., *Introduction to Singular Perturbations*, Academic Press, New York, 1974.

[6] V.R. Saksena, J. O'Reilly and P.V. Kokotovic, Singular perturbations and time-scales methods in control theory: Survey 1979-1983, *Automatica* 20 (1984), pp. 273-293.

On Some Optimal Control Problems of Differential Equations with Additional Equality Constraints

Urszula Ledzewicz-Kowalewska, Professor, Department of Mathematics and Statistics, Southern Illinois University at Edwardsville, Edwardsville, Illinois

ABSTRACT

In the paper the problems of optimal control of ordinary differential equations and of linear partial differential equations with additional equality constraints are considered. Extentions of the Dubovitskii-Milyutin method for the case of n equality constraints given by Fréchet and by Gâteaux differentiable operators are presented. For these various optimal control problems the local extremum principles obtained by using these specifications of the Dubovitskii-Milyutin theorem are given.

I. FORMULATION OF PROBLEM I. ON SOME SPECIFICATION OF THE DUBOVITSKII-MILYUTIN METHOD.

First, let us consider the following problem I: minimize the functional

$$I(x,u) = \int_0^1 f^0(x,u,t)dt \qquad (1)$$

under the constraints

$$\dot{x} = f(x,u,t) \ , \ x(0) = 0; \qquad (2)$$

$$x(1) = x_1; \qquad (3)$$

$$g(x,u,t) = 0; \qquad (4)$$

$$u \in U; \qquad (5)$$

where $x \in \overline{W}^n_{11}(0,1)$, $u \in L^r_\infty(0,1)$; the set $U = \{u \in L^r_\infty(0,1) : u(t) \in M \subset R^r\}$; functions

---

*on leave from: *Institute of Mathematics, University of Lodz, 90-238 Lodz, Poland*

$f^0 : R^n \times R^r \times R \to R$, $f : R^n \times R^r \times R \to R^n$, $g : R^n \times R^r \times R \to R^k$; x_1 is a fixed point of the space R^n.

We assume that

- there exist derivatives f^0_x, f^0_u, f_x, f_u, g_x, g_u which are bounded for any $(x,u) \in \overline{W}^n_{11} \times L^r_\infty$;
- the functions f^0, f, f^0_x, f^0_u, f_x, f_u are continuous with respect to (x,u) for any $t \in [0,1]$ and measurable with respect to t, (g_x, g_u) is continuous with respect to (x,u) in the topology of the space $L(\overline{W}^n_{11} \times L^r_\infty, L^k_\infty)$;
- the set M is closed, convex and possesses a nonempty interior.

Remark 1.1. In Problem I we can add the inequality operator constraints

$$h_j(x,u,t) \leq 0, \qquad j = 1, 2, ..., l. \tag{6}$$

but the problem will be of similar difficulty. It is considered in [9].

Remark 1.2. The space $\overline{W}^n_{11}(0,1)$ is a subspace of the space $W^n_{11}(0,1)$ which contains functions satisfying the condition $x(0) = 0$.

The space $W^n_{11}(0,1)$ is the space of absolutely continuous functions whose derivatives $\dot{x} \in L^n_1$. The norm in this space is of the form

$$\| x \| = | x(0) | + \int_0^1 | \dot{x}(t) | \, dt.$$

Remark 1.3. This problem I, but without the nonoperator constraint $u \in U$, where U-some set, was investigated by Boltianski, Makowski, Anorow, Dzgarkawa in [3], [16], [2] and [4], respectively, by making use of the variational method.

Necessary condition for Problem I cannot be proved by using the Dubovitskii-Milyutin method because this method can be applied to obtain the extremum principle for problems of optimal control with only one equality constraint.

This follows from the fact that the condition

$$\Big(\bigcap_{i=1}^{n} C_i\Big)^* = \sum_{i=1}^{n} C_i^* \quad \text{(where } C_i \text{ are cones, } C_i^* \text{ are cones dual to them)}$$

holds only for convex and open cones which correspond to the inequality constraints (cf. [5], Lemma 5.7).

For the convex and closed cones which correspond to the equality constraints,this condition has the form

$$\Big(\bigcap_{i=1}^{n} C_i\Big)^* = \overline{\sum_{i=1}^{n} C_i^*} \qquad \text{(cf. [5], Lemma 5.8)}$$

where by $^-$ we denote the weak * closure.

This means that, in order to apply the Dubovitskii-Milyutin method to problems of optimal control with more than one equality constraint, we must find the assumptions under which the algebraic sum of closed sets is closed.

We know that the algebraic sum of two closed subspaces is closed when one of them is of finite dimensionality (cf. [5], Lemma 2.6). This fact was applied in Girsanov's monograph (cf. [5]) to obtain the extremum principle for a problem of optimal control with the fixed right point of the trajectory.

In [17] Walczak used some properties of cones in normed spaces to get the conditions under which the algebraic sum of weakly * closed cones is weakly * closed. Then, by using these results, he obtained some generalization of the Dubovitskii-Milyutin method for the case of n equality constraints in arbitrary, even nonoperator form under some assumptions about the cones. This generalization was applied in [11], [18] and [19] to problems of optimal control with additional equality constraints.

In [7] some specifications of the Dubovitskii-Milyutin method is given without any additional assumptions about the cones but for the case of n equality constraints in the operator form.

The following lemma is proved.

LEMMA 1.1. ([7]) If the equality constraints Z_i, $i = 1, 2, \ldots, n$, are given in the operator form i.e. $Z_i = \{x \in X : F_i(x) = 0\}$ for $i = 1, 2, \ldots, n$, where $F_i : X \to Y_i$ are Fréchet differentiable operators at the point $x_0 \in \bigcap_{i=1}^{n} Z_i$, X, Y_i, $i = 1, 2, \ldots, n$ are Banach spaces, the operator $F : X \to Y_1 \times \ldots \times Y_n$ in the form

$$F(x) = \big(F_1(x), F_2(x) \ldots F_n(x)\big)$$

is regular at the point x_0, then the condition

$$\Big(\bigcap_{i=1}^{n} C_i\Big)^* = \sum_{i=1}^{n} C_i^* \qquad \text{holds,}$$

where $C_i = TC(Z_i, x_0)$, i.e. by C_i we denote a cone tangent to the set Z_i at the point x_0.

By making use of Lemma 1.1 some specification of the Dubovitskii-Milyutin lemma and the Dubovitskii-Milyutin theorem is obtained in [7] for the case of n equality constraints given in the operator form.

THEOREM 1.1. ([7]) Let

1° functional $G(x)$ attain its local minimum on the set $Z = \bigcap_{i=1}^{n+m} Z_i$ at the point $x_0 \in Z$;

2° $G(x)$ decrease regularly at the point x_0 with the cone of decrease C_0;

3° the inequality constraints $Z_i \subset X$ for $i = 1, 2, \ldots, n$ be regular at the point x_0 with feasible cones C_i, $i = 1, 2, \ldots, n$;

4° the equality constraints Z_i, $i = n + 1 \ldots n + m$, be in the form

$$Z_i = \{x \in X \ : \ F_i(x) = 0\}$$

where $F_i : X \to Y_i$ for $i = n + 1 \dots n + m$ are regular operators at the point $x_0 \in X$ with tangent cones C_i, $i = n + 1 \dots n + m$; the operator $F : X \to Y_{n+1} \times Y_{n+2} \times \dots \times Y_{n+m}$ given by the formula

$$F(x) = \left(F_{n+1}(x) \dots F_{n+m}(x)\right)$$

be such that $\operatorname{Im} F'(x_0)$ is closed in $Y_{n+1} \times \dots \times Y_{n+m}$,

then there exist linear and continuous $f_i \in C_i^*$, $i = 0, 1 \dots n + m$, not vanishing simultaneously and such that

$$f_0 + f_1 + \dots + f_n + f_{n+1} + \dots + f_{n+m} = 0.$$

Remark 1.4. The last equation is a specification of the Euler-Lagrange equation for the case of n equality constraints in the operator form.

Remark 1.5. By using the same Lemma 1.1 some specifications of the Dubovitskii-Milyutin sufficient condition for convex problems with n equality constraints in the operator form is obtained in [10].

II. THE EXTREMUM PRINCIPLE.

Above specification of the Dubovitskii - Milyutin method has been applied to problems of optimal control of ordinary differential equations in [8],[9] and of partial differential equations in [10],[12] with additional equality constraints.

First, we can obtain the extemum principle for problem I.

THEOREM 2.1. ([8]) Local extremum principle. If

1° (x^0,u^0) is an optimal process in problem I,

2° an operator A: $\overline{W}^n_{11} \times L^r_\infty \to \overline{W}^n_{11} \times L^k_\infty \times R^n$ given by the formula

$$A(\bar{x},\bar{u})(t) = \left(\bar{x}(t) - \int_0^t (f_x\bar{x} + f_u\bar{u})dt,\ g_x\bar{x} + g_u\bar{u},\ \bar{x}(1)\right)$$

is such that its image is closed in $\overline{W}^n_{11} \times L^k_\infty \times R^n$,

then there exist $\lambda_0 \geq 0$, $\psi \in L^n_\infty(0,1)$, $\omega \in b^k.\ a(0,1)$, $a \in R^n$ such that

1° $\|\ \omega\ \| + |\lambda_0| + \|\psi\| + \|a\| > 0,$

2° $-\lambda_0 \int_0^1 f^0_x\bar{x}dt + \int_0^1(\dot{\bar{x}} - f_x\bar{x})\psi(t)dt + \int_0^1 g_x\bar{x}d\omega + (a,\bar{x}(1)) = 0$ for any $\bar{x} \in \overline{W}^n_{11}$,

3° $\lambda_0 \int_0^1 f^0_u u^0(t)dt + \int_0^1 f_u u^0(t)\psi(t)dt - \int_0^1 g_u u^0(t)d\omega = \min_{u\in U}(\lambda_0 \int_0^1 f^0_u u(t)dt +$

$+ \int_0^1 f_u u(t)\psi(t)dt - \int_0^1 g_u u(t)d\omega).$

Remark 2.1. The form of the proposition of Theorem 2.1 is rather difficult in applications. This follows from the fact that space dual to $L^k_\infty(0,1)$ is the space $b.^k a(0,1)$.

If we assume that the function $\omega = (\omega_1, \omega_2, \dots \omega_k) \in b.^k a(0,1)$ from Theorem 2.1 has the property that the functions ω_i, i = 1, 2, ..., k, are measures, we can rewrite the proposition of this theorem in a simpler form: there exists $\lambda_0 \geq 0$, $v \in L_1^k$ and an absolutely continuous function $\phi : [0,1] \to R^n$, not vanishing simultaneously, satisfying the adjoint equation

$$\dot{\phi} = \lambda_0 f_x^0 - f_x^* \phi - g_x^* v \tag{7}$$

and such that

$$\left(\lambda_0 f_u^0 - f_u^* \phi - g_u^* v,\ u - u^0(t)\right) \geq 0 \tag{8}$$

for any $u \in U$ and $t \in [0,1]$ a.e.

Remark 2.2. In [9] it is proved, without the assumption that ω_i, i = 1, 2, ..., k, are measures, that from condition 2° of the proposition of Theorem 2.1 and the properties of linear and continuous functionals follows the adjoint equation (7).

Really let us consider problem I with additional inequality constraints given by (6) where $h_j : R^n \times R^r \times R \to R$, j = 1, 2, ..., l. (We shall call it problem II).

In addition to the previous assumptions of Problem I we have to assume that

- there exist derivatives h_{ix}, h_{iu}, i = 1, 2, ..., l which are bounded for any $(x,u) \in \overline{W}_{11}^n \times L_\infty^r$;
- h_{ix}, h_{iu} for i = 1, 2, ..., l are continuous with respect to (x,u) for any t and measurable with respect to t.

We also introduce

Definition 2.1. Let X be a Banach space of measurable, essentially bounded functions defined on the interval [0,1]. Each functional f defined on X of the form

$$f(x) = \int_0^1 \tilde{f}(t)x(t)dt \quad \text{where } \tilde{f} \in L_1(0,1)$$

will be called absolutely continuous.

LEMMA 2.1. Any linear and continuous functional f can be written in the form $f = f^a + f^s$, where f^a - an absolutely continuous functional, f^s - a singular functional. This representation is unequivalent.

In [9] the following extremum principle is proved for problem II.

THEOREM 2.2. ([9]) If

1° (x^0,u^0) is an optimal process in problem PII,

2° the operator $A : \overline{W}_{11}^n \times L_\infty^r \to \overline{W}_{11}^n \times L_\infty^k \times R^n$ in the form

$$A(\bar{x},\bar{u})(t) = \Big(\bar{x}(t) - \int_0^t \big(f_x(x^0,u^0,t)\bar{x}(t) + f_u(x^0,u^0,t)\bar{u}(t)\big)dt,$$
$$g_x(x^0,u^0,t)\bar{x}(t) + g_u(x^0,u^0,t)\bar{u}(t), \bar{x}(1)\Big)$$

is such that its image is closed in $\overline{W}^n_{11} \times L^k_\infty \times R^n$,

3° $h_{ix}\left(x^0(t),u^0(t),t\right) \neq 0$ or $h_{iu}\left(x^0(t),u^0(t),t\right) \neq 0$ for any $t \in R_i$

where

$$R_i = \{t \in [0,1] \ : \ h_i\left(x_0(t),u_0(t),t\right) = 0\} \text{ for any } i = 1, 2, ..., l$$

then there exist $\lambda_0 \geqq 0$ and functions $\psi \in L^n_\infty(0,1)$, $\omega \in b\,.^k a(0,1)$, $\gamma_i \in b.\ a(0,1)$ concentrated on the sets R_i for $i = 1, 2, ..., l$, such that

1° $||\omega|| + |\lambda_0| + ||\psi|| + \sum_{i=1}^{l} ||\gamma_i|| > 0,$

2° $\dot{\phi} = -\lambda_0 f^0_x(x^0,u^0,t) - f^*_x(x^0,u^0,t)\phi + g^*_x(x^0,u^0,t)\widetilde{f}_\omega - \sum_{i=1}^{l} h^*_{ix}(x^0,u^0,t)\widetilde{f}_{\gamma_i}$ where $\phi = -\psi$, $\widetilde{f}_\omega$, $\widetilde{f}_{\gamma_i}$ are the integrable functions corresponding to the functionals f_ω and f_{γ_i}, respectively, in the sense of Lemma 2.1 and Definition 2.1, functionals f_ω and f_{γ_i}, $i = 1, 2, ..., l$ are given in the form

$$f_\omega(v) = \int_0^1 v(t)d\omega, \qquad f_{\gamma_i}(s) = \int_0^1 s(t)d\gamma_i, \ i = 1, 2, ..., l.$$

for $s \in L^1_\infty$, $v \in L^k_\infty$.

3°
$$\lambda_0 \int_0^1 f^0_u(x^0,u^0,t)u^0(t)dt + \int_0^1 f_u(x^0,u^0,t)u^0(t)\phi(t)dt -$$

$$- \int_0^1 g_u(x^0,u^0,t)u^0(t)d\omega + \sum_{i=1}^{l} \int_0^1 h_{iu}(x^0,u^0,t)u^0(t)d\gamma_i =$$

$$= \min_{u \in U} \Big(\lambda_0 \int_0^1 f^0_u(x^0,u^0,t)u(t)dt + \int_0^1 f_u(x^0,u^0,t)u(t)\phi(t)dt -$$

$$- \int_0^1 g_u(x^0,u^0,t)u(t)d\omega + \sum_{i=1}^{l} \int_0^1 h_{iu}(x^0,u^0,t)u(t)d\gamma_i\Big).$$

III. FORMULATION OF PROBLEM III. THE EXTREMUM PRINCIPLE.

Proceeding in the similar way, we can apply the specification of the Dubovitskii-Milyutin method to problems of optimal control in Sobolev spaces, as well. We can consider such problems of partial differential equations with additional equality constraints in the operator form.

The example of it is the optimal control problem of parabolic differential equations with terminal condition considered in [10], i.e. problem III of the form:

minimize the functional

$$I(y,u) = \int_Q f(x,t,y,u)dxdt \tag{9}$$

under the constraints

$$\frac{\partial y}{\partial t} + A(t)y = u, \; x \in \Omega, \; t \in (0,T), \tag{10}$$

$$y(x,t) = 0, \quad x \in \Gamma, \quad t \in (0,T), \tag{11}$$

$$y(x,0) = y_0(x), \tag{12}$$

$$y(x,T) = y_1(x), \tag{13}$$

$$u \in U \tag{14}$$

where $f : R^n \times R^1_+ \times R^n \times R \to R$, $\Omega \subset R^n$ is a bounded open set with boundary Γ which is a C^∞ manifold of dimension $(n - 1)$. Locally, Ω is totally on one side of Γ, U is a closed, convex, with nonempty interior subset of the space $L^2(Q)$ where $Q = \Omega \times (0,T)$. The functions $y_0(x)$ and $y_1(x)$ are some fixed functions of the space $L^2(\Omega)$.

The operator A(t) has the form

$$A(t)y = -\sum_{ij=1}^{n} \frac{\partial}{\partial x_i} a_{ij}(x,t) \frac{\partial y(x,t)}{\partial x_j} \tag{15}$$

where $a_{ij}(x,t) \in C^1(Q)$ for i,j = 1, 2, ..., n and satisfy the condition

$$\sum_{ij=1}^{n} a_{ij}(x,t)\xi_i\xi_j \geq \alpha \sum_{i=1}^{n} \xi_i^2 > 0 \text{ for any } (x,t) \in Q \tag{16}$$

and $\xi_i \in R$, i=1,2, ..., n.

We consider problem III under the assumptions

(i) there exist derivatives f_y and f_u which are bounded for any $(y,u) \in L^2(O,T; H_0^1(\Omega)) \times L^2(Q)$.

(ii) the functions f, f_y, f_u are continuous with respect to (y,u) for any $(x,t) \in Q$ and measurable with respect to (x,t).

We denote by $Y = L^2(O,T ; H_0^1(\Omega))$ the space of states, by $Y^* = L^2(0,T ; H^{-1}(\Omega))$ the space of adjoint states.

Problem III includes two equality constraints in the operator form: conditions (10), (11), (12) form the first, condition (13) is the second.

Making use of Theorem 1.1 (a specification of the Dubovitskii-Milyutin theorem in the case of n equality constraints in the operator form), we can obtain the extremum principle for problem III.

THEOREM 3.1. ([10]) **Local extremum principle.** If

1° (y^0,u^0) is an optimal process in problem III,

2° an operator $A : Y \times L^2(Q) \to L^2(0,T ; H^{-1}(\Omega)) \times L^2(\Omega) \times L^2(\Omega)$ given by the formula

$$A(\bar{y},\bar{u})(t) = \left(\frac{\partial \bar{y}}{\partial t} + A\bar{y} - \bar{u},\ \bar{y}(x,0),\ \bar{y}(x,T)\right)$$

is such that its image is closed in $L^2(0,T\ ;\ H^{-1}(\Omega)) \times L^2(\Omega) \times L^2(\Omega)$,

then there exist $\lambda_0 \geq 0$ and a function $p(\cdot)$ satisfying the equation

$$\frac{\partial p}{\partial t} = -\ \lambda_0 f_y + A^*(t)p \qquad \text{for} \qquad x \in \Omega,\ t \in (0,T)$$

with a boundary condition

$$p(x,t) = 0 \qquad \text{for} \qquad x \in \Gamma,\ t \in (0,T),$$

and such that the maximum condition

$$\int_Q (p + \lambda_0 f_u)(u - u^0)dxdt \geq 0 \text{ for any } u \in U \text{ holds.}$$

Remark 3.1. Making use of the Dubovitskii-Milyutin sufficient condition (cf. Remark 1.5) in [10], we obtained the sufficient condition of optimality for problem III under the assumptions of the convexity and continuity of the function f.

Remark 3.2. Similar optimal control problem of parabolic system, but with different cost functional (9) and without terminal condition (13) was considered by Lions in [15].

Remark 3.3. Another problem of optimal control with additional equality constraint has been considered in [12]. There is an optimal control problem of second order hyperbolic system with initial condition, boundary condition and with additional equality constraint in the form of terminal condition $\frac{\partial y(x,T)}{\partial t} = 0$. The local extremum principle for this problem has been proved in [12].

IV. CASE OF OPTIMAL CONTROL PROBLEM WITH GÂTEAUX DIFFERENTIABLE OPERATOR CONSTRAINT.

Specification of the Dubovitskii-Milyutin method from [7] has been extended to the case of n equality constraints given by Gâteaux differentiable operators in [13]. The method of contractor directions has been applied and some extension of the Dubovitskii-Milyutin theorem has been proved.

THEOREM 4.1. ([13]) Let

1° a functional G(x) attain its local minimum on the set $Z = \bigcap_{i=1}^{n+m} Z_i$ at the point $x_0 \in Z$;

2° G(x) decrease regularly at the point x_0 with the cone of decrease C_0;

3° the inequality constraints $Z_i \subset X$ for $i = 1, 2, \ldots, n$ be regular at the point x_0 with the feasible cones C_i, $i = 1, 2, \ldots, n$;

4° the equality constraints Z_i, $i = n + 1, \ldots, n + m$, be given in the form

$$Z_i = \{x \in X : F_i(x) = 0\}$$

where $F_i : X \to Y_i$ for $i = n + 1, \ldots, n + m$ are Gâteaux-differentiable operators in some neighborhood $S(x_0,r)$ of the point $x_0 \in Z$ and G-regular at this point, the operator $F : X \to Y_{n+1} \times \ldots \times Y_{n+m}$ given by the formula $F(x) = (F_{n+1}(x), \ldots, F_{n+m}(x))$ is such that $F|_{\overline{S(x_0,r)}}$ is a closed, nonlinear operator and $\mathrm{Im}\, \nabla F(x_0)$ is closed in $Y = Y_{n+1} \times \ldots \times Y_{n+m}$. In addition, the following condition is satisfied: there is a constant C which has the property that, for every $x \in S(x_0,r)$, $y \in \mathrm{Im} \nabla F(x)$, there exists an element $h \in X$ such that

$$\nabla F(x)h = y \quad \text{and } ||h|| \leq C\, ||y||; \tag{17}$$

5° the cones C_i be of the form

$$C_i = \{x \in X : \nabla F_i(x_0)\, x = 0\};$$

then there exist linear continuous functionals $f_i \in C_i^*$, $i = 0, 1, \ldots, n + m$, not vanishing simultaneously and such that

$$f_0 + f_1 + \ldots + f_n + f_{n+1} + \ldots + f_{n+m} = 0$$

(the Euler-Langrange equation).

Proof. We shall consider two cases.

(1) First, let us assume that $\mathrm{Im}\nabla F(x_0) = Y$. In this case, it is enough to show that

$$\bigcap_{i=0}^{n+m} C_i = \emptyset.$$

Suppose to the contrary that there exists some $h \in \bigcap_{i=0}^{n+m} C_i$.

Since $h \in C_i$ for $i = 0, 1, 2, \ldots, n$, therefore from the definitions of a feasible direction and the direction of decrease of the functional (cf. [5]) and from the fact that the number of sets Z_i, $i = 1, 2, \ldots, n$, is finite it follows that there exist a neighborhood $O(h)$ of the vector h, some $\alpha < 0$ and some $\varepsilon_0 > 0$ such that, for any $\bar{h} \in O(h)$ and $\varepsilon \in (0, \varepsilon_0)$, we have that $x_0 + \varepsilon\bar{h} \in \bigcap_{i=1}^{n} Z_i$ and $G(x_0 + \varepsilon\bar{h}) \leq G(x_0) + \varepsilon\alpha$. Now, let us consider the fact that $h \in C_i$ for $i = n + 1, \ldots, n + m$. From assumption 5 it follows $\nabla F_i(x_0)\, h = 0$ for $i = n + 1, \ldots, n + m$ i.e. we have that $\nabla F(x_0)\, h = 0$. It is easy to check that the operator $F|_{\overline{S(x_0,r)}}$ satisfies all the assumptions of Corollary 1.1 from [1]. Making use of this corollary we get that $h \in TC\left(\bigcap_{i=n+1}^{n+m} Z_i, x_0\right)$. Let consider element $x(\varepsilon) = x_0 + \varepsilon h + r(\varepsilon) \in \bigcap_{i=n+1}^{n+m+1} Z_i$ as in the definition of the tangent direction (cf. [4], Section 5) and let ε_1 be such that $\bar{h}(\varepsilon) = h + \frac{1}{\varepsilon}\, r(\varepsilon) \in 0(h)$ for all $0 < \varepsilon < \varepsilon_1$.

Now, let us put $\tilde{\varepsilon} = \min(\varepsilon_0, \varepsilon_1,)$. Then it is easily noticed that, for any $\varepsilon \in (0, \tilde{\varepsilon})$

$x_0 + \varepsilon\bar{h}(\varepsilon) \in Z$ and $G(x_0 + \varepsilon\bar{h}(\varepsilon)) < G(x_0)$. This contradicts assumption 1. We have thus shown that $\bigcap_{i=0}^{n+m} C_i = \emptyset$. All the assumptions of Lemma 1.2 (1) from [13] are satisfied, so, by making use of part 1 of this lemma, we obtain the Euler-Lagrange equation.

(2) Now, assume that $\mathrm{Im}\nabla F(x_0) \neq Y$. In this case, from assumption 4 we have that $\mathrm{Im}\nabla F(x_0)$ is a proper closed subspace of Y. It is easy to notice that all the assumptions of Lemma 1.2 (2) from [13] are satisfied; then, applying this lemma, we obtain the proposition in this case, too.

This extension has been also applied to optimal control problems in [14].

Let us consider our optimal control problem II, but under weaker assumptions about differentiability.

- there exist Gâteaux derivatives of the functions f^0, f, g, h_i, i = 1, 2, ..., l, with respect to x and with respect to u, denoted ∇f_x^0, ∇f_u^0, ∇f_x, ∇f_u, ∇g_x, ∇g_u, ∇h_{ix}, ∇h_{iu}, for i = 1, 2, ..., l respectively.
- functions f^0, f, h_i, i = 1, 2, ..., l satisfy the Lipschitz condition with respect to (x,u) for any bounded (x,u) uniformly with respect to t and measurable with respect to t for any (x,u);
- function g is continuous with respect to (x,u) and measurable with respect to t for any (x,u), ∇h_{ix}, ∇h_{iu}, i = 1, 2, ..., l are continuous with respect to (x,u,t);
- for any function f^0, f, g one of the following conditions is satisfied: Gâteaux derivative with respect x is continuous with respect to u for any t or Gâteaux derivative with respect to u is continuous with respect to x for any t;
- set U is closed, convex and possesses a nonempty interior in $L_1^r(0,1)$.

Problem II under above weaker assumptions will be called problem II′.

By using Theorem 4.1 the local extremum principle for problem II′ has been proved in [14].

THEOREM 4.2 ([14]). Let

1° (x^0,u^0) be an optimal process for problem II′;

2° there exist a ball $S((x^0,u^0),r)$ with centre at (x^0,u^0) and radius r and a constant C such that for any $(x,u) \in S((x^0,u^0); r)$, $y = (y_1,y_2,y_3) \in \overline{W}_{11}^n \times L_1^k \times R^n$ exists $(\bar{x}, \bar{u}) \in \overline{W}_{11}^n \times L_1^k$ such that

$$y = \left(\bar{x}(t) - \int_0^t (\nabla f_x(x,u,t)\,\bar{x} + \nabla f_u(x,u,t)\bar{u})dt,\ \nabla g_x(x,u,t)\,\bar{x} + \nabla g_u(x,u,t)\,\bar{u},\ \bar{x}(1)\right)$$

and

$$||\dot{\bar{x}}|| + ||\bar{u}|| \leq C(||\dot{y}_1|| + ||y_2|| + ||y_3||) \quad \text{for } t \in [0,1] \text{ a.e.};$$

3° the operator A: $\overline{W}^n_{11} \times L^r_1 \to \overline{W}^n_{11} \times L^k_1 \times R^n$ in the form

$$A(\bar{x},\bar{u})(t) = \Big(\bar{x}(t) - \int_0^t (\nabla f_x(x^0,u^0,t)\,\bar{x}(t) + (\nabla f_u(x^0,u^0,t)\,\bar{u}(t))dt,$$

$$\nabla g_x(x^0,u^0,t)\,\bar{x}(t) + \nabla g_u(x^0,u^0,t)\,\bar{u}(t),\ \bar{x}(1)\Big)$$

has a closed image in $\overline{W}^n_{11} \times L^k_1 \times R^n$;

4° $\nabla h_{ix}(x^0(t), u^0(t),t) \neq 0$ or $\nabla h_{iu}(x^0(t), u^0(t), t) \neq 0$ for any $t \in R_i$, where

(*) $$R_i = \{t \in [0,1] : h_i(x^0(t), u^0(t), t) = 0\} \quad \text{for } i = 1, 2, ..., l;$$

then there exist $\lambda_0 \geq 0$ and functions $\psi \in L^n_1$, $\omega \in L^k_1$, $\gamma_i \in L_1$ for i = 1, 2, ..., l, such that

1° $|\lambda_0| + ||\psi|| + ||\omega|| + \sum_{i=1}^{l} ||\gamma_i|| > 0,$

2° $\dot{\varphi} = \lambda_0 \nabla f^0_x(x^0,u^0,t) - \nabla f^*_x(x^0,u^0,t)\varphi - \nabla g^*_x(x^0,u^0,t)\omega + \sum_{i=1}^{l} \nabla h^*_{ix}(x^0,u^0,t)\gamma_i$

where $\varphi = -\psi$,

3° $h_i(x^0, u^0,t)\gamma_i(t) = 0$ for i= 1,2,..., l,

4° $$\lambda_0 \int_0^1 \nabla f^0_u(x^0,u^0,t)\,u^0(t)dt + \int_0^1 \nabla f_u(x^0,u^0,t)\,u^0(t)\psi(t)dt -$$

$$- \int_0^1 \nabla g_u(x^0,u^0,t)\,u^0(t)\omega(t)dt + \sum_{i=1}^{l} \int_0^1 \nabla h_{iu}(x^0,u^0,t)\,u^0(t)\gamma_i(t)dt =$$

$$= \min_{u \in U} \Big(\lambda_0 \int_0^1 \nabla f^0_u(x^0,u^0,t)\,u(t)dt + \int_0^1 \nabla f_u(x^0,u^0,t)\,u(t)\psi(t)dt -$$

$$- \int_0^1 \nabla g_u(x^0,u^0,t)\,u(t)\omega(t)dt + \sum_{i=1}^{l} \int_0^1 \nabla h_{iu}(x^0,u^0,t)\,u(t)\gamma_i(t)dt \Big).$$

References

1° Altman M.,"An application of the method of contractor directions to nonlinear programming",Numer. Funct. Anal.Optim.1, No.6, (1979).

2° Anorow B. P. , "Princip maksimuma dla processov c organiceniami obscevo vida," Avtomatika i Telemakhanika, No. 3, (1967), pp. 5-15, No. 4, (1967), pp. 5-15.

3° Boltianski W. , Gamkerlidze S., "Matematyczna teoria procesow optymalnych," Warszawa, (1971).

4° Dzgarkawa D.T., "O niedokhodimykh uslovjakh optimalnosti w zadacakh so smiesannymi organiceniami," Biuletin Akademii Nauk Gruzinskoj SSR, 97, No. 3, (1980).

5° Girsanov I. W. , "Lectures on mathematical theory of extremum problems," Lecture Notes in Economics and Mathematical Systems, 67, Springer-Verlag, (1972).

6° Joffe A.D. , Tikhomirov W.M. , "Theory of extremal problems," Academic Press, New York, New York, (1979).

7° Ledzewicz-Kowalewska U. , "On some specification of the Dubovitskii-Milyutin method," Nonlinear Analysis, Vol. 10, No. 12, (1986), pp. 1367-71.

8° Ledzewicz-Kowalewska U . , "Application of some specification of the Dubovitskii-Milyutin method to problems of optimal control," Nonlinear Analysis, Vol. 12, No. 2, (1988), pp. 101-108.

9° Ledzewicz-Kowalewska U. , "The necessary condition for the problem of optimal control with equality and inequality constraints," Control and Cybernetics, Vol. 14, No. 4, (1985), pp. 351-360.

10° Ledzewicz-Kowalewska U. , "The necessary and sufficient condition for the problems of optimal control of parabolic systems with equality constraints," Bull. de la Soc. des Scien. at des Lettr. de Lodz, Vol. XXXVI, 30, No. 40, (1986), pp. 1-12.

11° Ledzewicz-Kowalewska U. , "The extremum principle for problems of optimal control with mixed constraints," Acta Universitatis Lodziensis, Folia Matematica 2, (1987), pp. 37-59.

12° Ledzewicz-Kowalewska U. , "Optimal control problems of hyperbolic systems with operator and nonopeator equality constraints," Applicable Analysis, Vol. 27, pp. 199-215.

13° Ledzewicz-Kowalewska U. , "Application of the method of contractor directions to the Dubovitskii-Milyutin formalism," Journal of Mathematical Analysis and Applications, Vol. 125, No. 1, (1987), pp. 174-184.

14° Ledzewicz-Kowalewska U. , "On the optimal control problems with Gâteaux differentiable operator constraints," Journal of Mathematical Analysis and Applications, Vol. 130, No. 2, (1988), pp. 301-315.

15° Lions I.L. , "Optimal control of systems governed by partial differential equations," Springer-Verlag, New York, New York, (1971).

16° Makowski K. and Neustadt L. W. , "Optimal control phase variable equality and inequality constraints," SIAM Journal Control, Vol. 12, No. 2, (1974).

17° Walczak S. , "On some properties of cones in normed spaces and their applications to investigating extremal problems," Journal of Optimization Theory and Applications, Vol. 42, No. 4, (1984), pp. 561-581.

18° Walczak S. , "On some control problem," Acta Universitatis Lodziensis, Folia Matematica 1, (1984).

19° Walczak S. , "The local extremum principle for problems with constraints upon phase coordinates," Bull. de la Soc. des Scien. at des Lettr. de Lodz, Vol. XXXII, No. 3, (1982).

Differential Inclusions, Linear Selections, and an Optimal Control Problem

S. J. Lee, Department of Mathematics, University of South Florida, Tampa, Florida

M. Z. Nashed, Department of Mathematical Sciences, University of Delaware, Newark, Delaware

Abstract

This paper examines some aspects of the role of algebraic and proximinal selections in differential and operator inclusions. Several examples of such inclusions are given. An optimal control problem for a linear differential inclusion involving a partial differential operator is analyzed in the framework of algebraic selections and least-squares problems.

§1. Differential and Operator Inclusions

An early impetus for the study of differential inclusions arose from control theory and differential equations with discontinuous functions, specifically from situations where the velocity $x'(t)$ is not uniquely determined by the state–dynamics of the system or where $x'(t)$ incurs discontinuities. More generally, differential and operator inclusions arise from mathematical models of dynamical and physical systems for which we have an incomplete description or a loose mode of dependence of their generalized coordinates.

The study of multivalued mappings, operator inclusions, and optimization problems involving multivalued mappings has gained considerable importance in recent years due to occurrence of such situations in various areas of analysis, differential equations, continuum mechanics and control theory. Multivalued mappings and inclusions provide also a natural framework for the study of certain issues arising from problems involving single–valued operators. Usually we encounter single–valued operators which are (at least) densely defined in a relevant Banach space, say X. In many cases, however, the mathematical study of these operators leads to the introduction of operators or quadratic forms whose domain is no longer dense in X. Typical examples from physics and analysis include spectral analysis of Hamiltonian perturbed by finite–rank operators, scattering theory, Trotter–type approximations of semigroups and approximation of Schrödinger operators. Also in the presence of constraints, characterizations of solutions of many classes of optimization problems usually involve a mixture of operator equations and inclusions.

For motivation we consider in this section examples and situations (some rather classical) that lead to multivalued mappings and inclusions in the context of differential equations, control theory, and optimization.

Example 1. The implicit differential equation $f(t, x(t), x'(t)) = 0$ can be regarded as a differential inclusion

$$x'(t) \in F(t, x(t)),$$

where F is defined by

$$F(t, x) := \{z : f(t, x, z) = 0\}.$$

Example 2. The dynamical system

$$x'(t) = f(t, x(t), u(t)), \; x(0) = x_0$$

controlled by $u \in U$, can be rewritten in the form of a differential inclusion

$$x'(t) \in F(t, x(t)), \; x(0) = x_0,$$

where $F(t, x) := \{f(t, x, u) : u \in U\}$. Note that in this formulation the control does not appear explicitly in the inclusion, and is "embedded" in the multivalued map F. In the particular case of the linear control system

$$x'(t) = Ax(t) + Bu(t), \; x(0) = x_0,$$

$$F(t,x) := \{Ax + Bu : u \in U\}$$

defines an affine manifold if U is a subspace.

Example 3. Consider the continuous steepest–descent equation

$$x'(t) = -\text{grad}\,\psi(x(t)),\ x(0) = x_0$$

for minimizing a Fréchet differentiable convex functional ψ on a normed space X. If ψ is a lower semicontinuous convex functional that is not necessarily differentiable, then this equation is replaced by the *subgradient inclusion*

$$x'(t) \in -\partial\psi(x(t)),\ x(0) = x_0,$$

where $\partial\psi(x)$ is the subdifferential of ψ at x.

More generally, the operator inclusions

$$x'(t) \in A(x(t)),\ x(0) = x_0,$$

where A is a maximal monotone or maximal accretive (multivalued) mapping have been extensively studied.

Example 4. As usual, constraints play a major role in differential inclusions arising in optimal control and optimization theory. For example, in viability theory, one seeks a solution of the differential inclusion

$$x'(t) \in F(t, x(t))$$

that satisfies

$$x(t) \in V \quad \text{for all} \quad t,$$

where V is a given closed set, the so–called viability set. This imposes a constraint on the time evolution of the trajectories. This problem is extensively studied in Aubin and Cellina [1].

Example 5. In some optimal control problems of differential inclusions one is interested in the so–called relaxed trajectories, i.e. maps x on an interval into $\mathbb{R}^n$ such that

$$x'(t) \in co\,F(t, x(t)) \quad \text{almost everywhere},$$

where $co\,F$ denotes the convex hull of F. See Clarke [2].

Example 6. Differential inclusions provide an effective framework for the study of the differential equation

$$x'(t) = f(t, x(t)),\ x(0) = x_0$$

with a discontinuous right–hand side. The idea is to embed $f(t, x(t))$ into a set–valued map $F(t, x)$ which has sufficient regularity so that the associated differential inclusion $x'(t) \in F(t, x(t))$ has trajectories closely related to the trajectories of the discontinuous differential equation.

We now focus on simple examples of multivalued linear mappings and linear inclusions. Let V and W denote real or complex vector spaces. A subspace M of $V \times W$ is called a *linear relation.* We identify M with the graph of a *multivalued* linear mapping $M(\cdot)$, whose domain is

$$\operatorname{Dom} M := \{x \in V : (x, y) \in M \quad \text{for some} \quad y \in W\},$$

and

$$M(x) := \{y : (x, y) \in M\}.$$

The nullspace and range of M are denoted by $\operatorname{Null} M$ and $\operatorname{Range} M$, respectively. For a Banach space Z with dual $Z^\star$, the natural pairing of $Z \times Z^\star$ is denoted by $\langle \cdot, \cdot \rangle$.

Example 7. Let X and Y be Banach space and let $M \subset X \times Y$ be a linear relation. The *adjoint* (or *adjoint subspace*) of M is the linear relation $M^\star \subset Y^\star \times X^\star$ defined by

$$M^\star = \{(y, x) : (x, -y) \in M^\perp\}, \tag{1.1}$$

where $M^\perp$ denotes the annihilator of M. Thus $M^\star(\cdot)$ is a multivalued linear mapping from $Y^\star$ to the power set of $X^\star$ defined by the conditions:

$$\begin{cases} x^\star \in M^\star(y^\star) & \text{if and only if } x^\star \in X^\star,\ y^\star \in Y^\star \text{ and} \\ \langle y, y^\star \rangle = \langle x, x^\star \rangle & \text{for all } y \in M(x),\ x \in \operatorname{Dom} M. \end{cases} \tag{1.2}$$

Of course, if M is the graph of a single–valued densely defined linear operator, then $M^\star$ is the graph of a single–valued operator, namely the unique adjoint operator in this case.

Adjoint subspaces play an important role in the study of differential equations with general boundary conditions, optimization and control problems and operator equations with nondense domain (see Coddington and Dijksma [3], Lee [4], Lee and Nashed [5], [6], [7] and references cited therein).

Example 8. Let A be a (single–valued) operator with domain in V and range in W. The mapping B which assigns to each $y \in \operatorname{Range} A$, the set $B(y)$ of all solutions of $Ax = y$ is a simple, but important, example of a multivalued operator. In the case when A is linear, (single–valued) selections of B are inner inverses of A restricted to the range of A, and the set $B(y)$ is easily characterized in terms of any of these selections: $B(y) = A^{\#}y + \operatorname{Null} A$, where $A^{\#}$ is any selection of the multivalued mapping A^{-1}.

Example 9. Let A be a linear operator on a vector space V into a Hilbert space H_2 and let $y \in H_2$. An element $u \in V$ is called a least–squares solution (LSS) of the equation $Ax = y$ if

$$\|Au - y\| = \inf\{\|Ax - y\| : x \in V\}.$$

It follows immediately that u is a LSS if and only if $Au - y$ is orthogonal to the range of A. If A has a dense domain in a Hilbert space H_1, then this condition is equivalent to $A^\star(Au - y) = 0$, where $A^\star$ is the adjoint of A. If A is bounded, then the last equation can be written in the traditional form of the "normal equations": $A^\star Au = A^\star y$. However, in general, Au and y need not be in the domain of $A^\star$. Thus a characterization of LSS in the case when $\operatorname{Dom} A$ is dense in H_1 but A is not necessarily bounded takes the form of an inclusion:

$$Au - y \in \operatorname{Null} A^\star.$$

This situation is more complicated when $\operatorname{Dom} A$ is not dense since the adjoint of A is then multivalued.

Example 10. Let X, Y, Z be Hilbert spaces and let $A : X \to Y$ and $L : X \to Z$ be bounded linear operator. Assume that A and L have nontrivial nullspaces. For a given $z \in Z$, let S_z denote the set of all least–squares solutions of $Lx = z$ (we assume S_z is nonempty, i.e. we take $z \in R(L) + R(L)^{\perp}$). It is easy to show that an element $w \in S_z$ that minimizes $\|Ax - y\|$, where y is a fixed element in Y, is characterized by the conditions

$$A^{\star}Aw - A^{\star}y \in (\text{Null}\, L)^{\perp} \quad \text{and} \quad L^{\star}Lw = L^{\star}z.$$

§2. Algebraic, Topological, and Proximinal Selections of Multivalued Linear Mappings

Let V and W be two (nonempty) sets and let F be a multivalued mapping on V into the power set of W. A *selection* of F is a single–valued mapping $f : V \to W$ with $f(v) \in F(v)$ for all $v \in V$. Selections provide an important methodological tool for reducing multivalued mappings, or certain aspects thereof, to single–valued mappings which can be studied more easily. They allow us to reduce inclusions to equations involving (single–valued) operators, or to other inclusions that are easier to handle. Often we are interested in the existence of selections with additional properties (e.g., selections that are continuous, or measurable, or minimal) and there are many selection theorems that guarantee the existence of such selections under suitable assumptions on the multivalued mapping (see Aubin and Cellina [1]).

We are interested here in *selections for multivalued linear mappings.* Selection theorems for nonlinear mappings provide very little information in this case. There are many properties, issues and subtleties in the case of linear selections that have no counter part in the general setting of multivalued mappings. Algebraic, topological, and proximinal selections of multivalued linear mappings, and related topics in normed linear relations have been thoroughly investigated by the authors in [8] and [9]. We give here only a few definitions and elementary properties that are useful for Sections 3–5 of this paper.

Definition 2.1. An *algebraic selection* (or *algebraic operator part*) R of a linear relation M in $V \times W$ is a single–valued linear selection of M with $\text{Dom}\, R = \text{Dom}\, M$ such that $R \subset M$.

The subspace $M_\infty := \{0\} \times M(0)$ is called the "*infinite part*" or the "*purely multivalued*" *part* of M. Clearly M is (the graph of) a single-valued linear operator if and only if $M(0) = 0$. The following useful characterizations of algebraic operator parts follow immediately from Definition 2.1 and serve as important tools in the analysis of multivalued linear mappings:

- Any algebraic complement R to M_∞ in M is an algebraic operator of M; conversely any algebraic selection R of M is an algebraic complement to M_∞. Thus $M = M_\infty \dot{+} R$ (algebraic sum).

- R is an algebraic selection of M if and only if there exists an *algebraic projector* (i.e. a linear idempotent) $\mathbb{P}$ of M onto $M_\infty = \{0\} \times M(0)$ with $R = \operatorname{Null} \mathbb{P}$.

Definition 2.2. Let R be an algebraic selection of M, so $R = \operatorname{Null} \mathbb{P}$ as above. If $\mathbb{P}(x_1, y) = \mathbb{P}(x_2, y)$ for all $(x_1, y), (x_2, y) \in M$, then we say that R is a *principal* algebraic selection of M.

The graph inverse of a linear relation M is denoted by M^{-1}. Often we are interested in algebraic, topological, and proximinal selections of M^{-1}. We shall state most of our results on selections for M^{-1} because of frequent occurrence of inverse mappings. Similar results hold for M by replacing M^{-1} by M and $\operatorname{Null} M$ by $M(0)$.

There are marked differences in properties of these selections for M^{-1} depending on whether or not M itself is (the graph) of a single-valued mapping. In particular, if M is single-valued, then all algebraic selections of M^{-1} are principal.

The following results are established in [8]:

Theorem 2.1. (a) Every principal algebraic selection of M^{-1} is the composition of the relation Graph $(I - P)$ with the relation M^{-1}:

$$R = (\operatorname{Graph}(I - P))M^{-1},$$

where P is an algebraic projector of $\operatorname{Dom} M$ onto $\operatorname{Null} M$.

(b) If M is (the graph of) a single–valued operator, then every algebraic selection of M^{-1} is principal, but not conversely.

Theorem 2.2. Let R be an algebraic selection of M^{-1}. Then

(a) $M = MRM$ and $RMR = (\{0\} \times (RM)(0)) \dotplus R$. In particular R is an algebraic selection of RMR.

(b) RM is (the graph of) a single–valued operator if and only if R is principal.

For problems of analysis and applications to differential and integral equations, control theory, and best approximations for linear inclusions, algebraic selections of a multivalued mappings do not suffice. Such problems often involve normed space settings and it is important that the associated projectors be continuous. This motivates the following definition.

Definition 2.3. Let $M \subset X \times Y$ be a linear relation. A subspace R of M^{-1} is a *topological* selection (or *topological*) operator part of M^{-1} if there exists a continuous projector from M^{-1} onto $\{0\} \times \text{Null}\, M$ such that $R = \text{Null}\, I\!P$. A topological operator part of M^{-1} is called a *principal* topological operator part (selection) if $I\!P(y, x)$ is independent of y.

It follows form Theorem 2.1 that every principal topological selection of M^{-1} is of the form $R = (\text{Graph}\,(I - P))M^{-1}$ for some continuous projector P of $\text{Dom}\, M$ onto $\text{Null}\, M$. A principal topological operator part R of M is a subspace of the form $R = (\text{Graph}\,(I - Q))M$, where Q is a continuous projector of $\text{Range}\, M$ onto $M(0)$.

Definition 2.4. A principal topological selection R of M^{-1} defined by $R = (\text{Graph}(I - P))M^{-1}$ is called a *proximinal selection* (or *proximinal operator part*) of M^{-1} if $\|I - P\| = 1$, where P is a continuous projector of $\text{Dom}\, M$ onto $\text{Null}\, M$.

Proximinal selections have a "best approximation" property. In Section 3 we will consider proximinal selections only in a Hilbert space context, and in Section 5 we apply them to an optimal control problem for a linear differential inclusion. Unlike algebraic selections, topological or proximinal selections do not always exist.

§3. Constrained Least–Squares Solutions of Linear Inclusions

Definition 3.1. Let H_1 and H_2 be Hilbert spaces and let M be a linear manifold in $H_1 \times H_2$. Let A be a set in H_1 and $g \in H_2$. Then $u \in H_1$ is called a *restricted least-squares solution* (briefly, a restricted LSS) of the inclusion $g \in M(x)$ *with respect to* A if

(i) $u \in (\operatorname{Dom} M) \cap A$, and

(ii) there exists $t \in M(u)$ such that

$$\|g - t\| = d(g, \operatorname{Range}(M|_A)),$$

where $M|_A$ is the restriction of M to A, and

$$d(g, W) = \inf\{\|g - x\| : x \in W\}.$$

Note that if M is (the graph of) a linear operator and A is an algebraic subspace of H_1, then this definition coincides with the usual definition of a least-squares solution (LSS) of an operator equation. In the case when A is a vector subspace, notions of LSS of a linear relation were introduced and investigated in [5].

The following theorems provide equivalent characterizations and conditions for existence of restricted least–squares solutions of linear inclusions, and demonstrate the role of selections in the study of inclusions. For proofs and related results, as well as applications to singular control problems, see [7].

Theorem 3.1. Let H_1 and H_2 be Hilbert spaces. Let $L \subset H_1 \times H_2$ and $N \subset H_1$ be linear manifolds. Let $L_{s,P}$ be an arbitrary, but fixed algebraic operator part of L corresponding to an algebraic projector P of H_1 onto $L(0)$, i.e. $L_{S,P} = (\operatorname{Graph}(I - P))L$. Let

$$S := g \dot{+} N, \quad \text{and} \quad M := L|_N,$$

where $g \in \operatorname{Dom} L$. Then for fixed $h \in H_2$, the following statements are equivalent:

(i) w is a restricted least-squares solution of the inclusion $h \in L(x)$ with respect to S.

(ii) $k := g - w$ is a least-squares solution of

$$L_{s,P}(g) - h \in M(x).$$

(iii) $w \in S \cap \operatorname{Dom} L$ and

$$L_{s,P}(w) - h \in L(0) \dotplus \operatorname{Null} M^\star.$$

(iv) $w \in S \cap \operatorname{Dom} L$ and there exists $z \in L(0)$ such that

$$\langle L_{s,P}(w) - h - z, L_{s,P}(u) + k \rangle = 0$$

for all $k \in L(0)$ and $u \in (\operatorname{Dom} L) \cap N$.

Remark 3.1. The three equivalent characterizations of a restricted LSS of the inclusion $h \in L(x)$ with respect to S are expressed in terms of algebraic selections (operator parts) and adjoint subspaces. In characterization (ii) of Theorem 3.1, the restricted LSS is equivalent to an unrestricted, but modified, LSS. Characterization (iii) is a generalized form of the normal equations; we will call

$$L_{s,P}(w) - h \in L(0) + \operatorname{Null} M^\star$$

the "*normal inclusion*" for the given inclusion $h \in L(x)$. In the case of a single-valued operator L with domain H_1, and with $g = 0, N = H_1, M = L$, the "normal inclusion" reduces to $M(w) - h \in \operatorname{Null} M^\star$, which gives the normal equations.

Theorem 3.2. Let the assumptions of Theorem 3.1 hold. Then:

(i) The inclusion $h \in L(x)$ has a restricted least-squares solution with respect to S if and only if

(3.1). $$L_{s,P}(g) - h \in \operatorname{Range} M \dotplus \operatorname{Null} M^\star$$

In particular, if Range M is closed, then a restricted least-squares solution exists for each $h \in H_2$.

(ii) Assume that (3.1) holds and that Null M is closed. Then the set of all restricted least–squares solutions of $h \in L(x)$ with respect to S, denoted by Ω_h, is given by

$$\Omega_h = [g - M^{\#}(L_{s,P}(g) - h)] \dot{+} \text{Null } M, \tag{3.2}$$

where $M^{\#}$ is the orthogonal generalized inverse of the multivalued mapping M.

§4. Applications of Linear Selections to Differential Inclusions

In this section we indicate some applications of the theory of linear selections to differential inclusions. We consider three types of differential inclusions.

Example 4.1. Let X be a normed space and let I be an interval. Let $M \subset X \times X$ be a linear relation. Let $t_0 \in I$, $x_0 \in X$ be fixed. Consider the differential inclusion

$$\begin{cases} x'(t) \in M(x(t)), & \text{almost everywhere (a.e.)} \\ x(t_0) = x_0. \end{cases} \tag{4.1}$$

Here $M(x(t)) := \{y \in X : (x(t), y) \in M\}$. By a solution of (4.1) we mean an X–valued map $x(t)$ of $t \in I$ such that the strong derivative $x'(t)$ exists for almost all $t \in I$ and satisfies (4.1).

As an example of the space, take $X := L_2(\Omega)$ where Ω is an open subset of $I\!R^n$. Then, in this case x is a real–valued function of $n+1$ scalar variables: $t \in I$, and $s_1, \ldots, s_n \in I\!R^n$. In this case, (4.1) is a partial differential equation. Let R be an algebraic selection of M. Then

$$M = R \dot{+} (\{0\} \times M(0)).$$

Thus (4.1) is equivalent to

$$\begin{cases} x'(t) - R(x(t)) \in M(0), & \text{a.e. for } t \in I \\ x(t_0) = x_0. \end{cases} \tag{4.1$'$}$$

Thus if M is purely multivalued, i.e., $M = \{0\} \times M(0)$, and if $x_0 \neq 0$, then (4.1) is not solvable; while if $M = X \times X$, then (4.1) has infinitely many solutions, in sharp contrast to the case when M is (the graph of) a single–valued linear operator.

Example 4.2. Let I be an interval and let X be a normed space of $I\!R^n$-valued functions x defined on I. Let $t_0 \in I$ and $x_0 \in I\!R^n$ be given. Let M be a linear relation in $X \times X$. Consider

$$x' \in M(x), \quad x(t_0) = x_0. \tag{4.2}$$

By a solution of (4.2) we mean an $I\!R^n$-valued function $x(t)$ of $t \in I$ such that x is absolutely continuous on I, $x(\cdot) \in X$, and that $x'(\cdot) \in M(x)$ and $x(t_0) = x_0$. Here

$$M(x) := \{y \in X : (x, y) \in M\}.$$

Let R be an algebraic selection of M. Thus $M = R \dot{+} (\{0\} \times M(0))$. Then (4.2) is equivalent to

$$\begin{cases} x' - R(x) \in M(0) \\ x(t_0) = x_0. \end{cases} \tag{4.2)'$$

Notice that $M(0)$ in this setting is a function space.

Example 4.3. Let X be a normed space of $I\!R^n$-valued functions defined on an interval I. Let $x_0 \in I\!R^n$ and $t_0 \in I$ be given. Let M be a linear relation in $X \times X$. Consider

$$\begin{cases} x'(t) \in (M(x))(t) \quad \text{a.e.} \\ x(t_0) = x_0. \end{cases} \tag{4.3}$$

Here $x(t) \in I\!R^n$, and

$$\begin{aligned} (M(x))(t) &:= \{y \in X : (x, y) \in M\}(t) \\ &:= \{y(t) \in I\!R^n : (x, y) \in M\}. \end{aligned}$$

Let R be any algebraic selection of M. Then (4.3) is equivalent to

$$\begin{cases} x'(t) - (R(x))(t) \in (M(0))(t) \\ x(t_0) = x_0. \end{cases} \tag{4.3)'$$

As in Example 4.2, $M(0)$ in this setting is a function space, namely,

$$(M(0))(t) := \{y(t) \in I\!R^n : (0, y) \in M\}.$$

Remark 4.1. Notice that if M is single–valued, then since $M(0) = \{0\}$, it follows that the inclusions (4.2) and (4.3) reduce to equivalent equations. Otherwise, they are not equivalent notions. For example, let $M := \{(x, x + \alpha y) : x \in X, \alpha \in I\!R\}$, where y is a nonzero element of X. Then

$$x'(t) \in (M(x))(t) \quad \text{a.e.}$$

if and only if

$$x'(t) \in \{x(t) + \alpha y(t) : \alpha \in I\!R\} \quad \text{a.e.}$$

Equivalently,

$$x'(t) - x(t) = \alpha(t)y(t)$$

for some real–valued function $\alpha(t)$.

On the other hand,

$$x' \in M(x)$$

if and only if there exists a *constant* α such that

$$x'(t) - x(t) = \alpha y(t) \quad \text{a.e. for} \quad t \in I.$$

§5. An Application to a PDE Control Problem

In this section we will consider a PDE control problem involving a linear relation.

Let I be an interval and let Ω be a region in $I\!R^n$. Let $L_2(I \times \Omega)$ be the Hilbert space of real–valued function $x(t, \omega)$ of $(t, \omega) \in I \times \Omega$ such that

$$\|x\| := \left(\int_{I \times \Omega} |x|^2 dA \right)^{1/2} < \infty.$$

Let M be a linear relation in $L_2(I \times \Omega) \times L_2(I \times \Omega)$ such that if $x \in \operatorname{Dom} M$ then for almost all $\omega \in \Omega$ fixed, $x(t, \omega)$ as a function of $t \in I$ is locally absolutely continuous on I, and $\frac{\partial x}{\partial t} \in L_2(I \times \Omega)$. Consider the problem of minimizing

$$C(x, u) := \|x\|^2 + \|u\|^2$$

subject to

$$u \in L_2(I \times \Omega), \quad x \in \operatorname{Dom} M, \tag{5.1}$$

$$\frac{\partial x}{\partial t} + Vu \in M(x) \tag{5.2}$$

$$B(x)\Big|_{\partial(I \times \Omega)} = \phi. \tag{5.3}$$

Here V is a linear operator on $L_2(I \times \Omega)$ into $L_2(I \times \Omega)$; ϕ is a real–valued function defined on the boundary, $\partial(I \times \Omega)$, of the region $I \times \Omega$; B is a linear operator on $\operatorname{Dom} M$ into $L_2(I \times \Omega)$; $y\Big|_{\partial(I\times\Omega)}$ denotes the restriction of y to the boundary $\partial(I \times \Omega)$.

In this section we will show that, using algebraic selections, we can reduce the above constrained minimization problem to a least–squares problem. Let $\mathbb{B}$ be the operator on $\operatorname{Dom} M$ defined by

$$\mathbb{B}(x) := B(x)\Big|_{\partial(I\times\Omega)}.$$

Let $M^{\P}$ be an algebraic selection of M. Define an operator T on $\operatorname{Dom} M$ by

$$T(x) := \frac{\partial x}{\partial t} - M^{\P}(x), \quad x \in \operatorname{Dom} T = \operatorname{Dom} M.$$

Let $T^{\#}$ be an algebraic selection of T^{-1}, the graph inverse of T. Define an operator $\hat{\mathbb{B}}$ by

$$\operatorname{Dom} \hat{\mathbb{B}} := \operatorname{Null} T \cap \operatorname{Dom} \mathbb{B},$$

$$\hat{\mathbb{B}}(x) := \mathbb{B}(x), \quad x \in \operatorname{Dom} \hat{\mathbb{B}}.$$

Let $\hat{\mathbb{B}}^{\#}$ denote an algebraic selection of $\hat{\mathbb{B}}^{-1}$.

Finally, let $\mathcal{D}$ denote the set of ordered triples (u, h, k_0) such that

$$u \in L_2(I \times \Omega),\ h \in M(0),\ h - Vu \in \operatorname{Range} T, \tag{5.4}$$

(5.5) $$k_0 \in \text{Null}\, \hat{\mathbb{B}},\ T^{\#}(h - Vu) \in \text{Range}\, \hat{\mathbb{B}}.$$

Then it is clear that $\mathcal{D}$ is a subspace.

Theorem 5.1. Assume that

$$\phi \in \text{Range}\, \hat{\mathbb{B}}.$$

Then $u \in L_2(I \times \Omega)$, $x \in L_2(I \times \Omega)$ satisfy (5.1), (5.2) and (5.3) if and only if there exists $(u, h, k_0) \in \mathcal{D}$ such that

$$x = T^{\#}(h - Vu) + (\hat{\mathbb{B}})^{\#}\phi - \hat{\mathbb{B}}^{\#}(T^{\#}(h - Vu)) + k_0.$$

Proof. Assume that (5.1)–(5.3) are satisfied for some $u, x \in L_2(I \times \Omega)$. Then, since

$$M(x) = M^{\P}(x) \dot{+} M(0),$$

the condition (5.2) becomes

$$T(x) + Vu \in M(0).$$

Equivalently, there exists $h \in M(0)$ such that

$$T(x) + Vu = h.$$

Thus h, u satisfy the condition (5.4) and

(5.6) $$x = T^{\#}(h - Vu) + k$$

for some $k \in \text{Null}\, T$.

Since x satisfies (5.3), we see that

$$\hat{\mathbb{B}}(k) = \phi - T^{\#}(h - Vu).$$

This implies that $\phi - T^{\#}(h - Vu) \in \text{Range}\, \hat{\mathbb{B}}$ and so $T^{\#}(h - Vu) \in \text{Range}\, \hat{\mathbb{B}}$ as it was assumed that $\phi \in \text{Range}\, \hat{\mathbb{B}}$.

Therefore

$$k = (\hat{\mathbb{B}})^{\#}(\phi) - (\hat{\mathbb{B}})^{\#}(T^{\#}(h - Vu)) + k_0 \tag{5.7}$$

for some $k_0 \in \text{Null}\,\hat{\mathbb{B}}$.

Thus u, h, k_0 satisfy (5.5). Substituting k in (5.7) into (5.6) we see that x is given as in the theorem. The "if" part of the theorem is easy to check. ■

Define a linear operator $\mathcal{S}$ from $\mathcal{D}$ into $L_2(I \times \Omega) \times L_2(I \times \Omega)$ by

$$\mathcal{S}(u, h, k_0) := (u, T^{\#}(h - Vu) - (\hat{\mathbb{B}})^{\#}(T^{\#}(h - Vu)) + k_0).$$

We will say that a pair (u^+, x^+) is optimal if (u^+, x^+) satisfies (5.1), (5.2), (5.3) and it minimizes the cost functional $\mathcal{C}$ over all (u, x) satisfying (5.1), (5.2), (5.3).

Theorem 5.2. Assume that $\phi \in \text{Range}\,\hat{\mathbb{B}}$. Then (u^+, x^+) is an optimal pair if and only if

$$x^+ = T^{\#}(h^+ - Vu^+) + (\hat{\mathbb{B}})^{\#}(\phi - T^{\#}(h^+ - Vu^+)) + k_0^+$$

for some $(u^+, h^+, k_0^+) \in \mathcal{D}$ such that (u^+, h^+, k_0^+) is a least-squares solution of the operator equation

$$\mathcal{S}(u, h, k_0) = -(0, (\hat{\mathbb{B}})^{\#}(\phi)).$$

Proof. This follows immediately from the above theorem.

Remark 5.1. If $B(x) := x$, then the condition (5.3) becomes a standard boundary condition

$$x(t, \omega) = \phi(t, \omega), \quad (t, \omega) \in \partial(I \times \Omega).$$

Thus, if we assume further that $M = L_2(I \times \Omega) \times \{0\}$, then the assumption that $\phi \in \text{Range}\,\hat{\mathbb{B}}$ becomes: ϕ depends only on ω.

Remark 5.2. By using the concept of algebraic selection we have reduced a control problem involving a linear relation (multivalued mapping) to a single-valued control problem. We have not worked out characterizations of the optimal control pair (u^+, x^+). Obviously the control problem is too general to give an explicit characterization of (u^+, x^+) due to the general nature of M, V and B. Notice that if $Mx := \frac{\partial x}{\partial t}$, then the state inclusion (5.2) becomes an operator equation. The techniques used in [7] can be easily extended to obtain explicit characterizations of (u^+, x^+) in concrete cases.

References

[1] J. P. Aubin and A. Cellina, *Differential Inclusions*, Springer–Verlag, Berlin, Heidelberg, New York, 1984.

[2] F. H. Clarke, *Optimization and Nonsmooth Analysis*, Wiley–Interscience, New York, 1983.

[3] E. A. Coddington and A. Dijksma, Adjoint subspaces in Banach spaces, with applications to ordinary differential subspaces, *Ann. Mat. Pura. Appl.* **118** (1978), 1–118.

[4] S. J. Lee, Multi–response quadratic control problems, *SIAM J. Control Optim.* **24** (1986), 771–788.

[5] S. J. Lee and M. Z. Nashed, Least–squares solutions of multi–valued linear operator equations in Hilbert spaces, *J. Approx. Theory* **38** (1983), 380–391.

[6] S. J. Lee and M. Z. Nashed, Gradient method for nondensely defined closed unbounded linear operators, *Proc. Amer. Math. Soc.* **33** (1983), 429–435.

[7] S. J. Lee and M. Z. Nashed, Constrained least–squares solutions of linear inclusions and singular control problems in Hilbert spaces, *Appl. Math. Optim.* (1988), (in press).

[8] S. J. Lee and M. Z. Nashed, Algebraic and topological selections of multi–valued linear relations (1988), submitted for publication.

[9] S. J. Lee and M. Z. Nashed, Normed linear relations (1988), submitted for publication.

Optimal Feedback

Stanislaw Lojasiewicz, Jr., Professor, Department of Mathematics, University of Rhode Island, Kingston, Rhode Island

W. Szpunar, Department of Mathematics, Rochester Institute of Technology, Rochester, New York

The problem that we have shortly named in the title of this article is as old as the mathematical theory of optimal control. Moreover, since the classical field theory of the calculus of variations may be considered as a method for constructing optimal feedback, we can say that the problem is twice as old as the modern optimal control.

The notion of optimal feedback called also (after engineers) closed-loop optimal control or optimal synthesis (whenever we think rather about the procedure than about the object) seems familiar to everybody engaged in the classical optimization. It indicates the importance of this notion. However, we will risk the following provocative statement. The 50% of the readers of this article would be in trouble, whenever asked about providing an intuitively complete and mathematically rigorous definition of the optimal feedback for a reasonably general optimal control problem. Moreover, only few mathematicians would be able to provide such a definition together with a proof of the existence (in a large class of problems) of the object that they have defined.

Let us describe a reasonably general optimal control problem that seems accessible to the dynamic programming method.

We consider control system:

$$\dot{x} = f(t,x,u), \quad \text{with} \quad (t,x) \in R \times R^n \quad \text{and} \quad u \in \Omega \subset R^m \tag{1}$$

with target $M \subset R \times R^n$ and cost functional

$$J(x,u) = g(t_1,x(t_1)) + \int_{t_0}^{t_1} h(t,x(t),u(t))dt.$$

Our problem is to minimize $J(x,u)$ over all solutions of (1) satisfying $x(t_0) = x_0$, $(t_1,x(t_1)) \in M$ where $(t_0,x_0) \in R \times R^n$ is prescribed and $t_1 \geq t_0$ is free. Keeping f, Ω, M, g and h fixed, we get a family of minimization problems indexed by initial data (t_0,x_0). A minimizing pair (x,u) is called optimal, x is called optimal arc and u optimal control.

Let us assume for a moment, that for every initial data there exists at most one optimal pair (x,u) (we identify any two controls equal almost everywhere) and moreover that f and u are continuous. For every $(t_0,x_0) \in R \times R^n$ such that an optimal pair (x,u) beginning at (t_0,x_0) exists put $v(t_0,x_0) = u(t_0)$. Then, since a left-hand side restriction of an optimal pair is an optimal pair (Bellman's principle of optimality) we get $u(t) = v(t,x(t))$ and therefore x satisfies O.D.E.

$$\dot{x} = f(t,x,v(t,x)). \tag{2}$$

If, moreover (2) would have the uniqueness of solutions, we could say that (2) characterizes optimal arcs. In such a case there is a consensus to call the map v optimal feedback. So, where is the problem? For several simple but important control systems, even if there exists unique natural candidate v for our optimal feedback, this candidate is not regular enough (frequently discontinuous) to guarantee the required uniqueness of solutions in (2). A tempting way to surmount this difficulty is to prove that, despite that v may be discontinuous, v is piecewise smooth in some sense; so that f(t,x,v(t,x)) will be piecewise smooth and then to define the meaning of a solution of (2) in such a way that (2) has the uniqueness (to the right) of solutions and that optimal arcs satisfy (2).

This type of program was first proposed in the early sixties by V. Boltyansky [1] (with however different emphasis) under the name of regular synthesis.

The first theorem about the existence of regular synthesis that covers a natural class of systems rather than some particular examples was done only in the late seventies by P. Brunovsky [2]. Then this program was intensively continued by H. Sussmann with the most impressive results in [6] and [7].

Until now, the existence theorems were obtained mainly for some systems with analytic data and piecewise analytic value function. Let us disregard some negative aspects of this program: very complicated definitions, long and difficult proofs, limited applicability, see [5]. A real drawback is that almost all definitions of regular synthesis are different (despite that they are similar). Why? Because there is no unique natural way to define solutions to an O.D.E. with piecewise smooth right-hand side in such a way that the uniqueness would be guaranteed. So there is no hope for a universal definition of the regular synthesis.

Now, let us mention an important dilemma. If for an initial data (t_0,x_0) there is more than one optimal arc beginning at (t_0,x_0), we cannot preserve two requirements at the same time: uniqueness (to the right) of solutions in (2) and every optimal arc satisfies (2).

If we choose uniqueness, we face the following nontrivial problem. Whenever is it possible to select from the family of all optimal arcs a subfamily with the uniqueness to the right property, and such that it covers all initial data for which an optimal arc exists?

If we try to fulfill the second requirement, it seems hopeless to expect that all optimal arcs will satisfy (2) with some v in a reasonable sense, in general. This suggests to replace single-valued feedback v by a multivalued feedback, i.e. to replace O.D.E. (2) by a differential inclusion.

We are going to present this approach as developed in [3] and [4]. First we will reformulate our general control problem in the language of differential inclusions.

Instead of (1) we will consider

$$\dot{x} \in F(t,x) \quad \text{where} \quad F(t,x) = \{v \in R^n: v = f(t,x,u),\ u \in \Omega\} \tag{3}$$

with the same target M and the modified cost functional

$$I(x) = g(t_1,x(t_1)) + \int_{t_0}^{t_1} L(t,x(t),\dot{x}(t))dt$$

defined for x satisfying (3) and $(t_1,x(t_1)) \in M$ where $L(t,x,v) = \inf\{h(t,x,u): v = f(t,x,u),\ u \in \Omega\}$.

So, now our "system" is the quadruple (F,M,g,L). The optimal cost, frequently called value function $V: R \times R^n \to [-\infty,+\infty]$ is defined by $V(t_0,x_0) = \inf\{I(x): x(t_0) = x_0\}$ and optimal arcs are characterized by $I(x) = V(t_0,x(t_0)) < +\infty$. In [3] we have introduced the class of terminal systems. Most reasonable systems, including all with the uniqueness to the right of optimal solutions are terminal and all systems considered below in this paper are terminal, too.

Let (F,M,g,L) be a terminal system. Consider differential inclusion

$$\dot{x} \in G(t,x) \tag{4}$$

where G is a multivalued selection of F i.e., $G(t,x) \subset F(t,x)$ holds for $(t,x) \in R \times R^n$.

DEFINITION. G is called optimal feedback iff

(i) every optimal arc satisfies (4);

(ii) every arc satisfying (4) is a restriction of an optimal arc.

REMARK. If an optimal feedback exists, then it is uniquely determined provided that we agree to identify two differential inclusions with the same sets of all solutions.

Now, we are going to present two important classes of terminal systems in which optimal feedback always exists. The existence is an immediate consequence of Theorems 2 and 3 in [4], but the direct proofs should not be long.

For both classes we assume: $F(x)$ are nonempty, compact, convex for $x \in R^n$, F is locally lipschitzian (in Hausdorff metric) the growth condition, i.e., the existence of $c > 0$ such that

$$yx \leq c(1 + x^2) \quad \text{holds for} \quad y \in F(x), \quad x \in R^n.$$

First class: $M = \{T\} \times R^n$ with $T \in R$, $L \equiv 0$ and g locally lipschitzian,

Second class: $M = R \times \{x_1\}$ with $x_1 \in R^n$, $g \equiv 0$ and $L \equiv 1$ and

$$0 \in \text{int } F(x_1). \tag{5}$$

The optimal feedbacks are provided by effective formulas: For the first class,

$$G(t,x) = \{y \in F(x): D_{(1,y)}V(t,x) \text{ exists and equals } 0\}$$

for $t < T$, $x \in R^n$ and $G(t,x) = \emptyset$ otherwise for the second class:

$$G(x) = \{y \in F(x): D_{(1,y)}V(x) \text{ exists and equals } -1\}$$

for $x \in R^n$ such that $0 < V(x) < +\infty$ and $G(t,x) = \emptyset$ otherwise where $D_{(1,y)}V(t,x)$ is the symmetric directional derivative of V, i.e.

$$D_{(1,y)}V(t,x) = \lim_{\varepsilon \to 0} \frac{V(t+\varepsilon,\ x + \varepsilon y) - V(t,x)}{\varepsilon}$$

(we do not assume $\varepsilon > 0$). The second class with $F(x) = Ax + \Omega$ where A is $n \times n$ matrix and with condition (5) disregarded, is the oldest problem in modern optimal control theory: linear time optimal control problem.

It is not difficult to see that for this class an optimal feedback (in the sense of our definition) may not exist due to the discontinuities of the value function V.

Condition (5) is equivalent to the lipschitzianity of V. But what happens if the value function is merely continuous? The lack of any answer to this question in both regular synthesis and differential inclusions approaches, clearly indicates how poor is our knowledge in the classical linear control theory. We will finish our considerations by directly constructing the optimal feedback for the time optimal problem with $x_1 = 0$, $A = 0$, i.e., for systems with constant differential inclusion $\dot{x} \in F$. Let us remark, that even in such a simple case the value function may be discontinuous. For simplicity, we will assume $n = 2$.

Let $(t_0, x_0) \in R \times R^2$ and $t \geq 0$ be done. The set of all points of the form $x(t_0 + t)$ where x is a solution of (3) defined on $[t_0, t_0+t]$ satisfying $x(t_0) = x_0$ is explicitely provided by $x_0 + tF$. Then the origin may be reached from x_0 in time t by solutions of $\dot{x} \in F$ iff $0 \in x_0 + tF$. Hence, the value function for our problem is done by

$$V(x_0) = \inf\{t \geq 0: 0 \in x_0 + tF\}. \tag{6}$$

Obviously $V(x_0) = 0$ iff $x_0 = 0$. Put $C = \{x \in R^2: 0 < V(x) < +\infty\}$. Then C is the cone generated by -F (with the origin removed if $0 \in F$). Let T denote the restriction of V to $C \cup \{0\}$. Then (6) may be strenghthened to

$$T(x) = \min\{t \geq 0: -x \in tF\}. \tag{7}$$

For $x \in C$ let us define $g(x) = -x/T(x)$. Then $T(\alpha x) = \alpha x$ and $g(\alpha x) = g(x) \in F$ holds for $\alpha > 0$. Hence, an optimal arc beginning at $x_0 \in C$ at the moment t_0 can be defined by

$$x(t) = x_0 + (t - t_0)g(x_0) \quad \text{for} \quad t \in [t_0, t_0 + T(x_0)] . \tag{8}$$

Moreover this arc satisfies the following O.D.E.:

$$\dot{x}(t) = g(x(t)) \quad \text{in} \quad [t_0, t_0 + T(x_0)]. \tag{9}$$

However, optimal arcs need not satisfy (9) in general. Thus we will enlarge (9) to a differential inclusion. For this purpose we are going to study a certain partition of the cone C related to the position of g(x) in F. This partition will allow us to propose a formula for G(x). Then, we will prove that this G is indeed an optimal feedback for our problem.

Rewriting (7) for $x \neq 0$ as $\frac{1}{T(x)} = \max\{\alpha > 0: \alpha(-x) \in F\}$ we get the following geometric description of g:

PROPERTY 1. g(x) is the farest point from the origin, in the intersection of F with the halfline starting at the origin and passing through -x. Let C_{ext} be the cone at all $x \in C$ such that g(x) is an extremal point of F. We will prove that the cone $C \setminus C_{ext}$ is open and will describe its connected components. Let $x_0 \in C \setminus C_{ext}$. Then $g(x_0)$ is not an extremal point of F and since $g(x_0) \in \partial F$ (we recall n = 2), there exists unique line L_0 such that

$$g(x_0) \in ri(F \cap L_0) \tag{10}$$

where ri stands for the relative interior. Moreover, the line L_0 does not contain the origin and F is contained in the half-plane containing the origin and bounded by L_0 (otherwise we would have violated property 1). Let C_0 be the cone generated by $-ri(F \cap L_0)$ Then C_0 is an open, proper, convex cone and $x_0 \in C_0 \subset C$. Moreover we have:

$$g(x) \in ri(F \cap L_0) \quad \text{for} \quad x \in C_0 \tag{11}$$

$$g(x) \in F \cap L_0 \quad \text{for nonzero} \quad x \in cl\, C_0, \text{ and} \tag{12}$$

$$F \cap L_0 = cl\{g(x): x \in C_0\} . \tag{13}$$

From (11) we get $C_0 \subset C \setminus C_{ext}$ and therefore $C \setminus C_{ext}$ is open. We claim that C_0 is a connected component of $C \setminus C_{ext}$. All we need is to prove: $(C \setminus C_{ext}) \cap \partial C_0 = \emptyset$. Suppose the contrary, i.e., that there exists $x_1 \in (C \setminus C_{ext}) \cap \partial C_0$. Proceeding with x_1 similarly as with x_0, we get a line L_1 and an open cone C_1 such that $x_1 \in C_1$ and $g(x) \in ri(F \cap L_1)$ for $x \in C_1$. But $x_1 = \lim x_k$ with $x_k \in C_0$. Hence $x_k \in C_1$

for k large enough and therefore we get $g(x_k) \in \operatorname{ri}(F \cap L_0) \cap \operatorname{ri}(F \cap L_1)$. Hence $L_0 = L_1$ (otherwise we would get $g(x_k) \in \operatorname{int} F$), and therefore $x_1 \in C_1 = C_0 = \operatorname{int} C_0$ that contradicts $x_1 \in \partial C_0$.

Construction of G. We put:

1) $G(x_0) = \emptyset$ if $x_0 \notin C_0$;
2) $G(x_0) = \{g(x_0)\}$ if $x_0 \in C_{ext}$;
3) $G(x_0) = \operatorname{cl}\{g(x): x \in C_0\}$ where C_0 is a connected component of $C \setminus C_{ext}$ such that $x_0 \in C_0$.

Now we are going to prove (i) from the definition of the optimal feedback. So, let x be an optimal arc. Without loss of generality, we may assume: $x(0) = x_0$ and $T(x_0) > 0$. Then $0 \neq x(t) \in C$ for $t \in [0,T_0[$ where $T_0 = T(x_0)$. For any nonempty, compact convex subset A of R^2 and $p \in R^2$ we put

$$s(A,p) = \max\{pa: a \in A\} \quad \text{and} \quad S(A,p) = \{a \in A: pa = s(A,p)\}.$$

Fix $p \neq 0$ such that

$$g(x_0) \in S(F,p). \tag{14}$$

Since $g(x_0) \in \partial F$, such a p exists. We claim that

$$\dot{x}(t) \in S(F,p) \quad \text{a.e. in} \quad [0,T_0]. \tag{15}$$

Indeed, from $-x_0 = x(T_0) - x(0) = \int_0^{T_0} \dot{x}(t)dt$ we get:

$$g(x_0) = \frac{1}{T_0}\int_0^{T_0} \dot{x}(t)dt,$$

hence

$$\frac{1}{T_0}\int_0^{T_0} s(F,p)dt = s(F,p) = pq(x_0) = \frac{1}{T_0}\int_0^{T_0} p\dot{x}(t)dt.$$

Therefore $\int_0^{T_0} s(F,p) - p\dot{x}(t)dt = 0$. Since $\dot{x}(t) \in F$ we have $s(F,p) - p\dot{x}(t) \geq 0$ a.e. in $[0,T_0]$. Hence, $s(F,p) = p\dot{x}(t)$ a.e. in $[0,T_0]$ and so we have proved (15).

We will consider two cases:

First case: $g(x_0)$ is an extremal point of F. Choose p such that (14) holds and put $\hat{F} = S(F,p)$. Then by (15), x is also an optimal for a new time optimal problem with F replaced by $\hat{F}$, $\hat{T}(x_0) = T(x_0)$ and $\hat{g}(x_0) = g(x_0)$ hold where $\hat{T}$ and $\hat{g}$ stands for T and g in the new problem. Since $\hat{g}(x_0) = g(x_0)$ is an end point of the possibly degenerated segment $S(F,p) = \hat{F}$ we can choose $\hat{p} \neq 0$ such that $S(\hat{F},\hat{p}) = \{\hat{g}(x_0)\}$. This means that (14) is satisfied with the new data and hence the new (15) becomes $\dot{x}(t) = \hat{g}(x_0)$ a.e. in $[0,T_0]$. Consequently, $x(t) = x_0 + tg(x_0) = (1 - \frac{t}{T_0})x_0$. Therefore x(t) is a positive multiple at x_0, so $g(x(t)) = g(x_0)$, $G(x(t)) = G(x_0)$, and finally $\dot{x}(t) \in G(x(t))$ holds a.e. in $[0,T_0]$.

Second case: $g(x_0)$ is not an extremal point of F. Pick L_0 and p such that (10) and (14) hold. Then $F \cap L_0 = S(F,p)$. Moreover, by (13) we have $F \cap L_0 = G(x_0)$. Hence (15) becomes: $\dot{x}(t) \in G(x_0)$ a.e. in $[0,T_0]$. Let t_1 be the smallest $t \in [0,T_0]$ such that $x(t) \notin C_0$. Then $G(x(t)) = G(x_0)$ for $t \in [0,t_1[$ and therefore

$$\dot{x}(t) \in G(x(t)) \quad \text{a.e. in} \quad [0,t_1]. \tag{16}$$

So if $t_1 = T_0$ we are done. If $t_1 < T_0$ we have $0 \neq x(t_1) \in C$ and therefore $x(t_1) \in C_{ext}$. Since the arc y defined on $[0, T_0-t_1]$ by $y(t) = x(t_1 + t)$ is still optimal and $g(y(0))$ is an extremal point of F, the first case applied to y leads to $\dot{y}(t) \in G(y(t))$ a.e. in $[0, T_0 - t_1]$. Coming back to x and recalling (16) we get $\dot{x}(t) \in G(x(t))$ a.e. in $[0,T_0]$. So we have proved (i).

For the converse, let x be a nontrivial arc of (4) defined on $[\alpha,\beta]$ with $\alpha < \beta$. Since (8) permits to extend optimally any arc ending in C, all we need is to prove that

$$T(x(\beta)) - T(x(\alpha)) = \alpha - \beta. \tag{17}$$

Let C_1 be a connected component of $C \setminus C_{ext}$ and $t_2 \in [\alpha,\beta]$. We claim:

$$x(t_2) \in C_1 \quad \text{implies} \quad x(t) \in C_1 \quad \text{for} \quad t \in [\alpha,t_2]. \tag{18}$$

Suppose the contrary. Then, since C_1 is open we could find $t_1 \in [\alpha,t_2[$ such that $x(t_1) \in \partial C_1$ and $x(t) \in C_1$ for $t \in]t_1,t_2]$. Pick an outward normal p to the convex cone cl C_1 at the point $x(t_1)$. Since $C_1 = \text{int cl } C_1$ we have

$$px(t_2) < px(t_1) = 0. \tag{19}$$

But $\dot{x}(t) \in G(x(t)) \subset -\text{cl } C_1$ holds, a.e. in $[t_1,t_2]$. Hence $p\dot{x}(t) \geq 0$, and therefore the mapping $t \to px(t)$ is nondecreasing in $[t_1,t_2]$ which contradicts (19). So (18) is true. Put $t_1 = \min\{t \in [\alpha,\beta]: x(t) \notin C\setminus C_{ext}\}$ if the last set is nonempty and $t_1 = \beta$ otherwise. In the case $t_1 < \beta$ (18) leads to $x(t) \notin C\setminus C_{ext}$ for $t \in [t_1,\beta]$. Then (9) holds a.e. in $[t_1,\beta]$. Hence $x(t)\dot{x}(t) < 0$ a.e. in $[t_1,\beta]$ and therefore $x(t) \neq 0$ for $t \in [t_1,\beta[$. Put $y(t) = \frac{x(t)}{|x(t)|}$ and $r(t) = |x(t)|$. By (9) we get $\dot{y}(t) = 0$ a.e. in $[t_1,\beta]$ and therefore

$$x(t) = r(t)y(t_1) \quad \text{holds in} \quad [t_1,\beta]. \tag{20}$$

Since $x(t) \in C$ a.e. in $[\alpha,\beta]$ (20) gives $y(t_1) \in C$. Hence $T(x(t)) = r(t)T(y(t_1))$ holds in $[t_1,\beta]$ and so the map $t \to T(x(t))$ is absolutely continuous. Again, by (9) and (20) we get $\dot{r}(t) = \frac{-1}{T(y(t_1))}$, hence $\dot{\overline{T(x(t))}} = -1$ a.e. in $[t_1,\beta]$ and therefore

$$T(x(\beta)) - T(x(t_1)) = t_1 - \beta. \tag{21}$$

In the case $\alpha < t_1$ put $x_0 = x(\alpha)$. Pick L_0 such that (10) holds. Then applying (18) to the corresponding C_0 we get $x(t) \in C_0$ for $t \in [\alpha,t_1[$. Let m be the point in L_0 closest to the origin. Then by (12) $g(a) \in L_0$ for nonzero $a \in \text{Cl } C_0$ and hence $T(a) = \frac{-ma}{m^2}$ for $a \in \text{Cl } C_0$. Therefore the mapping $t \to T(x(t)) = -\frac{mx(t)}{m^2}$ is absolutely continuous on $[\alpha,t_1]$; by (13) we get $\dot{x}(t) \in L_0$ a.e. in $[\alpha,t_1]$. Then $m\dot{x}(t) = m^2$ and $\dot{\overline{T(x(t))}} = -1$ a.e. in $[\alpha,t_1]$. Hence $T(x(t_1)) - T(x(\alpha)) = \alpha - t_1$ and therefore (21) gives (17). So the proof of (ii) is complete, and thus we have proved that G is optimal feedback.

REFERENCES

[1] V. Boltyanski, Sufficient conditions for optimality and the justification of the dynamic programming method, *SIAM J. Control, 4* (1966), pp. 326-361.

[2] P. Brunovsky, Every normal linear system has a regular time optimal synthesis, *Math. Slovaca, 28* (1978), pp. 81-100.

[3] S. Lojasiewicz, Jr., The existence of optimal solutions in nonlinear control systems, *Bull. Pol. Ac. Sci. Math. 32* (1984), pp. 485-499.

[4] S. Lojasiewicz, Jr., The existence of optimal feedback in nonlinear control systems, *Bull. Pol. Ac. Sci. Math. 32* (1984), pp. 501-514.

[5] S. Lojasiewicz, Jr. and H. Sussmann, Some examples of reachable sets and optimal costs functions that fail to be subanalytic, *SIAM J. Control, 23* (1985), pp. 584-598.

[6] H. Sussmann, Regular synthesis for time optimal control of single-input real analytic systems in the plane, *SIAM J. Control 25* (1987), pp. 1145-1162.

[7] H. Sussmann, Subanalytic sets and regular synthesis, to appear.

Bang-Bang Control of Quasi-Differential Equations

Lawrence Markus, Professor, Department of Mathematics, Institute of Technology, University of Minnesota, Minneapolis, Minnesota and Department of Mathematics, University of Warwick, Coventry, England

1. THIRTY YEARS: MAIS, OÙ SONT LES NEIGES D'ANTAN?

The initial impetus to the modern theory of optimal control arose around three decades ago in the investigations of the control dynamics of linear differential systems with constant coefficients, and more specifically in the bang-bang regulation of the linear harmonic oscillator [1,2]

$$\ddot{x} + k^2 x = u(t).$$

Shortly thereafter these studies were developed into a general theory of optimal control of nonlinear systems [6,7,10,11], and presented accordingly in organized treatises [5,8,12].

In this note we propose an extension of the theory of minimal-time optimal control to a class of linear differential systems with time-dependent coefficients, and demonstrate that the bang-bang principle remains valid. This important class of systems, defined later as quasi-differential equations, will contain the Mathieu equation

$$\ddot{\chi} + (\alpha + \beta \cos t)\chi = u(t)$$

and more generally the scalar n-th order linear ordinary differential equations

$$\chi^{(n)} - a_{n-1}(t)\chi^{(n-1)} - \cdots - a_0(t)\chi = b(t)u, \qquad S)$$

with the state

$$x = \begin{pmatrix} x^1 \\ x^2 \\ \vdots \\ x^n \end{pmatrix}$$

in $\mathbb{R}^n$ (as usual, $\chi = \chi^{(0)} = x^1$; $\chi^{(1)} = \dot{x}^1 = x^2, \cdots, \chi^{(n-1)} = x^n$), and with the real scalar controller u(t) as some suitably chosen function of the time t on an open interval $I \subset \mathbb{R}$. The real coefficients $a_0(t), a_1(t), \cdots, a_{n-1}(t)$ and b(t) are piecewise continuous---or perhaps only measurable and suitably integrable on I, as specified later.

Of course, the control dynamics of S) in $\mathbb{R}^n$ is analysed in terms of the linear system

$$\dot{x} = \begin{pmatrix} 0 & 1 & 0 & 0 & \cdots & 0 \\ 0 & 0 & 1 & 0 & \cdots & 0 \\ \vdots & & & & & \\ 0 & 0 & 0 & 0 & \cdots & 1 \\ a_0(t) & a_1(t) & & & \cdots & a_{n-1}(t) \end{pmatrix} x + \begin{pmatrix} 0 \\ 0 \\ \vdots \\ 0 \\ 1 \end{pmatrix} u(t)$$

with the state vector

$$x(t) = \begin{pmatrix} x^1(t) \\ x^2(t) \\ \vdots \\ x^n(t) \end{pmatrix} = \begin{pmatrix} x^{(0)}(t) \\ x^{(1)}(t) \\ \vdots \\ x^{(n-1)}(t) \end{pmatrix}$$

at each time t. The particular format of these coefficient matrices motivate the definition of the class of Zettl matrices, and the concept of quasi-differential equations, described in Section 3.

2. FULLY CONTROLLABLE LINEAR DIFFERENTIAL SYSTEMS

Consider a linear control system

$$\dot{x} = A(t)x + B(t)u \qquad (\text{note: } \dot{x}(t) = \frac{dx}{dt}) \tag{L}$$

with state vector

$$x = \begin{pmatrix} x^1 \\ x^2 \\ \vdots \\ x^n \end{pmatrix} \in \mathbb{R}^n, \quad \text{and control vector} \quad u = \begin{pmatrix} u^1 \\ u^2 \\ \vdots \\ u^m \end{pmatrix} \in \mathbb{R}^m,$$

at each time $t \in I$, where I is a prescribed open interval in $\mathbb{R}$. Here the coefficient matrices $A(t) \in L^1_{loc}(I, \mathbb{R}^{n\times n})$ and $B(t) \in L^1_{loc}(I, \mathbb{R}^{n\times m})$, that is, the corresponding entries

$$a_{ij}(t) \quad \text{for} \quad 1 \leq i, j \leq n, \text{ and } \quad b_{ij}(t) \quad \text{for} \quad 1 \leq i \leq n, 1 \leq j \leq m$$

are each integrable on every compact subinterval of I.

For each initial state $x_0 \in \mathbb{R}^n$, and controller $u(t) \in L^\infty_{loc}(I_1, \mathbb{R}^{m\times 1})$ for some subinterval $I_1 = [t_0, t_1] \subset I$, there exists a unique (absolutely continuous) solution or response $x(t)$ on I_1:

$$x(t) = X(t)\, x_0 + X(t) \int_{t_0}^{t} X(s)^{-1} B(s) u(s)\, ds.$$

Here $X(t)$ is the usual fundamental solution matrix defined by

$$\dot{X}(t) = A(t)X(t), \quad X(t_0) = I.$$

Definition. The linear control system

$$\dot{x} = A(t)x + B(t)u, \tag{L}$$

for state vector $x \in \mathbb{R}^n$ and controllers $u(t) \in L^\infty_{loc}(I_1, \mathbb{R}^{m\times 1})$, is called *fully controllable in* $\mathbb{R}^n$ *starting at instant* $t_0 \in I$ in case:

> for each pair of initial state x_0 and target state $x_T \in \mathbb{R}^n$ and for each positive duration $I_1 = [t_0,T] \subset I$, there exists a controller u(t) on $t_0 \leq t \leq T$ such that the corresponding response x(t) from $x(t_0) = x_0$ is controlled or steered to $x(T) = x_T$.

If L) is fully controllable in $\mathbb{R}^n$, starting at each initial instant $t_0 \in I$, then L) is called *fully controllable on* I.

We recall that the set of attainability from x_0, on the duration $[t_0,T] \subset I$ is the subset of $\mathbb{R}^n$

$$K_{x_0}(t_0,T) = \{X(T)x_0 + \int_{t_0}^{T} X(T)X(s)^{-1}B(s)u(s)ds \mid \text{all controllers } u(s) \text{ on } t_0 \leq s \leq T\}$$

The corresponding set of attainability from the origin $x_0 = 0$, say starting at initial time $t_0 = 0$ in I, is denoted $K_0(T)$. In this notation the linear system L) is fully controllable in $\mathbb{R}^n$, starting at the instant $t_0 = 0$, in case

$$K_0(T) = \mathbb{R}^n, \text{ for each } T > t_0 \text{ in } I.$$

In the autonomous case, where $A(t) = A$ and $B(t) = B$ are constant, and where $I = \mathbb{R}$, the classical necessary and sufficient condition [8] for L) to be fully controllable on I is

$$\text{rank } [B, AB, A^2B, \ldots, A^{n-1}B] = n.$$

However in the case of time-dependent coefficient matrices A(t) and B(t), no similar criterion is available---unless these coefficient matrices are assumed to be highly differentiable [5,8].

In this note we present an appropriate sufficiency test for the full controllability of L), under the hypothesis that (A(t), B(t)) have the format of a Zettl-Everitt pair [3,4,9], and hence L) is equivalent to a scalar quasi-differentiable equation $\mathcal{Q}$), as will be described in the next Section 3. An extensive treatment of the general control theory of quasi-differential equations can be found in [3], but our treatment seems more direct and elementary. In Section 4 we present new results on bounded controllers and on the bang-bang principle for minimal-time optimal controllers.

3. CONTROLLABILITY OF QUASI-DIFFERENTIAL EQUATIONS

One of the first problems of a first course in control theory concerns the controllability of the real scalar linear differential equation

$$\chi^{(n)} - a_{n-1}(t)\chi^{(n-1)} - a_{n-2}(t)\chi^{(n-2)} - \dots - a_0(t)\chi = b(t)u, \quad \mathcal{S})$$

In this paper we shall provide the control theory for the corresponding linear system with time-varying coefficients

$$\dot{x} = A(t)x + B(t)u, \qquad \mathcal{L})$$

in fact for a more extensive and generalized type of linear system where the coefficient matrices (A(t), B(t)) constitute a Zettl-Everitt pair, as defined next. For simplicity of exposition *we henceforth assume that* $t_0 = 0$ *lies in* I*, and that only scalar controllers* $u(t) \in L_{loc}(I, \mathbb{R})$ *are involved, that is,* $m = 1$.

Definition. The pair of matrices, defined on the open interval $I \subset \mathbb{R}$,

$$A(t) \in L^1_{loc}(I, \mathbb{R}^{n\ n}) \quad \text{and} \quad B(t) \in L^1_{loc}(I, \mathbb{R}^{n\ 1})$$

constitute a Zettl-Everitt pair in case they display the format

$$A(t) = \begin{pmatrix} * & a_{12}(t) & 0 & 0 & 0 & \dots & 0 & 0 \\ * & * & a_{23}(t) & 0 & 0 & \dots & 0 & 0 \\ & & & & & & & \\ * & * & & & & & & a_{n-1,n}(t) \\ * & * & & & & & & * \end{pmatrix},$$

$$B(t) = \begin{pmatrix} 0 \\ 0 \\ \vdots \\ b(t) \end{pmatrix}.$$

In greater detail the conditions are:

$$a_{i,i+1}(t) \neq 0 \quad \text{a.e. on } I, \text{ for } 1 \leq i \leq n-1$$

and

$$a_{ij}(t) \equiv 0 \quad \text{a.e. on } I, \quad \text{for } 1 \leq i \leq n,\ j \geq i+2.$$

Remarks.

1) We say that $A(t)$ is a Zettl matrix, $A(t) \in Z_n(I)$, and note that this format provides a generalization of the companion-form matrix arising from the scalar differential equation *S*).

2) In the notation of the quasi-derivatives $\chi_A^{[s]}(t)$ for $0 \leq s \leq n$, for a suitably differentiable real scalar function $\chi(t)$, we can describe such a Zettl-Everitt control system *L*) by a scalar quasi-differential control equation

$$\chi_A^{[n]} = b(t)u. \qquad \mathcal{Q})$$

Here the quasi-derivatives of $\chi(t)$, constructed relative to a prescribed matrix $A(t) = (a_{ij}(t)) \in Z_n(I)$, are defined as follows:

$$\chi_A^{[0]}(t) = \chi(t)$$

(which we also denote by $\chi_A^{[0]}(t) = x^1(t)$)

$$\chi_A^{[1]}(t) = a_{12}(t)^{-1}\{\dot{\chi}_A^{[0]}(t) - a_{11}(t)\chi_A^{[0]}(t)\}$$

(which we also note by $\chi_A^{[1]}(t) = x^2(t)$, in accord with the first row of the differential system *L*) $\dot{x} = A(t)x + b(t)u$ with Zettl-Everitt pair $(A(t),B(t))$, namely,

$$\dot{x}^1 = a_{11}x^1 + a_{12}x^2 \quad \text{or} \quad x^2 = a_{12}^{-1}\{\dot{x}^1 - a_{11}x^1\}).$$

Continue to define

$$\chi_A^{[2]}(t) = a_{23}^{-1}\{\dot{\chi}_A^{[1]} - (a_{21}\chi_A^{[0]} + a_{22}\chi_A^{[1]})\}$$

(which we also denote by $\chi_A^{[2]}(t) = x^3(t)$, in accord with

$$\dot{x}^2 = a_{21}x^1 + a_{22}x^2 + a_{23}x^3).$$

and so forth

$$\vdots$$

$$\chi_A^{[n-1]}(t) = x^n(t).$$

We observe that if A(t) is the companion matrix corresponding to the scalar differential equation *S*), then these quasi-derivatives are merely the classical derivatives, $\chi_A^{[s]}(t) = \chi^{(s)}(t)$ for $0 \leq s \leq n-1$.

However we define $\chi_A^{[n]}(t)$, as based on a general matrix $A(t) \in Z_n(I)$, to include the information of the last row of the linear system *L*). Thus we write

$$\chi_A^{[n]}(t) = \dot{\chi}_A^{[n-1]}(t) - \sum_{s=1}^{n} a_{ns}(t)\chi_A^{[s-1]}(t)$$

or equally well

$$\chi_A^{[n]}(t) = \dot{x}^n - a_{n1}(t)x^1 - a_{n2}(t)x^2 - \cdots - a_{nn}(t)x^n.$$

In this notation the first (n-1)-rows of the linear system

$$\dot{x} = A(t)x + B(t)u \qquad L)$$

merely define the quasi-derivatives $\chi_A^{[s]}(t)$ for $0 \leq s \leq n-1$, and the last row can be written as the scalar quasi-differential control equation

$$\chi_A^{[n]} = b(t)u. \qquad \mathcal{Q})$$

In this sense the two differential control equations *L*) and $\mathcal{Q}$) are merely different notations for the same control system in $\mathbb{R}^n$ (see [3] for further details). In this paper, we shall usually employ the more familiar notation of the matrix differential system in $\mathbb{R}^n$

$$\dot{x} = A(t)x + B(t)u. \qquad L)$$

with the Zettl-Everitt pair (A(t),B(t)).

Theorem 1. Consider the linear control system in $\mathbb{R}^n$
L) $\dot{x} = A(t)x + B(t)u$ *on the open interval* $I \subset \mathbb{R}$, *where*

$$A(t) \in Z_n(I), \quad B(t) = \begin{bmatrix} 0 \\ 0 \\ \vdots \\ 0 \\ b(t) \end{bmatrix} \in L^1_{loc}(I),$$

so (A(t), B(t)) *constitute a Zettl-Everitt pair, and the controllers* $u(t) \in L^{\infty}_{loc}(I_1)$ *for intervals* $I_1 \subset I$.

If $b(t) \neq 0$ a.e. *on* I, *then* L) *is fully controllable in the state space* $\mathbb{R}^n$, *starting at any* $t_0 \in I$.

The proof of Theorem 1 (see [3,9] for details) shows that $\dim K_0(T) < n$ implies that there exists a nontrivial solution $\eta(t)$ of the adjoint equation

$$\dot{\eta} = -\eta A(t)$$

such that

$$\eta = (\eta_1, \ldots, \eta_n)$$

satisfies

$$\eta(t)B(t) \equiv \eta_n(t)b(t) \equiv 0 \quad \text{a.e. on} \quad [t_0, T] \subset I.$$

But the special format of the Zettl matrix A(t) shows that $\eta_n(t) \equiv 0$ a.e. implies $\eta(t) \equiv 0$ a.e., which is impossible.

In the notation of quasi-differential equations the conclusion of Theorem 1 takes the following form.

Corollary. The scalar quasi-differential equation

$$x_A^{[n]} = b(t)u \quad \text{on} \quad I \subset \mathbb{R} \qquad \mathcal{Q})$$

with $A(t) \in Z_n(I)$ *and* $b(t) \neq 0$ a.e. *on* I, *is fully controllable in the state space* $\mathbb{R}^n$, *starting at any initial instant* $t_0 \in I$.

4. MINIMAL-TIME OPTIMAL CONTROL OF QUASI-DIFFERENTIAL EQUATIONS

In this final section we turn to the problem of minimal-time optimal control of an initial state $x_0 \neq 0$ to the origin in $\mathbb{R}^n$, with bounded controllers u(t) in the linear system

$$\dot{x} = A(t)x + B(t)u. \qquad L)$$

Again we assume that the coefficient matrices

$$A(t) \in Z_n(I), \quad B(t) = \begin{bmatrix} 0 \\ 0 \\ \vdots \\ 0 \\ b(t) \end{bmatrix} \in L^1_{loc}(I) \quad (b(t) \neq 0 \text{ a.e. on } I)$$

constitute a Zettl-Everitt pair on the open interval $I \subset \mathbb{R}$. Also for simplicity we assume $t_0 = 0$ lies in I, and the control restraint is defined by

$$|u(t)| \leq 1 \quad (\text{so } u(t) \in L^\infty_{loc}(I)).$$

Under these circumstances it is known [8] that the set of attainability, from the origin in duration $[0,T]$, is a compact convex subset of $\mathbb{R}^n$. We still denote this set by $K_0(T)$ and note that it varies continuously with $T > 0$ (say, in terms of the Hausdorff metric). In addition the controllability asserted in Theorem 1 guarantees that $K_0(T)$ has non-empty interior, so $K_0(T)$ is a compact convex-body on $\mathbb{R}^n$.

For the given linear control system $L)$ fix an initial state $x_0 \in \mathbb{R}^n$ and consider the corresponding set of attainability $K_{x_0}(T)$, from x_0 on $0 \leq t \leq T$, using the controllers restrained by $|u(t)| \leq 1$. Then

$$K_{x_0}(T) = X(T)x_0 + K_0(T) \quad (\text{vector sum}),$$

so $K_{x_0}(T)$ is merely a translate of $K_0(T)$ in $\mathbb{R}^n$.

If a controller $u(t)$ on $0 \leq t \leq T$, with $|u(t)| \leq 1$, steers x_0 to a point $x(T)$ on the boundary $\partial K_{x_0}(T)$, then $u(t)$ is called an extremal controller, and the corresponding response $x(t)$ on $0 \leq t \leq T$, is an extremal solution of the control system $L)$, see [8]. A necessary and sufficient condition that $u(t)$ on $0 \leq t \leq T$ is an extremal controller is that the Pontryagin Maximal Principle holds - namely: there exists a nontrivial solution $\eta(t)$ of

$$\dot{\eta} = -\eta A(t)$$

for which

$$\eta(t)B(t)u(t) = \max_{|u|\leq 1} \eta(t)B(t)u \quad \text{a.e. on} \quad [0,T].$$

But since $b(t) \neq 0$ a.e. on I, we note that

$$\eta(t)B(t) = \eta_n(t)b(t) \neq 0 \quad \text{a.e. on} \quad [0,T],$$

for otherwise (because of the format of the adjoint equation) the vector $\eta(t) \equiv 0$ on some set of positive measure.

Theorem 2 (Bang-bang). Consider the linear control system in $\mathbb{R}^n$ *L)* $\dot{x} = A(t)x + B(t)u$, *on open interval* $I \subset \mathbb{R}$, *where*

$$A(t) \in Z_n(I), \quad B(t) = \begin{bmatrix} 0 \\ 0 \\ \vdots \\ 0 \\ b(t) \end{bmatrix} \in L^1_{loc}(I), \quad (b(t) \neq 0 \text{ a.e. on } I),$$

so $(A(t),B(t))$ *forms a Zettl-Everitt pair. We use controllers* $u(t) \in L^{\infty}_{loc}(I_1)$ *for intervals* $I_1 \subset I$, *subject to the restraint* $|u(t)| \leq 1$.

If a controller $u(t)$ *on* $0 \leq t \leq T$ *steers an initial state* $x_0 \in \mathbb{R}^n$ *to the boundary* $\partial K_{x_0}(T)$, *then there exists a nontrivial solution* $\eta(t)$ *of*

$$\dot{\eta} = -\eta A(t)$$

so that the extremal controller $u(t)$ *satisfies*

$$u(t) = \operatorname{sgn}[\eta_n(t)b(t)] \quad \text{a.e. on} \quad 0 \leq t \leq T.$$

Thus $|u(t)| = 1$ *a.e., which means that* $u(t)$ *is of bang-bang type.*

Proof. The proof of Theorem 2 follows immediately from the Pontyagin Maximal Principle. Since

$$u(t) = \operatorname{sgn}[\eta_n(t)b(t)] \quad \text{a.e. on} \quad 0 \leq t \leq T,$$

and since $\eta_n(t)b(t) \neq 0$ a.e., we conclude that

$$|u(t)| = 1 \quad \text{a.e. on} \quad [0,T].$$

This asserts that an extremal controller u(t) must be of bang-bang type. □

For an extremal controller u(t) on $0 \le t \le T$, with

$$u(t) = \operatorname{sgn}[\eta_n(t)b(t)]$$

there can be an infinite number of switches for u(t) between the values +1 and -1. But if, in addition, the coefficient matrices A(t), B(t) are real analytic on I, then there are only a finite number of switches for the extremal controller u(t) on $0 \le t \le T$.

We next apply these considerations to the minimal-time optimal control problem, where we seek to steer $x_0 \neq 0$ to the origin in $\mathbb{R}^n$, in the minimal possible time $T^* > 0$. By familiar arguments [8] we conclude that a minimal-time optimal controller $u^*(t)$ on $0 \le t \le T^*$ must be an extremal controller.

Theorem 3. Consider the linear control system in $\mathbb{R}^n$

$$\dot{x} = A(t)x + B(t)u, \quad \textit{on open interval} \quad I \subset \mathbb{R}, \tag{L}$$

where (A(t),B(t)) *form a Zettl-Everitt pair with* $b(t) \neq 0$ a.e. *on* I, *as before. The controllers* $u(t) \in L^\infty_{loc}(I_1)$, *for intervals* $I_1 \subset I$, *satisfy the restraint*

$$|u(t)| \le 1.$$

If $x_0 \neq 0$ *can be steered to the origin of* $\mathbb{R}^n$ *by some admissible controller, then there exists a minimal-time optimal controller* $u^*(t)$ *on* $0 \le t \le T^*$, *for the minimal-time* $T^* > 0$. *Moreover* $u^*(t)$ *is unique* (a.e.), *and* $u^*(t)$ *is necessarily of bang-bang type.*

The next two corollaries give the interpretation of Theorem 3 for scalar differential, and quasi-differential control systems --- which are merely notational reformations of the system L).

Corollary 1. Consider the scalar control system

$$\chi^{(n)} - a_{n-1}(t)\chi^{(n-1)} - \dots - a_0(t)\chi = b(t)u \tag{S}$$

with coefficients $a_i(t)$, $b(t)$ *in* $L^1_{loc}(I)$ *and* $b(t) \neq 0$ a.e. *on the open interval* $I \subset \mathbb{R}$. *Controllers* $u(t) \in L^\infty_{loc}(I_1)$, *on various intervals* $I_1 \subset I$, *satisfy the restraint* $|u(t)| \leq 1$.

If an initial state $(x_0^1 = \chi(0), \ldots, x_0^n = \chi^{(n-1)}(0))$ *can be steered to the zero state in* $\mathbb{R}^n$, *then there exists a unique* (a.e.) *minimal-time optimal controller* $u^*(t)$ *on* $0 \leq t \leq T^*$, *and* $|u^*(t)| = 1$ a.e., *so* $u^*(t)$ *is of bang-bang type.*

Corollary 2. Consider the scalar quasi-differential equation

$$\chi_A^{[n]} = b(t)u \tag{2}$$

where $A(t) \in Z_n(I)$ *and* $b(t) \neq 0$ a.e. *is in* $L^1_{loc}(I)$, *on the open interval* $I \subset \mathbb{R}$. *Controllers* $u(t) \in L^\infty_{loc}(I_1)$, *on various intervals* $I_1 \subset I$, *satisfy the restraint* $|u(t)| \leq 1$.

If an initial state $(\chi_A^{[0]}(0), \ldots, \chi_A^{[n-1]}(0))$ *can be steered to the zero state in* $\mathbb{R}^n$, *then there exists a unique* (a.e.) *minimal-time optimal controller* $u^*(t)$ *on* $0 \leq t \leq T^*$, *and* $|u^*(t)| = 1$, *so* $u^*(t)$ *is of bang-bang type.*

As interesting applications of Theorem 3, we can relate our theories of controllability to some particular scalar differential equations, such as the Mathieu equation. We refer to the more extensive paper [9] for these details.

ACKNOWLEDGEMENT

This research was partially supported by NSF Grant DMS 86-07687 and AFOSR-1SSA-860088.

REFERENCES

1. R. Bellman, I. Glicksberg, O. Gross, On the Bang-Bang Control Problem, *Quart. Appl. Math. 14*, 11-18 (1956).
2. D. Bushaw, Optimal Discontinuous Forcing Terms, *Contrib. to the Theory of Diff. Eqs. Vol. 4*, Princeton Univ., 29-52 (1958).
3. W. N. Everitt, Linear control theory and quasi-differential equations, to appear in *Math. Zeitschrift*.

4. W. N. Everitt and L. Markus, Nonlinear quasi-differential control systems, to appear.

5. H. Hermes and J. P. LaSalle, *Functional Analysis and Time Optimal Control*, Academic Press, New York, 1969.

6. J. P. LaSalle, Basic principle of the Bang-Bang Servo, *Bull. Am. Math. Soc.*, *60*, 54 (]959).

7. E. B. Lee and L. Markus, Optimal control of nonlinear processes, *Arch. Rat. Mech.*, *8*, 36-58 (]961).

8. E. B. Lee and L. Markus, *Foundations of Optimal Control Theory*, Wiley Publ., New York, 1967, Republ. by Krieger, 1986.

9. L. Markus, Control of quasi-differential equations, *Annales Polon. Math.*, to appear (1988).

10. L. Markus and N. Wagner, On stability theorems for nonlinear servomechanisms, *J. Math. Mech.*, *6*, 393-500 (1957).

11. L. S. Pontryagin, Optimal control processes, *Usp. Mat. Nauk*, *14*, 3-20 (1959).

12. L. S. Pontryagin, G. Boltyanski, R. V. Gamkrelidze, and E. F. Mischenko, *The Mathematical Theory of Optimal Processes*, Interscience Publ., New York, 1962.

Towards Dynamic Control Optimization of Flexible Manufacturing Cells

Pablo Spinadel, Flexible Automation Group, Seibersdorff, Austria, Technical University, Vienna, Austria, and University of Buenos Aires, Buenos Aires, Argentina

A B S T R A C T

In the last years some researchers have begun to point out that: "... we can obtain considerable gains in efficiency by the adoption of appropriate methods for factory organization, planning and scheduling and production control...".

If we want to reach these objectives, it will be necessary to adapt already existing optimal control techniques or to develop new ones.

As an example, within Flexible Manufacturing Systems, the assembly environment needs an optimal control mechanism with enough flexibility to solve not only the static optimization problem, as a result of management constraints and production conditions, but also the real time problems of machine break-downs or changes in priority of orders or unscheduled express orders. As practical application and study prototype, a HIgh Flexible Assembly Cell (HIFAC) being erected at the Austrian Research Centre Seibersdorf (ARCS) in our laboratory is outlined in this paper. After a discussion of cell configuration requirements, the necessity of a DYnamic Control Optimization (DYCO) mechanism is shown and the developments at ARCS are outlined.

INTRODUCTION

Traditional batch manufacturing systems were criticized because of the elevated levels of **Work-In-Process (WIP)** and because the jobs spent most of the time "**waiting** for something to happen to them" **(WFS-Time)**, (Buzacott & Yao, 1983). These two elements depend strongly on logistics and planning and scheduling **(P&S)** and also on the hardware and its layout.

Some very interesting advances were made in order to reduce the WFS-time for example in the machining area, a better understanding of the internal and external setup times {see **SMED** or **POKA-YOKE** (Shingo, 1985 and Shingo, 1986)} or the **CAD/CAM** coupling.

In the same way techniques like **JIT** (Voss, 1987 and Mortimer, 1986), or the new possibilities opened by the simulation Programs (Heginbotham, 1985 and Miller, 1987), give ways to reduce the WIP levels.

These new techniques provoked developments of FMS models, using operational research theories like **QUEUEING NETWORKS** (Buzacott & Yao, 1986) for performance evaluation and programming, and control theory for the development of scheduling and operating procedures (Buzacott & Yao, 1986, and Gershwin et al., 1984; see also Kusiak, 1986; Mamdani, 1986 and Mc Goldrick, 1986). Practically all these models were developed for machining systems, the reason being probably, the high costs of such systems that enable financial support by the industries.

This holds for most of the investigated problems, like machine balancing, aggregate planning, resource grouping, etc. all of them applied to large industries only. In other words, if we define a FMS like a group of **F**lexible **M**anufacturing **C**ells **(FMC)**, the problems solved were the interrelations between FMC and not the internal problems within a FMC.

At this point it is important to make a differentiation between different kinds of FMC's, because the development of new machining centers have progressed so far that all the cell work is done inside the machining centers. This transforms the cell problems into internal machine problems.

This is the reason for us to limit our study to assembly FMC's: **F**lexible **A**ssembly **C**ells **(FAC)**. Also we will refer to machining as FMC-cells unless specified otherwise.

The study of **FACs** obliges us to use other kinds of techniques different from that ones mentioned above because:

a) In most cases, assembly times are much shorter than machining times. This requires normally real-time solutions.

b) We need not only a static optimization of resource utilization but also a dynamic optimization that will allow us to reach the levels of flexibility the actual situation requires.

This paper consists of 4 parts:

Part A: General characteristics of a **FAC**, like scheduling, flexibility and control.

Part B: Description of the **HIFAC** project.

Part C: Overview of the proposed System Structure and of the model being developed therefore at **ARCS-FLAG.**

Part D: Some features of the Control Support and Decision Module.

PART A

CHARACTERISTICS OF A FAC

1.- INTRODUCTION

The most important factor in industrial research over the years has been the aim how to reach higher productivity. Flexibility as an industrial factor has varied cyclically over the years, accompanied by a constant increase in the production levels. **Without a new flexibility increase the industry will not be able to reach a new increase in the productivity** (Spinadel, 1987).

Modern industrial elements are already very flexible, so it will be very difficult to increase their inherent flexibility. More effort must be made to reach a better synchronisation, between elements and between elements and products, the reason being that in this field an increase of flexibility seems to be possible. To reach a better interrelation, we could change the physical and the temporal distribution of the resources. In this paper we will discuss basically the temporal distribution.

Systems are continously varying, so it will be very difficult to talk about static optimization and this condition forces us to analyze what kind of problems a development of a DYCO mechanism for a FAC will have.

2.- ASSEMBLY AS A PART OF PRODUCTION PROCESS

Assembly obtains a central position in production if short processing times and at the same time low in-process supplies are to be realized.

Technical conditions for obtaining these aims can be provided by means of flexible assembly that can be regarded as the bottleneck in the manufacturing process. By integrating flexible assembly into this process, linking areas like computer aided design, production planning, parts manufacturing and quality control with the assembly area, organizational conditions are provided.

Assembly parts can arrive at subcells via different paths, at optional sequences and frequences. This requirement can be accomplished by "equipping" single parts carriers with specific information. Equipped with its information, a carrier will "find" the right way to various stations (subcells) within the cell. Arrived at a subcell, specific operational work, requested by the data storage device at the parts carrier, will be made. As a result, **every single assembly technology needs to be installed only one time** (in one station), but on condition that these technologies can be applied in a **highly flexible** way.

3.- REQUIREMENTS OF FAC

The task of the control system for the described operating sequences can be outlined as follows:

According to orders from the production planning system (e.g. assembling of a certain amount of (different) variants of products within a limited period) the optimal sequence of single assembly steps has to be determined **provisionally.** This static determination is the basis for starting operating sequences with certain parts.

The fundamental condition for the function of the cell is the application of a **dynamic** control system. By this system, assembly sequences, paths and duration of assembly cycles are to be controlled **on-line.** The previously determined sequences can be corrected while assembly is progressing in the cell. In addition error detection and correction is a necessary feature of the control system.

Probably the most important elements of a **FAC** to reach the above mentioned features are Planning and Scheduling **(P&S)**, Flexibility **(F)** and Control **(C)**.

3.1.- PLANNING AND SCHEDULING

The traditional **P&S** method is to calculate always a goal schedule. This goal schedule is normally of high complexity and normally always arrives at suboptimal solutions that require an enormous computational effort (Kusiak, 1987). This kind of method produces an inherent limitation in industry that is: the goal schedule will only be recalculated if the variation is very great.

So the traditional system works normally at an **"Adapted" (inferior) level of a suboptimal solution**; this shows the necessity of new P&S structures.

The first step will be to define two types of scheduling:

LONG TIME SCHEDULING (LTS) and **SHORT TIME SCHEDULING (STS)**

3.1.1.- LONG TIME SCHEDULING

This kind of P&S is based usually on the required work the industry knows with a great time anticipation. This comes from the batch style systems, where the producer fixes the minimum time before an order must be solicitated or he works on stock. One of the changes the FMS must introduce where applied is to reduce drastically this time period. So the unit of **LTS** will be reduced from years and months to weeks and days.

Most of the industries using computer support today use the so-called Production Planning Systems **(PPS)**. There are a lot of different PPS's that more or less allow a LTS. So we do not want to develop a new PPS but will clearly define a small interface to such systems that allows the use of most of the existing PPS.

3.1.2.- SHORT TIME SCHEDULING

This is the level where today we need a fast reaction when the internal or external variables change or even the model parameters change.

The only way to reach the solution by a fast reaction, is not using all the information we have in a centralized way, the reason being that computational efforts will surely not be enough. In most of the cases the information that could be treated locally is very limited because it is not easy to know the way it affects other areas. A possible solution to this could be the use of expert systems.

Expert systems **(ES)** allow to find rapidly an information structure that permits us to distinguish between fundamental variables that must be sent for a central analysis and those whose influence could be analyzed using local heuristic knowledge in order to send a kind of resumed variable only to the central control. Also the ES allows the early detection of failure causes (for example to find a degraded function of a robot, based on the power consumption of a motor), which enables a reduction in the number of errors tending to the ideal **Zero Quality Control** (Shingo, 1986).

3.2.- FLEXIBILITY

The intuitive concept of flexibility is normally associated with the capacity of a system to "adapt correctly" to alternative constellations.

In Spinadel, 1988, we discuss some flexibility formulations that already exists and we introduce a new one that will allow us to broaden the definition introducing time analysis so as to be able to distinguish between a **STATIC FLEXIBILITY (SF)** and a **DYNAMIC FLEXIBILITY (DF).**

3.2.1.- STATIC FLEXIBILITY (SF)

The SF is associated with a correct functioning of the system in different constellations, that is, it does not take into account transition from one state to another of the same constellation or of a different constellation. Also it does not show what happens during transitions between constellations. We define the SF of system i in constellation k as:

$$SF_{ik} = \frac{(EA_{ik} - EP_k)}{(EI_k - EP_k)} \quad (1)$$

where **EA** = actual, **EI** = ideal, **EP** = worst } efficiency

3.2.2.- DYNAMIC FLEXIBILITY (DF)

We will associate the DF with the correct "reaction" of a system to any variation that occurs.

So DF regards the transition of one constellation to another and also between states of the same constellation. We define the **DF** as:

$$DF_{ik} = \frac{dSF}{dt}_{ik} = \left. \frac{a* \frac{dEA}{dt} - b* \frac{dEI}{dt}}{(EI - EP)^2} \right|_{ik} \qquad (2)$$

where $a > 0$; $b > 0$ and $\frac{dEP}{dt} = 0$

3.2.3.- ANALYSIS OF FLEXIBILITIES

From equation (1) we see that **SF** represents a measure of how the system works on each constellation. If we know all the constellations that conform the system world, we only need to optimize the resource utilization in all of them. This optimization is principally limited by economic factors like cost of resources. So if we introduce a new constellation in the system, the SF of the system will be fixed at the planning stage and will only depend on a good utilization of resources.

From all the possible variations we are interested only in the case of some problem or breakdown of a machine and in the normal transition from the assembly of one product to the assembly of another one.

A good system selection can be prove by studying what occurs with the Ideal Efficiency **EI** and the Actual Efficiency **EA** when the system change to another state (Spinadel, 1988).

More about flexibility in FMS can be found in Kumar, 1986; Zelenovic, 1982 and Buzacott, 1982.

3.3.- CONTROL

Within a FAC we must define two fundamental types: those where a new development could be made on-line (using simulation, automatic program generation and others) and those where we could only develop off-line.

In our case the tasks needed for new developments are planned and scheduled like any other job. This normally decreases the flexibility of the system, because although we do not stop the process we always need to disconnect the resource being programmed. For the system it looks like a total breakdown of this resource.

For this kind of systems we must define different levels of control:

* **Electronic control**
* **Hardware control**
* **Communication control**
* **General control**

We will shortly describe each of them, but will concentrate on the last two.

3.3.1.- ELECTRONIC CONTROL

This is the level where we do the direct control of the electronic. It has no intelligence by itself, but a high reaction velocity, that we must try to control as much as we can at this level. Normally this control is done by devices like PLC's.

As an example, in a robot this level would control the motors and input/output signals.

3.3.2.- HARDWARE CONTROL

Based on a limited logic and a group of control algorithms here the first decision will take place. The most important feature of this level is to realize a good filtering of the variables to pass only that one useful to the upper control level. Normally this control is done by the dedicated controllers included in each machine.

As an example with a robot where we know the point to reach and also perhaps the way to reach it, the controller must resolve which movements must be done by each axis and transduce this to the necessary signal for the motor.

3.3.3.- COMMUNICATION CONTROL

This is the first level where we need some intelligence and also the first one affected by the P&S. At this level there is a great part of the flexibility of the system. The function of the controller here is to control the resources. This means, for example, to load and unload programs from local memories, or to follow the elements being moved and also to secure a correct transfer of information without collision problems or time problems.

For the controller the elements will no more be robots or transport systems or some specific devices, they must be only resources with a determined group of characteristics. This will no more define hardware details but a set of programs that could be done, the time they need to do them, how much each costs, etc.

At this level we normally find micro- or mini-computers.

As an example we could talk about a micro-computer distributing programs from the central memory unit to all the machines with no collisions, or by requirement of the upper level showing the state of each resource and each job being done.

3.3.4.- GENERAL CONTROL

This is the fundamental level of the system where functions must be done like the following:

* control of a consistent functioning of the resources (each one must have what they need)
* resolve dynamically the fail problems that may appear.
* realize the alternative analysis in case of a variation on the goal schedule
* control the communication of the cell with the "world". In order to secure the readyness of elements it will need to continue functioning.

In this level we will find most of the parts of the cell like expert systems, simulation modules, management decision support, etc.

This level also could be defined as one where all modes differing from the automatic mode, like starting mode, degraded mode, ending mode, debugging mode, etc., must be analyzed and "well done".

Some authors like Corbell et al., 1985, propose a multilevel structure where one level determines in which mode the system is working and one or more other levels realize the control of each mode. This requires to know all the possible states of the system or, as we will propose, a basic model at which we will change the parameters with the help of expert systems.

From now on we will refer to this control level when we talk about the controller unless otherwise specified.

PART B

DESCRIPTION OF HIFAC PROJECT

1.- THE ARCS-FLAG

The Austrian Research Centre Seibersdorf (**ARCS**) is Austria's largest non university research institution. Proceeding from the special know-how acquired over many years in developing (since it foundation in 1956), designing and building equipment for handling radioactive materials, in 1985 the working programme for the new department "Flexible automation and production engineering" was introduced. Activities within this area have been concentrated on research and development in the field of assembly automation, building of special purpose machines and components (e.g. sensor guided sorting machines or robot grippers) and the transfer of problem solutions for production processes to industry.

2.- EUROPEAN STATUS

Connected with the EUREKA-Project EU72 "**FAMOS**" the ARCS has submitted the project "Flexible assembly cell and communications network for manufacturing of precision gearboxes". Industrial companies and research centres will cooperate in this project. The aim of the project is the development of new assembly technologies which can be applied by means of a communications network. FAMOS-projects are "industry led projects" - accordingly, problem solutions have to be adjusted to the needs of an industrial manufacturer.

The main objective of this project is to get a better understanding of the problems inside the cell as well as the problems of its interaction with the exterior, in order to obtain a generic system of dynamic control.

3.- HIFAC ELEMENTS

The **HIFAC** contains two robots (BOSCH, ASEA), a flexible conveyer-belt system with dot memory data carriers, robot-guided screwing and glueing devices, sensor-controlled press and robot-gripper, quality-control station and several feeding/supplying devices. One of these devices applies a vision system for flexible feeding; bar code readers are used for identification of certain assembly products. The assembly cell now being set up at the **ARCS** is the central component of a planned pilot plant, which also includes machine tools with CAD/CAM, a production planning system, stocks and automated guided vehicles. **Figure 1** shows the situation of the assembly cell, integrated into the general layout.

4.- MAIN OBJECTIVES OF THE HIFAC PROJECT

a) **Independence of the product:** The system must be capable of adapting itself with very small modifications to the assembly of different types of products.

b) **Independence of the used hardware and software:** The system must be independent from software and hardware elements with which it interacts. This means that at the system level we can only talk about resources with determined economic and temporal characteristics and not about determined elements.This requires to define clear interfaces, so simple as to be able to communicate without too much extra effort with elements of different suppliers.

c) **Practically demonstrable:** The system must be in a position to assemble some types of product. After a study of specific needs of the Austrian market, we decided to take precision gear boxes as assembly element that allow easily to define different types as well as different variations for each type.

d) **Flexible software:** The software must be easily expandable and, therefore, its design is modular.

e) **Flexible hardware:** The system to be controlled must be easily expandable. This has, as a consequence the use of a net of PC compatible computers inside the cell, permitting a high local flexibility. For the global management at the shop floor level, there is a protocol independent optical fiber net **(TOPNET)** developed at ARCS, which communicates to a Host.

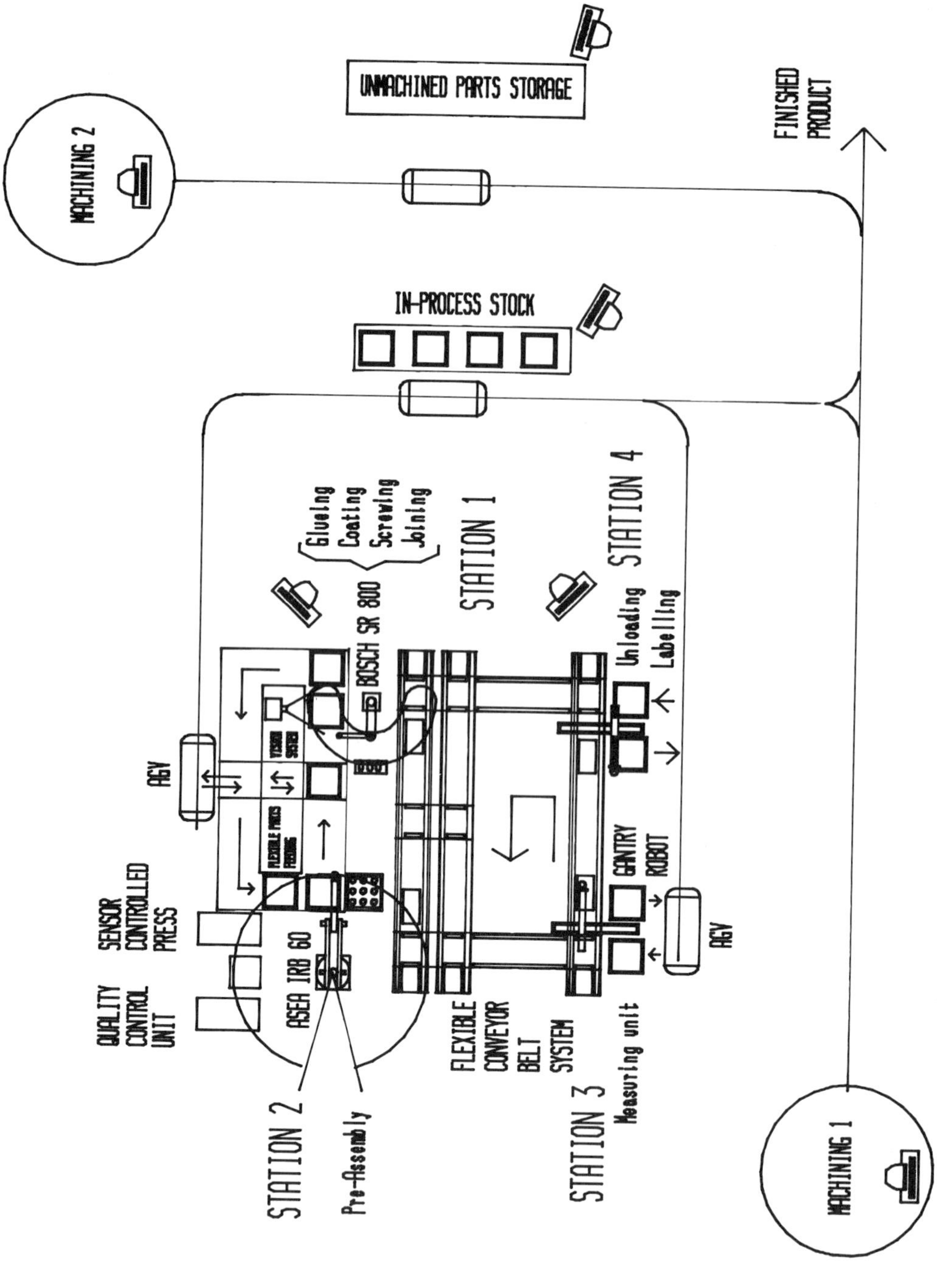
MACHINING 2
UNMACHINED PARTS STORAGE
FINISHED PRODUCT
IN-PROCESS STOCK
Glueing
Coating
Screwing
Joining
STATION 1
STATION 4
BOSCH SR 800
Unloading
Labelling
VISION SYSTEM
AGV
FLEXIBLE PARTS FEEDING
GANTRY ROBOT
AGV
SENSOR CONTROLLED PRESS
QUALITY CONTROL UNIT
ASEA IRB 60
FLEXIBLE CONVEYOR BELT SYSTEM
Measuring unit
STATION 2
Pre-Assembly
STATION 3
MACHINING 1

FIGURE 1

PART C

STRUCTURE OF THE MODEL OF HIFAC

1.- INTRODUCTION

Given a set of variables and the relations between them, the objective of a mathematical model is to obtain a set of values which we could define as solutions or responses. If we include the time in the model, this in general becomes an iterative model.

Taking into account the proposed solution and the model reactions it is possible to recalculate a set of variables which allow to generate a new set of solutions beginning the cycle again. Not all model variables are influenced by internal changes, some of them are determined in a static or dynamic way by elements external to the model. These last ones will be called "external variables", to distinguish them from the others, which will be called "internal variables".

As we have previously noted there exist definite relations between the system variables that include a group of parameters which are functions of the system elements.

If we represent this concept we get the model in **Figure 2.**

This model, largely used as a base, satisfies the requisites necessary for the so-called automatic state of the system, that is, for a static system where only processed elements are changing. It possesses also the necessary characteristic of a feedback system which permits to stabilize its functioning.

Using the common terminology in flexibility, this model will be suitable for a "constellation universe" with only one level in each one of them, that is, with a lonely basic state.

2.- SYSTEMS VARIATIONS

In any other system work-mode different from the automatic one, for example, a degrading mode, we are obliged to make a variation on the model. This variation is not referred to as the internal variables of the

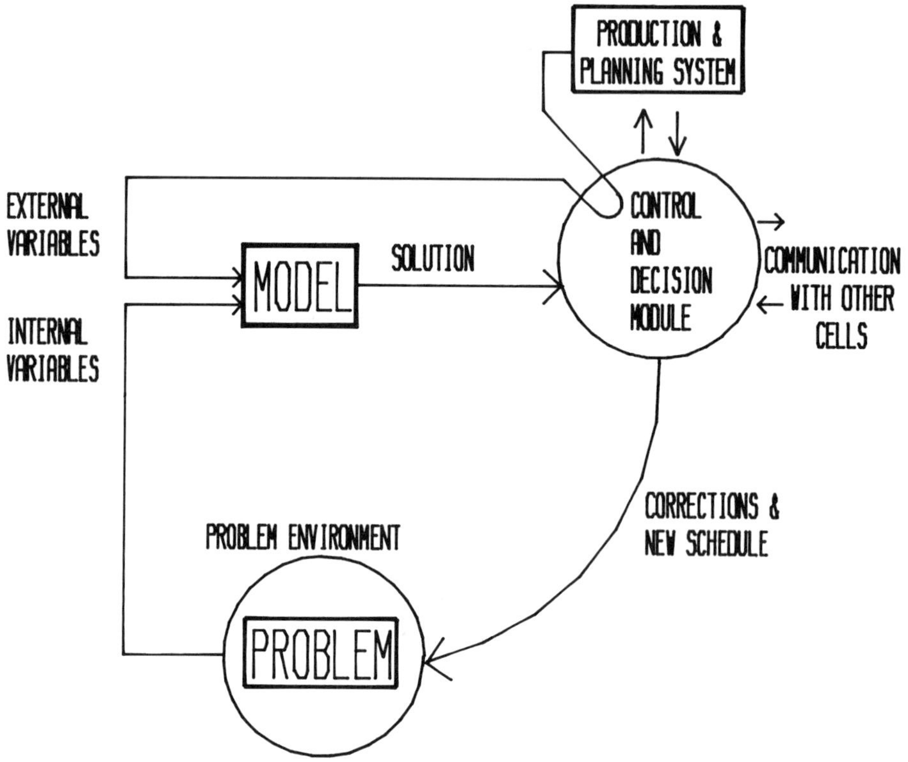
PRODUCTION &
PLANNING SYSTEM
EXTERNAL
VARIABLES
MODEL
SOLUTION
CONTROL
AND
DECISION
MODULE
COMMUNICATION
WITH OTHER
CELLS
INTERNAL
VARIABLES
CORRECTIONS &
NEW SCHEDULE
PROBLEM ENVIRONMENT
PROBLEM

FIGURE 2

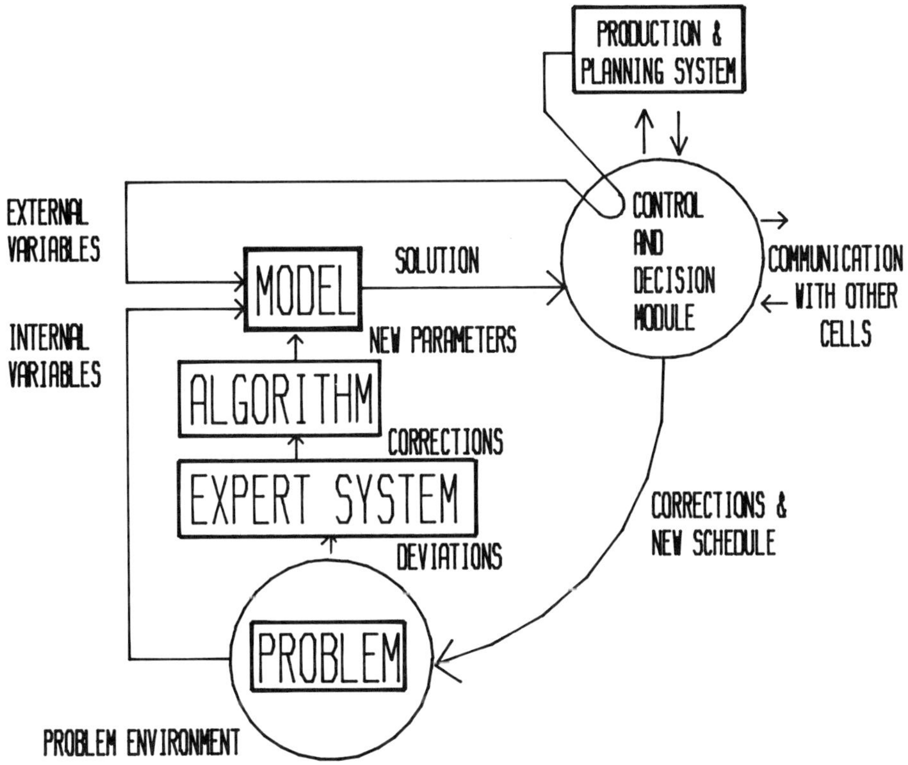
PRODUCTION & PLANNING SYSTEM
CONTROL AND DECISION MODULE
COMMUNICATION WITH OTHER CELLS
EXTERNAL VARIABLES
INTERNAL VARIABLES
MODEL
SOLUTION
NEW PARAMETERS
ALGORITHM
CORRECTIONS
EXPERT SYSTEM
DEVIATIONS
CORRECTIONS & NEW SCHEDULE
PROBLEM
PROBLEM ENVIRONMENT

FIGURE 3

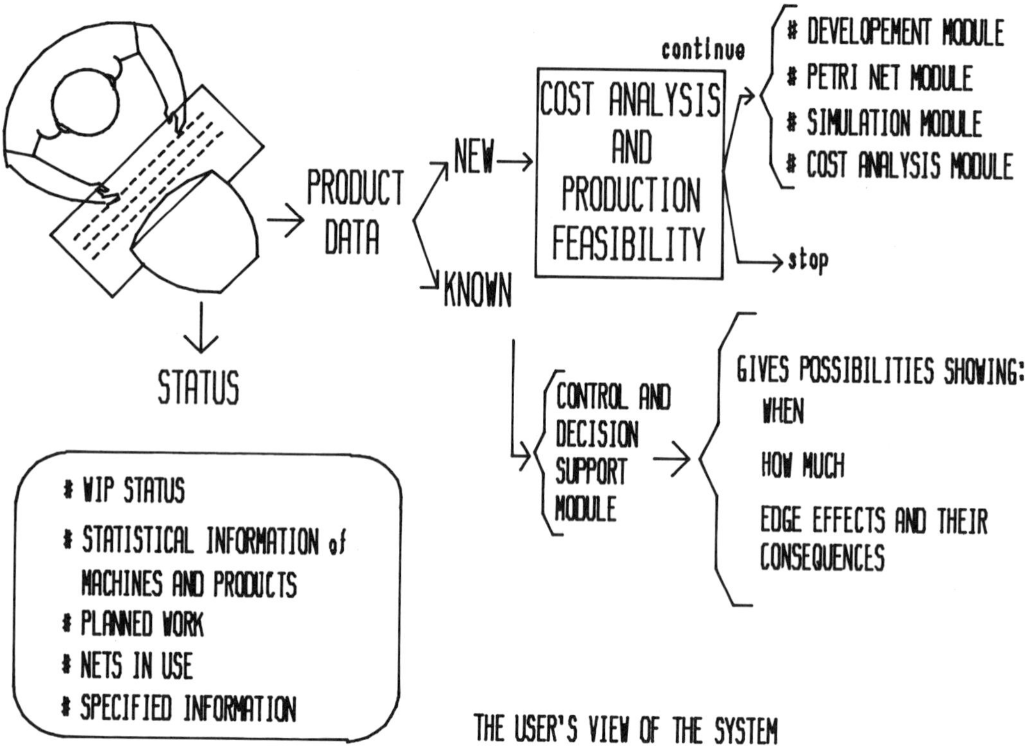
PRODUCT DATA
NEW
KNOWN
continue
COST ANALYSIS AND PRODUCTION FEASIBILITY
DEVELOPEMENT MODULE
PETRI NET MODULE
SIMULATION MODULE
COST ANALYSIS MODULE
stop
STATUS
WIP STATUS
STATISTICAL INFORMATION of MACHINES AND PRODUCTS
PLANNED WORK
NETS IN USE
SPECIFIED INFORMATION
CONTROL AND DECISION SUPPORT MODULE
GIVES POSSIBILITIES SHOWING:
WHEN
HOW MUCH
EDGE EFFECTS AND THEIR CONSEQUENCES
THE USER'S VIEW OF THE SYSTEM

FIGURE 4

system, because they represent the successive temporal states of elements processed in the model. They refer to the parameters that relate variables between them. Then we need a feedback of the parameters value to be able to distinguish different levels of a constellation.

This feedback is not easy to analyze because in general, a variation of the model directly affects only a limited set of variables (sometimes more) but indirectly it affects practically all them. By the same reason, the influence that is working on the system through a change in variables is very complicated to analyze, even more so if we pretend to optimize not only the degraded functioning of the system but also the way it returns to the basic state.

To analyze the influence of system degradation over each of the parameters and also its temporal variation to permit the return to the basic state using an optimal way, we need not only special algorithms translating the actual system reality to the mathematical model, but also heuristic knowledge and a data structure to get the best possible return.

We shall use the model in **Figure 3.**

We notice in this model that besides the external feedback we have also an internal one that in a way could be identified as a feedback for the model, to distinguish it from the other one, which will be a feedback for the system.

This feedback allows us to define only a basic model of the cell and to represent its variations with the help of flexible parameters. This will enable us to use net theory for the model without reaching complexity levels which cannot be analyzed.

The main concept of this **DYCO** is shown in **Figure 4.**

PART D

CONTROL SUPPORT AND DECISION MODULE

1.- FEATURES

Some of the features it must have are the following:

a) To permit an easy following of the cell functioning and of the elements composing the cell.

b) To facilitate the analysis of alternatives in planning.

c) To communicate with other cells.

d) To control the disposal of elements by the developing module.

e) To determine the frontiers of automatic decision.

f) To offer a decision support to management.

We will now discuss some of these characteristics:

2.1.- INTERFACES

Points a), b), c) and d) are related with the cell communication. This communication can be external, i.e. cell with environment, or internal between modules of the cell. As only a part of a whole system, the way in which the cell communicates with the other parts is the key to the successful functioning of the system. This way strongly depends on how interfaces are defined. We will discuss external interface characteristics by differentiating between three types:

TYPE I: PRODUCTION PLANNING SYSTEM

The long time strategy of an industry is normally defined at top management level, this means also the PPS level. By this interface the cell will receive a global schedule with earlier and later times for each job. In the other way, the PPS will receive from the cell information about possible alternative schedules and about the actual state of each planned job.

TYPE II: FLEXIBLE MACHINING CELL

The communication through this interface is basically directed by the FAC. The way this is done is to "pull up" the production (Buzacott & Yao, 1983) in order to receive parts comming from the FMC in JIT sense (Shingo, 1985 and Shingo, 1986).

The communication will be:

FAC -> FMC :
* Solicitation of status from elements needed in the future.
* Indication of received element characteristics in order to analyze possible failure causes.

FMC -> FAC :
* Indication of required status.
* Sending of the information belonging to elements being transferred from the FMC to the FAC.
* Preview message of time retards on FMC.

TYPE III: STOCK

This interface is similar to that above. The only difference is in a higher rigidity that comes from the strong interdependence between stock and the external, not controllable world.

A number of studies have shown the enormous interdependence between system functioning and external world. The long time suppliers require to sell products could annul all the efficiency of a FMS. A large stock is also not a solution because its cost would also annul the decrease achieved by use of FMS.

Perhaps the only solution will be in the future to generate a higher level that includes relations among the industrial participants.

The future could most probably require a JIT-production and management system that includes all parts and even raw material suppliers, linked together by a communication network.

2.2.- AUTOMATIC CONTROL

An automatic control is not always the optimal control. The automatic control has always boundary conditions that give a rigid context to it. As an example, let us suppose that an industry receives an express order which exeeds the disposable resources. It is possible that the system is working JIT and so has no possibility to include this new order in a new schedule. But it is also possible that some factor (for example economic or strategic) makes it convenient for the industry to pay the recharges of finishing later other works in order to schedule the new, more profitable express order.

The function of this module is to determine, in conjunction with the management, which boundary conditions could be changed in an automatic way and which must be asked for. To give more flexibility to this function we will define priority levels for the characteristics of the jobs. We call this function "determination of the frontiers of automatic decision".

2.3.- ALTERNATIVE MODELS

The point f) requires an user-friendly environment to allow the management an easy comparison of alternative possibilities and also an easy change of all the parameters that can be changed by him. This allowance is established by the system manager. This assures a hierarchical structure of the decisions. In order to allow a correct analysis of the distinct possibilities of the system, a simulation module will be necessary, enabling the analysis of not only the utilization of resources possibilities but also the economic side of each of them.

CONCLUSIONS AND OUTLOOK

The central part of the assembly cell, the flexible transport system and two stations (station 1 and 2) with peripheral equipment (press, glueing and screwing devices), were installed at the beginning of 1988. At present (June 1988), variants of different gearboxes can be assembled, provided that parts feeding is achieved via fixed pallets.

The current project phase comprises the installation of subcell components like vision systems for flexible feeding or a measuring unit for tolerance compensation, as well as hardware for the information system of the cell (8 kB data carriers, barcode readers). In addition the remaining two stations are equipped with computer hardware (PC AT/XT) and linked with the host computer of the cell (via ethernet-interface linked with other areas of production such as planning and parts manufacturing).

The third project phase concentrates on the development of the **DYCO** as a fundamental condition for attaining high flexibility, not only with single subcells, but also for the assembly area as a whole and in connection with preeceding areas of production.

Parallel to the first two phases, we are looking for specific techniques that allow dynamic optimization within a FAC.

We shall use extended generalized Petri nets for the construction of the control model and for simulation purposes.

Also the optimization will take place at the decision level and we intend to apply Evolution Strategies for this purpose.

A justification of this decision can be found in Spinadel, 1988a.

An introduction to Petri Nets can be found in Brauer, 1979 and Reisig, 1986 and for further studies Rozemberg, 1987; Favrel, 1985 and Kamath, 1986.

We have already pointed out that optimal control techniques are able to play an important role in the automation area. Also some problems will appear which probably need new techniques that must be developed in an interdisciplinary work. This work can only be done if mathematic researchers begin to be interested to work in this field.

So the main objective for sending this paper to the conference of Modern Optimal Control was to introduce the Automation field and especially the Flexible Assembly problem to mathematic researchers.

REFERENCES

1. BRAUER, W. editor (1979). Net theory and applications. Lecture notes in computer science; Springer Verlag.
2. BUZACOTT, J.A.; SHANTHIKUMAR, J.G. (1980). Models for understanding flexible manufacturing systems. a) Dept. of Ind. Eng., Univ. of Toronto, Ontario, Canada. b) AIIE Transactions - Dec.1980.
3. BUZACOTT, J.A. (1982). The fundamental principles of flexibility in manufacturing systems. Proceedings of the first conference in flexible manufacturing systems. North Holland. Amsterdam.
4. BUZACOTT, J.A. & YAO,D.D. (1983). Flexible manufacturing systems: a review of analytical models. Management Science - Vol. 32, No.7, July 1986.
5. BUZACOTT, J.A. AND YAO, D.D. (1986). On queueing network models of flexible manufacturing systems. J.C. BALTZER a.g. Scientific Publishing Company.
6. CORBEEL, D.; GENTINA, J.C. AND VERCAUTER, C. (1985). Application of an extension of Petri nets to modelization of control and production processes. Proceedings of the 6th european workshop on applications and theory of Petri nets.
7. FAVREL, J. AND LEE, K.H. (1985). Modelling, analyzing, scheduling and control of fms's by Petri nets. Modelling production management systems. FALSTER,P. & MAZUMDER,S.(ED)
8. GERSHWIN, S.B. ; HILDEBRANDT, R.R. ; SURI, R. AND MITTER, S.K. (1984) A control theorist's perspective on recent trends in manufacturing systems. Proceedings of the IEEE 23rd. Conference on Decision and Control, Las Vegas, NV.
9. HEGINBOTHAM, W.B. editor (1985); Simulation in Manufacturing. IFS (publications) Ltd, UK.
10. KAMATH, M. AND VISWANADHAM, N. (1986). Applications of Petri nets based models in the modelling and analysis of flexible manufacturing systems. Proceedings of the 1986 IEEE conference on robotic and automation.
11. KUMAR, VINOD (1986). On measurements of flexibility in flexible manufacturing systems: an information-theoretic approach. Proceedings of the second ORSA/TIMS conference on flexible manufacturing systems: Operations Research models and applications. Edited by STECKE, K.E. AND SURI, R.- Elsevier Science Publishers B.V.
12. KUSIAK, A. (1986). Application of operational research models and techniques in fms's. Dept. of Ind. & Mech. Eng., Manitoba Univ., Winnipeg, Canada.
13. KUSIAK, A. editor (1986). Flexible manufacturing systems: Methods and Studies. North Holland. Amsterdam.
14. KUSIAK, A. (1987). Artificial Intelligence and Operational Research in FMS. Dept. of Mech. & Ind. Eng., Manitoba Univ., Winnipeg, Canada.
15. MAMDANI, A. AND EFSTATHIOU, J. editors (1986). Expert systems and optimization in process control. UNICOM seminars Ltd.
16. MC GOLDRICK, P.F. editor (1986). Advances in manufacturing technology. Proceedings of the first national conference on production research. Univ. of Nottingham - Kogan Page Ltd.

17. MILLER, R.K. (1987); Manufacturing simulation.
S.E.A.I. Technical Publications and Technical Insights.
18. MORTIMER, J. editor (1986); Just-in-time an executive briefing.
IFS (publications) Ltd, UK. Springer Verlag.
19. REISIG, W. (1986). Petrinetze eine einführung. Springer Verlag.
20. ROZENBERG, G. editor (1987).
Advances in Petri nets 1987. Springer Verlag.
21. SHINGEO SHINGO (1985). A revolution in manufacturing: the SMED system.
Productivity Press. Stamford, Connecticut and Cambridge, Massachusetts
22. SHINGEO SHINGO (1986).
Zero Quality Control: source inspection and the POKA-YOKE system.
Productivity Press. Stamford, Connecticut and Cambridge, Massachusetts
23. SPINADEL, P. (1987).
Historical relation between flexibility and productivity.
ARCS. Flexible Automation Group. Working Paper. Nr. 011087.
24. SPINADEL, P. (1988). Dynamic and static flexibility.
ARCS. Flexible Automation Group. Working Paper. Nr. 010288.
25. SPINADEL, P. (1988a). HIFAC: why to use Petri nets.
ARCS. Flexible Automation Group. Working Paper. Nr. 020288.
26. STECKE, K.E.; DUBOIS, D.; SETHI, S.P.; RATHMILL, K. (1985).
Classification of flexible manufacturing systems: evolution towards the automated factor. Graduate school of business administration, University of Michigan, Ann Arbor. Working paper No. 363.
27. The Flexible Manufacturing System Handbook.(1982).
The Charles Stark Draper laboratory, Inc., Cambridge, Massachusetts.
28. VOSS, C.A. editor (1987); Just-in-time Manufacture.
IFS (publications) Ltd, UK. Springer Verlag.
29. ZELENOVIC, D.M. (1982).
Flexibility as a condition for effective production systems.
International journal on production research. No. 20. Pages 319-337.

Differential Games and Optimal Control Theory

Vera W. de Spinadel, Professor, University of Buenos Aires, Buenos Aires, Argentina

I. A BIT OF HISTORY

A differential game is a mathematical model designed to solve a conflicting situation that changes with time. Differential games need the help of Game theory, of the Calculus of Variations and of the theory of Optimal Control.

Davis (1962), in his classical book affirms that the origin of the problem of pursuit-evasion lies as far as the XV century with Leonardo da Vinci. However, the problem began to awake some interest among mathematicians in 1732, when P. Bouguer, in a memory published by the Academie Royale des Sciences, was able to find the pursuit curve, i.e., "the curve which follows a vessel that approaches another moving in a straight line, assuming that the velocities of both vessels are in a constant ratio." Years after, other authors considered the more difficult problem in which the pursuit curve was a circumference. These previous works had all the characteristic features that only one player, the pursuer, had freedom to move, while the other, the evader, was constrained to a fixed trajectory.

In 1954-1956, R. Isaacs in four memoranda published by the Rand Corp., was the first to consider the pursuit-evasion problem, where both players move freely. He solved this problem using the well-known Hamilton-Jacobi equations from Calculus of Variations. But, until 1965, when he published his famous book "Differential Games," the problem arose no special interest. Meanwhile, in 1959, Roxin and I found the so-called 'Bang-Bang' control problem through the paper of Bellman, Glicksberg and Gross, 1956, in Buenos Aires, Argentina, almost 30 years ago!

Since then, the enthusiasm followed uninterruptedly. American as well as Soviet engineers and mathematicians got dedicated to the formal study of differential games and the analysis of illustrative examples.

In the Soviet school, in particular, Pontryagin et al., in their classical book appeared in 1962, used similar methods as Isaacs, as their control problem is that of a differential game with only one player. In 1961, Kelendzeridze extended this method to two players and his work is a continuation of Isaacs'.

Most recently, in 1988, Krasovskii and Subbotin published their book "Game-Theoretic Control Problems," which is a revised version of their monograph from 1974. In this book, control problems are stated as follows:

Given a control system, whose dynamics is described by a system of ordinary differential equations, it is necessary to construct a strategy, i.e. a control based on the feedback principle which will assure a definite quality for the process controlled under specified conditions when the system is subject to uncertain factors. These factors could be a perturbation or a controlling action from the opposite side. In the frame of differential game theory, this type of problem is considered jointly with the inverse problem. The analysis of both inverse problems is useful, not only in situations in which two players control the system but also in cases in which the controlled system (by only one player) is subject to an uncertain perturbation.

American contributions, on the other hand, were oriented to the logical foundations of the subject. Berkovitz (1961) applied the Calculus of Variations to one player's strategy, when the opposite strategy remained fixed beforehand. Fleming (1957), in turn, introduced a continuous strategy as limit of two discrete ones, considering probability distributions which are expected to tend to certainty when the game becomes continuous.

At the same time, optimal control theory, which in appearance had nothing to do with a differential game, began to develop. Optimal control theory was introduced principally by Bellman (1956), LaSalle (1960), Bushaw (1958), Gamkrelidze (1958), Krasovskii (1957), and others. Optimal control theory is equivalent to a differential game with only one player and control problems can be extended to differential games adding an opponent. Besides, the analytic methods used to solve a differential game are applicable to control problems, suppressing one player.

Finally, we mention the very important class of differential games where the players must base their decisions on an incomplete information. For these games, results from stochastic process theory and from the general probability theory, are applied.

2. EQUATIONS OF MOTION

Let $x(t)$ be an n-dimensional phase vector which indicates the state of the controlled system at time t. Its dynamics is described by a system of n differential equations

$$\frac{dx(t)}{dt} = f(t,x(t),u(t),v(t)) \tag{1}$$

with the initial condition

$$x(t_0) = x_0 \tag{2}$$

u(t) is a p-dimensional vector which characterizes the controlling action at time t. The w-dimensional vector v(t) corresponds to the disturbance acting at time t. Besides

$$u(t) \in U \subset R^p; \quad v(t) \in V \subset R^q \tag{3}$$

where U and V are compact sets. If we deal with differential games, vectors u(t) and v(t) are called the controls of u- and v-player, respectively. It will be assumed that u(.) and v(.) are measurable functions, a condition which in general is necessary for system (1) and (2) to make sense.

In place of system (1) and (2), one can use a differential inclusion (Clarke, (1986)) which does not involve a specific disturbance but only indicates the bounds for a possible disturbance. In this case, the motion of the controlled system must satisfy the differential inclusion

$$\frac{dx(t)}{dt} \in \{f(t,(x),u(t),v): v \in V\} \tag{4}$$

and, as a rule, the set of solutions will be infinite.

The game begins at a given $x(t_0)$ and ends when x(t) reaches a given terminal closed manifold M. Let T be the first $t > t_0$ for which $x(t) \in M$.

Let the *payoff* be given by a functional

$$\mu = \mu[T,x(t),u(t),v(t): t_0 \leq t \leq T] \tag{5}$$

This is the quantity v-player has to pay to u-player.

More commonly, the payoff consists of two terms

$$P(u,v) = g(T,x(T)) + \int_0^T h(t,x(t),u(t),v(t))dt \tag{6}$$

where g(.) is the *terminal payoff* and the second term is the *integral payoff*.

u-player's objective will be to maximize P(.) while v's will be to minimize it.

With respect to the *strategy* to follow, this must indicate a player how to play the game, that is, how to choose its own control at time t. It must determine the value of the control variable based on the knowledge of the past state and the possible present state of the game. It is worthwhile to notice that a player cannot predict his opponent's future behavior. Besides, the set of strategies must be such that once they are chosen by both players, the corresponding controls u(.) and v(.) exist and are uniquely determined. Even in very simple cases, this does not always happen as, for instance, when the selection of u(.) depends from that of v(.) for the same t and reciprocally. In a real game, a certain delay in the transmission and processing of the information, together with some proper inertia, help to soften these incompatible situations, but for the formulation of the mathematical model, they are very important.

Summing up, the basic problem for each player is to optimize over the set of *admissible strategies*.

3. ADMISSIBLE STRATEGIES

R. Isaacs uses the concept of strategy as a function of the state variable x. For example, a strategy for u is a function $\hat{u}(t,x)$ such that the choice of the control be $u(t) = \hat{u}(t,x(t))$. This corresponds to a *feedback control*, very useful in many important practical games. Despite this fact, from a mathematical point of view, there may appear some situations of incompatibility or even indeterminancy. What is worst, there is no guarantee that for each choice of $\hat{u}(t,x(t))$ and $\hat{v}(t,x(t))$, the differential system (1)-(2) has a solution nor that this solutions is unique. That is why, when the possible choice of u(.) depends from the choice of v(.) or reciprocally, it may be difficult to define precisely the set of admissible paths.

Very simple examples show that optimal controls (and optimal paths also) possess discontinuities. The existence of *singular surfaces* is one of the key points in the formulation of the theory.

To avoid the difficulties inherent to the admissible strategies introduced by Isaacs, Varaiya (1967) and Friedman (1970), had developed a more abstract point of view. This conception does not invalidate Isaacs' results but attempts to give the theory more solid foundations. It is related with the ideas of Fleming and Berkovitz of approaching the continuous game by some discretization mechanism, but in this case there is no need to allow stochastic choices for each player. Varaiya's approach is not really a discretization but an introduction of time delays. Nevertheless, both are based on the idea of using maps in functional spaces to approximate and, in the limit, define the strategies.

This way, Friedman proves that under suitable assumptions, the differential game defined by (1)-(2)-(3)-(6) has a value V which is Lipschitz continuous with respect to t and x.

The result is valid for differential games with terminal condition $t = T$ fixed. In the general case when $(T,x(T)) \in M$, some difficulties arise. The first one is the problem of *controllability:* can a player force the termination of the game against his opponent's will? This is the clue question in a pursuit-evasion game. Following Isaacs, this is considered as a *game of kind*, where the result is yes or no, or in terms of the payoff, $P(u,v) = +1$ or -1. In a certain sense, it can be replaced by a *game of degree*, considering *capture time* as payoff for a pursuit game: $P(u,v) = T$. Under appropriate conditions, this optimal time problem has a pretty continuous solution in the so-called *region of controllability*, where the pursuer may catch the evader; outside this region, the Value of the game is $+\infty$. That means a discontinuity, and, in fact, the game may present discontinuities (of the Value and of the optimal strategies) even when capture time remains finite.

Analogous problems arise when we consider more general payoffs of the type (6). In this case, the notion of controllability must be considered more carefully. From the beginning, it is not easy to know which player (if any) wishes to terminate the game as soon as possible. And this may change from one region to another.

Isaacs introduced the notion of *useable part* of the terminal set. This part is determined by those points of the boundary ∂M where it is possible that the game (reasonably played) terminates. This means that from every point x_0 of some neighborhood of such limit point of M, any of the players may carry x(t) to the terminal set M in a certain time interval that tends to zero with the distance between x_0 and M.

The basic emphasis in the functional approach has been on the proof of existence theorems for the Value of the game. In differential games without value, Roxin (1970) has proved that the game becomes extremely sensible to infinitesimal changes in the information each player receives. In some particular cases, it is possible to get an estimation of the Value of the game using Liapunov functions and some adequate differential inequalities.

4. ECONOMIC APPLICATIONS

Almost all the published results in differential games have been restricted to two-person, zero-sum games in which the possibility of mutual interest between the conflicting partners, is ruled out. In order to apply the theory of differential games to the analysis of economic competition, it is necessary to state an N-persons, non-zero sum formulation. Ciletti and Starr (1970) have modelled the situation of N firms, each manufacturing a single product. In the market these N products are interchangeable but not identical. Even making very strong simplifying assumptions, it is not easy to overcome all the analytical difficulties. And the question remains: How useful might a simplified model be in increasing our understanding of oligopolistic competition?

Most recently, A. Mehlmann (1988) has analyzed the subject of non-zero sum differential games and their applications, devising simple types which reflect structural properties of some qualitative solvable differential games. Among the selected applications, the dynamics of oligopoly, with finite horizon and infinite horizon, is considered.

5. LINEAR DIFFERENTIAL GAMES

Let us consider a linear differential game defined by

$$x' = f(x,u,v) = Ax + Bu + Cv \tag{7}$$

where $x \in R^n$; u and v are control variables such that

$$u(t) \in U \subset R^p; \quad v(t) \in V \subset R^q \tag{8}$$

The game is over when x reaches a certain analytic manifold $M \subset R^n$. We are looking for a strategy for u such that the game is finished in the least possible time, assuming we know the state x and the value of v at a certain time t.

Following Pontrjagin's method we may study the controllability of system (7) in a similar way as we should for the corresponding control system

$$x' = Ax + Bu.$$

For that purpose, we shall assume M to be linear, for example, the plane $x_1 = 0$ and the control subject to restrictions

$$||u|| \leq 1; \quad ||v|| \leq 1$$

where $||u|| = \max|u_i|$. Like in optimal control theory, we introduce a covariant vector $\psi = (\psi_1, \psi_2, \ldots, \psi_n)$ and form the hamiltonian

$$H(x,\psi,u,v) = \psi f(x,u,v) = \psi Ax + \psi Bu + \psi Cv \tag{9}$$

Let us assume that u wants to minimize time and v wants to maximize it. Then the optimal controls will be

$$\bar{u} : H \text{ maximum} : \psi Bu \text{ maximum} \qquad \bar{u} = \text{sgn}\ (\psi B)^T \tag{10}$$

where $\text{sgn}\ x = \begin{pmatrix} \text{sgn}\ x_1 \\ \vdots \\ \text{sgn}\ x_n \end{pmatrix}$ and T indicates transposed matrix.

Correspondently,

$$\bar{v} : H \text{ minimum} : \psi Cv \text{ minimum} \qquad \bar{v} = -\text{sgn}\ (\psi C)^T \tag{11}$$

The canonical equations are the following:

$$x' = \frac{\partial H}{\partial \psi} = Ax + Bu + Cv \tag{12}$$

$$\psi' = -\frac{\partial H}{\partial x} = -\psi f_x = -\psi A \tag{13}$$

Integrating the last equation, we get

$$\psi(t) = \psi(0)\; e^{-At} \tag{14}$$

and together with (14) and (15) we are able to calculate $\bar{u}$ and $\bar{v}$ and finally integrate (16) to get x(t), assuming x(0) known and that *the optimal solution exists.*

We may generate optimal solutions beginning at the terminal point, on the manifold M. The "transversal condition" between M and the path is reduced, in the case of minimum time, to $\psi(T)$ being orthogonal to M, where T indicates final time. In fact, this coincides with Isaacs' condition because in this case the value of the game $V = V(t,x(t))$ is reduced to $V(x)$ and

$$\psi = \nabla V = \frac{\partial V}{\partial x} \tag{15}$$

being $V(x) = 0$ for $x \in M$. Then the vector $\psi(T) = \partial V(x(T))/\partial x$ is orthogonal to M.

If the path family covers the whole plane, then every point is a possible initial point to begin the path to reach M.

EXAMPLE 1. Let us consider the control system

$$\begin{cases} x_1' = x_2 + 2u \\ x_2' = -x_1 + v \end{cases} \tag{16}$$

where $|u| \leq 1$; $|v| \leq 1$. Let M be the plane $x_1 = 0$ and assume that u wants to minimize the time to reach M and v wants to maximize it. The hamiltonian results

$$H = \psi_1 x_2 - \psi_2 x_1 + 2\psi_1 u + \psi_2 v \tag{17}$$

from where

$$\begin{cases} \psi_1' = -\dfrac{\partial H}{\partial x_1} = \psi_2 \\[2ex] \psi_2' = -\dfrac{\partial H}{\partial x_2} = -\psi_1 \end{cases} \tag{18}$$

By integrating we get

$$\begin{cases} \psi_1(t) = K\cos(t+\phi) \\ \psi_2(t) = -K\sin(t+\phi) \end{cases} \tag{19}$$

To fix the final condition we may assume that the final time is 0 being the initial time some negative value of t. Let us assume

$$x_1(0) = 0 \qquad x_2(t) \text{ arbitrary}$$

$$\psi_2(0) = 0 \qquad \psi_1(0) \text{ arbitrary, i.e. } \psi_1(0) = 1$$

Replacing these in (19) we obtain: $\phi = 0$ or $\phi = \pi$; $K = 1$. The optimal strategies are

$$\left.\begin{aligned} \bar{u} = \operatorname{sgn}\psi_1(t) &= \operatorname{sgn}\cos t \\ &= \operatorname{sgn}\cos(t+\pi) \end{aligned}\right\} \text{ for H max} \tag{20}$$

$$\left.\begin{aligned} \bar{v} = -\operatorname{sgn}\psi_2(t) &= \operatorname{sgn}\sin t \\ &= \operatorname{sgn}\sin(t+\pi) \end{aligned}\right\} \text{ for H min} \tag{21}$$

The path equations are as follows

$$\begin{aligned} x_1' &= x_2 + 2\operatorname{sgn}\cos t \\ &= x_2 + 2\operatorname{sgn}\cos(t+\pi) \\ x_2' &= -x_1 + \operatorname{sgn}\sin t \\ &= -x_1 + \operatorname{sgn}\sin(t+\pi) \end{aligned} \tag{22}$$

Let us consider now which of the two possibilities $\phi = 0$ or $\phi = \pi$ is better. Just before the ending of the game, that is, for t negative small, u tries to accelerate and v to delay the termination. And so the point x must be near the axis $x_1 = 0$, for example, x_1 would be at the right: $x_1 > 0$. From $x_1' = x_2 + 2u$ it results that u wants to make x_1 as negative as possible and chooses $u = -1$. In this case, in (20) it is better to take

$$\bar{u} = \operatorname{sgn}\cos(t+\pi)$$

This happens if $x_2 < 2$ because then u is able to push the point towards the x_2-axis from the right. We may assure that the half-axis

$x_2 < 2$ is reachable by optimal paths from the right. Symmetrically, $x_2 > -2$ is reachable by optimal paths from the left (choosing $\phi = 0$).

Geometrically, in this example, u and v act in opposite directions. However, the control system with $u(v = 0)$ or $v(u = 0)$ is controllable. If both players use maximum force, the control system oscillates around a shifted centre whose coordinates are

$$
\begin{array}{llll}
u = 1\,; & v = 1 & : x_{2c} = -2\,; & x_{1c} = 1 \\
u = -1; & v = -1 & : x_{2c} = 2\,; & x_{1c} = -1 \\
u = 1\,; & v = -1 & : x_{2c} = -2\,; & x_{1c} = -1 \\
u = -1; & v = 1 & : x_{2c} = 2\,; & x_{1c} = 1.
\end{array}
$$

In Fig. 1 we have drawn the four circles with center (1,-2), (-1,2), (-1,-2) and (1,2). Taking into account that the oscillations are isochronous and the angles are equivalent to time, we may analyze which are the best choices for u and v. For example, starting from the half-axis $x_1 = 0$ for $t = 0$, $x_2 < 2$ for $t < 0$ with $u = -1$, $v = 1$. In this way, the region A is covered. After rotating 90° we must change v to -1. The switching points lie over the line $x_2 = 3$. With center in (-1,-1), the region B is covered but also region C. From a point of the shaded zone there exist two ways of reaching the coordinate axis: 1) $u = -1$, $v = 1$; 2) $u = -1$, $v = -1$.

In this example, integrating backwards the path equations, we get not only the optimal strategy but also another path which satisfies the optimality conditions.

If the manifold M is of lower dimension, some problems may appear, as will be seen in the following example.

EXAMPLE 2. Let the control system be defined by

$$
\begin{cases} x_1' = 2u_1 + v_1 \\ x_2' = 2u_2 + v_2 \end{cases} \tag{23}
$$

where $u_1^2 + u_2^2 \leq 1$; $v_1^2 + v_2^2 \leq 1$. The terminal manifold is the origin $x_1 = x_2 = 0$.

The hamiltonian results

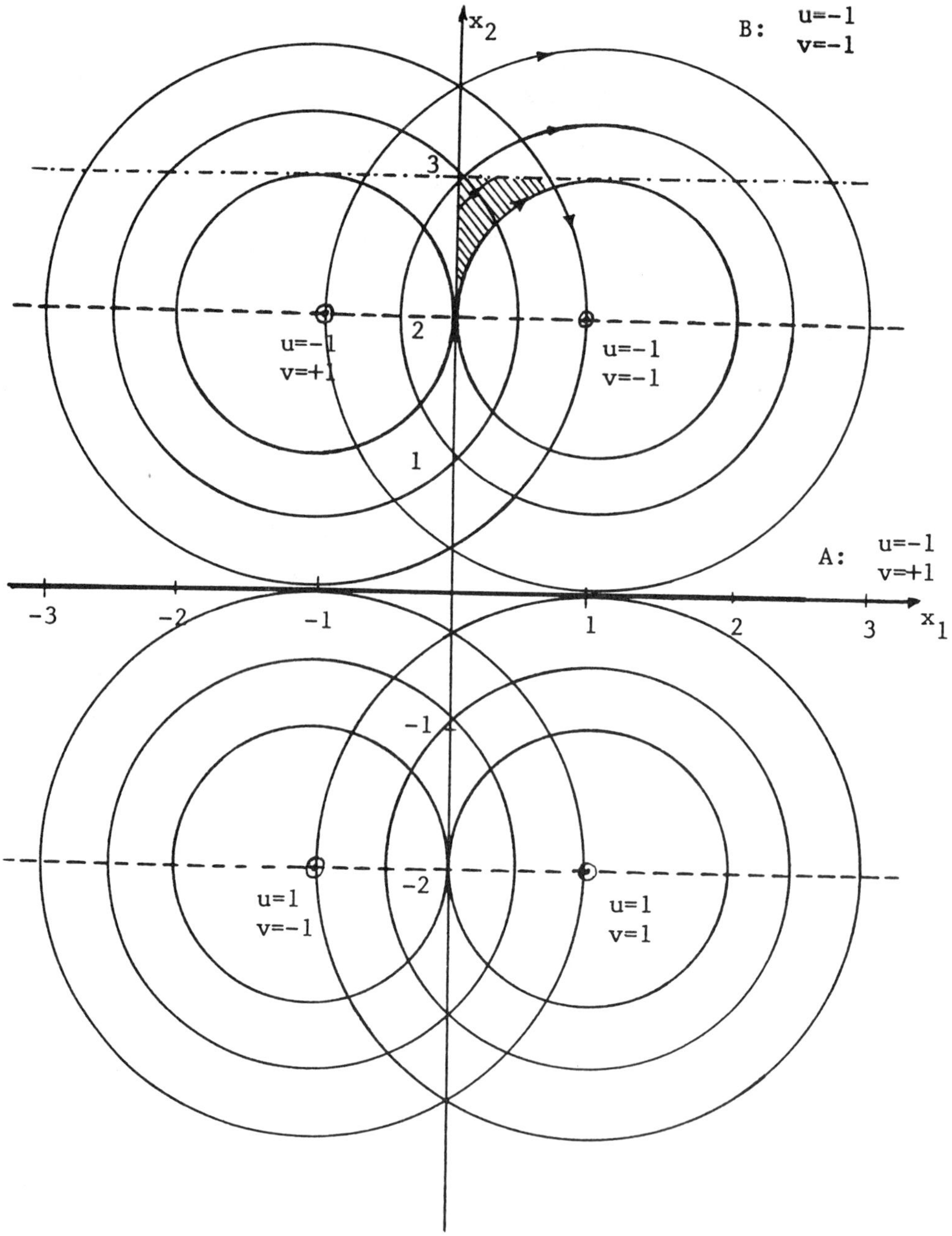

FIG. 1 *For* $x_1 > 0$: *Region A*, u = -1; v = +1; *Region B and C*, u = -1; v = -1. *For* $x_1 < 0$: *the solutions are symmetrical.*

$$H = 2\psi_1 u_1 + \psi_1 v_1 + 2\psi_2 u_2 + \psi_2 v_2$$

from where $\psi_1' = \psi_2' = 0$ which means $\psi_1 = \psi_2 = $ const. The optimal strategies are

$$\begin{cases} \bar{u} = \operatorname{sgn} \psi & \text{for H max} \\ \bar{v} = -\operatorname{sgn} \psi & \text{for H min} \end{cases}$$

The vectorial path equation may be written

$$x' = 2\bar{u} + \bar{v} = 2\bar{u} - \bar{u} = \bar{u} = \operatorname{sgn} \psi.$$

As a consequence, to reach the origin, ψ must lie in the direction of the origin. v has an opposite strategy but the only thing he gets is to produce a delay in the velocity. However, we shall prove that there exists a strategy for v which avoids that u reaches the origin. Being v's action weaker than u's, what is advisable for v is to play in a random way. Then u can only respond "with delay" to v's action. Mathematically, this is the "Brownian movement", that is, a stochastic process like a particle diffusion. It is obvious that u may get $x_1 = 0$ and also $x_2 = 0$ but the probability that both conditions are simultaneously satisfied, is zero. In general, when the stochastic path cuts the x_1-axis, will not intersect the x_2-axis and vice versa. As u does not know instantaneously v's action, he cannot maintain for example $x_1 = 0$ during a finite timer interval. The conclusion is that the differential game never ends and Pontrjagin's solution seems false! However, if we write the dynamics in polar form

$$\rho' = 2u_\rho + v_\rho \qquad \theta' = \rho^{-1}(2u_t + v_t)$$

where u_ρ and u_t are the radial and tangential components of vector u, and similar for v. Assuming that we may always choose

$$u_\rho = -1; \quad u_t = 0$$

that is, u pointing toward the origin, it results

$$\rho' = -2 + v_\rho \qquad \theta' = \rho^{-1} v_t$$

Being $|v_\rho| \leq 1$ it follows that $\rho' \leq 1$ and ρ decreases to zero, reaching the origin in finite time. Even when this result is mathematically correct, it is evident that physically it makes no sense, since it is impossible to maintain u always in radial direction because near the origin it would imply big changes of direction in very short times. The explanation is that the problem of reaching the origin is relatively artificial and does not occur in practical differential games. For differential games, the typical problem is to reach a manifold of dimension n - 1. For this, one has to replace the origin by a circulat neighborhood of radius ε, reducing it to the previous case. The important fact would be the "ε-controllability", which means reaching the "ε-neighborhood" of the origin in finite time.

EXAMPLE 3. Let us consider the control problem

$$\begin{cases} x_1' = x_1 + 2u \\ x_2' = v \end{cases}$$

where $|u| \leq 1$; $|v| \leq 1$ and M is the plane $x_1 = 0$. Then

$$H = \psi_1 x_1 + 2\psi_1 u + \psi_2 v$$

being

$$\begin{cases} \psi_1' = -\psi_1 & \psi_1(t) = K\cos(t + \pi) \\ \psi_2' = 0 & \psi_2(t) = \text{const.} \end{cases}$$

Taking as final conditions

$$x_1(0) = 0 \qquad x_2(0) \text{ arbitrary}$$

$$\psi_2(0) = 0 \qquad \psi_1(0) = 1$$

it results: $k = 1$; $\phi = 0$ or $\phi = \pi$.

The optimal strategies are the following:

$$\bar{u} = \text{sgn}\,\psi_1(t) = \text{sgn}\cos t = \text{sgn}\cos(t + \pi), \qquad (H \max)$$

$$\bar{v} = -\text{sgn}\,\psi_2(t) = -\text{sgn}\,\psi_2, \qquad (H \min)$$

These are two uncoupled equations and it is really a control problem for u (to reach $x_1 = 0$ in minimum time) and v cannot interfere. In this case, it is proved that using Pontrjagin's equations for $x_1 < 2$, the optimal strategy is not useful to reach the origin. If we assume $x_1 > 0$, then u wishes to have x_1 as negative as possible and chooses u = -1 and then $x_1 < 2$ and u is able to force the point towards the vertical axis from the right.

EXAMPLE 4. Let the control system be

$$\begin{cases} x_1' = x_2 + 2u \\ x_2' = -x_1 + v \end{cases}$$

where $|u| \leq 1$; $|v| \leq 1$ being M the plane $x_2 = 0$. The hamiltonian is

$$H = -\psi_1 x_2 - \psi_2 x_1 + 2\psi_1 u + \psi_2 v$$

being

$$\psi_1' = \psi_2 \qquad \psi_2' = -\psi_1$$

from where

$$\psi_1(t) = K\cos(t + \phi) \qquad \psi_2(t) = -K\sin(t + \phi)$$

with final conditions

$$x_2(0) = 0\ ; \quad x_1(0) = \text{arbitrary}$$

$$\psi_1(0) = 0\ ; \quad \psi_2(0) = \text{arbitrary, for example, } \psi_2(0) = 1.$$

It results: $K = 1$; $\phi = \pi$ or $\phi = 3\pi/2$.

The optimal strategies are the following

$$\left.\begin{aligned} \bar{u} = \operatorname{sgn} \psi_1(t) &= \operatorname{sgn} \cos(t + \tfrac{\pi}{2}) \\ &= \operatorname{sgn} \cos(t + \tfrac{3\pi}{2}) \end{aligned}\right\} \text{ for H max}$$

$$\left.\begin{aligned} \bar{v} = -\operatorname{sgn} \psi_2(t) &= \operatorname{sgn} \sin(t + \tfrac{\pi}{2}) \\ &= \operatorname{sgn} \sin(t + \tfrac{3\pi}{2}) \end{aligned}\right\} \text{ for H min}$$

and the path equations are

$$x_1' = x_2 + 2\operatorname{sgn}\cos(t + \tfrac{\pi}{2}) = x_2 + 2\operatorname{sgn}\cos(t + \tfrac{3\pi}{2})$$

$$x_2' = -x_1 + \operatorname{sgn}\sin(t + \tfrac{\pi}{2}) = -x_1 + \operatorname{sgn}\sin(t + \tfrac{3\pi}{2})$$

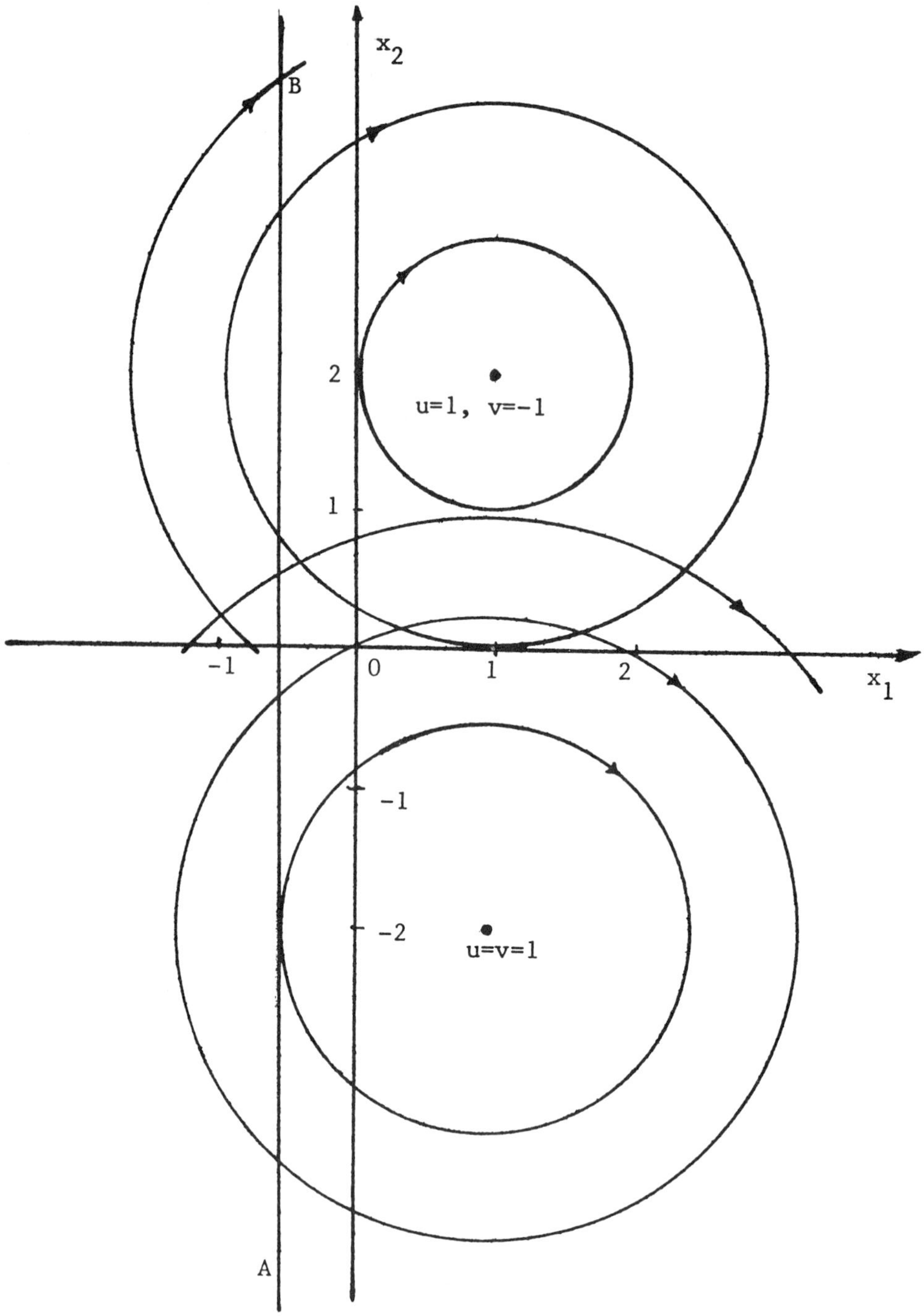

FIG. 2 *The circumference arc in each region has almost length π/2. This condition determines the position of the line AB.*

Let us analyze which of the two possibilities $\phi = \pi/2$ or $3\pi/2$ is better. Before ending the game, if $x_2 > 0$ or $x_2' < 0$, v intends to postpone the end and chooses $v = 1$. Then the game can be over from above ($x_2 > 0$) if and only if $-x_1 + v = -x_1 + 1 < 0 \Rightarrow x_1 > 1$. Similarly, the game can be over from below only if $x < -1$. The optimal paths are represented in Fig. 2. In the quadrant $x_1 > -1$, $x_2 > 0$ is $u = v = 1$ and the paths are circles of center (1,-2). These paths reach the x_1-axis for $x_1 > 1$. In the quadrant $x_1 < -1$, $x_2 > 0$ is $u = 1$, $v = -1$ and the paths are circles of center (1,2). Such paths begin over the x_1-axis and join the paths of the previous case. In the half-plane $x_2 < 0$, the behavior is similar.

CONCLUSIONS

The analysis of these two-dimensional linear differential games applying optimal control methods, shows that the mutual enrichment of both fields can be fruitful, but one must be very careful in translating methods and results from one to the other.

REFERENCES

1. Davis, H.T. (1962). *Introduction to Nonlinear Differential and Integral Equations*, Dover Publications, Inc., New York, pp. 113-127.

2. Bouguer, P. (1732). Sur de nouvelles courbes ausquelles on peut donner de nom de Lignes de Poursuite, Histoire de l'Academie Royale des Sciences, Paris, pp. 1-14.

3. Isaacs, R. (1953-1956). Differential Games, I, II, III, IV, Rand Corp. Research Memorandum RM-1391, 1399, 1411, 1468.

4. Isaacs, R. (1965). *Differential Games*, John Wiley and Sons.

5. Bellman, R.I., Glicksberg and Gross, O. (1956). On the Bang-Bang Control Problem, *Quart. Appl. Math.*, *14*, pp. 11-18.

6. Rixin E. and Spinadel, V. (1959). Reachable Zones in Autonomous Differential Systems, *Contr. to Diff. Eqs.*, *1*, No. 3, pp. 275-315.

7. Pontrjagin, et al. (1962). *The Mathematical Theory of Optimal Processes*, Interscience, New York.

8. Kelendzeridze, D.L., Theory of an Optimal Pursuit Strategy, *Dokladi Akad. Nauk USSR* (translated in Sov. Math., 1961).

9. Krasovskii, N.N. and Subbotin, A.I. (1988). *Game-Theoretic Control Problems*, Springer-Verlag.

10. Berkovitz, L.D. (1961). A Variational Approach to Differential Games, Rand Report RM-2772.

11. Fleming, W.H. (1957). A Note on Differential Games of Prescribed Duration, Contr. to the Theory of Games III, Annals of Mathem. Study 39, Princeton Univ. Press, Princeton, N.J.

12. LaSalle, J.P. (1960). The Time Optimal Control Problem, Contr. to the Theory of Nonlinear Oscillations, V, pp. 1-24, Princeton.

13. Bushaw, D.W. (1958). Optimal Discontinuous Forcing Terms, Contr. to the Theory of Nonlinear Oscillations, IV, Princeton.

14. Gamkrelidze, R.V. (1958). Theory of Time-Optimal Processes for Linear Systems, *Izv. Akad. Nauk. USSR Ser. Math., 22*, pp. 449-474.

15. Krasovskii, N.N. (1957). Concerning the Theory of Optimal Control, *Avtomat. i Telemeh., 18*, pp. 960-970.

16. Roxin, E. (1969). Axiomatic Approach in Differential Games, *Journ. of Optimiz. Theory and Appl., 3*, 3.

17. Clarke, F.H. and Loewen, P.D. (1986). The Value Function in Optimal Control: Sensitivity, Controllability and Time-Optimality, *SIAM J. Control. and Optimiz., 24*, No. 2, pp. 243-263.

18. Varaiya, P.P. (1967). The Existence of Solutions to a Differential Game, *J. Soc. Ind. Appl. Math., Ser. A, Control 5*, pp. 153-162.

19. Friedman, A. (1970). On the Definition of Differential Games and the Existence of Value and Saddle Points, *J. Diff. Eq., 7*, pp. 69-91.

20. Friedman, A. (1970). Existence of Value and Saddle Points for Differential Games of Pursuit and Evasion, *J. Diff. Eq., 7*, pp. 92-110.

21. Friedman, A. (1970). Existence of Value and Saddle Points for Differential Games of Survival, *J. Diff. Eq., 7*, pp. 111-125.

22. Roxin, E. (1970). On Differential Games without Value, Symposium on the American Automatic Control Council.

23. Ciletti, M.D. and Starr, A. W. (1970). Differential Games, a Critical View. Symposium on the American Automatic Control Council.

24. Mehlmann, A. (1988). *Applied Differential Games*, Plenum Publish. Corp.

25. Pontrjagin, L.S. (1964). On Some Differential Games (translated in *Soviet Math., 5*, No. 3.

Thirty Years of Optimal Control

Was the Path Unique?

Hector J. Sussmann, Professor, Department of Mathematics, Rutgers University, New Brunswick, New Jersey

1 INTRODUCTION

In this note I intend to take seriously the title of this conference —"Thirty years of modern optimal control"— and regard it as an opportunity to speculate on one aspect of the past and present of our discipline. Specifically, I want to look at the evolution of our field, regarded as a path in the space of conceptual structures, and at the point in this space where we find ourselves now, and I want to ask to what extent the path that has been followed was the one that we were *bound* to follow, and to what extent it is a special path, perhaps a product of peculiar initial conditions and historical circumstances. In asking this question —and arguing for a particular answer— I will proceed in a spirit similar to D. Ruelle's recent Gibbs lecture (cf. [17]). Ruelle asked: "*How natural is our human 20th-century mathematics?* ... How much does the existing construct depend on human nature and condition?" This is more or less equivalent to asking whether, if we were to establish contact with "little green men from outer space, ... [they] would have a similar central core for their mathematics; ... in their own language they would have a theorem saying that the image of a compact set by a continuous map is compact."

A "yes" answer to this question essentially means that mathematical concepts force themselves upon us by virtue of their internal logical structure and the logical relations among them, that the evolution of mathematical knowledge follows a "natural" path which is essentially determined by this structure. A "no" answer means that, just because mathematics has developed in a particular way, it does not follow that this is

Partially supported by NSF grant DMS83-01678-01, and by the CAIP Center, Rutgers University, with funds provided by the New Jersey Commission on Science and Technology and by CAIP's industrial members.

how it inexorably had to develop. It also means that the development *may have taken a wrong turn*, in that important concepts could have been missed and unimportant ones may have become the focus of everybody's attention.

For the specific case of Optimal Control Theory, I want to argue that there is a set of mathematical concepts, such as *coordinate-free properties, Lie brackets, real analyticity, integral manifolds, Lie algebras of vector fields*, that most civilizations of "little green men from outer space" would probably have come across during the very earliest stages of their development of the field, and immediately incorporated into the central core of their Control Theory, which would, for that reason, look quite different from ours. In most cases, the little green men would have been able to borrow these concepts from other areas of their mathematics, where they would have already been discovered (invented?) for different purposes. But, even in those planets where other mathematicians failed to hit upon the Lie bracket, the control theorists would have discovered it, exactly as on Earth we control theorists would have discovered the variations-of-constants formula if it hadn't been known before.

This is not to deny that the importance of these ideas has also been recognized on Earth, as shown by the tremendous amount of activity in Optimal Control in recent years that makes systematic use of them. However, in spite of this activity, these concepts are still often regarded by other control theorists as belonging to a somewhat esoteric fringe of our field called "differential geometric methods in control," or "differential geometric control theory." (I believe this name to be inappropriate, because *Differential Geometry* usually starts when one studies differentiable manifolds *with some extra structure*, such as a Riemann or Lorenz metric, or at least a connection, and this is seldom what one encounters in Optimal Control.) The implication of such terminology is the belief that this esoteric subfield somehow asks its own esoteric questions and then introduces esoteric concepts in order to solve them. Lie brackets presumably occur in Control Theory because one asks questions about Lie brackets, subanalytic sets occur because one wants to know whether certain reachable sets are subanalytic, etc. I will argue that this belief is false. My main thesis is that *the concepts that I listed above are forced upon almost anyone who asks the most basic, most natural questions in Optimal Control.* In particular, I will mostly focus on the concept of the *Lie bracket*, and argue that it is as entitled to belonging to the mainstream of our field as, say, concepts that we all accept as basic, such as the exponential of a matrix, Lagrange multipliers, convexity, or weak* convergence. (Those readers who think this is so obvious and well known that it does not need proof should look at the large number of mathematical books on Optimal Control where *the Lie bracket is not mentioned.* Cf., e.g., [1], [11], [12], [13], [16], [37].)

2 AN ALTERNATIVE SCENARIO: EXTRATERRESTRIAL OPTIMAL CONTROL

Our little green men's thoughts might have evolved as follows. The mathematical object that describes the evolution of a physical system is an *ordinary differential equation*, i.e. a *vector field* or *dynamical system*. When such a system evolves, its future path is uniquely determined by its present state. Suppose now that we want to study systems whose evolution can be affected by a controlling agent, i.e. we want to incorporate the element of *choice*. We should then look at systems of two or more vector fields. If, as is natural, we want to start by studying the simplest possible case, then it clear that we have to begin our development of control theory by studying the systems arising from *pairs of vector fields*. (We shall refer to these as *two-vector-field systems*.)

What questions would the little green men ask about such pairs? If X and Y are our two vector fields, then a *control* is clearly specified by giving a time interval $[a, b]$, and specifying the subset S_X of $[a, b]$ during which we follow X (or, equivalently, the subset $S_Y = [a, b] - S_X$ during which we follow Y). An obvious question is to what extent the point q that we get at the end —starting from a point p— depends on the set S_X. For instance, it might be that all we need to know in order to know q is *how much time was spent controlling in each of the two directions*, i.e. *the Lebesgue measure of the sets* S_X, S_Y. Or perhaps we need more information, i.e. it is not enough to know how much X-control and how much Y-control we have applied, and we also need to know the *order* in which these two controls were applied. So let us consider the most extreme situation, and compare two controls for which the sets S_X and S_Y have exactly the same size, but such that, in one case, all the X-controlling is done first, whereas in the other case the Y-controlling is done first. For these two controls, let us ask whether the point q will be the same.

To formulate this mathematically, let us introduce some notations. For a smooth (i.e. C^∞) vector field X on $\mathbb{R}^n$ (or on an open subset Ω of $\mathbb{R}^n$), let us use exponential notation for the flow, so that $t \to e^{tX}p$ denotes the integral curve of X that goes through p at time 0. If X and Y are two smooth vector fields, then the question we have asked above is, simply, whether the identity

$$e^{tX}e^{\tau Y}p = e^{\tau Y}e^{tX}p \tag{2.1}$$

holds for all choices of t, τ and p. If this happens, we say that the flows of X and Y *commute*. To be precise, we have to take into account the possibility that the integral curves of X or Y may fail to be defined for all times, i.e. that the flows of X or Y may exhibit explosions. The precise definition of commuting flows is as follows. We say that X and Y *have commuting flows* if for every $p \in \Omega$ there is an $\varepsilon > 0$ such that 2.1 holds whenever $t \in \mathbb{R}$, $\tau \in \mathbb{R}$, $|t| < \varepsilon$, $\tau < \varepsilon$.

One can then show that,

A. *If X and Y have commuting flows, then*

  a. *ε can be chosen to be independent of p as long as p stays in a compact set, and*

  b. *2.1 actually holds for all t and τ, with no restrictions, if both X and Y are complete.*

(Recall that a vector field X is *complete* if the integral trajectories of X are defined for all times.)

Our question now is how we can tell whether X and Y have commuting flows. But, before we answer this, it is worth showing that commutation has many interesting consequences. To begin with, we can broaden our analysis and study the two-vector-field system

$$\dot{x} \in \mathrm{co}(X(x), Y(x)) \, , \tag{2.2}$$

i.e.

$$\dot{x} = vX(x) + (1-v)Y(x) \, , \; 0 \leq v \leq 1 \tag{2.3}$$

or, equivalently (if we write $f = \frac{1}{2}(X+Y)$, $g = \frac{1}{2}(Y-X)$ and $u = 1-2v$),

$$\dot{x} = f(x) + ug(x) \, , \; -1 \leq u \leq 1 \, , \tag{2.4}$$

where the controls are required to be measurable, but we allow the values of u to be not just 1 and -1 but an arbitrary number in $[-1,1]$. (Our little green men would presumably have noticed, as we have on Earth, that it is always nice to have a closed reachable set, so that chattering controls should be permitted, and in particular the control set $\{-1,1\}$ might as well be replaced by the interval $[-1,1]$.)

Assume now —for the moment— that X and Y are complete. Then the assertion that the flows of X and Y commute is easily seen to be exactly equivalent to the statement that, if $\gamma : [a,b] \to \Omega$ is a bang-bang trajectory corresponding to a control $\bar{u}(\cdot) : [a,b] \to [-1,1]$, and α, β are, respectively, the total amount of time during which the control u is equal to 1 and -1 (i.e $\alpha = |\{t : \bar{u}(t) = 1\}|$ and $\beta = b - a - \alpha$, where $|\ldots|$ denotes Lebesgue measure) then

$$\gamma(b) = e^{\alpha Y} e^{\beta X} \gamma(a) \, . \tag{2.5}$$

Since $\alpha = \frac{1}{2}(b - a + \int_a^b \bar{u}(t)\,dt)$, we can rewrite 2.5 as

$$\gamma(b) = e^{\frac{1}{2}(b-a+\int_a^b \bar{u}(t)\,dt)Y} e^{\frac{1}{2}(b-a-\int_a^b \bar{u}(t)\,dt)X} \gamma(a) \, . \tag{2.6}$$

It then follows easily from the continuity of the integral as a functional of the control with respect to the weak topology, together with the fact that, in this topology, the bang-bang controls are dense in the class of all controls, that Formula 2.6 actually holds for an arbitrary control. What this says is that, when the flows commute, then all that matters is how much control you apply in each direction, and the order in which these controls are applied makes no difference. In particular, *the point $\gamma(b)$ to which a control $\bar{u}(\cdot)$ steers the initial point $p = \gamma(a)$ depends on $\bar{u}(\cdot)$ through two parameters only, namely, the total time and the integral of the control.* Formula 2.5 implies that $\mathcal{R}(p) \subseteq S$, where $S = \{e^{\tau Y} e^{tX} p : t \in \mathbb{R}, \tau \in \mathbb{R}, t \geq 0, \tau \geq 0\}$, so that S is the image of a subset of the plane under a smooth map. Therefore

B. *If X and Y have commuting flows, then the reachable set $\mathcal{R}(p)$ from a point p is at most two-dimensional.*

The precise meaning of "at most two dimensional" is that $\mathcal{R}(p) \subseteq S$, where S is the image of a two-dimensional manifold under a smooth map. With this interpretation we have actually established **B** when X and Y are complete. However, with a little bit more work one can generalize this to the noncomplete case, and **B** is still valid, and in fact the result is even stronger, because we can interpret "$\mathcal{R}(p)$ is at most two dimensional" to mean "$\mathcal{R}(p)$ is a subset of a two-dimensional connected submanifold of Ω."

Also, if we assume that X and Y are linearly independent at a point p, then the map $(\tau, t) \to e^{\tau Y} e^{tX} p$ is one-to-one for small enough τ, t, by the Implicit Function Theorem. This implies, in particular, that α and β can be recovered from a knowledge of $q = e^{\alpha Y} e^{\beta X} p = \gamma(b)$. Since $b - a = \alpha + \beta$, it follows that the time $b - a$ can be recovered from the knowledge of q, i.e. *all trajectories going from p to q take exactly the same time*, at least as long as this time is small. In particular, this means that

(#) *there is an $\varepsilon > 0$ such that all trajectories that start at p and take time $\leq \varepsilon$ are time-optimal.*

If $X(p)$ and $Y(p)$ are linearly independent, then Condition (#) is equivalent to

(##) *there exists a neighborhood N of p such that every trajectory through p which is entirely contained in N is time-optimal.*

The implication (##) $\Rightarrow$ (#) is trivial. To show the converse, pick a smooth function φ such that the derivatives of φ in the X- and Y- direction are > 0 at p. (This is possible because of the linear independence.) Then we can pick a neighborhood N' such that the X- and Y-derivatives of φ are $> c$ throughout N', where c is some positive constant. Let $N = N' \cap \{q : \varphi(q) < \varphi(p) + c\varepsilon\}$. Then every trajectory through p that is contained in N must take time $\leq \varepsilon$.

So we have established:

C. *if X and Y have commuting flows and $X(p)$, $Y(p)$ are linearly independent, then the minimum time problem associated with 2.4 is degenerate at p,*

where we define a minimum time control problem to be *degenerate* at a point p if (##) holds, and to be *degenerate* if it is degenerate at every p. In order words, except for the minor detail that degeneracy, as defined, only refers to "small time trajectories," a degenerate system is one for which all trajectories are time-optimal. A minimum-time problem for such a system is such that all trajectories that satisfy the constraints of interest are solutions. Clearly, the question whether a particular system is degenerate is the most basic one we can pose about the minimum time problem, in the same way as, for the problem of minimizing a function, the first basic question one should ask is whether the function is a constant. Any theorem that enables us to characterize degenerate problems will be about as basic as the Calculus theorem that says that a function on an open connected set is a constant if and only if it is everywhere differentiable and the gradient vanishes. Such a characterization will be described later and, not surprisingly, Lie brackets will be involved.

The hypothesis that $X(p)$ and $Y(p)$ are linearly independent is crucial, as can be seen by considering the system $\dot{x} = u$, $x \in \mathbb{R}$, $|u| \leq 1$. The restriction to small-time trajectories is also unavoidable, even if X and Y are complete. (This can easily be seen from the following example. Let $\Omega = \mathbb{R}^2 - \{(0,0)\}$, and let X, Y have components $(y, -x)$, (x, y), respectively. Then the trajectories of X are periodic with period 2π. Also, it is clear that X and Y have commuting flows. But a trajectory of X that goes from a point p back to p in time 2π is not time-optimal.)

Finally, one can strengthen **C** to

D. *If X and Y have commuting flows, $K \subseteq \Omega$ is compact, and $X(p)$ and $Y(p)$ are linearly independent for every $p \in K$, then there is an $\varepsilon > 0$ such that all trajectories that start at a point $p \in K$ and take time $\leq \varepsilon$ are time-optimal.*

3 THE LIE BRACKET

Now that our little green men have established that the commutation of the flows is of paramount control-theoretic significance, they should naturally turn to the question of how to give an *infinitesimal characterization* of this property. If we write X and Y as *column-vector-valued functions*, we can define the *Lie bracket* of X and Y to be the vector field $[X,Y]$ given by

$$[X,Y] = DY \cdot X - DX \cdot Y \ , \tag{3.1}$$

where DX is the square-matrix-valued dunction whose rows are the gradients of the components of X.

It then turns out that

E. *X and Y have commuting flows if and only if $[X,Y] \equiv 0$.*

More generally, using a Taylor expansion up to order two, one can prove the following *geometric characterization of the Lie bracket*:

$$[X,Y](p) = \lim_{t \to 0+\,,\,\tau \to 0+} \frac{1}{t\tau}\left[e^{\tau Y} e^{tX} e^{-\tau Y} e^{-tX} p - p\right] \ . \tag{3.2}$$

From 3.2 one can easily derive the "if" part of **E**. To derive the "only if" part, one shows the following result, which is important in its own right:

F. *if X and Y satisfy $[X,Y] \equiv 0$ on some neighborhood of p, and the vectors $X(p)$, $Y(p)$ are linearly independent, then there exists a smooth change of coordinates near p such that, in the new coordinates, X and Y are constant vector fields.*

Since constant vector fields obviously have commuting flows, the "if" part of **E** follows from **F**, at least near points p where X and Y are independent. The general case can be reduced to this one by the following procedure. Work in $\Omega \times \mathbb{R} \times \mathbb{R}$, i.e. add two new variables t and τ. Let $\tilde{X}(x,t,\tau)$ have components $(X(x),1,0)$. Similarly, let $\tilde{Y}$ have components $(Y(x),0,1)$. Then one can easily verify that $[\tilde{X},\tilde{Y}] \equiv 0$ if $[X,Y] \equiv 0$. Moreover, if $\pi : \Omega \times \mathbb{R} \times \mathbb{R} \to \Omega$ denotes the projection, then $\pi(e^{t\tilde{X}}(p,s,\sigma)) = e^{tX}(p)$, and similarly for Y. Therefore, if $\tilde{X}$ and $\tilde{Y}$ have commuting flows, the same is true for X and Y. This establishes **E** in general.

The Lie bracket of two vector fields therefore has the advantage of being an object that can be *explicitly computed from the components of the vector fields by means of algebraic operations* (without having to solve differential equations), and yields information about a property of the solutions. So it is a natural object to associate to a pair of vector fields.

4 LOCAL CLASSIFICATION AND COORDINATE FREE OBJECTS

While some of our little green men are busy asking some specific questions of interest about two-vector-field systems, other may take a broader point of view and ask for a *classification theory* of such pairs. Following the model of single vector fields, one may

(a) look first for a *local classification*. i.e. a classification of pairs of vector fields on a neighborhood of a given point p,

(b) try to start by looking at the least degenerate cases, and progress towards the more degenerate ones,

(c) wonder what *group* of transformations should be used in the classification theory,

(d) look for *canonical forms*.

The point of *(c)* is that, as is the case elsewhere in Mathematics, to each problem there is attached a group, namely, the group of transformations that preserve the properties of interest for the problem. It is then often convenient, for a particular problem, to look for a transformation in the group that will put the problem in a simpler form, from which it might be easy to see the solution. In our case, if we ask questions such as the ones discussed above (whether the minimum time control problem is degenerate, or whether the point q just depends on α and β as above), or many other related questions (such as whether the reachable sets have nonvoid interior, or whether all time-optimal trajectories are bang-bang, or whether the reachable set from a point p contains a neighborhood of p, or whether there is a smooth feedback control defined near p that makes p an asymptotically stable equilibrium, or whether the boundary of the reachable sets is a finite union of submanifolds), *the appropriate group is that of smooth tranformations of the state space*, variously referred to as *diffeomorphisms* or *coordinate changes*, depending on the point of view. So the classification theory one wants to develop is the local classification of pairs of vector fields under *smooth coordinate changes.*

For the case of a *single* vector field X, the "least degenerate" case is that of a point p that is not an equilibrium, i.e. a p such that $X(p) \neq 0$. The classification problem for this case is solved by the most basic result in the qualitative theory of ordinary differential equations, namely, the existence of a "flow box":

G. *if X is a smooth vector field near p, and $X(p) \neq 0$, then there exists a smooth change of coordinates on a neighborhood of p such that in the new coordinates X becomes a constant vector field.*

(Notice that, by making a further *linear* change of coordinates we could equally well have stated the conclusion of **G** to be that, in the new coordinates, $X = (1, 0, \ldots, 0)$.)

In the qualitative theory of differential equations, one then proceeds to the "degenerate case" when $X(p) = 0$, and within that one studies first the "nondegenerate case" of a *hyperbolic equilibrium* (obtaining the theory of *stable and unstable manifolds*), and then the more degenerate case when the linearization has some pure imaginary eigenvalues (so that one has to study *center manifolds*).

For our two-vector-field systems, Statement **F** is none other than the extension to the case of two vector fields of the flow-box theorem **G**. However, in this case the simultaneous "flow-boxing" of two vector fields is not possible unless their Lie bracket vanishes. Hence *the Lie bracket is the obstruction that prevents simultaneous flow-boxing.* When the Lie bracket does not vanish, *there is no classification theorem analogous to* **G**, *even in the "nonequilibrium" case.* (Naturally, the two-vector-field analogue of "nonequilibrium" is the case when the vector fields are linearly independent at a point p.) So, while the problem of the local classification of ordinary differential equation is completely solved by the flow-box theorem, and therefore does not require any further analysis, *the analogous problem for pairs of vector fields is very difficult and far from solved.*

5 RESULTS INVOLVING LIE BRACKETS

So far, our little green men have stumbled upon the Lie bracket, but most of our discussion has not made use of all the information it contains. We have only focused on the implications of the *vanishing* of the Lie bracket. However, once we have discovered this important object, it is possible to derive immediately a plethora of results that solve basic control-theoretic problems in terms of properties of Lie brackets. We list just a few of them, without proof.

First, let us ask the question when the minimum time problem for a two-vector-field system is degenerate. As explained above, the condition that the Lie bracket vanishes is *sufficient* for this. It's not hard to see that this is too strong a condition. The appropriate necessary and sufficient condition turns out to involve Lie brackets as well. Moreover, this condition turns out to be exactly the same for general control systems, i.e. systems involving more than two vector fields. To state the condition, let us first use $\mathcal{V}(\Omega)$ to denote the class of all C^∞ vector fields on Ω. If we define addition and multiplication by scalars in the obvious way, and then let $[\ldots,\ldots]$ be the Lie bracket as defined above, it turns out that $\mathcal{V}(\Omega)$ is a *Lie algebra.* (In general, a Lie algebra is a linear space V endowed with a bilinear map $V \times V \ni (x, y) \to [x, y] \in V$ such that $[x, y] = -[y, x]$ and $[x, [y, z]] + [y, [z, x]] + [z, [x, y]] = 0$ for all $x, y, z \in V$.) If $\mathcal{F} \subseteq \mathcal{V}(\Omega)$ is a set of smooth vector fields, let us use $\Lambda(\mathcal{F})$ to denote the smallest Lie subalgebra of $\mathcal{V}(\Omega)$ that contains $\mathcal{F}$. Also, let us use $\Lambda_0(\mathcal{F})$ to denote the smallest ideal of $\Lambda(\mathcal{F})$ that contains all the differences $X - Y$, $X \in \mathcal{F}$, $Y \in \mathcal{F}$. (A *Lie subalgebra* of a Lie algebra L is a linear subspace of L that is closed under the map $[\ldots,\ldots]$, and an *ideal* is a linear subspace I of L such that $x \in L$, $y \in I$ imply $[x, y] \in I$.) Also, we let $\mathcal{F}(p) = \{X(p) : X \in \mathcal{F}\}$. Call $\mathcal{F}$ *infinitesimally degenerate* at a point p if $\Lambda_0(p) \neq \Lambda(p)$. (One can easily see that

$\Lambda_0(p)$ is a subspace of $\Lambda(p)$ of codimension 0 or 1, so infinitesimal degeneracy is the condition that the codimension is 1.)

To a control system $\Sigma_f : \dot{x} = f(x,u)$, $u \in U$, in which f is assumed to be of class C^∞ in x for each fixed $u \in U$, we can associate the family $\mathcal{F}(f) = \{f(\cdot,u) : u \in U\}$, i.e. the set of all vector fields that correspond to all the possible constant values of u.

Let us call a control problem Σ_f *analytic* if all the vector fields $f(\cdot,u)$ are real analytic. We then have the following (cf. [24]):

Theorem 5.1. *An analytic control problem Σ_f is degenerate at a point p if and only if the corresponding family $\mathcal{F}(f)$ is infinitesimally degenerate at p.*

So, at least for the analytic case, we have a complete characterization of degeneracy in terms of Lie brackets. To understand what this means, and how this generalizes **C**, let us return to the two-vector-field case. In this case, we can write all the *iterated* brackets, i.e. f, g, $[f,g]$, $[f,[f,g]]$, $[g,[f,g]]$, $[f,[f,[f,g]]]$, $[g,[f,[f,g]]]$, $[[f,g],[f,[f,g]]]$, ... etc. Degeneracy will occur, in this case, if and only if, for each point p, the vector $f(p)$ does not belong to the linear hull of the values at p of all the other brackets. In particular, this includes **C** because $[X,Y] = 2[f,g]$, so that, if the hypothesis of **C** holds, then the only brackets that might be $\neq 0$ are f and g. Therefore, if $f(p)$ and $g(p)$ are independent, then $f(p)$ cannot belong to linear hull of the "other" brackets, and Theorem 5.1 applies.

Now let us turn to the problem of understanding what the reachable sets look like. Let $\mathcal{R}_f(p)$ denote the reachable set from p for the system Σ_f, and let $\mathcal{R}_f^T$, $\mathcal{R}_f^{\leq T}$ denote, respectively, the time T and the time $\leq T$ reachable sets. We could ask whether these sets are smooth submanifolds, but trivial examples show that this need not be so. The next best thing is if the sets are "close" to being submanifolds in some sense. For our purposes, let us call a subset S of a smooth manifold M *thick in* M if $S \subseteq \operatorname{Clos}\operatorname{Int} S$, where both the closure and the interior are taken relative to M. (This implies, among other things, that the set has *integer dimension* in any reasonable dimension theory, and cannot consist of a piece of some dimension together with some smaller dimensional "tail" sticking out. A convex set is thick in its affine hull.) Let us call a smooth submanifold M of Ω an *integral manifold* of a set $\mathcal{F}$ of smooth vector fields if M is connected and the tangent space T_xM of M at x equals $\mathcal{F}(x)$ for every $x \in M$. A *maximal* integral manifold is one that is maximal with respect to inclusion. We then have (cf. [6], [15]):

Theorem 5.2. *If $\mathcal{F}$ is a Lie algebra of real-analytic vector fields on Ω, then for every $p \in \Omega$ there exists a unique maximal integral manifold M of $\mathcal{F}$ such that $p \in M$.*

If we use $I(\mathcal{F},p)$ to denote the maximal integral manifold of $\mathcal{F}$ through p, then the following holds (cf. [24]):

Theorem 5.3. *For an analytic system Σ_f, the sets $\mathcal{R}(p)$, $\mathcal{R}_f^{\leq T}(p)$ (for $T > 0$) are thick subsets of the integral manifold $I(\Lambda(\mathcal{F}(f)),p)$. If the vector fields in $\mathcal{F}(f)$ are complete (or, more generally, if $\mathcal{R}_f^T(p)$ is connected), then $\mathcal{R}_f^T(p)$ is a thick subset of the submanifold $I(\Lambda_0(\mathcal{F}(f)),q)$, if q is any point of $\mathcal{R}_f^T(p)$.*

In particular, Theorem 5.3 can be used to give a Lie bracket characterization of the *accessibility property* (AP) from a point p, i.e. the property that the interior of $\mathcal{R}_f(p)$ is nonvoid (cf. [24]):

Theorem 5.4. *For an analytic system Σ_f, on Ω, with Ω open in $\mathbb{R}^n$, the AP holds from a point p iff $\Lambda(p) = \mathbb{R}^n$.*

One may then ask whether, if the AP from p holds, one might not be able to conclude something stronger, namely, that not only is the interior of $\mathcal{R}_f(p)$ nonvoid, but the point p itself is in the interior. (This property is called *local controllability from p*.) This seems to be a very difficult problem, and no complete answer is known, but a number of results have been proved. Let us limit ourselves to systems of the form

$$\dot{x} = f(x) + \sum_{i=1}^{m} u_i g_i(x) \ , \ |u_i| \leq 1 \ , \ i = 1, \ldots, m \ . \tag{5.1}$$

One can then see that the obstruction to local controllability comes from brackets such as $[g_i, g_j]$, $[g_i, [f, g_j]]$, $[[g_i, [f, g_j]], [g_r, g_s]]$, etc., i.e. iterated brackets with an even number of g's. It is then easy to prove:

Theorem 5.5. *Consider a system 5.1 such that the vector fields f, $g_1, \ldots, g_m$ are real analytic. Assume that the AP from p holds. Assume that all the iterated brackets with an even number of g's vanish at p. Then, for every $T > 0$, p is an interior point of the time T reachable set from p.*

(Notice that f itself is a bracket with 0 g's, and 0 is even, so the hypothesis of Theorem 5.5 implies in particular that $f(p) = 0$, i.e. p is an equilibrium point. In particular, the time T and time $\leq T$ reachable sets from p coincide.)

It turns out that the vanishing of the brackets with an even number of g's is a very interesting property. Actually, one can show that these systems can be put in a very useful canonical form by changing coordinates.

Theorem 5.6. *If a system 5.1 satisfies the hypotheses of Theorem 5.5, then one can make a real analytic change of coordinates on a neighborhood of p, that sends p to 0, such that in the new coordinates the system has the form*

$$\dot{x} = Ax + \sum_{i=1}^{m} u_i b_i \ , \tag{5.2}$$

where A is a square matrix and the b_i are vectors.

Such systems are called *linear systems*. For our little green men, a linear system is a "canonical form," under nonlinear changes of coordinates, of a general system for which the brackets with an even number of g's happen to vanish.

Theorem 5.5 is far from being the best one can prove. A lot of effort has recently been devoted to proving more general sufficient conditions (cf. [2], [5], [7], [8], [9], [10], [21], [22],[23], [25], [27], [29], [33], [36]). The common feature of all these results is that they involve some kind of *bracket neutralization condition*. That is, rather than require that certain brackets be zero, all that one asks is that these brackets be "neutralized," by being expressible, at p, as linear combinations of brackets that are "smaller" in some sense. For instance, one can prove (cf. [27]):

Theorem 5.7. *Consider a system 5.1 such that the vector fields f, $g_1, \ldots, g_m$ are real analytic. Assume that the AP from p holds. Assume that every iterated bracket with an even number of g's is equal, at p, to a linear combination of brackets with fewer g's. Then, for every $T > 0$, p is an interior point of the time T reachable set from p.*

Another interesting control problem, that the little green men would have rightly regarded as one of the most natural questions to be asked at the very beginning of the development of Control Theory, is the question of the *structure of optimal trajectories and optimal controls*, i.e. *what do time-optimal controls look like?* Assuming that our little green men have discovered the Pontryagin Maximum Principle, let us explore how this result can be used to obtain information about optimal controls. For simplicity, let us return now to two-vector-field systems, written in the form 2.4. To study optimal trajectories, we apply the Maximum Principle and end up concluding that one has to study the *switching function* $t \to \varphi(t)$ corresponding to an extremal γ, where

$$\varphi(t) = \Big\langle \lambda(t), g(\gamma(t)) \Big\rangle \,, \tag{5.3}$$

and λ is the adjoint variable. It then follows that the optimal control will be equal to 1 whenever $\varphi(t) < 0$, and to -1 when $\varphi(t) > 0$. This appears to suggest that optimal controls will tend to be "bang-bang," in the sense that their values will be 1 or -1. The switchings will occur at the zeros of the switching function. "Singular trajectories" can occur also, provided we manage to make φ vanish identically on an interval. And we should not forget about the possibility of even worse pathology, e.g. a switching function whose set of zeros is a Cantor-like set of positive measure.

In any case, the simplest possibility is for the zeros of the switching function to be isolated. Whenever this happens, we will have a "bang-bang theorem." The easiest way to make sure that the zeros of φ are isolated is to differentiate φ. If we can show that $\varphi(t)$ and $\dot\varphi(t)$ never vanish simultaneously, then the zeros of φ will be isolated, as desired. If we compute the derivative of the switching function we find ... a Lie bracket once again! Precisely, we get

$$\dot\varphi(t) = \Big\langle \lambda(t), [f,g](\gamma(t)) \Big\rangle \,. \tag{5.4}$$

This already makes it possible to get *something*. If we are *in the plane*, and the vectors $g(x)$ and $[f,g](x)$ are linearly independent at every point x, then the conditions $\varphi(t) = 0$ and $\dot\varphi(t) = 0$ can never hold simultaneously, because $\lambda(t) \neq 0$. So we have:

Theorem 5.8. *For a system 2.4 in the plane, if $g(x)$ and $[f,g](x)$ are linearly independent for every x, then every time-optimal trajectory is bang-bang.*

If we want to prove a similar result in higher dimensions, then we have to keep differentiating the switching function. Unfortunately, we obtain

$$\frac{d^2\varphi}{dt^2}(t) = \Big\langle \lambda(t), \big([f,[f,g]] + u(t)[g,[f,g]]\big)(\gamma(t)) \Big\rangle \,, \tag{5.5}$$

which has the drawback that the control appears explicitly. The easiest way to get around this problem is to assume that the coefficient of the control —which happens to be a bracket with an even number of g's— vanishes identically. If we keep on differentiating, we see that the simplest case to study is that when all brackets with an even number of g's vanish. In that case, the successive derivatives of the switching function are of the form

$$\frac{d^j\varphi}{dt^j} = \left\langle \lambda(t), \mathrm{ad}_f^j(g)(\gamma(t)) \right\rangle , \tag{5.6}$$

where ad_f is the operator $h \to [f,h]$. Eventually, one of the $\mathrm{ad}_f^j(g)$ will be a linear combination of the preceding ones, and therefore φ will satisfy a linear homogeneous differential equation (possibly with nonconstant coefficients). This implies, in particular, that its zeros are isolated, so we get bang-bang trajectories once again. The conclusion is none other than the *bang-bang theorem for linear systems.* However, for our little green men this is just the simplest, crudest result that you can prove by just assuming that every bracket that bothers you actually vanishes. This, of course, is too much. As in the case of local controllability, one can see that it is not necessary for all the offending brackets to vanish. Using this, one can prove genuinely *nonlinear* bang-bang theorems (cf. [26], [18], [3]).

Clearly, bang-bang theorems are only the beginning of the study of the general problem of the structure of optimal controls. If, as one can easily see from trivial examples, time-optimal controls are not always bang-bang, then what can be said about them? Are they, for instance, finite concatenations of "bang-bang" and "singular" arcs? Can they have a more complicated structure involving, e.g., an uncountable set of points of discontinuity?

It turns out that, for this question, the Lie brackets play a fundamental role as well. Moreover, *the difference between C^∞ and real-analytic systems is crucial.* To see this, let us observe first that, in order to study these questions, the obvious thing to do is to apply the Pontryagin Maximum Principle, or whatever other necessary conditions for optimality might be available, in order to prove something about our optimal controls. As we have seen in our discussion of bang-bang theorems, the Lie brackets show up as soon as you write down what the Maximum Principle gives, and try to use it. To understand more complicated possibilities, let us go back to the case of systems in the plane. As explained above, for such systems the time-optimal trajectories are bang-bang, *as long as we stay away from the set $S = \{x : \Delta(x) = 0\}$*, where Δ is the determinant of the matrix whose columns are g and $[f,g]$. To understand "singular" extremals one needs to understand the set S better. For general C^∞ systems one cannot say much about this set, other than the not very useful fact that it is closed. For real-analytic systems, the picture is much more interesting. The set S is the zero set of a *real analytic function* and therefore it has a very special structure. There are two possibilities (assuming that Ω is connected): either $S = \Omega$, i.e. $\Delta \equiv 0$, or S is a locally finite union of real-analytic arcs and branch points. The first case turns out to be exactly that of a *degenerate problem* in the sense defined earlier, so that in this case all trajectories are time-optimal, at least locally. In the second case, the "singular trajectories" will have to live in the arcs that make up the set S, and this imposes severe restrictions on what may happen. One can also analyze the junctions between singular and bang-bang arcs, and carry the analysis to completion (cf. [30], [31], [32]), proving that

Theorem 5.9. *For a nondegenerate real-analytic system 2.4 in an open connected subset Ω of the plane, every time-optimal control is a finite concatenation of real-analytic controls that are either "bang" (i.e. $\equiv 1$ or $\equiv -1$) or singular.*

One can attempt to pursue a similar analysis in higher dimensions. It turns out that the problem becomes much more difficult, but a very rich structure is beginning to emerge, in which the Lie brackets naturally play a central role. Significant results on this problem have recently been obtained by Bressan (cf. [4]) and Schättler (cf. [18], [19], [20]).

We conclude our list of examples of Lie bracket results by stating a theorem that explains, at least in principle, the surprising ubiquitousness of the Lie bracket. Suppose we are given a collection $\mathcal{F}$ of smooth vector fields. A *Lie bracket relation* (LBR) at p among the vector fields in $\mathcal{F}$ is a formal linear combination R of iterated formal brackets of symbols —i.e. indeterminates— that stand for the vector fields in $\mathcal{F}$ (technically, an element of the free Lie algebra generated by indeterminates $\{X_f : f \in \mathcal{F}\}$) that vanishes when the actual vector fields of $\mathcal{F}$ are substituted for the symbols representing them. (For instance, if $f \in \mathcal{F}$ and $g \in \mathcal{F}$, and F, G are the corresponding indeterminates, then $F+[F,G]-3[G,[F,G]]$ is a LBR at p iff $f(p)+[f,g](p)-3[g,[f,g]](p) = 0$.) Assume now that we are given two families $\mathcal{F}^i = \{f_a^i : a \in A\}$, $i = 1, 2$ of vector fields, *indexed by the same set* A, and points $p^i \in \Omega^i$. Use the same indeterminates F_a for both systems. Let us say that System $\mathcal{F}^1$ near p^1 is *equivalent* to System $\mathcal{F}^2$ near p^2 if there exists a diffeomorphism Φ from a neighborhood N^1 of p^1 onto a neighborhood N^2 of p^2 that transforms f_a^1 to f_a^2 for each a. In other words, the two systems are equivalent if they really are the same up to a change of coordinates. Then one has the following (cf. [28]):

Theorem 5.10. *Assume that the systems $\mathcal{F}^i$ as above are analytic and satisfy the accessibility property from p^i. Then $\mathcal{F}^1$ near p^1 is equivalent to $\mathcal{F}^2$ near p^2 if and only if both systems have exactly the same LBR's at the correponding points p^i.*

This theorem says that, at least for the case of analytic systems, all local properties at a point are determined by the LBR's. This explains why in so many cases the results are naturally stated in terms of Lie brackets.

6 THE ROLE OF REAL ANALYTICITY

Our main emphasis so far has been on the role of the Lie bracket. However, we have also encountered at various points the concept of real analyticity. It should come as no surprise that this matters. Take, for instance, Theorem 5.3. This says, in particular, that the reachable set $\mathcal{R}_f(p)$ is entirely contained in a submanifold whose dimension is that of $\Lambda(\mathcal{F}(f))(p)$. But $\Lambda(\mathcal{F}(f))(p)$ is an object that can be computed by just looking at the vector fields $f(\cdot, u)$ and their derivatives of all orders *at the point p*. So we are saying, among other things, that if we are in, say, $\mathbb{R}^4$, but the dimension of $\Lambda(\mathcal{F}(f))(p)$

turns out to be 3, then it follows that there is no way that one can fill up a 4-dimensional set by means of trajectories starting at p, *even if one is allowed to go far away.* So the values of the vector fields and their derivatives at a point p impose severe restrictions on what these vector fields can do even far away from p. Clearly, this sort of thing cannot possibly happen for general C^∞ systems, and it is not surprising that real analyticity is needed to derive it.

Another striking illustration of the difference between the C^∞ and the real-analytic case is provided by the following two facts (cf. [34] and [35]).

H. *Let $a < b$, and let $\eta : [a,b] \to [-1,1]$ be an arbitrary measurable function. Then there exist a single-input C^∞ system of the form 2.4 in $\mathbb{R}^3$, and points p, q in $\mathbb{R}^3$, such that η steers p to q and no other control does.*

K. *Consider a real analytic system of the form 2.4. Then, given any pair of points p, q such that there is a control η that steers p to q in time T, it follows that there is a (possibly different) $\tilde{\eta}$ that also steers p to q in time T and is analytic on an open dense subset J of its interval of definition.*

The contrast between **H** and **K** is quite remarkable. For C^∞ systems, **H** says that there is no general *a priori* regularity theorem for optimal controls. If we know that a control η is the unique time-optimal control steering p to q for some C^∞ problem, then there is nothing we can infer about η from this information, since every η can arise in this way for some problem. (Notice that the η of **H** *is* time-optimal, since it is the only control that steers p to q.) On the other hand, for *real-analytic problems* we can infer *something*. It is not much, of course, but it is enough to show that there is a fundamental difference between the two cases. At the moment, it is still a mystery how big this difference is. For instance, it does not seem to be known whether we can strengthen the conclusion of **K** to make the complement of J a null set, or even a countable one. (That it cannot be taken to be finite follows from "Fuller's problem," cf. [14].)

7 CONCLUSION

In the planet of the little green men, the development of control theory began by studying the simplest control systems where there really is some control (i.e. two-vector-field system) and asking the most natural questions, such as

1. when is the minimum-time problem degenerate?
2. when does the effect of a control that involves alternation between two different actions actually depend on the order in which these actions are carried out rather than just how much time has been alotted to each one?
3. when does a bang-bang theorem hold?
4. what do reachable sets look like, and how can one calculate their dimension?
5. when does the reachable set have a nonvoid interior?
6. when does local controllability hold?

7. what are optimal controls like?

Having asked these questions, they immediately found that the effort to answer them leads right away to the Lie bracket, and to drawing a sharp line of demarcation between C^∞ and real analytic systems, the latter having many interesting properties that do not hold in the C^∞ case. This is why all their textbooks on Control Theory start with the study of vector fields, Lie brackets, Lie algebras of vector fields, integral manifolds, and real analytic systems. This path seems to me to be much more "natural" than the one that has been followed on Earth. So the path we have followed is not the only one we could have followed, and perhaps not the one we *should* have followed. However, we are free to make choices now that will affect our future path. We can (and probably should) make an effort to incorporate the missing elements (geometry and topology) into the mainstream of our field, so as to make the next thirty years even more exciting than the first ones.

REFERENCES

1 L. D. Berkovitz, *Optimal Control Theory*, Springer-Verlag, New York, 1974.

2 R. M. Bianchini and G. Stefani, Sufficient conditions of local controllability, *Proc. 25th. I.E.E.E. Conf. Decision and Control*, Athens, Greece (December 1986).

3 A. Bressan, A high-order test for optimality of bang-bang controls, *S.I.A.M. J. Control and Optimization*, 23: 38-48 (1985).

4 A. Bressan, The generic local time-optimal stabilizing controls in dimension 3, *S.I.A.M. J. Control and Optimization*, 24: 177-190 (1986).

5 P. Brunovsky, Local controllability of odd systems, *Banach Center Publications* 1: 39-45, Warsaw, Poland (1974).

6 R. Hermann, On the accessibility problem in control theory, in *International Symposium on Nonlinear Differential Equations and Nonlinear Mechanics*, J.P. LaSalle and S. Lefschetz Eds., Academic Press, New York (1963).

7 H. Hermes, On local and global controllability, *S.I.A.M. J. Control* 12: 252-261 (1974).

8 H. Hermes, Lie algebras of vector fields and local approximation of attainable sets, *S.I.A.M. J. Control and Optimization* 16: 715-727 (1974).

9 M. Kawski, A new necessary condition for local controllability, in *Proceedings of the Conference on Differential Geometry*, San Antonio, 1986.

10 M. Kawski, High-order small-time local controllability, to appear in *Nonlinear Controllability and Optimal Control*, H.J. Sussmann ed., to be published by M. Dekker, Inc.

11 E. B. Lee and L. Markus, *Foundations of Optimal Control Theory*, J. Wiley, New York, 1967.

12 G. Leitmann, *An Introduction to Optimal Control*, McGraw-Hill, New York, 1966.

13 J. Macki and A. Strauss, *Introduction to Optimal Control Theory*, Springer-Verlag, New York, 1982.

14 C. Marchal, Chattering arcs and chattering controls, *J. Optim. Theory and Applic.*, 11: 441-468 (1973).

15 T. Nagano, Linear differential systems with singularities and an application to transitive Lie algebras, *J. Math. Soc. Japan* 18: 398-404 (1966).

16 L. S. Pontryagin, V. G. Boltyanskii, R. V. Gamkrelidze, E. F. Mischenko, *The Mathematical Theory of Optimal Processes* (translated by K. N. Trigoroff, L. W. Neustadt, editor), J. Wiley, 1962.

17 D. Ruelle, Is our mathematics natural? The case of equilibrium statistical mechanics, *Bull. Amer. Math. Soc.*, 19: 259-268 (July 1988).

18 H. Schättler, On the local structure of time-optimal bang-bang trajectories in $\mathbb{R}^3$, *S.I.A.M. J. Control and Optim.*, 26, No. 1: 186-204 (1988).

19 H. Schättler, The local structure of time-optimal trajectories in dimension 3 under generic conditions, *S.I.A.M. J. Control and Optim.*, 26, No. 1: 186-204 (1988).

20 H. Schättler and H. J. Sussmann, On the regularity of optimal controls, *Journal of Applied Mathematics and Physics (ZAMP)* 38: 292-301, (1987).

21 G. Stefani, Local controllability of nonlinear systems: an example, *Systems and Control Letters*, 6, No. 2: 123-125 (1985).

22 G. Stefani, On the local controllability of a scalar-input control system, in *Theory and Applications of Nonlinear control Systems*, C.I. Byrnes and A. Lindquist eds., Elsevier Science Publ. (1986).

23 G. Stefani, Polynomial approximations to control systems and local controllability, *Proc. 24th. I.E.E.E. Conference on Decision and Control*, Ft. Lauderdale, Fla. (1986).

24 H. J. Sussmann and V. Jurdjevic, Controllability of nonlinear systems, *J. Diff. Equations* 12: 95-116 (1972).

25 H. J. Sussmann, A sufficient condition for local controllability, *S.I.A.M. J. Control* 16, No.5: 790-802 (1978).

26 H. J. Sussmann, A bang-bang theorem with bounds on the number of switchings, *S.I.A.M. J. Control and Optimization* 17, No. 5: 629-651 (1979).

27 H. J. Sussmann, Lie brackets and local controllability: a sufficient condition for scalar input systems, *S.I.A.M. J. Control and Optimization* 21, No. 5: 686-713 (1983).

28 H. J. Sussmann, Lie brackets, real analyticity and geometric control theory, in *Differential Geometric Control Theory*: 1-115, R.W. Brockett, R.S. Millman and H.J. Sussmann Eds., Birkhäuser Boston Inc. (1983),

29 H. J. Sussmann, A general theorem on local controllability, *S.I.A.M. J. Control and Optimization* 25, No. 1: 158-194 (1987).

30 H. J. Sussmann, The structure of time-optimal trajectories for single-input systems in the plane : the C^∞ nonsingular case, *S.I.A.M. J. Control and Optimization* 25: 433-465 (1987).

31 H. J. Sussmann, The structure of time-optimal trajectories for single-input systems in the plane: the general real-analytic case, *S.I.A.M. J. Control and Optimization* 25, No. 4: 868-904, (1987).

32 H. J. Sussmann, Regular synthesis for time-optimal control of single-input real-analytic systems in the plane, *S.I.A.M. J. Control and Optimization* 25, No. 5: 1145-1162 (1987).

33 H. J. Sussmann, A general theorem on symmetries and local controllability, in *Proceedings of the 24th I.E.E.E. Conference on Decision and Control*: 27-32, Fort Lauderdale, Florida, (1985).

34 H. J. Sussmann, A weak regularity theorem for real analytic optimal control problems, *Revista Matemática Iberoamericana* 2, No. 3: 307-317 (1986).

35 H. J. Sussmann, Recent developments in the regularity theory of optimal trajectories, in *Linear and Nonlinear Mathematical Control Theory*, Rendiconti del Seminario Matematico, Università e Politecnico di Torino, Fascicolo Speciale 1987, 149-182.

36 K. Wagner, Uber den Steuerbarkeitsbegriff bei nichtlinearen Kontrollsystemen, *Arch. Math.*, 47: 29-40 (1986).

37 J. Warga, *Optimal Control of Differential and Functional Equations*, Academic Press, New York (1972).

Exact Controllability in the Presence of Damping

Roberto Triggiani, Professor, Applied Mathematics Department, University of Virginia, Charlottesville, Virginia

1. PROBLEM AND STATEMENT OF RESULTS

Let Ω be an open bounded domain in R^n with sufficiently smooth boundary Γ which is divided as $\Gamma = \Gamma_0 \cup \Gamma_1$, with Γ_0 possibly empty and Γ_1 non-empty and relatively open. In Ω we consider the following non-homogeneous problem of Dirichlet type for the wave equation in the presence of 'damping' in the solution w (t,x)

$$\begin{cases} w_{tt} = \Delta w - Bw_t & \text{in } (0,T] \times \Omega = Q & (1.1a) \\ w(0,\cdot) = w^0; \ w_t(0,\cdot) = w^1 & \text{in } \Omega & (1.1b) \\ w|_{\Sigma_0} = 0 & \text{in } (0,T] \times \Gamma_0 = \Sigma_0 & (1.1c) \\ w|_{\Sigma_1} = u & \text{in } (0,T] \times \Gamma_1 = \Sigma_1 & (1.1d) \end{cases}$$

with control function $u \in L^2(0,T; L^2(\Gamma_1)) = L^2(\Sigma_1)$. The operator B is taken, for the time being, to be a bounded operator on $L^2(\Omega)$. The (optimal) regularity theory (with minor modifications) of [L-T.1,2], [L.1] [L-L-T.1] then gives that $u \in L^2(\Sigma_1) \to \{w, w_t\} \in C([0,T]; L^2(\Omega) \times H^{-1}(\Omega))$, continuously, for any $0 < T < \infty$. When $B = 0$, the exact controllability problem for (1.1) on the natural state space $L^2(\Omega) \times H^{-1}(\Omega)$ has been solved only recently [L-T.3], [L.1] [H.1], [T.1] on the basis of techniques used first in the regularity theory mentioned above. As a consequence a spur of activities on the exact controllability/regularity theory of other dynamics has followed: wave equation with Neumann control, [L-T.4], [L.1], various cases of plate equations [L.1], [L-L.1], [L-T.5], [L-T.6] etc.

The proofs of these results make definite use of the property that the corresponding homogeneous problem is either 'energy-conservative' in the natural norms which arise in the application of suitable multipliers [H.1], [L.1], [L.2], [L-T.3], [L-T.6], [T.1], or else are 'energy-conservative' in norms which are equivalent to natural norms arising in the application of multipliers [L-T.5]. This feature plainly fails in the case of 'damped' problems such as (1.1). In additions, (1.1a) displays an additional energy term $B\phi_t$ which complicates the analysis. On the other hand, 'too much' damping can drastically change the original character of the dynamics by making its solutions smoother and thus affect adversely exact controllability (to and from the origin) over the original (less smooth) spaces of the undamped case. For instance, consider the abstract second order equation: $\ddot{x} + Ax + B_1 \dot{x} = 0$ on a Hilbert space X which, once written as a first order equation in [x,x] on the space $E = \mathcal{D}(A^{1/2}) \times X$, gives rise to the operator $\mathcal{A}_{B_1} = \begin{vmatrix} 0 & I \\ -A & -B_1 \end{vmatrix}$. Here, A is a positive selfadjoint operator on X and B_1 is also positive selfadjoint, with $\mathcal{D}(A^{\alpha_2}) \subset \mathcal{D}(B_1) \subset \mathcal{D}(A^{\alpha_1})$, $0 < \alpha_1 \le \alpha_2$. The following results are established in [C-T.1], [C-T.2], [C-T.3].

(i) Let $\rho_1 A^{\alpha_1} \le B$ on $\mathcal{D}(B)$ (1.2)

$$B \le \rho_2 A^{\alpha_2} \text{ on } \mathcal{D}(A^{\alpha_2})$$

for some constants $0 \le \rho_1, \rho_2 < \infty$; $1/2 \le \alpha_1 \ \alpha_2 \le 1$. Then the strongly continuous semigroup of contractions $\exp(\mathcal{A}_{B_1} t)$ generated on E by the operator $\mathcal{A}_{B_1}$ (once closed) is also analytic (holomorphic) on a triangular sector of $\mathbb{C}$ containing R^+.
(ii) Let now (1.2) hold for $0 < \alpha_1 \le \alpha_2 < 1/2$. Then $\exp(\mathcal{A}_{B_1} t)$ is not analytic on E, but is differentiable here for all $t > 0$.
The above results justify why in (1.1) we have restricted the damping operator B to be bounded (case $\alpha_1 = \alpha_2 = 0$ above).

In this note, we address the case of exact controllability under damping, which was recently raised to us. While space limitations force us to consider only problem (1.1), the new ideas presented here - in addition to adaptation of techniques employed in the non-damped cases mentioned above - are applicable to other boundary conditions and/or to plate equations, etc.

First, that exact controllability is preserved under 'small' perturbations B is most directly established through the following argument. Exact controllability of (1.1) on [0,T], 0 < T, $0 < T < \infty$ (to the origin) is the property that for any initial data $\{w^0, w^1\}$ in $L^2(\Omega) \times H^{-1}(\Omega)$ there exists a suitable $u \in L^2(0,T; L^2(\Gamma_1))$ such that $w(T,\cdot) = w_t(T,\cdot) = 0$. By reversing time, this property is equivalent to saying that the zero initial data are 'steered' to all of $L^2(\Omega) \times H^{-1}(\Omega)$ at time T by means of $L^2(\Sigma_1)$ - controls along the dynamics obtained from (1.1a) by replacing B with -B, see (3.2) below. Equivalently stated, the operator $\mathcal{L}_T$ defined in (3.1) below has to be onto $L^2(\Omega) \times H^{-1}(\Omega)$. But if $\mathcal{L}_T$ is surjective, then so is $\mathcal{L}_T + P$, for all suitably small bounded perturbations P, as one sees by considering, equivalently, that if $\mathcal{L}_T^*$ has a continuous inverse, see (3.3), so does $\mathcal{L}_T^* + P^*$, for all $\| P^* \| = \| P \|$ suitably small. But B bounded on $L^2(\Omega)$ in (1.1a) produces a bounded perturbation to $\mathcal{L}_T$. Thus we have the following qualitative result.

THEOREM 1.1 ('openness property of E.C.') Let problem (1.1) with B = 0 be exactly controllable in [0,T], see [L.1], [T.1], etc. Then there exists a constant $\delta > 0$ such that: if $\| B \| < \delta$, then problem (1.1) is exactly controllable on [0,T]. □

One may give a more precise statement of exact controllability for problem (1.1) than Theorem 1.1, for general $B \in \mathcal{L}(L^2(\Omega))$, but still of local character, i.e. with $\| B \|$ sufficiently small, see Appendix A. Plainly, the openness property in Theorem 1.1 above holds around each operator $B_0 \in \mathcal{L}(L^2(\Omega))$, not only $B_0 = 0$, such that the corresponding problem (1.1) is exactly controllable for $B = B_0$. In the text of the present paper, we specialize B to be a multiplication operator

$$(Bf)(x) = \gamma(x)\, f(x), \tag{1.3}$$

$\gamma(x)$ a scalar function on Ω with appropriate smoothness. In our first result, Lemma 1.1, we provide an identity for the solution ϕ of the homogeneous problem (1.4) below, followed by an inequality of the sought after type for $\gamma(x)$ variable. Next, we restrict to the case $\gamma(x) = \gamma =$ constant (of any size and any size) and obtain from Lemma 1.1 a global exact controllability result for (1.1) for all T > some T_0 (computable) > 0. As a corollary, the perturbation argument behind the proof of the openness property yields exact controllability for (1.1) for the same T, with $\gamma(x) = \gamma_0 + B_1$, where γ_0 is an arbitrary real constant and B_1 lies in a suitably small sphere of $\mathcal{L}(L^2(\Omega))$, centered at the origin and with radius depending on γ_0 and T.

In the sequel we shall let ϕ be the solution of the corresponding homogeneous problem

$$\begin{cases} \phi_{tt} = \Delta\phi - B\phi_t & \text{in } Q & (1.4a) \\ \phi|_{t=0} = \phi^0; \ \phi_t|_{t=0} = \phi^1 & \text{in } \Omega & (1.4b) \\ \phi|_{\Sigma} = 0 & \text{in } \Sigma & (1.4c) \end{cases}$$

LEMMA 1.2 Let B be as in (1.3) with $\gamma(x)$ smooth. Let $h(x) = [h_1(x), \cdots h_n(x)]$ be a $C^2(\overline{\Omega})$ - vector field on Ω with transpose of its Jacobian given by the $n \times n$ matrix $H = H(x)$

$$H(x) = \begin{vmatrix} \dfrac{\partial h_1}{\partial x_1} & \cdots & \dfrac{\partial h_1}{\partial x_n} \\ \cdot & & \cdot \\ \cdot & & \cdot \\ \cdot & & \cdot \\ \dfrac{\partial h_n}{\partial x_1} & \cdots & \dfrac{\partial h_n}{\partial x_n} \end{vmatrix} \tag{1.5}$$

a) Then the following identity holds for the solution ϕ of problem (1.4)

$$½ \int_\Sigma \left[\frac{\partial \phi}{\partial \nu}\right]^2 e^{\gamma(x)t}\, h \cdot \nu \, d\Sigma = \int_Q H\nabla\phi \cdot \nabla\phi \, e^{\gamma(x)t} \, dQ + \int_Q (h \cdot \nabla\phi)(\nabla\phi \cdot \nabla\gamma)(t\, e^{\gamma(x)t})\, dQ$$

$$- ½ \int_Q \phi\phi_t \, e^{\gamma(x)t}\, h \cdot \nabla\gamma \, dQ + ½ \int_Q \phi\nabla\phi \cdot \nabla(\mathrm{div}(e^{\gamma(x)t} h))\, dQ + \beta_{0,T} \tag{1.6a}$$

$$\beta_{0,T} = ½ \left[(\phi_t, \phi \, \mathrm{div}(e^{\gamma(x)t} h))_\Omega\right]_0^T + \left[(e^{\gamma(x)t}\phi_t, h \cdot \nabla\phi)_\Omega\right]_0^T \tag{1.6b}$$

b) Next, assume that the matrix $H(x)$ in (1.5) and the matrix $M_{\gamma h}$ defined by

$$M_{\gamma h}(x) = \begin{vmatrix} \gamma_{x_1} h_1 & \cdots & \gamma_{x_n} h_1 \\ \cdot & & \cdot \\ \cdot & & \cdot \\ \cdot & & \cdot \\ \gamma_{x_1} h_n & \cdots & \gamma_{x_n} h_n \end{vmatrix} \tag{1.7}$$

satisfy the following conditions for all $v \in [L^2(\Omega)]^n$

$$H(x)\, v(x) \cdot v(x) \geq \rho_1 \, |v(x)|^2_{R^n} \quad \text{for some constant } \rho_1 > 0, \ \forall x \in \Omega \tag{1.8}$$

$$M_{\gamma h}(x)\, v(x) \cdot v(x) \geq \rho_2 \, |v(x)|^2_{R^n} \quad \text{for some constant } \rho_2, \ \forall x \in \Omega \tag{1.9}$$

$$\rho_1 > 0; \ m_{\rho T} > 0, \ \text{where } m_{\rho T} = \min_{0 \leq t \leq T} (\rho_1 + t\rho_2) = \begin{cases} \rho_1 & \text{if } \rho_2 \geq 0 \\ \rho_1 + T\rho_2, & \text{if } \rho_2 < 0 \end{cases} \tag{1.10}$$

Then the following inequality holds true for the solution ϕ of problem (1.4) for any $\varepsilon > 0$

$$1/2 \int_\Sigma \left[\frac{\partial \phi}{\partial \nu}\right]^2 e^{\gamma(x)t}\, h \cdot \nu \, d\Sigma + \mathrm{const}_{T,\gamma,h,\varepsilon,\rho_1,\rho_2} \|\phi\|^2_{C([0,T];\, L^2(\Omega))} \geq$$

$$\left[\frac{\rho_1}{2} - \varepsilon c_{\gamma,T,\rho,h}\right] \int_0^T \int_\Omega e^{\gamma(x)t} (\phi_t^2 + |\nabla\phi|^2)\, d\Omega \, dt$$

$$- (K_h + \varepsilon) \left\{ \int_\Omega e^{\gamma(x)T} (|\phi_t(T)|^2 + |\nabla\phi(T)|^2)\, d\Omega + E(0) \right\} \tag{1.11}$$

where $c_{\gamma,T,\rho,h} = 0$ if $\gamma(x) \equiv \text{const}$, and $K_h = \max |h|/2$ over $\overline{\Omega}$

$$E(t) = \int_\Omega \phi_t^2 + |\nabla\phi|^2 \, d\Omega \quad \square \tag{1.12}$$

For the case $\gamma(x) \equiv$ constant (of any size and any sign), whereby the matrix $M_{\gamma h} = 0$ in (1.7) and (1.10) is satisfied with $\rho_2 = 0$, exact controllability of (1.1) is obtained from Lemma 1.1 for T sufficiently large.

THEOREM 1.3 Let B be as in (1.3) with $\gamma(x) \equiv$ constant γ, γ being an arbitrary real number. Let there exists a vector field $h(x) \in C^2(\bar{\Omega})$ such that:

$$h \cdot \nu \le 0 \text{ on } \Gamma_0; \text{ and (1.8) is satisfied} \tag{1.13}$$

Then, for T sufficiently large (as in the proof below), problem (1.1) is exactly controllable on $L^2(\Omega) \times H^{-1}(\Omega)$ $\square$

By perturbation argument as in Theorem 1.1 we obtain

COROLLARY 1.4 Let $B = B_0 + B_1$, with $B_0 = \gamma_0 I$, γ_0 an arbitrary real constant and $B_1 \in \mathcal{L}(L^2(\Omega))$. Let the vector field condition (1.13) hold true, so that Theorem 1.3 applies for $\gamma(x) = \gamma_0$ on [0,T], T some suitable large. Then, there exists a constant $\delta > 0$ (depending on γ_0 and T) such that if $\| B_1 \| < \delta$, then problem (1.1) is exactly controllable on [0,T] on $L^2(\Omega) \times H^{-1}(\Omega)$ $\square$

REMARKS J. L. Lions has also obtained a 'local' result while Zuazua has also obtained a global result for γ = constant (private communications). Their methods are different from ours; in particular they do not use the behavior of E(t) given by (A.15) or (3.11) below, which is a new feature of the problem in question. Rather, in proving the crux of the matter, inequality (3.4) below, J. L. Lions' method consists in comparing the damped ϕ - problem (1.4) with the undamped ϕ - problem (1.4) with B = 0; this succeeds for small damping. Zuazua uses the same idea of J. L. Lions after the preliminary classical change of variable $v = e^{\gamma/2t}\phi$ of the ϕ - problem (1.4) with $\gamma(x)$ = const into a v - problem which, as is well known, does not contain v_t but the lower order term v. We thank J. Lagnese for raising to us the problem discussed here. $\square$

2. PROOF OF LEMMA 1.2

Part a) We multiply (1.4a) by $e^{\gamma(x)t}(h \cdot \nabla\phi)$ as in [L-T.3] and integrate first in [0,T] and then in Ω. This produces a cancellation of the damping term $\gamma(x)\phi_t$, as desired. Proceeding as, say, in [T.1], [L-T.3] we obtain (computations are omitted)

$$\tfrac{1}{2}\int_\Sigma \left[\frac{\partial\phi}{\partial\nu}\right]^2 e^{\gamma(x)t} h \cdot \nu \, d\Sigma = \int_Q H\nabla\phi \cdot \nabla\phi \, e^{\gamma(x)t} dQ + \tfrac{1}{2}\int_Q (\phi_t^2 - |\nabla\phi|^2)\, \mathrm{div}\,(e^{\gamma(x)t}h)\, dQ$$
$$+ \int_Q (h \cdot \nabla\phi)\, \nabla\phi \cdot \nabla\gamma\, (te^{\gamma(x)t}) dQ + [(e^{\gamma(x)t}\phi_t, h \cdot \nabla\phi)_\Omega]_0^T \tag{2.1}$$

where we have used the identity

$$\nabla\phi \cdot \nabla(h \cdot \nabla\phi\, e^{\gamma(x)t}) = H\nabla\phi \cdot \nabla\phi\, e^{\gamma(x)t} + \tfrac{1}{2} h \cdot \nabla(|\nabla\phi|^2)\, e^{\gamma(x)t}$$
$$+ (h \cdot \nabla\phi)\, \nabla\phi \cdot \nabla\gamma\, (t\, e^{\gamma(x)t}) \tag{2.2}$$

(ii) We now multiply (1.4a) by $\phi\, \mathrm{div}\,(e^{\gamma(x)t}h)$ and integrate first over [0,T], then over Ω. Thus again produces a cancellation of the damping term $\gamma(x)\,\phi_t$. Proceeding as in [T.1] we obtain (computations are omitted)

$$\int_Q (\phi_t^2 - |\nabla\phi|^2)\, \mathrm{div}\,(e^{\gamma(x)t}h)\, dQ = [(\phi_t, \phi\, \mathrm{div}(e^{\gamma(x)t}h))_\Omega]_0^T$$
$$+ \int_Q \phi\, \nabla\phi \cdot \nabla(\mathrm{div}(e^{\gamma(x)t}h))\, dQ - \int_Q \phi\, \phi_t\, e^{\gamma(x)t} h \cdot \nabla\gamma\, dQ \tag{2.3}$$

Thus inserting (2.3) into (213) produces (1.6).

Part b) Step 1. We have

$$(\nabla\phi \cdot \nabla\gamma)h = M_{\gamma h}\,\nabla\phi \tag{2.4}$$

with $M_{\gamma h}(x)$ the matrix defined in (1.7). Thus, with reference to the first two terms on the right hand side of Eq (1.6a), we obtain by virtue of (2.4) and assumptions (1.8), (1.9)

$$\int_Q H\nabla\phi \cdot \nabla\phi\, e^{\gamma(x)t}\, dQ + \int_Q t(h \cdot \nabla\phi)\,(\nabla\phi \cdot \nabla\gamma)\, e^{\gamma(x)t}\, dQ \geq$$

$$\rho_1 \int_Q |\nabla\phi|^2 e^{\gamma(x)t}\, dQ + \rho_2 \int_Q t\,|\nabla\phi|^2\, e^{\gamma(x)t}\, dQ \tag{2.5}$$

For any $\varepsilon > 0$, we estimate the remaining two terms in Q in (1.6a) by using $2a\,b \leq \varepsilon\, a^2 + \frac{1}{\varepsilon}\, b^2$. We obtain

$$1/2\left|\int_Q \phi\nabla\phi \cdot \nabla(\operatorname{div} e^{\gamma(x)t}\, h)\, dQ\right| \leq \varepsilon \int_Q |\nabla\phi|^2 e^{\gamma(x)t}\, dQ$$

$$+\frac{1}{\varepsilon}\operatorname{const}_{T,h,\gamma}\|\,\phi\,\|^2_{C([0,T];\, L^2(\Omega))} \tag{2.6}$$

$$1/2\left|\int_Q \phi\, \phi_t\, e^{\gamma(x)t}\, h \cdot \nabla\gamma\, dQ\right| \leq \varepsilon\mu_{\gamma T} \int_Q \phi_t^2\, e^{\gamma(x)t}\, dQ + \frac{T}{\varepsilon}\,\mu_{\gamma T}\,\|\,\phi\,\|^2_{C([0,T];\, L^2(\Omega))} \tag{2.7}$$

where $\mu = \mu_{\gamma,T,h} = \sup |h \cdot \nabla\gamma|/4$ over $\overline{\Omega} \times [0,T]$. (Note that $\mu = 0$ if $\gamma(x) \equiv$ const. In estimating terms, we shall pull out explicitly constants which vanish when $\gamma(x) =$ const). Using estimates (2.5)-(2.7) in the right hand side of EQ (1.6a), we obtain

$$1/2 \int_\Sigma \left[\frac{\partial\phi}{\partial\nu}\right]^2 e^{\gamma(x)t}\, h \cdot \nu\, d\Sigma + \operatorname{const}_{T,h,\gamma,\varepsilon}\|\,\phi\,\|^2_{C([0,T];\, L^2(\Omega))} \geq$$

$$(\rho_1 - \varepsilon)\int_Q |\nabla\phi|^2\, e^{\gamma(x)t}\, dQ + \rho_2 \int_Q t\,|\nabla\phi|^2\, e^{\gamma(x)t}\, dQ$$

$$-\,\varepsilon\,\mu_{\gamma,T,h}\int_Q \phi_t^2\, e^{\gamma(x)t}\, dQ + \beta_{0,T} \tag{2.8}$$

Step 2 We multiply Eq (1.4a) by $e^{\gamma(x)t}\phi$, respectively by $t\,e^{\gamma(x)t}\,\phi$ (This latter step is not needed when $\gamma(x)$ = const, for in this case $M_{\gamma h} = 0$ and $\rho_2 = 0$) and integrate by parts as in obtaining (2.3). We get

$$\int_Q (\phi_t^2 - |\nabla\phi|^2)\, e^{\gamma(x)t} dQ = \left[\int_\Omega \phi\, \phi_t\, e^{\gamma(x)t}\, d\Omega\right]_0^T + \int_Q \phi\, \nabla\phi \cdot \nabla(e^{\gamma(x)t})\, dQ \tag{2.9}$$

$$\int_Q t(\phi_t^2 - |\nabla\phi|^2)\, e^{\gamma(x)t}\, dQ = T\int_\Omega \phi(T)\, \phi_t(T)\, e^{\gamma(x)T}\, d\Omega$$

$$+\int_Q t\,\phi\, \nabla\phi \cdot \nabla(e^{\gamma(x)t})\, dQ - \int_Q \phi\, \phi_t\, e^{\gamma(x)t}\, dQ \tag{2.10}$$

With reference to the first two terms in Q at the right hand side of Eq (2.8), we split each integral in two halves $\int_Q = 1/2\int_Q + 1/2\int_Q$, express in one of them $|\nabla\phi|^2$ over Q in terms of ϕ_t^2 over Q via (2.9), respectively (2.10), and obtain respectively

$$\int_Q |\nabla\phi|^2 e^{\gamma(x)t}\, dQ = 1/2\int_Q (\phi_t^2 + |\nabla\phi|^2) e^{\gamma(x)t}\, dQ - 1/2\int_Q \phi\nabla\phi \cdot \nabla(e^{\gamma(x)t}) dQ$$

$$- 1/2[\int_\Omega \phi\, \phi_t\, e^{\gamma(x)t}\, d\Omega]_0^T \tag{2.11}$$

$$\int_Q t\, |\nabla\phi|^2 e^{\gamma(x)t}\, dQ = 1/2 \int_Q t(\phi_t^2 + |\nabla\phi|^2)\, e^{\gamma(x)t}\, dQ - 1/2 \int_Q t\, \phi\, \nabla\phi \cdot \nabla(e^{\gamma(x)t})\, dQ$$

$$+ 1/2 \int_Q \phi\, \phi_t\, e^{\gamma(x)t}\, dQ - T \int_Q \phi(T)\, \phi_t(T)\, e^{\gamma(x)T}\, d\Omega \tag{2.12}$$

We insert (2.11), (2.12) in the right hand side (R.H.S.) of (2.8) and obtain

$$\text{R.H.S. of (3.8)} = \frac{(\rho_1-\varepsilon)}{2} \int_Q (\phi_t^2 + |\nabla\phi|^2)\, e^{\gamma(x)t}\, dQ + \frac{\rho_2}{2} \int_Q t\, (\phi_t^2 + |\nabla\phi|^2)\, e^{\gamma(x)t} dQ$$

$$- \varepsilon\, \mu_{\gamma,T,h} \int_Q \phi_t^2\, e^{\gamma(x)t}\, dQ$$

$$- \frac{(\rho_1-\varepsilon)}{2} \int_Q \phi\, \nabla\phi \cdot \nabla(e^{\gamma(x)t})\, dQ - \frac{\rho_2}{2} \int_Q t\, \phi\, \nabla\phi \cdot \nabla(e^{\gamma(x)t})\, dQ$$

$$+ \frac{\rho_2}{2} \int_Q \phi\, \phi_t\, e^{\gamma(x)t}\, dQ + b_{0,T} \tag{2.13}$$

$$b_{0,T} = \beta_{0,T} - \frac{(\rho_1-\varepsilon)}{2} [\int_\Omega \phi\, \phi_t\, e^{\gamma(x)t}\, d\Omega]_0^T - \rho_2\, T \int_\Omega \phi(T)\, \phi_t(T)\, e^{\gamma(x)T}\, dQ \tag{2.14}$$

Note that if $\gamma(x) \equiv$ const, then $\rho_2 = \mu_{\gamma,T,h} = \nabla(e^{\gamma(x)t}) = 0$ and (2.13), (2.14) become drastically simpler.
Step 3 We estimate as usual the integrals in (2.13) with lower order terms

$$|\int_Q t^\alpha\, \phi\, \nabla\phi \cdot \nabla(e^{\gamma(x)t})\, dQ| \le \varepsilon\, g_{\gamma,T} \int_Q |\nabla\phi|^2\, e^{\gamma(x)t}\, dQ + \text{const}_{\gamma,T,\varepsilon,\alpha} \|\phi\|^2_{C([0,T];\, L^2(\Omega))} \tag{2.15}$$

with $\alpha = 0, 1$ and $g_{\gamma,T} = \sup |\nabla(e^{\gamma(x)t})|$ over $\overline{\Omega} \times [0,T]$. Note that $g_{\gamma,T} = 0$ if $\gamma(x) \equiv$ const.

$$|\int_Q \phi\, \phi_t\, e^{\gamma(x)t} dQ| \le \varepsilon \int_Q \phi_t^2\, e^{\gamma(x)t}\, dQ + \text{const}_{\gamma,T,\varepsilon} \|\phi\|^2_{C([0,T];\, L^2(\Omega))} \tag{2.16}$$

Using estimates (2.15), (2.16) in (2.13) and recalling (2.8) we obtain for any $\varepsilon > 0$:

$$1/2 \int_\Sigma \left[\frac{\partial\phi}{\partial\nu}\right]^2 e^{\gamma(x)t}\, h \cdot \nu\, d\Sigma + \text{const}_{T,h,\gamma,\varepsilon,\rho_1,\rho_2} \|\phi\|^2_{C([0,T];\, L^2(\Omega))}$$

$$\ge \left[\frac{\rho_1}{2} - \varepsilon\, c_{\gamma,T,\rho_1,\rho_2,h}\right] \int_Q (\phi_t^2 + |\nabla\phi|^2)\, e^{\gamma(x)t} dQ$$

$$+ \frac{\rho_2}{2} \int_Q t\, (\phi_t^2 + |\nabla\phi|^2)\, e^{\gamma(x)t} dQ + b_{0,T}$$

where $c_{\gamma,T,\rho_1,\rho_2,h} = 0$ if $\gamma(x) = $ const. Recalling (1.6b) and (2.14), we readily obtain

$$|b_{0,T}| \le (K_h + \varepsilon) \int_\Omega \{e^{\gamma(x)T}\, (\phi_t^2(T) + |\nabla\phi(T)|^2) + \phi_1^2 + |\nabla\phi_0|^2\}\, d\Omega$$

$$+ \text{const}_{\gamma,h,T,\rho} \|\phi\|^2_{C([0,T];\, L^2(\Omega))} \tag{2.18}$$

with $K_h = \max |h|/2$ over $\overline{\Omega}$. Then, recalling (1.10), we obtain the desired conclusion (1.11) of part b) by (2.17), (2.18) □

3. PROOF OF THEOREM 1.3 ($\gamma(x)$ = constant):
Steps 0, 1 apply to any B.
Step 0 By reversing time, exact controllability in [0,T] of problem (1.1) on the space $X = L^2(\Omega) \times H^{-1}(\Omega)$ is equivalent to the property that the operator $\mathcal{L}_T$ (bounded from $L^2(0,T; L^2(\Gamma_1)) \to X$, see [L-L-T.1])

$$u \in L^2(\Sigma_1) \to \mathcal{L}_T u = \{y(T), y_t(T)\} \tag{3.1}$$

corresponding to problem (1.1) with B replaced by -B

$$\begin{cases} y_{tt} = \Delta y + B y_t & \text{in } Q & (3.2a) \\ y(0,\cdot) = 0 = y_t(0,\cdot) = 0 & \text{in } \Omega & (3.2b) \\ y|_{\Sigma_0} = 0 & \text{in } \Sigma_0 & (3.2c) \\ y|_{\Sigma_1} = u & \text{in } \Sigma_1 & (3.2d) \end{cases}$$

be onto X; equivalently that the Hilbert space adjoint $\mathcal{L}_T^*$ have a continuous inverse; i.e. there exists $C_T > 0$ such that

$$\| \mathcal{L}_T^*\{z_0, z_1\} \|_{L^2(\Sigma_1)} \geq C_T \| \{z_0, z_1\} \|_X \tag{3.3}$$

Step 1 By arguing as, say, in [T.1], one may show that following P.D.E. characterization of (3.3). Details are omitted.

Lemma 3.1. Condition (3.3) - and hence exact controllability of problem (1.1) on $X = L^2(\Omega) \times H^{-1}(\Omega)$ within the class of $L^2(0, T; L^2(\Gamma_1))$ - controls - is equivalent to the following: there exists $C'_T > 0$ such that

$$\int_0^T \int_{\Gamma_1} \left[\frac{\partial \phi}{\partial \nu}\right]^2 d\Sigma_1 \geq C'_T \| \{\phi^0, \phi^1\} \|^2_{H_0^1(\Omega) \times L^2(\Omega)} \tag{3.4}$$

where ϕ solves the homogeneous problem (1.4) □

Step 2 Lemma 3.2. Let the assumptions of Theorem 1.3 be fulfilled. Then inequality (3.4) holds true. □

Proof. Step 1. Now $B = \gamma I$, with γ arbitrary real constant. Eq (1.11) of Lemma 1.2b specializes in this case to (we write $\rho_1 = \rho$, since $\rho_2 = 0$ in this case)

$$1/2 \int_\Sigma \left[\frac{\partial \phi}{\partial \nu}\right]^2 e^{\gamma t}\, h \cdot \nu \, d\Sigma + \text{const}_{T,\gamma,h,\varepsilon,\rho} \| \phi \|^2_{C([0,T]; L^2(\Omega))}$$

$$\geq \frac{\rho}{2} \int_0^T e^{\gamma t} E(t)\, dt - (K_h + \varepsilon)[e^{\gamma T} E(T) + E(0)] \tag{3.5}$$

(we have repeatedly noted that the computations in the proof of Lemma 1.2b become markedly simpler when $\gamma(x)$ = const as several terms vanish; in particular the constant $c_{\gamma,T,\rho,h}$ is zero in this case, see below (1.11)).
Step 2 Problem with $B = \gamma I$, γ real constant, can be written abstractly as $\ddot{\phi} = -A\phi - \gamma\dot{\phi}$ on $L^2(\Omega)$ with solution given by

$$\begin{vmatrix} \phi(t) \\ \dot{\phi}(t) \end{vmatrix} = e^{\mathcal{A}_\gamma t} \begin{vmatrix} \phi_0 \\ \phi_1 \end{vmatrix}; \quad \mathcal{A}_\gamma = \begin{vmatrix} 0 & 0 \\ -A & -\gamma I \end{vmatrix} : \quad \begin{matrix} \text{generator of a s.c. group} \\ \text{on } Y = \mathcal{D}(A^{1/2}) \times L^2(\Omega) \end{matrix} \tag{3.6}$$

$$E(t) = \| \{\phi(t), \dot{\phi}(t)\} \|^2_Y = \| e^{\mathcal{A}_\gamma t} [\phi^0, \phi^1] \|^2_Y \tag{3.7}$$

where $Af = -\Delta f$, $\mathcal{D}(A) = H^2(\Omega) \cap H_0^1(\Omega)$, positive definite on $L^2(\Omega)$. Let $\{\mu_n\}$ be the eigenvalues of A, $\mu_n > 0$, $\mu_n \to +\infty$ and let $\{e_n\}$ be the corresponding eigenvectors in $L^2(\Omega)$. The eigenvalues $\lambda_n^{+,-}$ of $\mathcal{A}_\gamma$ with corresponding eigenvectors $\psi_n^{+,-}$ on Y are

$$2\lambda_n^{+,-} = -\gamma \pm \sqrt{\gamma^2 - 4\mu_n}\,; \quad \psi_n^{+,-} = [e_n, \lambda_n^{+,-} e_n] \tag{3.8}$$

say, normalized in Y: $\| \psi_n^{+,-} \|_Y^2 = 2\mu_n \| e_n \|^2 = 1$. Let N be the integer such that

$$\gamma^2 - 4\mu_n \geq 0,\ n=1,\cdots N; \quad \gamma^2 - 4\mu_n < 0,\ \ n=N+1, N+2, \cdots$$

so that the $\lambda_n^{+,-}$ satisfy for $n \geq N+1$

$$\lambda_n^{+,-} = -\gamma/2 \pm i\sqrt{\mu_n - (\frac{\gamma}{2})^2} \quad n = N+1, \cdots \tag{3.9}$$

We next invoke some crucial spectral properties of $\mathcal{A}_\gamma$ noted in [T.2] (see also [C-T.1.2] for a more general operator). These a fortiori imply that the eigenvectors $\{\psi_n^{+,-}, n \geq N+1\}$ form a Riesz basis on the infinite dimensional subspace $Y_n^\infty = \overline{sp}\,\{\psi_n^{+,-}, n \geq N+1\}$ of Y; and that moreover

$$Y = Y_N^+ \oplus Y_N^- \oplus Y_N^\infty \tag{3.10}$$

$$Y_N^+ = sp\{\psi_n^+,\ 1 \leq n \leq N\}, \quad Y_n^- = sp\{\psi_n^-,\ 1 \leq n \leq N\}.$$

Step 3 We shall first show that inequality (3.4) holds true for initial conditions $\{\phi^0, \phi^1\} \in Y_N^\infty$. In this case, the Riesz basis property on Y_N^∞ mentioned above implies that the corresponding solution $\{\phi(t), \dot{\phi}(t)\}$ of problem (1.4) satisfies

$$E(t) \sim e^{-\gamma t}E(0), \text{ meaning } \ c_N\, e^{-\gamma t}E(0) \leq E(t) \leq C_N e^{-\gamma t}E(0) \tag{3.11}$$

for two constants $0 < c_N < C_N < \infty$ independent of the initial data. Property (3.11) is the new feature over the conservative cases of undamped problems mentioned in the introduction. Using (3.11) on the right hand side (R.H.S.) of (3.5) yields

$$\text{R.H.S. of (3.5)} \geq [\frac{\rho}{2}\, c_N\, T - (K_h + \varepsilon)\,(C_N + 1)]\, E(0) \tag{3.12}$$

where the term in [ ] is positive provided

$$T > \frac{2\,K_h}{\rho}\,\frac{C_N + 1}{c_N}, \quad 2\,K_h = \max\,|h|\ \text{ over } \overline{\Omega} \tag{3.13}$$

Step 4 As to the left hand side of (3.5) one uses assumption (1.13) on h plus a Lemma like Lemma A.2 in the Appendix to absorb $\| \phi \|^2_{C([0,T];\, L^2(\Omega))}$ by $\int_{\Sigma_1} \left(\frac{\partial \phi}{\partial \nu}\right)^2 d\Sigma_1$ (note that the uniqueness property (A.4) of Holmgren type required in the argument holds true automatically if γ = constant). This proves inequality (3.4) for $\{\phi^0, \phi^1\} \in Y_N^\infty$.

Step 5 The analysis of the right hand side of (3.5) for $\{\phi^0, \phi^1\} \in Y_N^+ \oplus Y_N^-$, is direct since this is a finite dimensional eigenspace. □

APPENDIX A: a local result for general $B \in \mathcal{L}(L^2(\Omega))$

In the case of a general bounded operator B on $L^2(\Omega)$, not necessarily of the form (1.3), one may in principle extend the approach of section 2 by applying first the operator $(h \cdot \nabla\phi)\, e^{Bt}$ to both sides of Eq (1.4a), integrate by parts in t (thereby producing the desired cancellation of the damping term $B\phi_t$) etc. to obtain the counterpart of Eq (2.1); next, one uses the multiplier ϕ div $((e^{B^*t} 1_\Omega)h)$ on Eq (1.4a), 1_Ω being the identically one function on Ω, which plays the role of ϕ div $(e^{\gamma(x)t}\, h)$ in section 2. It seems, however, that in order to obtain readable results, one needs more information on B, e.g. to compute terms like $\nabla(e^{B^*t}(h \cdot \nabla\phi))$. This Appendix retains a general B; but the result is local, though a quantitatively more precise result than Theorem 1.1.

THEOREM A.1 Let $h(x) \in C^2(\overline{\Omega})$ be a vector field satisfying the vector field condition:

(i) $h \cdot \nu \leq 0$ on Γ_0; ν = outward unit vector (A.0)

(ii) there exists a constant $\rho > 0$ such that for all $v(x) \in [L^2(\Omega)]^n$

$$\int_\Omega H(x)v(x)\cdot v(x)\,d\Omega \geq \rho \int_\Omega |v(x)|^2_{R^n}\,d\Omega \tag{A.1}$$

Moreover, let the damping operator $B \in \mathcal{L}(L^2(\Omega))$ satisfy

$$0 \leq \text{Re}\,(Bx, x)_\Omega \leq C_B \|x\|^2,\ x \in L^2(\Omega) \tag{A.2}$$

for some $C_B \leq \|B\|$, and let

$$\|B\| < \frac{\rho}{5M_h},\ M_h = \max |h(x)| \text{ over } \overline{\Omega} \tag{A.3}$$

Finally, let the corresponding homogeneous problem satisfy the uniqueness property (Holmgren-type) [H.2, chapter 5]

$$\begin{cases} \phi_{tt} = \Delta\phi - B\phi_t \\ \phi|_\Sigma = 0 \text{ and } \frac{\partial\phi}{\partial\nu}|_{\Sigma_1} = 0 \Rightarrow \phi \equiv 0 \\ \text{T sufficiently large} \end{cases} \tag{A.4}$$

Then, for T sufficiently large, problem (1.1) is exactly controllable on $L^2(\Omega) \times H^{-1}(\Omega)$ (Assumption (A.5) may be eliminated, as in [T.1], by trading it with another assumption requiring $\max|\nabla(\text{div } h)|$ (over $\overline{\Omega}$) sufficiently small with respect to ρ, which in particular holds true for radial vector fields).

Proof of Theorem A.1. We must show inequality (3.4) for problem (1.4) in ϕ. As in the ´undamped' case $B = 0$, we multiply Eq (1.4a) by $h \cdot \nabla\phi$ and ϕ div h, see e.g. [T.1], and integrate by parts. We obtain the following identity, see e.g. [T.1], after using the B.C. (1.4c).

$$1/2 \int_\Sigma \left[\frac{\partial\phi}{\partial\nu}\right]^2 h\cdot\nu\, d\Sigma = \int_Q H\nabla\phi\cdot\nabla\phi\, dQ + \int_Q (B\,\phi_t)\, h\cdot\nabla\phi\, dQ$$

$$+ 1/2 \int_Q (B\,\phi_t)\,\phi \text{ div } h\, dQ + \int_Q \phi\,\nabla(\text{div } h)\cdot\nabla\phi\, dQ + \beta'_{0,T} \tag{A.5}$$

$$\beta'_{0,T} = [(\phi_t, h\cdot\nabla\phi)_\Omega]_0^T + 1/2[(\phi_t, \phi \text{ div } h)_\Omega]_0^T \tag{A.6}$$

which reveals the presence of two extra integral terms over the 'undamped' case [T.1]. Of these, it is the term involving both ϕ_t and $\nabla\phi$ (energy terms) that will impose restrictions on the damping B. (This problem is avoided by using the multipliers at the opening paragraph of this Appendix, in particular as in section 2 for B as in (1.3)). Recalling E(t) from (1.12), we split $\int_q |\nabla\phi|^2\, dQ = (1/2+1/2) \int_Q |\nabla\phi|^2\, dQ$, use

$$\int_Q |\nabla\phi|^2\, dQ = \int_Q \phi_t^2\, dQ - \int_Q (B\phi_t)\,\phi dQ - [(\phi_t, \phi)_\Omega]_0^T \tag{A.7}$$

from the multiplier ϕ and thus obtain after using (A.1)

$$\text{R.H.S. of (A.5)} \geq \frac{\rho}{2}\int_0^T E(t)dt + \int_Q (B\phi_t)\, h\cdot\nabla\phi\, dQ$$

$$+ 1/2 \int_Q (B\phi_t)\,\phi \text{ div } h\, dQ - \frac{\rho}{2}\int_Q (B\phi_t)\,\phi\, dQ$$

$$+ \int_Q \phi\,\nabla(\text{div } h)\cdot\nabla\phi\, dQ + \beta_{0,T} \tag{A.8}$$

$$\beta_{0,T} = \beta'_{0,T} - \frac{\rho}{2}[(\phi_t, \phi)_\Omega]_0^T = [(\phi_t, h\cdot\nabla\phi)_\Omega]_0^T + 1/2\,[(\phi_t, \phi \text{ div } h)_\Omega]_0^T - \frac{\rho}{2}[(\phi_t, \phi)]_0^T \tag{A.9}$$

We recall M_h in (A.3) and estimate the bilinear terms

$$|\int_Q (B\phi_t)\, h\cdot\nabla\phi\, dQ| \le \frac{\|B\|}{2} M_h \int_0^T E(t)\,dt,\quad |[(\phi_t, h\cdot\nabla\phi)_\Omega]_0^T| \le \frac{M_h}{2}[E(T)+E(0)] \qquad (A.10)$$

while for the 3rd, 4th, 5th integrals on the right of (A.8) with lower order terms we have

$$\left|\begin{matrix}\text{lower order integral terms}\\ \text{in Q in (A.8)}\end{matrix}\right| \le \varepsilon \int_0^T E(t) + \text{const}_{B,\varepsilon,T} \|\phi\|^2_{C([0,T];\, L^2(\Omega))} \qquad (A.11)$$

Thus, from (A.10), (A.11) and from

$$|[1/2(\phi_t, \phi \,\text{div}\, h)_\Omega - \frac{\rho}{2}(\phi_t, \phi)_\Omega]_0^T| \le \varepsilon[E(T)+E(0)] + \text{const}_{\rho,h,\varepsilon}\|\phi\|^2_{C([0,T];\, L^2(\Omega))} \qquad (A.12)$$

we obtain after using assumption (A.0) as well

$$\int_{\Sigma_1} \left[\frac{\partial\phi}{\partial\nu}\right]^2 h\cdot\nu\, d\Sigma_1 + \text{const}_{B,h,\varepsilon}\|\phi\|^2_{C([0,T];\, L^2(\Omega))} \ge a\int_0^T E(t)\,dt - b\,(E(T)+E(0)) \qquad (A.13)$$

$$a = \rho - \|B\|\, M_h - \varepsilon \qquad b = M_h + \varepsilon \qquad (A.14)$$

where, by assumption (A.4), the coefficient a is positive for ε sufficiently small. Integrate by parts (1.4) premultiplied by ϕ_t and recall (A.3) to have $0 \ge \frac{dE}{dt} = 2\,\text{Re}\,(B\phi_t, \phi_t)_\Omega \ge -2C_B\|\phi_t\|^2 \ge -2C_B E(t)$ and hence

$$E(0) \ge E(t) \ge e^{-2C_B t} E(0) \qquad (A.15)$$

the new relationship on E(t) which replaces the conservative behavior used in the undamped case, e.g. [T.1]. (A more refined analysis was used in the special case $B = \gamma I$ in Theorem 1.3) Since a in (A.14) is positive by assumption (with ε small) and b in (A.14) is positive, we use (A.15) in (A.13) and obtain

$$\text{R.H.S. of (A.13)} \ge f(T)\,E(0);\quad f(T) = a\frac{1-e^{-2C_B T}}{2C_B} - 2b. \qquad (A.16)$$

To guarantee, as desired, that f(T) which is strictly increasing in T, be positive for sufficiently large T, it suffices to require that its limit $a/2C_B - 2b$ be positive; i.e. that $a - 4\,C_B b > 0$. This is in turn achieved provided $\|B\| < \frac{\rho}{5M_h}$, as summed.

Thus, with $f(T) > 0$, we return to (A.13) and (A.16) and obtain

$$\int_{\Sigma_1} \left[\frac{\partial\phi}{\partial\nu}\right]^2 h\cdot\nu\, d\Sigma_1 + \text{const}_{B,T,\varepsilon}\|\phi\|^2_{C([0,T;\, L^2(\Omega))} \ge f(T)\,E(0) \qquad (A.17)$$

The final step to obtain inequality (3.4) is

Lemma A.2. Inequality (A.17) (with $f(T) > 0$) implies: there is $k_T > 0$ such

$$\|\phi\|^2_{C([0,T];\, L^2(\Omega))} \le k_T \int_0^T \int_{\gamma_1} \left[\frac{\partial\phi}{\partial\nu}\right]^2 d\Sigma_1 \qquad (A.18)$$

Proof As, in the undamped case [L.1], [L.2], [L-T.4], [L-T.5], [L-T.6] based on the uniqueness property (A.5) to obtain a contradiction. □

References

[C-T.1] Chen, S. and R. Triggiani, "Proof of Two Conjectures of G. Chen and D. L. Russell on Structural Damping for Elastic Systems: the case: $\alpha = ½$", Proceedings of the Seminar in Approximation and Optimization held at University of Habana, Cuba, January 12-14, 1987; Springer-Verlag Lectures Notes in Mathematics, 1988.

[C T.2] Chen, S. and R. Triggiani, "Proof of extensions of two conjectures on structural damping for elastic systems: the case $\frac{1}{2} \leq \alpha \leq 1$," *Pacific J. of Mathematics,* to appear.

[C-T.3] Chen, S. and R. Triggiani, "Differentiability of semigroups arising with gentle dissipation: the case $0 < \alpha < \frac{1}{2}$", to appear.

[H.1] Ho, L. F., "Observabilite frontiere de l'equation des ondes," CRAS, vol. 302, Paris 1986.

[H.2] Hormander, L., *Linear partial differential operators,* Springer-Verlag 1969.

[L.1] Lions, J. L., "Exact controllability, stabilization and perturbations," SIAM Review, 1986.

[L.2] Lions, J. L., *Controllability exacte, perturbations et stabilization de systemes distribues,* 1 and 2, Masson, 1988.

[L-L.1] Lagnese, J. and J. L. Lions, *Modeling, analysis and control of thin plates,* Masson, 1988.

[L-L-T.1] Lasiecka, I., J. L. Lions, and R. Triggiani, "Non homogeneous boundary value problems for second order hyperbolic operators," *J. Mathem. Pures at Appliquees,* vol. 69, pp. 149-192, 1986.

[L-T.1] Lasiecka, I. and R. Triggiani, "A cosine operator approach to modeling $L_2(0,T; L_2(\Gamma)$ - boundary input hyperbolic equations," *Applied Math and Optimiz.,* vol. 7, pp. 35-83, 1981.

[L-T.2] Lasiecka, I. and R. Triggiani, "Regularity of hyperbolic equations under $L_2(0,T; L_2(\Gamma)$ - boundary terms," *Applied Math and Optimiz., vol. 10, pp. 275-286, 1983.*

[L-T.3] Lasiecka, I. and R. Triggiani, "Uniform exponential energy decay of the wave equation in a bounded region with," *J. Diff. Eqts.,* vol. 66, pp. 340-390, 1987.

[L-T.4] Lasiecka, I. and R. Triggiani, "Exact controllability for the wave equation with Neumann boundary control," *Applied mathem. & Optimiz.,* to appear.

[L-T.5] Lasiecka, I. and R. Triggiani, "Exact controllability of the Euler-Bernoulli equation with controls in the Dirichlet and Neumann boundary conditions: a non-conservative case," *SIAM J. Control & Optimization,* to appear.

[L T.6] Lasiecka, I. and R. Triggiani, "Exact controllability for a class of Euler-Bernoulli equation with boundary controls for displacement and moments," *J. Mathem. Analysis & Applications,* to appear.

[T.1] Triggiani, R., "Exact boundary controllability on $L^2(\Omega) \times H^{-1}(\Omega)$ for the wave equation with Dirichlet control acting on a portion of the boundary, and related problems", *Applied Math. and Opt.,* vol. 18, pp. 241-277, 1988. Also Springer-Verlag Lectures Notes in Control Sciences, Vol. 102, pp. 292-332, Proceedings of 3rd International Conference, Vorau, Austria, July 6-12, 1986. Also in Recent Advances in Communication and Control Theory,honoring the sixtieth anniversary of A. V. Balakrishnan (R.E. Kalman and G. I. Marchuk, Eds.), pp. 262-286, Optimization Software (New York).

[T.2] Triggiani, R., *Improving Stability Properties of Hyperbolic Damped Equations by Boundary Feedback,* pp. 400-409, Springer-Verlag Lecture Notes LNCIS, 1985.

[Z.1] Zuazua, E., private communication.

Some Selected Problems of Optimal Control

Jack Warga, Professor, Department of Mathematics, Northeastern University, Boston, Massachusetts

1. INTRODUCTION.

At a conference celebrating "Thirty years of optimal control" it would seem appropriate to briefly review a number of problems and concepts that mark important stages in the development of the theory of deterministic optimal control. The selection is quite subjective and reflects some of the author's interests, past and present. However, most of the problems described below had at various times attracted the attention of many outstanding

---

Partially supported by N.S.F. grant DMS 88619002

researchers, and their importance seems undisputed.

The term "optimal control" seems to have originated with the Pontryagin school and to have at first applied to problems defined by ordinary differential equations. Typically, one considered the (vector) differential equation

$$x(t) = a + \int_{t_0}^{t} f(s,x(s),u(s))ds \quad \forall\ t \in T := [t_0, t_1] \qquad (1)$$

with associated endpoint (isoperimetric) restrictions, say of the form $h_1(x(t_1)) = 0 \in \mathbb{R}^m$. Given a (usually compact) set $U \subset \mathbb{R}^k$, the problem was to determine a (Lebesgue) measurable (control) function $u: T \to U$ and an absolutely continuous $x: T \to \mathbb{R}^n$ that minimize a cost functional, say of the form $h_0(x(t_1))$, on the set $\mathcal{A}$ of (admissible) couples (x,u) that satisfy the differential equation (1) and the endpoint restrictions. Later, much more general functional equations and restrictions were included in the area of research but, even today, the term "optimal control" is still associated in many people's minds with ordinary differential equations.

2. EXISTENCE OF OPTIMAL SOLUTIONS. RELAXED CONTROLS.

The first existence theorems in the calculus of variations were derived by L. Tonelli [30] around 1910-1911 (characteristically, only about ten years after Hilbert attacked the Dirichlet problem by use of weak solutions that, in another form, also turned out to hold the key to existence theorems in the calculus of variations and in optimal control). Much more general variational problems were considered in 1937 by L.C. Young [49] who introduced the concept of "generalized curves". In the context of optimal control (of ordinary differential equations), the first existence theorem was derived in 1959 by A.F. Filippov [11], then, independently, by E. Roxin [26], J. Warga [32] and T. Ważewski [48] in 1962 and A. Ghouila-Houri [15] in 1967. The critical step in all of these endeavors was the assumption of the convexity (or the convexification) of the set $f(t,x,U)$ related to the differential equation (1).

Once appropriate assumptions (concerning f, h_0, h_1, the compactness of U, and the uniform boundedness and equicontinuity of the solutions $x(\cdot)$ of (1)) are made, one can consider a minimizing sequence $((x_i,u_i))$ in $\mathcal{A}$, that is, such that

$$\lim_i h_0(x_i(t_1)) = \inf \{ h_0(x(t_1)) \mid (x,u) \in \mathcal{A} \}.$$

While the sequence (x_i) admits a uniformly convergent subsequence, many examples show that the functions u_i may become rapidly oscillating and have no subsequence that converges in any conventional sense. However, if some u_i "jumps" rapidly between values $r_1,\ldots,r_k$ in U over some small interval $[t,t+\eta]$ with relative frequencies $\alpha_1,\ldots,\alpha_k$ then

$$\int_t^{t+\eta} f(s,x(s),u_i(s))ds \cong \int_t^{t+\eta} \Sigma_j\, \alpha_j f(s,x(s),r_j)ds,$$

and thus $u_i(s)$ acts like the probability measure with masses α_j at r_j. This suggests that we replace the "original" control function $s \to u(s)$ by a "relaxed" control function $s \to \sigma(s)$, where $\sigma(s)$ is an element of the set rpm(U) of Radon probability measures on U, and that the differential equation (1) be replaced by the "relaxed" equation

$$x(t)= a + \int_{t_0}^{t} f(s,x(s),\sigma(s))ds \quad \forall\, t \in T, \tag{2}$$

where

$$f(s,x(s),\sigma(s)) := \int f(s,x(s),r)\sigma(s)(dr).$$

With an appropriate "weak norm" topology, the set $\mathcal{S}$ of relaxed controls is a convex and compact subset of a Banach space. We can embed the set $\mathcal{R}$ of original controls in $\mathcal{S}$ by identifying $u(s)$ with the Dirac measure $\delta_{u(s)}$ concentrated at $u(s)$ with probability 1. It turns out that $\mathcal{R}$ is dense in $\mathcal{S}$ and the solution $t \to x(t)$ of equation (2) depends continuously on σ. The continuity of $\sigma \to x(\sigma)(\cdot):\mathcal{S} \to C(T,\mathbb{R}^n)$ and the compactness of $\mathcal{S}$ immediately yield quite general existence theorems. In particular, the existence of an optimal solution is assured even if we add to the (finite-dimensional) endpoint condition a new (infinite-dimensional) "unilateral" condition of the form $h_2(t,x(t)) \in A \ \ \forall\ t \in T$, where $h_2: T \times \mathbb{R}^n \to \mathbb{R}^{m'}$ is continuous and $A \subset \mathbb{R}^{m'}$ closed.

Relaxed controls offer other advantages in addition to compactness and existence theorems. Linear variations of relaxed controls in the convex set $\mathcal{S}$ - analogous to the weak variations of the calculus of variations - yield necessary conditions for minimum (including a generalized Weierstrass E-condition = Pontryagin's maximum principle) that would otherwise require the much more complicated use of strong (needle) variations of ordinary control functions. Furthermore, the use of relaxed controls enables us to obtain controllability theorems and related necessary conditions for minimum in any "abundant" set of ordinary control functions. An abundant set $\mathcal{U}$ [36] is defined by the property that, for every choice of $k \in \{1,2,\dots\}$, $\sigma_0,\dots,\sigma_k \in \mathcal{S}$, and

$$\theta \in \mathcal{T}_k := \{\ (\theta_1,\dots,\theta_k)\ \mid\ \theta_j \geq 0,\ \Sigma_j\ \theta_j \leq 1\ \},$$

there exists a sequence $(u_n(\theta))$ in $\mathcal{U}$ such that

$$\lim_n\ u_n(\theta) = \sigma_0 + \Sigma_j\ \theta_j(\sigma_j - \sigma_0) \quad \text{uniformly for } \theta \in \mathcal{T}_k\ ,$$

$$\theta \to u_n(\theta): \mathcal{T}_k \to \mathcal{S} \quad \text{is continuous for each } n = 1,2,\dots\ .$$

One can show that, in particular, any set of controls closed under finite measurable concatenations ("splicings") is abundant. Thus the sets of all Lebesgue measurable or of all simple controls are abundant. Separate, and relatively easy, arguments show that other sets, such as those of all continuous or piecewise continuous controls, are also abundant.

The term "relaxed" was introduced by Warga [32] in 1962 but its fairly general acceptance is without doubt due to its subsequent adoption by McShane [24].

3. UNILATERAL PROBLEMS.

Since the early days of optimal control researchers had sought to generalize the maximum principle to unilateral problems (also referred to as problems with bounded phase coordinates or with state inequality restrictions). We have already defined these problems (for the control of ordinary differential equations) as the search for a minimum of $h_0(x(t_1))$ subject to equation (1) (respectively (2)) and the restrictions

$$h_1(x(t_1)) = 0 \in \mathbb{R}^m , \quad h_2(t,x(t)) \in A \quad \forall\ t \in T .$$

The first rigorous derivation of the maximum principle for such problems, in 1959, is due to R.V. Gamkrelidze [14]. However, this first derivation was based on the assumption that the optimal control satisfies certain regularity conditions but without providing any criteria for verifying this assumption. Without such regularity assumptions, the maximum principle for original unilateral problems was first derived in 1963-1965 by Dubovitskii and Milyutin [8,9] and for relaxed unilateral problems in 1964-1965 by Warga [33,34]. There were literally dozens of other derivations since that time, most of them for C^1 problems in which the functions $f(t,\cdot,u)$, h_0, h_1, and $h_2(t,\cdot)$ are assumed to be continuously differentiable.

The early derivations of the maximum principle for unilateral problems were long and complicated. We shall illustrate how much better the situation is now understood by proving, in a few lines, a simplified version of the principle for unilateral problems without the endpoint restriction $h_1(x(t_1)) = 0$. Specifically, let $\varphi_0(\sigma) := h_0(x(\sigma)(t_1))$, $\varphi_2(\sigma)(t) := h_2(t,x(\sigma)(t))$ $\forall\ \sigma \in \mathcal{S}$, $t \in T$, and let σ_0 yield the minimum of $\varphi_0(\sigma)$ on the set $\varphi_2^{-1}(C)$, where

$$C := \{ c \in C(T,\mathbb{R}^{m'}) \mid c(T) \subset A \}$$

and A is a convex subset of $\mathbb{R}^{m'}$ with a nonempty interior. We denote by C° the interior of C, by $C(T,\mathbb{R}^{m'})^*$ the topological dual of

$C(T,\mathbb{R}^{m'})$, and by $D\varphi_i(\sigma_0:\sigma-\sigma_0)$ the directional derivative

$$\lim \alpha^{-1}[\varphi_i(\sigma_0+\alpha(\sigma-\sigma_0)) - \varphi_i(\sigma_0)] \text{ as } \alpha \to 0+$$

which we assume to be linear and continuous in σ over $\mathcal{S}$.

THEOREM. There exists $\ell = (\ell_0,\ell_2) \in [0,\infty) \times C(T,\mathbb{R}^{m'})^*$ such that $\ell \neq 0$,

(a) $\ell\, D(\varphi_0,\varphi_2)(\sigma_0:\sigma-\sigma_0) \geq 0 \quad \forall\, \sigma \in \mathcal{S}$,

(b) $\ell_2\, \varphi_2(\sigma_0) = \max\{\, \ell_2\, c \mid c \in C \,\}$.

(Relation (a) is an "integrated" form of the maximum principle.)

PROOF. Let

$W := \{D(\varphi_0,\varphi_2)(\sigma_0:\sigma-\sigma_0) \mid \sigma \in \mathcal{S}\}, \quad Z := (-\infty,0) \times [C^\circ - \varphi_2(\sigma_0)].$

Then W and Z are convex, their closures contain 0, and Z° is nonempty. If $W \cap Z$ is empty then, by the convex separation theorem, there exists $\ell = (\ell_0,\ell_2) \in \mathbb{R} \times C(T,\mathbb{R}^{m'})^*$ such that

$$\ell \neq 0, \quad \ell_0 z_0 + \ell_2 z_2 \leq 0, \quad \ell w \geq 0 \quad \forall\ (z_0,z_2) \in \bar{Z}, \quad w \in \bar{W}.$$

This implies that $\ell_0 \geq 0$ and relations (a) and (b) are valid. If W and Z have a point in common then there exists $\sigma_1 \in \mathcal{S}$ such that

$$D\varphi_0(\sigma_0:\sigma_1-\sigma_0) < 0, \quad D\varphi_2(\sigma_0:\sigma_1-\sigma_0) \in C^\circ - \varphi_2(\sigma_0)$$

which implies that

$$\varphi_0(\sigma_0 + \alpha(\sigma_1-\sigma_0)) < \varphi_0(\sigma_0), \qquad \varphi_2(\sigma_0 + \alpha(\sigma_1-\sigma_0)) \in C^\circ$$

for all sufficiently small positive α, contradicting the assumption that σ_0 is optimal. Q.E.D.

We also observe that the above proof is applicable if σ_0 is optimal among original controls. Indeed, we may approximate $\sigma_0 + \alpha(\sigma_1-\sigma_0)$ by original controls, again showing contradiction. In the case when we do not discard endpoint restrictions, the proof is conceptually similar but technically much more involved.

4. OPTIMAL CONTROL OF FUNCTIONAL-INTEGRAL EQUATIONS.

The formalism of optimal control, and specifically the distinction between state and control functions, was introduced in the period following World War II in response to a number of engineering optimization problems which were most naturally expressed in this form. This was particularly the case because of the inequality restrictions that engineering reality imposed on the control functions. This new formalism was so natural and convenient that known theorems of the calculus of variations, due to Valentine [31], that could have applied to many of these problems, were ignored or forgotten. However, the outcome was another demonstration of the often made observation that "it helps sometimes to be a little ignorant". Because the problem was studied in its control form, an approach was developed that could be relatively easily extended to problems defined by more general functional equations. Leaving aside problems defined by partial differential equations (which require a very different and highly specialized approach in both variational and control formulations), the methods of the optimal control of ordinary differential equations were generalized to problems defined by controlled difference-differential and delay-differential equations, of which the former have the form

$$x(t)= a + \int_{t_0}^{t} f(s,x(s),x(s-\beta_1),\ldots,x(s-\beta_p),u(s))ds \quad \forall\ t \in T \qquad (4)$$

where $0 < \beta_1 < \ldots < \beta_p$ and $x(\cdot)$ is given on the interval $[t_0-\beta_p, t_0]$. In fact, an existence theory and first order necessary conditions entirely analogous to those described in sections 2 and 3 were developed [36, Chaps. VII and VIII] for problems in which the domain T of the state and control functions can be any compact metric space with a positive nonatomic Radon measure μ and equation (2) is replaced by

$$x(t) = \int f(t,s,\xi(x)(s),\sigma(s))\mu(ds) \quad \forall\ t \in T,$$

where x is an element of either $C(T, \mathbb{R}^n)$ or some $L^p(\mu,\mathbb{R}^n)$ and ξ is a C^1 function between two appropriate spaces among those just mentioned. This functional-integral equation includes the delay-differential equations as special cases.

In a number of problems a further complication is introduced, with delays or more general transformations applied to the original control functions as well as to state functions. For one class of such problems the old theory can be generalized [37]: these are problems with separated (i.e. additively coupled) delays in controls, e.g. when equation (4) is replaced by

$$x(t)= a + \int_{t_0}^{t} \Sigma_j f_j(s,x(s),x(s-\beta_1),\ldots,x(s-\beta_p),u(s-\beta_j))ds \quad \forall\ t \in T$$

If the delays are not separated but are all integral multiples of the same number (or composition powers of the same variable delay) then a transformation enables us to reduce the problem to a manageable form [35]. In other cases the task facing us is much more formidable and no existence theory has as yet crystallized.

5. PROBLEMS WITH CONFLICTING CONTROLS.

These are problems in which two players are involved, each controlling his own control function, and with conflicting goals. The best known class of such problems, first formulated and studied by R. Isaacs [18], are the *differential games* involving (for games of fixed duration) an ordinary differential equation of the form

$$x(t)= a + \int_{t_0}^{t} f(s,x(s),u(s),v(s))ds \quad \forall\ t \in T:= [t_0,t_1], \qquad (5)$$

where u is controlled by player I and v by player II. At any moment t each of the players knows the past history of x, u and v and must choose the current value of his control. Player I seeks to minimize the "payoff function", say $h_0(x(t_1))$, while II seeks to maximize it.

The theory of differential games is still in a state of flux, partly because of the difficulties in formulating the "rules of the game" in such a manner that equation (5) admit a solution. (In particular, how can we prevent a situation in which the choices of u(t) and v(t) at various times t result in the integrand in (5) being nonmeasurable?) In studying necessary conditions, one of the assumptions that most (or all?) of the researchers in the field (e.g. Isaacs [18,19], Berkovitz [1,2,3], Friedman [12], Elliott and Kalton [10], and others) have found to be indispensable (at least in the case of original controls) is the "separation" of u and v (essentially required by the "Isaacs condition"), i.e. the assumption that

$$f(s,x(s),u(s),v(s)) = f_1(s,x(s),u(s)) + f_2(s,x(s),v(s)),$$

an assumption analogous to one that we mentioned in connection with delayed controls.

Another class of problems with conflicting controls [36, Chaps. IX and X] involves, as a special case, equation (5) but with the flow of information favoring player II. In these problems, player I must disclose in advance the choice of u and, with this knowledge in hand, player II determines the choice of v. For problems of this kind, it is no longer necessary to assume that u and v interact through an ordinary differential equation. In fact, one might consider a functional-integral equation such as

$$x(t) = \int f(t,s,\xi(x)(s),u(s),v(s))\mu(ds) \quad \forall \quad t \in T ,$$

again with T an arbitrary compact metric space and μ nonatomic.

A general conflicting control problem of this kind can be defined by functions $\varphi_0(u)$, $\varphi_1(u)$, $\varphi_2(u,v)$ defined for all u and v in appropriate sets $\mathcal{U}$ and $\mathcal{V}$ of original controls, where φ_0 and φ_1 are defined, say, as in section 3 while φ_2 is similarly defined with u replaced by (u,v). The optimization problem is to minimize φ_0 on $\mathcal{U}$ subject to the restrictions

$$\varphi_1(u) = 0 \in \mathbb{R}^m, \quad \varphi_2(u,v) \in A \subset \mathbb{R}^{m'} \quad \forall\ v \in \mathcal{V} .$$

(A special case of this problem is the determination of $\inf_u \sup_v f(u,v)$). In the special case when the conflicting controls are separated, one can establish existence theorems based on embedding u and v in the corresponding sets of relaxed controls, and obtain first order conditions generalizing the maximum principle. Furthermore, the relaxed problem is then the "limit" of related original problems. However, simple examples show that the latter is no longer the case when u and v are nonadditively coupled, say, when we have terms involving the scalar product of u(s) and v(s). To handle this more complicated situation, that is, to appropriately compactify the sets of control functions so that the new "compactified" problem be a limit of original problems, we must introduce both a relaxed version of u and a "hyperrelaxed" version of v.

We shall use a simple example to illustrate the situation. Let

$$U = \{ (z_1, z_2) \in \mathbb{R}^2 \mid z \cdot z := z_1^2 + z_2^2 = 1 \}, \qquad T = [0,1],$$

and let $\mathcal{U} = \mathcal{V}$ be the set of all (Lebesgue) measurable $u: T \to U$. Our problem is to determine

$$a := \inf_u \sup_v \int_0^1 u(s) \cdot v(s)\, ds .$$

Clearly, for any choice of u and s, $v(s) = u(s)$ yields the maximum of $u(s) \cdot v(s)$, and thus $a = 1$. If we were to replace u by its relaxed version $\sigma: T \to \mathrm{rpm}(U)$ and choose

$$\sigma(s) = \frac{1}{2}\, \delta_{(0,1)} + \frac{1}{2}\, \delta_{(0,-1)} \qquad \forall\ s \in T,$$

where δ_z is the Dirac measure with probability 1 at z, then

$$\sup_v \int_0^1 ds \int r \cdot v(s)\ \sigma(s)(dr) = 0,$$

showing that the new problem is not a "limit case" of the old one. In replacing u(s) by the probability measure $\sigma(s)$ we have denied the player controlling v the knowledge of the exact location of u(s). In order to restore this lost advantage due to v, we must allow v(s) to vary with the variable r of integration with respect to $\sigma(s)$ and thus consider

$$a' := \inf_{\sigma} \sup_{v} f(\sigma, v),$$

where

$$f(\sigma, v) := \int_0^1 ds \int r \cdot v(s,r)\, \sigma(s)(dr).$$

Clearly, the maximizing value of $v(s,r)$ is r for all s and r, yielding $a' = a = 1$.

In a general situation, a point-valued $(s,r) \to v(s,r)$ may be unable to yield a maximum, and we must therefore be ready to replace v by a function $\pi: T \times U \to \mathrm{rpm}(V)$, where V is the range of the original function $v: T \to V$. We refer to such Borel measurable π as "hyperrelaxed controls". In this new context, with u embedded in its corresponding set of relaxed controls and v in the corresponding set of hyperrelaxed controls, we obtain existence theorems, show that the compactified problem is a limit case of the original problem, and derive first order necessary conditions for minimum that generalize the maximum principle and, incidentally, bear some formal resemblance to optimality conditions for differential games.

6. NONSMOOTH OPTIMAL CONTROL.

Let us return to the optimal control problem of minimizing $h_0(x(t_1))$ subject to

$$x(t) = a + \int_{t_0}^{t} f(s, x(s), u(s))ds \quad \forall\, t \in T := [t_0, t_1] \tag{6}$$

and the endpoint restriction $h_1(x(t_1)) = 0$. If we assume that h_0, h_1 and $f(t,\cdot,\cdot)$ are continuous, $f(\cdot, v, r)$ measurable, and the solutions of equation (6) that yield small values of $|\, h_1(x(t_1))|$ uniformly bounded and equicontinuous then the use of relaxed controls yields the existence of an optimal relaxed solution. If we assume, furthermore, that the solutions x of equation (6) are unique then we can show that relaxed controls can be "imitated" by original controls (and there are simple examples showing that uniqueness cannot be just dispensed with). However, in order to even formulate the classical form of the maximum principle, one must assume that $f(t,\cdot,r)$ is differentiable, and the proofs even assume that it is C^1. Since, in a number of important control problems, we cannot make the differentiability assumptions, and we wish to use necessary conditions for minimum in the search for candidates, the question arises whether the maximum principle cannot be replaced by alternate formulations that do not require derivatives. The first steps in this direction were made by researchers using the methods of convex analysis and dealing with "convex" problems characterized by the property that local minima are also global and first order necessary conditions are also sufficient. For "nondifferentiable and nonconvex" problems, the earliest results were derived in 1973 by E.L. Kugušev [22], F.H. Clarke [5], and J. Warga [38], using quite different approaches. Since that time a large number of researchers have contributed to the field of *nonsmooth analysis* which has expanded beyond optimal control to lay the basis of a calculus of nondifferentiable functions.

The basic problem and the first task of nonsmooth analysis is to define a substitute for the derivative when the latter does not exist. The reader's first reaction will be to observe that such generalized derivatives are, by now, an old tool of analysis in the form of distributional and weak derivatives. However, the latter reflect global properties of functions acting as operators while optimization deals largely with local properties. The first, and probably the simplest, example of such a local generalized derivative is the *subgradient* $\partial f(x)$ of a convex function f at $x \in \mathbb{R}^n$. It can be defined as the set of all p such that $y \to f(y) - p\cdot y$ is minimized by $y = x$. An equivalent definition of $\partial f(x)$ is as the convex hull of the set of all cluster points of derivatives $f'(y)$ as y converges to x in the set where $f'(y)$ are defined. Clarke [7] uses the latter formulation to define the *generalized Jacobian* $\partial f(x)$ of a locally Lipschitzian $f:\mathbb{R}^n \to \mathbb{R}^m$. A somewhat more general definition is that of Warga's *derivate containers* $\Lambda f(x)$ which, however, unlike the generalized Jacobian, are not uniquely defined for a given f and x. A derivate container is defined by $\Lambda f(x) = \bigcap_{\varepsilon>0} \Lambda^{\varepsilon} f(x)$, where $\Lambda^{\varepsilon} f(x)$ are sets such that $\Lambda^{\varepsilon} f(x) \subset \Lambda^{\varepsilon'} f(x)$ if $\varepsilon < \varepsilon'$ and there exists a sequence (f_i) of C^1 functions converging uniformly to f in some neighborhood of x with $f_i'(y) \in \Lambda^{\varepsilon} f(x)$ if $|y-x| \leq \varepsilon$ and $i \geq 1/\varepsilon$. A derivate container $\Lambda f(x)$ is compact but, unlike Clarke's generalized Jacobian $\partial f(x)$, it need not be convex, and some derivate containers may be properly contained in $\partial f(x)$. Furthermore, $\partial f(x)$ is itself a special derivate container. On the other hand, $\partial f(x)$ is contained in the convex hull of every derivate container [42].

Even more general definitions of generalized set-valued derivatives of functions between finite- or infinite-dimensional Banach spaces have been proposed and studied (see e.g. Clarke [6,7], Ioffe [17], Mordukhovič [25], Warga [45]). One of the main reasons for this lack of uniformity is that different generalized derivatives serve different purposes and provide the framework for sometimes more general and at other times more detailed theorems. However, many of these generalized derivatives $\Delta f(x)$ provide similar basic theorems of the nonsmooth calculus. Thus, if x minimizes f locally then $0 \in \Delta f(x)$, and if all the elements of $\Delta f(x)$ have uniformly bounded inverses then f is a local homeomorphism at x.

Starting with the early work of Kugušev, Clarke, and Warga, various derivations of the "nonsmooth" maximum principle and of related controllability conditions have been obtained for unilateral problems with controlled differential equations, differential inclusions, functional-integral equations, etc.

7. HIGHER ORDER CONTROLLLABILITY AND NECESSARY CONDITIONS.

We shall find it convenient to deal with a model defined by a convex subset Q of a topological vector space, an abundant subset $\mathcal{U}$ of Q, a convex set C with a nonempty interior in a topological vector space $\mathcal{Z}$, a function $\varphi = (\varphi_1, \varphi_2): Q \to \mathbb{R}^m \times \mathcal{Z}$, and a point $q_0 \in Q$ with $\varphi_2(q_0) \in C$. As an example of how this model applies to optimal control problems we may consider the unilateral control problem in which Q represents the set $\mathcal{S}$ of relaxed controls, $\mathcal{U}$ an abundant set of original controls, $x(\sigma)$ the solution of a differential or a functional-integral equation, and

$$\varphi_1(\sigma) = (h_0, h_1)(x(\sigma)(t_1)), \qquad \varphi_2(\sigma)(t) = h_2(t, x(\sigma)(t)),$$
$$C = \{ c \in C(T, \mathbb{R}^{m'}) \mid c(T) \subset A \}.$$

Controllability questions relate to the (generalized) *attainable* sets

$$\mathrm{Att}(\mathcal{V}, G) := \{ \varphi_1(q) \mid q \in \mathcal{V},\ \varphi_2(q) + G \subset C \}, \quad \mathrm{Att}(\mathcal{V}) = \mathrm{Att}(\mathcal{V}, \{0\}),$$

where $\mathcal{V} = \mathcal{U}$ or $\mathcal{V} = Q$ and G will henceforth denote a neighborhood of 0 in $\mathcal{Z}$. Some of these questions are:

(Controllability [44]) Is $\varphi_1(q_0) \in \mathrm{Att}(\mathcal{V}, G)^\circ$ for some G?

(Conical controllability [46]) Is some truncated open cone with vertex at $\varphi_1(q_0)$ and with axis parallel to a given vector w contained in $\mathrm{Att}(\mathcal{V}, G)$?

(Directional controllability [47]) Is some segment originating at $\varphi_1(q_0)$ and parallel to a given vector w contained in $\mathrm{Att}(\mathcal{V})$?

The first of these questions is closely related to the maximum principle and associated "transversality" conditions. In fact, the usual proofs of the maximum principle (whose basic ideas go back to McShane's seminal paper [23]) essentially show that if the maximum principle is not satisfied then $\varphi_1(q_0) \in \mathrm{Att}(\mathcal{V})^\circ$. However, as many examples demonstrate, the latter may still be the case even if the maximum principle holds, this principle being a necessary but not a sufficient condition for minimum. It is, therefore, natural to follow the lead of the finite-dimensional analysis and of the calculus of variations and to investigate second and higher order conditions for minimum or controllability. This had been done extensively for "free" optimal control problems (without endpoint or unilateral restrictions) [13], much less so for problems with endpoint restrictions (Jacobson [20], Krener [21], Warga [39-41]), and only recently for problems as modelled here, with both endpoint and unilateral restrictions (Bernstein [4], Gilbert and Bernstein [16], Warga [43,44,46,47]).

If we consider any control problem then it is clear that q_0 cannot be optimal if we have controllability at q_0. Thus the negation of a sufficient condition for controllability provides a necessary condition for minimum, and this is essentially the case with the maximum principle for first order conditions. In the latter case, the set V of variations of (φ_1, φ_2) for the relaxed problem is convex, and a sufficient condition for controllability is that the origin be an interior point of V. Since $0 \in V$, the negation of it is that 0 belongs to the boundary ∂V of V, and one can show then that there exists a Lagrange coefficient ℓ (a support functional to V at 0) that is required for the maximum principle. The existence of ℓ is thus equivalent to the statement $0 \in \partial V$. However, if we assume that the maximum principle is satisfied at q_0 and establish second or higher order conditions for controllability then the situation is different. The nonlinear set V of higher order variations is not generally convex and, in order to define a Lagrange coefficient (a support functional to a set of variations), one is constrained to choose convex subsets of V, thus sacrificing a great deal of information. This is effectively what happened in the early work on higher order conditions even though, as it appears, we were all unaware of it. The alternative is to apply controllability theorems not involving Lagrange coefficients. This is all the more justified by the way higher order necessary conditions are generally used: they do not serve to find new candidates for minimum but rather to discard some of those provided by lower order conditions, a goal that can also be achieved by applying controllability theorems.

We shall complete this section by providing an example of a

simple second order conical controllability criterion at q_0. Let $\mathbf{e}_1,\ldots,\mathbf{e}_k$ be the columns of the unit $k \times k$ matrix,

$$\mathcal{T}_k := \{ (\theta_1,\ldots,\theta_k) \mid \theta_j \geq 0, \Sigma_j \theta_j \leq 1 \}, \qquad y_1,\ldots,y_k \in Q - q_0,$$

$$\psi(\theta) := (\psi_1,\psi_2)(\theta) = (\varphi_1,\varphi_2)(q_0 + \sum_j \theta_j y_j) \qquad \forall\, \theta \in \mathcal{T}_k ,$$

and assume that ψ has a second order Taylor approximation at 0 (relative to $\mathcal{T}_k$), that is, there exist a linear operator $\psi'(0)$ and a bilinear symmetric operator $\psi''(0)$ such that

$$\lim |\theta|^{-2}\{ \psi(\theta) - [\psi(0) + \psi'(0)\theta + \psi''(0)\theta^2] \} = 0 \text{ as } \theta \to 0 \text{ in } \mathcal{T}_k.$$

THEOREM. Let $I := \{1,2,,\ldots,i_1\}$, $J := \{i_1+1,\ldots,k\}$, $w \in \mathbb{R}^m$, and assume that, for all $i \in I$, $j \in J$,

(a) $\psi'(0)\mathbf{e}_i \in \{0\} \times [C - \varphi_2(q_0)]$,

(b) $\psi''(0)(\sum_{i\in I} \mathbf{e}_i)^2 + 2\, \psi'(0) \sum_{j\in J} \mathbf{e}_j \in \{w\} \times [C - \varphi_2(q_0)]^\circ$,

(c) the vectors $\psi''(0)(\Sigma_i \mathbf{e}_i)\mathbf{e}_s$, $\psi'(0)\, \mathbf{e}_j$ for $s \in I$, $j \in J$ span $\mathbb{R}^m$.

Then there exist $\gamma_0 > 0$ and neighborhoods G_1, G_2 of 0 in $\mathbb{R}^m, \mathcal{Z}$ such that

$$\varphi_1(q_0) + \gamma(w + G_1) \subset \{ \varphi_1(u) \mid u \in \mathcal{U}, \quad \varphi_2(u) + \gamma G_2 \subset C \}$$

for all $\gamma \in (0,\gamma_0]$. In particular, if $w=0$, then $\varphi_1(q_0) \in \mathrm{Att}(\mathcal{U})^\circ$.

REFERENCES.

1. L. D. Berkovitz, Advances in Game Theory, Ann .Math.Stud.,No. 52, Princeton University Press, Princeton, N.J., 1964, 127-174

2. L. D. Berkovitz, Necessary conditions for optimal strategies in a class of differential games and control problems, SIAM J. Control 5 (1967),1-24

3. L. D. Berkovitz, The existence of value and saddle point in games of fixed duration, SIAM J. Control Optim.23 (1985),172-196; Erratum and addendum, ibid. 26 (1988), 740-742

4. D. S. Bernstein, A systematic approach to higher order necessary conditions in optimization theory, SIAM J. Control Optim. 22 (1984), 211-238

5. F. H. Clarke, Necessary conditions for nonsmooth problems in optimal control and calculus of variations, Ph.D. thesis, University of Washington, 1973

6. F. H. Clarke, Generalized gradients of Lipschitz functionals, Adv. Math. 40 (1981), 52-67

7. F. H. Clarke, Optimization and Nonsmooth Analysis, John Wiley and Sons, New York, 1983

8. A. Y. Dubovitskii and A. A. Milyutin, Extremum problems with constraints, Dokl. Akad. Nauk SSSR 149 (1963), 759-762 (Soviet Math. Dokl. 4 (1963), 452-455)

9. A. Y. Dubovitskii and A. A. Milyutin, Extremum problems in the presence of restrictions, Zh. Vychisl. Mat. i Mat. Fiz. 5 (1965), 395-453 (USSR Comp. Math. and Math. Phys. 5 (1965), No. 3, 1-80)

10. R. J. Elliott and N. J. Kalton, The existence of value in differential games, Mem. Amer. Math. Soc. 126 (1972)

11. A. F. Filippov, On certain questions in the theory of optimal control, Vestnik Moskov. Univ. Ser. Mat. Mekh. Astr. 2 (1959), 25-32 (SIAM J. Control 1 (1962), 76-84)

12. A. Friedman, Differential Games, John Wiley and Sons, New York, 1971

13. R. Gabasov and F. M. Kirillova, High order necessary conditions for optimality, SIAM J. Control 10 (1972), 127-168

14. R. V. Gamkrelidze, Optimum-rate processes with bounded phase coordinates, Dokl. Akad. Nauk 125 (1959), 475-478

15. A. Ghouila-Houri, Sur la généralisation de la notion dc commande d'un système guidable, Revue Française Inform. Recherche Operationnelle (1967), No. 4, 7-32

16. E .G. Gilbert and D. S. Bernstein, Second-order necessary conditions in optimal control: accessory-problem results without normality, J. Optim. Theory Appl. 41 (1983), 75-106

17. A. D. Ioffe, Nonsmooth analysis: differential calculus of nondifferentiable mappings, Trans. Amer. Math. Soc. 266 (1981), 1-56

18. R. Isaacs, Games of pursuit, Report 1-257, RAND Corporation, 1951

19. R. Isaacs, Differential Games, Wiley and Sons, New York, 1965

20. D. H. Jacobson, A new necessary condition of optimality for singular control problems, SIAM J. Control 7 (1969), 578-595

21. A. J. Krener, The high order maximal principle and its application to singular extremals, SIAM J.Control Optim. 15 (1977), 256-293

22. E. I. Kugušev, The maximum principle in problems of optimal control of systems with non-smooth right-hand side, Vestnik Moskov. Univ. Ser. I Mat. Meh. 28 (1973), No. 3, 103-117

23. E. J. McShane, Necessary conditions in generalized-curve problems of the calculus of variations, Duke Math. J. 7 (1940), 513-536

24. E. J. McShane, Relaxed controls and variational problems, SIAM J. Control 5 (1967), 438-485

25. B. Sh. Mordukhovič, Nonsmooth analysis with nonconvex generalized differentials and nonsmooth maps, Dokl. Akad. Nauk BSSR 24 (1980), 684-687

26. E. Roxin, The existence of optimal controls, Michigan Math. J. 9 (1962), 109-119

30. L. Tonelli, Sui massimi e minimi assoluti del calcolo delle variazioni, Rend. Circ. Mat. Palermo 32 (1911), 297-333 (in Opere Scelte 2, 10-66)

31. F. A. Valentine, The problem of Lagrange with differential inequalities as side conditions, Contributions to the Calculus of Variations 1933-1937, University of Chicago Press, 1937, 407-448

32. J. Warga, Relaxed variational problems, J. Math. Anal. Applic. 4 (1962), 111-128

33. J. Warga, Minimizing variational curves restricted to a preassigned set, Trans. Amer. Math. So. 112 (1964), 432-455

34. J. Warga, Unilateral variational problems with several inequalities, Michigan Math. J. 12 (1965), 449-480

35. J. Warga, The reduction of certain control problems to an "ordinary differential" type, SIAM Review 10 (1968), 219-222

36. J. Warga, Optimal Control of Differential and Functional Equations, Academic Press, New York, 1972

37. J. Warga, Optimal controls with pseudodelays, SIAM J. Control Optim. 12 (1974), 286-299

38. J. Warga, Necessary conditions without differentiability assumptions in optimal control, J. Diff. Eqs. 18 (1975), 41-62

39. J. Warga, A second order condition that strengthens Pontryagin's maximum principle, J. Diff. Eqs. 28 (1978), 284-307

40. J. Warga, A second order Lagrangian condition for restricted control problems, J. Optim. Theory Applic. 24 (1978), 465-473

41. J. Warga, A hybrid relaxed-Lagrangian second order condition for minimum, in Differential Games and Control Theory, Vol. 3, P.T. Liu and E. Roxin, eds., Marcel Dekker, New York, 1979

42. J. Warga, Fat homeomorphisms and unbounded derivate containers, J. Math.Anal. Applic. 81 (1981), 545-560; Erratum, ibid. 90 (1982), 582-583

43. J. Warga, Second order necessary conditions in optimization, SIAM J. Control Optim. 22 (1984), 524-528

44. J. Warga, Higher order conditions with and without Lagrange multipliers, SIAM J. Control Optim. 24 (1986), 715-730

45. J. Warga, Homeomorphisms and local C^1 approximations, Nonlinear Anal. TMA 12 (1988), 593-597

46. J. Warga, Higher order conditons for conical controllability, SIAM J. Control Optim., to appear

47. J. Warga, Global directional controllability, preprint

48. T. Ważewski, Sur une généralisation de la notion des solutions d'une equation au contingent, Bull. Acad. Polon. Sci. Ser. Sci. Math. Astron. Phys. 10 (1962), 17-21

49. L. C. Young, Generalized curves and the existence of an attained absolute minimum in the calculus of variations, C.R. Sci. Lettres Varsovie, C III 30 (1937), 212-234

Stopping Times for Stochastic Approximation

George Yin, Professor, Department of Mathematics, Wayne State University, Detroit, Michigan

Abstract. The problem of stopping times for multidimensional stochastic approximation is investigated in this paper. sequentially determined confidence ellipsoids and cuboids are constructed for the determination of appropriate stopping rules. The result in [1] is extended to the case of moving average noise models.

Keywords. stochastic approximation, moving average processes, confidence ellipsoid, confidence cuboid, stopping rules.

1. Introduction

One of the important problem in control and optimization is to find roots of a given function $f(\cdot)$. In many cases, $f(\cdot)$ is not explicitly known, or there is no sufficient information readily available concerning its value, so the usual numerical analysis procedures cannot be employed. However, the system might be simulated or observed based on sampled data. More often than not, one can only obtain noise or error corrupted observations. I.e., one observes $f(x)$+noise. In 1951, Robbins and Monro suggested a way to solve the aforementioned problem ([2]). Their method is known as stochastic approximation method (S.A.). Because of their wide range of applicability, the S.A. methods have been witnessed tremendous developments for the past three decades. Among those, we mention the work of Nevelson and Khasminskii [3], Ljung [4], Kushner and Clark [5], and Kushner [6].

To see the close connection between stochastic approximation and optimal control, we give one particular example. More examples and applications can be found in [3-6]. Consider the following optimal control of Markov decision problem. Let $\{X_n\}$ be a Markov chain, and the decisions be encoded in $\{U_n\}$. For every admissible control policy π, we wish to minimize the long-run average $J(\pi) = \lim_{n\to\infty} E^{\pi}\frac{1}{n}\sum_{i=1}^{n} c(X_i, U_i)$, where $c(\cdot,\cdot)$ denotes the instant cost. Let $\min_\pi J(\pi) = V$. By means of Lagrangian method, it can be shown that there exist two Markov stationary policies $\underline{g}$ and $\overline{g}$, such that $J(\underline{g}) < V < J(\overline{g})$. Simple memoryless randomization at each step between $\underline{g}$ and $\overline{g}$ produces an one-parameter family of Markov stationary polices $\{f^\eta; 0 \le \eta \le 1\}$. Under suitable conditions, there is a solution of the equation $J(f^\eta) = V$. The solution of this equation is hard to find in general. Nevertheless, this problem can be recast to a stochastic approximation problem. For more details, the readers are referred to [7] and the references therein.

Stochastic approximation algorithms are recursive schemes. In practice, with a prescribed estimation error, one often wishes to terminate the procedure if the desired precision is reached. Although convergence and rates of convergence have been treated extensively, the problem of stopping times for stochastic approximation has not been given enough attention. Three of the exceptions are [8-10]. In these papers, stopping rules were determined by means of constructing appropriate confidence intervals. However, the study was confined to one dimensional scalar problem. The noise involved was assumed to be

---

This research was supported in part by Wayne State University under the Wayne State University Research Award.

independent and identically distributed random variables only. From an application point of view, this is far from satisfactory. Recently, some preliminary results of stopping rules for multivariate stochastic approximation with dependent noise have been developed in [1]. The essential idea depends upon the construction of sequentially determined confidence regions. Asymptotic normality was proved via an application of the functional central limit theorem and the weak convergence method ([11-12]) for noise of martingale difference type. In this paper, our objective is to generalize our result in [1] and to build further on the insights for designing various confidence regions which will lead to reasonable stopping rules for multivariate stochastic approximation.

Let X, $w \in \mathbb{R}^r$. Consider the following S.A. algorithm:

$$Y_n = f(X_n) + w_n, \quad X_{n+1} = X_n + \frac{1}{n} Y_n. \tag{1.1}$$

Y_n is the observation taken at time n, and w_n is the unobservable noise.

We shall assume that the following conditions hold.

(A1) $f(\cdot)$ is a continuous function, such that

(a) $f(x) = 0$ has a unique root θ.

(b) $f(x) = f_x(\theta)(x-\theta) + O(|x-\theta|^{1+\gamma})$ for some $\gamma > 0$, and $A = \frac{I}{2} + f_x(\theta)$ is a stable matrix; i.e., all eigenvalues of A have negative real parts.

(c) $|f(x)| \le K(1+|x|)$, for some $K > 0$.

(A2) $\{w_n\}$ is a sequence of moving average process, i.e., for some positive integer p, $w_n = \sum_{j=0}^{p} C_j \psi_{n-j}$, where C_j, $0 \le j \le p$ are positive definite matrices, such that $\psi_j = 0$, for all $j \le 0$, and

(a) $\mathcal{F}_n$ is a sequence of increasing σ-algebras, such that $(\psi_n, \mathcal{F}_n)$ is a martingale difference for each n.

(b) For some $a > 2$, $\sup_n E(|\psi_n|^a | \mathcal{F}_{n-1}) < \infty$ w.p.1.

(c) There is a symmetric positive definite matrix $S \in \mathbb{R}^{r \times r}$, such that

$$E(\psi_n \psi_n' | \mathcal{F}_{n-1}) \to S \text{ w.p.1 as } n \to \infty.$$

(A3) There exists a twice continuously differentiable Liapunov function $v(x) \ge 0$, such that

(a) $v(x) > 0 \ \forall x \neq \theta$.

(b) $v(x) \to \infty$, as $|x| \to \infty$.

(c) $v_x'(x) f(x) < 0$, for all $x \neq \theta$.

(A1)-(A3) are the conditions on the function, the noise processes and the existence of a Liapunov function for the system respectively. The gain or step size is chosen to be $\{\frac{1}{n}\}$, other step size can be incorporated into our frame work. The mathematical analysis will be essentially the same. The assumption on C_i's can be weakened to require $\sum_{i=0}^{p} C_i$ to be non-singular. Adaptively identifying such matrices are possible.

The remainder of this paper is organized in the following way. Some preliminary results are given next. The determination of stopping rules via construction of appropriate ellipsoidal confidence regions is discussed in section 3. The asymptotic validity is established in section 4. An alternative stopping rule via determination of cuboidal confidence regions is presented in section 5. Finally, the paper is closed with some concluding remarks in the last section. In the sequel, K_i, $0 \le i \le 5$ stand for various positive constants.

2. Preliminaries

Lemma 1. *Under assumptions (A1)-(A3),*

1° *$\{X_n\}$ defined by (1.1) is strongly consistent. I.e., $X_n \to \theta$ w.p.1.*

2° *$X_n - \theta = o(n^\alpha)$ w.p.1, for $\alpha < \frac{1}{2}$.*

3° *$\sqrt{n}(X_n - \theta) \xrightarrow{D} N(0, \Sigma)$, with Σ given by*

$$\Sigma = \int_0^\infty e^{At} \bar{S} e^{A't} dt \tag{2.1}$$

where $A = I/2 + f_x(\theta)$, and $\bar{S} = \sum_{i=0}^p \sum_{j=0}^p C_i S C_j'$.

The proof can be found in [5] and [13]. In what follows, we shall construct a sequence of consistent estimates for Σ. To do so, we need to construct two sequences, one for estimating $f_x(\theta)$, the other one for estimating S.

2.1. Construction of D_n

We construct a sequence of estimates of $f_x(\theta)$ by means of one sided finite difference scheme. To start with, for each $i \le r$, let Z_n^i be the measurement at parameter value $X_n + \frac{1}{n^{\frac{1}{4}}} e^i$,

$$Z_n^i = f(X_n + \frac{1}{n^{\frac{1}{4}}} e^i) + \varsigma_n^i. \tag{2.2}$$

Where $\{e^1, \ldots, e^r\}$ is the standard basis of R^r.

Define D_n^i as

$$D_n^i = \frac{1}{n^{\frac{1}{4}}} (Z_n^i - Y_n) \tag{2.3}$$

with $Y_n = f(X_n) + w_n$.

Lemma 2. *Let (A1)-(A3) be satisfied, and let $\{\varsigma_n^i\}$, $i \le r$ either be a moving average type of noise or satisfy that $\sum \frac{1}{n} \varsigma_n^i$ converges w.p.1. Then $\{D_n^i\}$, $i \le r$ are consistent estimates of $f_x^i(\theta)$, for $i \le r$. Where $f_x^i(\theta)$ denotes the ith column of the gradient matrix $f_x(\theta)$.*

The proof relies on Lemma 1, and it is straight forward.

Remark: There is nothing special about $n^{\frac{1}{4}}$. In fact, any n^α with $\alpha < \frac{1}{2}$ will do. The important thing is to fulfill the convergence speed requirement 2^0 in Lemma 1.

2.2. Construction of S_n

A good candidate for $\{S_n\}$ is the "sample covariance". We define

$$S_n^0 = \frac{1}{n} \sum_{k=1}^n Y_k Y_k' \tag{2.4}$$

$$S_n^i = \frac{1}{n}\sum_{k=1}^{n}(Y_kY_{k-i}' + Y_{k-i}Y_k'),\ 1 \le i \le p \tag{2.5}$$

and

$$S_n = \sum_{i=0}^{p} S_n^i. \tag{2.6}$$

Lemma 3. *If (A1)-(A3) are satisfied, then $S_n \to \bar{S}$ w.p.1.*

Proof: We claim that

$$S_n^0 \xrightarrow{n} \sum_{l=0}^{p} C_l S C_l' \tag{2.7a}$$

$$S_n^i \xrightarrow{n} \sum_{l=0}^{p-i}(C_{l+i}SC_l' + C_lSC_{l+i}'),\ 1 \le i \le p. \tag{2.7b}$$

It is clear that Lemma 3 will follow from (2.7). We shall prove (2.7b) only. In fact, we only prove the case of $i = 1$. The rest of them can be verified in a similar fashion.

In view of (2.5) and by symmetry, we only need to examine the term

$$s_n^1 = \frac{1}{n}\sum_{k=1}^{n} Y_kY_{k-1}'.$$

Define

$$\phi_n = \sum_{l=0}^{p-1} C_{l+1}E(\psi_{n-l-1}\psi_{n-l-1}'|\mathcal{F}_{n-l-2})C_l',\ s^1 = \sum_{l=0}^{p-1} C_{l+1}SC_l$$

then

$$\begin{aligned} s_n^1 - s^1 &= (1-\frac{1}{n})(s_{n-1}^1 - s^1) + \frac{1}{n}f(X_n)f'(X_{n-1}) \\ &+ \frac{1}{n}f(X_n)w_{n-1}' + \frac{1}{n}w_nf'(X_{n-1}) + \frac{1}{n}(w_nw_{n-1}' - \phi_n) + \frac{1}{n}(\phi_n - s^1). \end{aligned} \tag{2.8}$$

To complete the proof, we shall utilize the following lemma.

Lemma 4. *(cf. [14]) Let a sequence $\{x_n\}$ be given by*

$$x_{n+1} = (1-\frac{1}{n})x_n + \frac{1}{n}b_n + \frac{1}{n}c_n \tag{2.9}$$

where $\sum \frac{1}{n}b_n$ converges and $c_n \to 0$ as $n \to \infty$, then $x_n \to 0$ as $n \to \infty$.

Define

$$x_n = s_n^1 - s^1$$

$$b_n = f(X_n)w_{n-1}' + w_n'f(X_{n-1}) + (w_nw_{n-1}' - \phi_n)$$

$$c_n = f(X_n)f'(X_{n-1}) + (\phi_n - s^1).$$

It can be shown (cf [3], [6]) that for n sufficiently large, $E|X_n - \theta|^2 = O(1/n)$. Owing to the representation (A1) (b) of $f(\cdot)$, for n large enough,

$$E|f(X_n)|^2 = O(1/n). \tag{2.10}$$

Equation (2.10), (A2) (b) together with Cauchy-Schwartz inequality yield

$$\sum_{n=2}^{\infty} \frac{1}{n} E|f(X_n)w'_{n-1}| \leq \sum_{n=2}^{\infty} \frac{1}{n} E^{\frac{1}{2}}|f(X_n)|^2 E^{\frac{1}{2}}|w_{n-1}|^2$$
$$\leq K_0 \sum_{l=0}^{p} \sum_{n=2}^{\infty} \frac{1}{n} E^{\frac{1}{2}}|f(X_n)|^2 E^{\frac{1}{2}}\left(E|\psi_{n-l}|^2|\mathcal{F}_{n-l-1}\right) < \infty.$$

Theorem 4.2.1 in [15] now implies $\sum \frac{1}{n} f(X_n)w'_{n-1}$ converges w.p.1. By the same token $\sum \frac{1}{n} w_n f'(X_{n-1})$ converges w.p.1.

By virtue of the local martingale convergence theorem in [16],

$$\sum \frac{1}{n}(w_n w'_{n-1} - \phi_n) \text{ converges w.p.1}$$

and hence $\sum \frac{1}{n} b_n$ converges w.p.1. In view of the strong consistency of X_n, the first term in c_n tends to 0, as $n \to \infty$. By virtue of (A2)(c), the second term also tends to 0. Thus, Lemma 4 yields that $s_n^1 - s^1 \to 0$ w.p.1, as $n \to \infty$. The proof of Lemma 3 is completed. □

Let $A_n = \frac{I}{2} + D_n$, with the constructions of D_n and S_n, we can define Σ_n as

$$\Sigma_n = \int_0^{\infty} e^{A_n t} S_n e^{A'_n t} dt.$$

Theorem 1. *Under (A1)-(A3),* $\Sigma_n \xrightarrow{n} \Sigma$ *w.p.1.*

Remark: Notice that Σ is symmetric and positive definite, hence, Σ_n are non-singular for all but finitely many n's. Without loss of generality, we shall assume Σ_n to be non-singular for all n.

3. Confidence ellipsoid

Since $\sqrt{n}(X_n - \theta)$ is asymptotically normal, we can define an ellipsoidal confidence region for θ as follows. Let

$$U_n = n(X_n - \theta)'\Sigma_n^{-1}(X_n - \theta). \tag{3.1}$$

Then,

$$E_n = \{\theta; U_n \leq c\} \tag{3.2}$$

is an ellipsoidal confidence region for θ.

Let $V(E_n)$ denote the volume of E_n. The following formula was derived in [17].

$$V(E_n) = \frac{\pi^{\frac{r}{2}} \left(\frac{c}{n}\right)^{\frac{r}{2}} (det\Sigma_n)^{\frac{1}{2}}}{\Gamma(\frac{r}{2}+1)} \tag{3.3}$$

It is clear that to make the estimates close to θ with high probability is equivalent to make θ inside of a confidence ellipsoid with volume small enough. Henceforth, instead of forcing $|X_n - \theta| \leq \epsilon$, we shall make $V(E_n) \leq \epsilon^r$.

$\Sigma_n^{-1} \to \Sigma^{-1}$ implies that

$$U_n - n(X_n - \theta)\Sigma^{-1}(X_n - \theta) \xrightarrow{P} 0 \tag{3.4}$$

and hence, $U_n \xrightarrow{D} \chi_r^2$, the Chi-square random variable with r degrees of freedom. Therefore, if $c = c_\alpha$, we have

$$P(\theta \in E_n) = P(U_n \leq c_\alpha) \xrightarrow{n} P(\chi_r^2 \leq c_\alpha) = 1 - \alpha. \tag{3.5}$$

For any $\epsilon > 0$, we can choose n large enough, such that $V(E_n) \leq \epsilon^r$.

Consequently,

$$\nu_{\epsilon,\alpha} = \frac{\pi c_\alpha (det\Sigma)^{\frac{1}{r}}}{\epsilon^2 \left(\Gamma(\frac{1}{2}r+1)\right)^{\frac{2}{r}}} \leq n. \tag{3.6}$$

Thus, for any $\epsilon > 0$, if a confidence ellipsoid with volume no larger than ϵ^r is desired, we impose the following stopping rule $\tau_{\epsilon,\alpha}$,

$$\tau_{\epsilon,\alpha} = \inf\{n; \nu_{\epsilon,\alpha}^n \leq n\} \tag{3.7}$$

where $\nu_{\epsilon,\alpha}^n = \frac{\pi c_\alpha (det\Sigma_n)^{\frac{1}{r}}}{\epsilon^2 \left(\Gamma(\frac{1}{2}r+1)\right)^{\frac{2}{r}}}$. Our main result is:

Theorem 2. *If (A1)-(A3) are satisfied, and Σ_n^{-1} exist, then the following equations hold.*

$$\lim_{\epsilon \to 0} \frac{\tau_{\epsilon,\alpha}}{\nu_{\epsilon,\alpha}} = 1 \ w.p.1 \tag{3.8}$$

and

$$\lim_{\epsilon \to 0} P\{\theta \in E_{\tau_{\epsilon,\alpha}} \text{ and } V(E_{\tau_{\epsilon,\alpha}}) \leq \epsilon^r\} = 1 - \alpha. \tag{3.9}$$

Equation (3.8) is readily verified. We need only prove (3.9). In order to do so, we need only prove the asymptotic normality for the stopped stochastic process. I.e, we need only show $\sqrt{\tau_{\epsilon,\alpha}}(X_{\tau_{\epsilon,\alpha}} - \theta) \xrightarrow{D} N(0, \Sigma)$.

4. Asymptotic normality for the stopped procedure

It is well-known (cf. [3]) that for iterates X_n given by (1.1), for some $m \geq 1$,

$$\sqrt{n+1}(X_{n+1} - \theta) = \sum_{k=m}^{n} \frac{1}{\sqrt{k}} \sqrt{1 + \frac{1}{k}} A_{nk} w_k + o(1) \tag{4.1}$$

where $o(1) \to 0$ in probability, and

$$A_{jk} = \begin{cases} \prod_{l=k+1}^{j}(I + A/l), & \text{if } j > k; \\ I, & \text{if } j = k. \end{cases}$$

Moreover, since $\sqrt{1+1/n} = 1 + O(1/n)$, (4.1) can be further reduced to

$$\sqrt{n+1}(X_{n+1} - \theta) = \sum_{k=m}^{n} \frac{1}{\sqrt{k}} A_{nk} w_k + o(1)$$

with $o(1) \to 0$ in probability.

By virtue of equation (3.8), $\tau_{\epsilon,\alpha} \to \infty$ and $\nu_{\epsilon,\alpha} \to \infty$ w.p.1, as $\epsilon \to 0$. We claim that the following lemma holds.

Lemma 5. *Under (A1)-(A3),*

$$\sqrt{\tau_{\epsilon,\alpha}+1}(X_{\tau_{\epsilon,\alpha}+1} - \theta) = \sum_{k=m}^{\tau_{\epsilon,\alpha}} \frac{1}{\sqrt{k}} A_{\tau_{\epsilon,\alpha},k} w_k + o(1). \tag{4.2}$$

Proof: Define $Q_n = \sqrt{n}(X_n - \theta)$. It follows from (1.1) and the assumptions,

$$Q_{n+1} = Q_n + \frac{A}{n} Q_n + \frac{1}{\sqrt{n}}\sqrt{1+\frac{1}{n}}\delta(X_n) + \frac{C_n}{\sqrt{n}}(X_n - \theta) + \frac{1}{\sqrt{n}}\sqrt{1+\frac{1}{n}} w_n. \tag{4.3}$$

Where $\delta(x) = O(|x-\theta|^{1+\gamma})$, $C_n = O(1/n)$.

It is easily seen that for some $m \geq 1$, we have

$$\begin{aligned} Q_{n+1} = A_{n,m-1}Q_m &+ \sum_{k=m}^{n} \frac{1}{\sqrt{k}} A_{nk}\left(\sqrt{1+\frac{1}{k}}\delta(X_k) + C_k(X_k - \theta)\right) \\ &+ \sum_{k=m}^{n} \frac{1}{\sqrt{k}}\sqrt{1+\frac{1}{k}} A_{nk} w_k. \end{aligned} \tag{4.4}$$

Equation (4.4) yields

$$Q_{\tau_{\epsilon,\alpha}+1} = A_{\tau_{\epsilon,\alpha},m-1}Q_m + \sum_{k=m}^{\tau_{\epsilon,\alpha}} \left(b_k^1 + b_k^2\right) + \sum_{k=m}^{\tau_{\epsilon,\alpha}} \frac{1}{\sqrt{k}}\sqrt{1+\frac{1}{k}} A_{\tau_{\epsilon,\alpha},k} w_k \tag{4.5}$$

where

$$b_k^1 = \frac{1}{\sqrt{k}}\sqrt{1+\frac{1}{k}} A_{\tau_{\epsilon,\alpha}k}\delta(X_k), \; b_k^2 = \frac{1}{\sqrt{k}} A_{\tau_{\epsilon,\alpha}k} C_k (X_k - \theta). \tag{4.6}$$

We shall show that as $\epsilon \to 0$, the first two terms on the right hand side of (4.5) tend to 0 in probability.

(A1) and the expression $A = I/2 + f_x(\theta)$, imply $|e^{At}| \le K_1 e^{-\lambda_1 t}$ for some $\lambda_1 > 0$. Also, $A_{nk} = (I + r_{nk})e^{A\ln(n/k)}$, $k \le n-1$, where $|r_{nk}| \to 0$, as $k \to \infty$ uniformly in $n > k$. It follows that

$$\begin{aligned} |A_{\tau_{\epsilon,\alpha},m-1}| &\le |I + r_{\tau_{\epsilon,\alpha},m-1}||e^{A\ln\left(\frac{\tau_{\epsilon,\alpha}}{m-1}\right)}| \\ &\le K_2 e^{-\lambda_1 \ln\left(\frac{\tau_{\epsilon,\alpha}}{m-1}\right)} \to 0 \text{ as } \epsilon \to 0. \end{aligned} \tag{4.7}$$

The last step above follows from the fact that $\tau_{\epsilon,\alpha} \to \infty$ w.p.1, as $\epsilon \to 0$.

By virtue of (2.10), we can choose m so that Q_m is bounded in probability. Moreover, for any $\eta > 0$,

$$\begin{aligned} &P\{|A_{\tau_{\epsilon,\alpha},m-1}Q_m| > \eta\} \\ \le & P\{|A_{\tau_{\epsilon,\alpha},m-1}||Q_m| > \eta; |Q_m| \le \frac{1}{\eta}\} + P\{|A_{\tau_{\epsilon,\alpha},m-1}||Q_m| > \eta; |Q_m| > \frac{1}{\eta}\} \\ \le & P\{|A_{\tau_{\epsilon,\alpha},m-1}| > \eta^2\} + P\{|Q_m| > \frac{1}{\eta}\}. \end{aligned} \tag{4.8}$$

Now, (2.10), (4.7) (4.8) yield $A_{\tau_{\epsilon,\alpha},m-1}Q_m \xrightarrow{P} 0$, as $\epsilon \to 0$.

Next, we consider $\sum_{k=m}^{\tau_{\epsilon,\alpha}} b_k^1$. Since

$$\begin{aligned} &P\{|\sum_{k=m}^{\tau_{\epsilon,\alpha}} \frac{1}{\sqrt{k}} A_{\tau_{\epsilon,\alpha}k}\delta(X_k)| > \eta\} \\ \le & P\{|\sum_{k=m}^{\tau_{\epsilon,\alpha}} \frac{1}{\sqrt{k}} A_{\tau_{\epsilon,\alpha}k}\delta(X_k)| > \eta; \frac{\tau_{\epsilon,\alpha}}{\nu_{\epsilon,\alpha}} \le 1\} + P\{\frac{\tau_{\epsilon,\alpha}}{\nu_{\epsilon,\alpha}} > 1\}. \end{aligned} \tag{4.9}$$

By virtue of (3.8), the second term on the right side of (4.9) tends to 0. Hence, we need only consider the first term. Note that for $\frac{\tau_{\epsilon,\alpha}}{\nu_{\epsilon,\alpha}} \le 1$,

$$\begin{aligned} &P\{|\sum_{k=m}^{\tau_{\epsilon,\alpha}} \frac{1}{\sqrt{k}} A_{\tau_{\epsilon,\alpha}k}\delta(X_k)| > \eta; \frac{\tau_{\epsilon,\alpha}}{\nu_{\epsilon,\alpha}} \le 1\} \\ \le & P\{\sum_{k=m}^{\nu_{\epsilon,\alpha}} \frac{1}{\sqrt{k}} |A_{\tau_{\epsilon,\alpha}k}||\delta(X_k)| > \eta; \frac{\tau_{\epsilon,\alpha}}{\nu_{\epsilon,\alpha}} \le 1\}. \end{aligned}$$

For the $\eta > 0$ given above, there is a $\Delta > 0$, such that $|\delta(x)| \le \eta^2|x|$, for $|x - \theta| < \Delta$, and $P\{\sup_{n\ge m}|X_n - \theta| < \Delta\} > 1 - \eta$.

In view of the above paragraph and (2.10), applying Chybeshev's inequality,

$$\begin{aligned} &P\{\sum_{k=m}^{\nu_{\epsilon,\alpha}} \frac{1}{\sqrt{k}} |A_{\tau_{\epsilon,\alpha}k}||\delta(X_k)| > \eta; \frac{\tau_{\epsilon,\alpha}}{\nu_{\epsilon,\alpha}} \le 1\} \\ \le & P\{\sum_{k=m}^{\nu_{\epsilon,\alpha}} \frac{1}{\sqrt{k}} |A_{\tau_{\epsilon,\alpha}k}||\delta(X_k)| > \eta; \sup_{k\ge m}|X_k - \theta| < \Delta\} + \eta \\ \le & P\{\sum_{k=m}^{\nu_{\epsilon,\alpha}} \frac{1}{\sqrt{k}} |A_{\tau_{\epsilon,\alpha}k}||X_k - \theta| > 1/\eta\} + \eta \le K_4\eta \sum_{k=m}^{\nu_{\epsilon,\alpha}} \frac{1}{k}\left(\frac{k}{\nu_{\epsilon,\alpha}}\right)^{\lambda_1} + \eta. \end{aligned} \tag{4.10}$$

Since

$$\lim_{\epsilon\to 0}\sum_{k=1}^{\nu_{\epsilon,\alpha}}\left(\frac{k}{\nu_{\epsilon,\alpha}}\right)^{\lambda_1-1}\frac{1}{\nu_{\epsilon,\alpha}}=\int_0^1 u^{\lambda_1-1}du=\frac{1}{\lambda_1} \tag{4.11}$$

$$P\{\sum_{k=m}^{\tau_{\epsilon,\alpha}}\frac{1}{\sqrt{k}}|A_{\tau_{\epsilon,\alpha}k}||\delta(X_k)|>\eta\}\le K_5\eta. \tag{4.12}$$

Consequently, $\sum_{k=m}^{\tau_{\epsilon,\alpha}}b_k^1\xrightarrow{P}0$, as $\epsilon\to 0$. Similar argument yields $\sum_{k=m}^{\tau_{\epsilon,\alpha}}b_k^2\xrightarrow{P}0$, as $\epsilon\to 0$. Moreover, it can be seen that

$$\sum_{k=m}^{\tau_{\epsilon,\alpha}}\frac{1}{\sqrt{k}}\sqrt{1+\frac{1}{k}}A_{\tau_{\epsilon,\alpha}k}w_k=\sum_{k=m}^{\tau_{\epsilon,\alpha}}\frac{1}{\sqrt{k}}A_{\tau_{\epsilon,\alpha}k}w_k+o(1).$$

Hence, the lemma is proved. □

Lemma 6. *Under the assumptions of Theorem 2,*

$$\sum_{k=m}^{\tau_{\epsilon,\alpha}}\frac{1}{\sqrt{k}}A_{\tau_{\epsilon,\alpha}k}w_k=\sum_{k=m}^{\nu_{\epsilon,\alpha}}\frac{1}{\sqrt{k}}A_{\nu_{\epsilon,\alpha}k}w_k+o(1) \tag{4.13}$$

with $o(1)\to 0$ *in probability as* $\epsilon\to 0$.

Proof: It can be seen

$$\sum_{k=m}^{\tau_{\epsilon,\alpha}}\frac{1}{\sqrt{k}}A_{\tau_{\epsilon,\alpha}k}w_k=\sum_{k=m}^{\nu_{\epsilon,\alpha}}\frac{1}{\sqrt{k}}A_{\nu_{\epsilon,\alpha}k}w_k+\sum_{k=m}^{\tau_{\epsilon,\alpha}}\frac{1}{\sqrt{k}}A_{\tau_{\epsilon,\alpha}k}w_k-\sum_{k=m}^{\nu_{\epsilon,\alpha}}\frac{1}{\sqrt{k}}A_{\nu_{\epsilon,\alpha}k}w_k. \tag{4.14}$$

For any $\eta>0$,

$$\begin{aligned}
&P\Big\{|\sum_{k=m}^{\nu_{\epsilon,\alpha}}\frac{1}{\sqrt{k}}A_{\nu_{\epsilon,\alpha}k}w_k-\sum_{k=m}^{\tau_{\epsilon,\alpha}}\frac{1}{\sqrt{k}}A_{\tau_{\epsilon,\alpha}k}w_k|>\eta\Big\}\\
\le &P\Big\{|\sum_{k=m}^{\nu_{\epsilon,\alpha}}\frac{1}{\sqrt{k}}A_{\nu_{\epsilon,\alpha}k}w_k-\sum_{k=m}^{\tau_{\epsilon,\alpha}}\frac{1}{\sqrt{k}}A_{\tau_{\epsilon,\alpha}k}w_k|>\eta;\frac{\tau_{\epsilon,\alpha}}{\nu_{\epsilon,\alpha}}\le 1\Big\}+P\Big\{\frac{\tau_{\epsilon,\alpha}}{\nu_{\epsilon,\alpha}}>1\Big\}\\
\le &P\Big\{|\sum_{k=m}^{\nu_{\epsilon,\alpha}}\frac{1}{\sqrt{k}}A_{\nu_{\epsilon,\alpha}k}w_k-\sum_{k=m}^{\tau_{\epsilon,\alpha}}\frac{1}{\sqrt{k}}A_{\nu_{\epsilon,\alpha}k}w_k|>\eta;\frac{\tau_{\epsilon,\alpha}}{\nu_{\epsilon,\alpha}}\le 1\Big\}\\
&+P\Big\{|\sum_{k=m}^{\tau_{\epsilon,\alpha}}\frac{1}{\sqrt{k}}A_{\tau_{\epsilon,\alpha}k}w_k||A_{\nu_{\epsilon,\alpha}\tau_{\epsilon,\alpha}}-I|>\eta;\frac{\tau_{\epsilon,\alpha}}{\nu_{\epsilon,\alpha}}\le 1\Big\}+P\Big\{\frac{\tau_{\epsilon,\alpha}}{\nu_{\epsilon,\alpha}}>1\Big\}.
\end{aligned} \tag{4.15}$$

The assertion follows form (4.15). Some detailed calculations are omitted. Hence, Lemma 6 holds. □

Now, we are in a position to complete the proof of Theorem 2. By Lemma 6, $\sum_{k=m}^{\tau_{\epsilon,\alpha}}\frac{1}{\sqrt{k}}A_{\tau_{\epsilon,\alpha}k}w_k$ and $\sum_{k=m}^{\nu_{\epsilon,\alpha}}\frac{1}{\sqrt{k}}A_{\nu_{\epsilon,\alpha}k}w_k$ have the same limit in distribution. In view of Lemma 1, we conclude

$$\sum_{k=m}^{\tau_{\epsilon,\alpha}}\frac{1}{\sqrt{k}}A_{\tau_{\epsilon,\alpha}k}w_k\xrightarrow{D}N(0,\Sigma).$$

As a consequence, Theorem 2 follows.

5. Confidence cuboid

In this section, we give an alternative stopping rule for multivariate stochastic approximation. This stopping rule is based on sequentially determined multidimensional confidence rectangular regions. We shall call such a region as a confidence cuboid. This is a natural extension of one dimensional confidence intervals.

Since the development is parallel to the previous sections, the method and results will be discussed in an informal manner so that the main ideas are clear.

In lieu of the vector process X_n, we consider its components. By virtue of (1.1),

$$X_{n+1}^i = X_n^i + \frac{1}{n}(f^i(X_n) + w_n^i) \tag{5.1}$$

where the superscript i stands for the ith component of appropriate processes. Lemma 1 implies that $X_n^i \to \theta^i$ w.p.1, $n^\alpha(X_n^i - \theta^i) \to 0$ for $0 \leq \alpha < \frac{1}{2}$. Furthermore, it is not hard to show that $\sqrt{n}(X_n^i - \theta^i) \xrightarrow{D} N(0, \sigma_i^2)$. In addition, sequentially determined estimates of σ_i^2 can be constructed in the light of the previous sections. For notational simplicity and brevity, we shall not go into the details.

We assume that $\{\sigma_{n,i}^2\}$, $i \leq r$ are sequences of consistent estimates for σ_i^2. If for each n, $\sqrt{n}(X_n^i - \theta^i)$ is normally distributed with mean zero and variance σ_i^2, and if c_{α^i} is the $100(1 - \frac{\alpha^i}{2})$-th percentile of a standard normal random variable, then

$$P\Big(\Big|\frac{\sqrt{n}(X_n^i - \theta^i)}{\sigma_i}\Big| \leq c_{\alpha_i}\Big) = 1 - \alpha^i. \tag{5.2}$$

A $100(1 - \frac{\alpha^i}{2})\%$ confidence interval for θ^i is given by

$$I_n^i = (X_n^i - c_{\alpha^i}\sigma_i, X_n^i + c_{\alpha^i}\sigma_i). \tag{5.3}$$

Recall that $\sqrt{n}(X_n^i - \theta^i)$ is asymptotically normal. In view of the above argument, if a $100(1 - \frac{\alpha^i}{2})\%$ confidence interval on θ^i of length 2ϵ is sought, the desired stopping rule for the ith component process can be defined as

$$\tau_{\epsilon,\alpha^i}^i = \inf\{n; \nu_n^i \leq n\} \tag{5.4}$$

with

$$\nu_n^i = c_{\alpha^i}^2 \sigma_{n,i}^2 / \epsilon^2. \tag{5.5}$$

By virtue of the analogous argument as in section 3, we can show that

$$\lim_{\epsilon} P(|X_{\tau_{\epsilon,\alpha^i}^i} - \theta^i| \leq \epsilon) = 1 - \alpha^i. \tag{5.6}$$

Next, we consider the cuboid or r-dimensional rectangle

$$I_n = \{\theta; \theta^i \in I_n^i, i \leq r\} \tag{5.7}$$

with I_n^i given by (5.3).

In view of the well-known Bonferroni's inequality,

$$P(\theta \in I_n) \geq 1 - \sum_{i=1}^{r} P(\theta^i \notin I_n^i). \tag{5.8}$$

Define the stopping rule for the stochastic approximation algorithm as

$$\tau_{\epsilon,\alpha} = \max_{i \leq r} \tau^i_{\epsilon,\alpha^i}. \tag{5.9}$$

By virtue of (5.6) (5.8),

$$\lim_{\epsilon \to 0} P(\theta \in I_{\tau_{\epsilon,\alpha}}) \geq 1 - \sum_{i=1}^{r} \lim_{\epsilon \to 0} P(\theta^i \notin I^i_{\tau^i_{\epsilon,\alpha}}) \geq 1 - \sum_{i=1}^{r} \alpha^i. \tag{5.10}$$

In particular, if we choose $\alpha^i = \frac{\alpha}{r}$, then I_n is a confidence cuboid with limiting confidence coefficient $1 - \alpha$.

6. Some discussions

1. Stopping times for multidimensional stochastic approximation have been developed in this paper. The essential idea is to construct appropriate confidence regions.

2. Although only the RM type algorithms have been discussed, nothing seems to prevent the analogue to the development for the KW type algorithms and for more general S.A. algorithms of the form $X_{n+1} = X_n + a_n f(X_n, w_n)$.

3. Stopping times for parallel stochastic approximation algorithms (cf. [18-20]) can be treated. The trick is in essence of last section. Stopping rules for each processor are determined first, and the stopping rule for the multi-processor procedure is then determined by taking the maximum of the stopping times for each individual processor. Finally, Bonferroni's inequality is employed for getting the limit.

4. For the sake of ease in presentation, simple probabilistic arguments are employed in this paper. An alternative approach–the weak convergence methods for multidimensional SA algorithms with MA and ϕ-mixing type of correlated noise processes can be worked out as well.

5. One of the important thing which we have not done is to investigate the empirical behavior of the stopping rules by means of Monte Carlo simulations. The small sample property should be very interesting to us.

Acknowledgement. I wish to thank Prof. Emilio Roxin, the organizer of this conference, for hospitality, various support and assistance.

References

[1] G. Yin, A stopped stochastic approximation algorithm, *Systems & Control Letters*, 11(1988), 107-115.

[2] H. Robbins and S. Monro, A stochastic approximation method, *Ann. Math. Statist.* 22(1951), 400-407.

[3] M.B. Nevelson and R.Z. Khasminskii, *Stochastic Approximation and Recursive Estimation*, Translation of Math. Monographs, v47, 1973, AMS.

[4] L. Ljung, Analysis of recursive stochastic algorithms, *IEEE Trans. on Automatic Control* AC-22, 551-575.

[5] H.J. Kushner and D.S. Clark, *Stochastic Approximation for Constrained and Unconstrained Systems*, Springer-Verlag, 1978, Berlin.

[6] H.J. Kushner, *Approximation and Weak Convergence Methods for Random Processes with Applications to Stochastic Systems Theory*, 1984, MIT Press, Cambridge, MA.

[7] A.M. Makowski and A. Shwartz, Implementation issues for Markov decision processes, *Stochastic Differential Systems, Stochastic Control Theory and Applications*, ed. W. Fleming and P.L. Lions, 323-339.

[8] R.H. Farrel, Bounded length confidence intervals for the zeros of a regression function, *Ann. of Math. Statist.* 33(1962), 237-247.

[9] R. Sielken, Stopping times for stochastic approximation procedures, *Z. Wahr.* 26(1973), 67-75.

[10] D.F. Stroup and H.I. Braun, On a new stopping rule for stochastic approximation, *Z. Wahr.* 60(1982), 535-554.

[11] S. Ethier and T.G. Kurtz, *Markov Processes*, 1986, J. Wiley & Sons.

[12] P. Billingsley, *Convergence of Probability Measures*, 1968, J. Wiley & Sons.

[13] H.F. Chen, Asymptotic normality of stochastic approximation under correlated noise, *J. Systems Science and Math. Sciences*, 4(1984), 136-159.

[14] Y.M. Zhu, Extensions of the relations of series to infinite products and the convergence of a class of recursive algorithms, *Mathematica Numerica Sinica*, 7(1985), 369-376.

[15] E. Lukacs, *Stochastic Convergence*, 1970, Academic Press.

[16] Y.S. Chow, Local convergence of martingales and the law of large numbers, *Ann. of Math. Statist.* 36(1965), 552-558.

[17] H. Cramer, *Mathematical Methods of Statistics*, 1946, Princeton Univ. Press, Princeton.

[18] H.J. Kushner and G. Yin, Asymptotic properties of distributed and communicating stochastic approximation algorithms, *SIAM J. Control and Optimization*, 25(1987), 1266-1290.

[19] H.J. Kushner and G. Yin, Stochastic approximation algorithms for parallel and distributed processing, *Stochastics*, 22(1987), 219-250.

[20] G. Yin and Y.M. Zhu, On w.p.1 convergence of a parallel stochastic approximation algorithm, to appear in *Probability in the Engineering and Informational Sciences*.

Remarks and Personal Reminiscences

Laurence C. Young, Professor, Center for the Mathematical Sciences, University of Wisconsin at Madison, Madison, Wisconsin

Occasionally, as in this meeting, it is important to go back in time. As presumably the oldest participant, I shall speak mainly of what came before the 30 years most of us have lived through. We then spoke of the Calculus of Variations, but Optimal Control is a better name.

Only by looking back, can we hope to learn what we would all like to understand, how concepts, theories, devices, algorithms, came into being, which all of a sudden made possible for a subject to be so much clearer, so much more streamlined.

Ah, but there is more to it than even that! It takes time, often many years, for us to appreciate something new. For years Fourier was unable even to publish. For years too, Galois' epoch-making manuscript on Groups languished somewhere. Then we have Riemann's inaugural lecture: it took 50 years for it to lead to Relativity, and 20 years after that this was still not accepted by the Nobel Prize Committee, who were careful to say, in their award to Einstein, that it was not awarded for Relativity!

Old I certainly am! Old enough to say with Felix Klein on his mere 50-th birthday, and with more justice,

> When everyone begins to speak well of you,
> you are on the downward path.

Still, some young people feel ahead in our changing world. I was young once ... I was all the more struck with Hilbert in 1932, when I heard him quote so much in his lectures Dirichlet, the friend of Jacobi. Young people tend to value a brilliant teacher. Universities give prizes for excellent teaching. But Felix Klein, whom the great Carathéodory ranked as the greatest teacher of all time, remarks in his Development of mathematics in the XIX-th century, that while Koenigsberg and Berlin could benefit from the great activity of Jacobi, his forcefulness and brilliant style, Goettingen had nothing similar -- only an unapproachable Gauss and a Riemann unappreciated by his students, with in between Dirichlet quietly lecturing for 4 brief years to small classes.

'Yet in the long run, such things matter less
than the depth to which man's thought can reach.'

We heard how Lefschetz, whom we are honouring with Lasalle, ran his Center. I could tell you similarly about Littlewood: you could certainly come to see him, but the next day in class, he would explain that he was unprepared, he had had a visitor ... (Needless to say, visitors were rare ...) There was, however, no noticeable difference between his unprepared and prepared lectures: both consisted in distributing printer's proofsheets and making trivial corrections, such as, on line so and so, page so and so, an asterisk should be a star; on Wednesdays this continued right into a waiting taxi! Yet Littlewood, like Lefschetz, built up a school of mathematicians. What was the secret? They, the pupils, were doing this work! They were even flattered by being expected to manage, this was an honour they much show themselves worthy of ...

By contrast, a young Fellow once came in great distress to my father: he was doing all he could, wearing himself out, and still his pupils were failing their examinations. How was this possible? Very simply, my father said, they should have done the work, not you ...

The method is the same in creative work, the main difference lies in the quality of the person overseeing it. Littlewood, Lefschetz, were giants. So was the quiet Dirichlet. We can all say with Newton,

If I have seen further than others, it is
because I have stood on the shoulders of giants.

I have known personally legendary figures we now think of as giants -- who not necessarily had a high reputation in their time, for they were not mainly teachers and solvers of past problems, they created for the future.

I feel close also to giants my parents knew, and who perhaps knew me when I was 2 or 3 years old. But even such pioneers as Galois, Cauchy, Gauss, right up to Lagrange and Euler -- who made of the Calculus of Variations no longer a collection of isolated problems but a theory -- are only, when we think of it, a few handshakes away. This is why, always, I advise students to go back to the sources, whether in Banach, in Zygmund, or all the way to Euler. Read by all means what is up-to-date, but also what started it all! Do today's great ones agree? Let me only mention a bust hanging in Fine Hall: Lefschetz made a special point of showing it to me. It was black, I did not recognize it. Finally Lefschetz told me it was Hilbert's.

It goes without saying that we should spread our reading not merely in time, to the past, but also in space, to cognate subjects. This is really self-evident in Optimal Control, which brings together, not only those we commonly call mathematicians, but also economists, physicists, biologists, engineers and others, who nowadays should count as mathematicians too. I am against artificial distinctions. At a meeting, my father said to a young man,

> You must be an Englishman.
> No, replied he, I'm a Dutchman.
> Same thing, said my father.

The young man was delighted, he told me the story himself. He was van der Corput, whom I got to know well.

As a matter of course, I urge graduate students to take an active part in all the Seminars they can manage, and not to concentrate solely on a thesis. Above all, they must not be afraid of unfamiliar ground. A pupil of Felix Klein was startled to have been recommended by him for a professorship of astronomy:

> But I know nothing of astronomy! he said.
> You can tackle it, Felix Klein replied,
> you are a mathematician.

He made a good astronomer. But in some cases the switch from one subject to another is actually desired! My friend Coulson was in this way allowed to change his professorship at Oxford, from one of mathematics, to one of Chemistry. But at least within what we still commonly call mathematics, whether pure or applied, we should not fear having to pass from one domain to another when a theory leads that way. Even for angels, the point of a needle should be no refuge from the unknown, nor from hard work.

Ah, yes! Hard work! Time and energy consuming! Yes, we are right to try to reduce it so as to devote ourselves to better things! But what if the hard work is learning a language? Prior to about 1450, for hundreds of years mathematicians had to know, not only Latin and Greek, but also Arabic! I honestly believe that even for them the linguistic difficulties were much smaller than the mathematical ones. However nowadays we are quite spoilt: people imagine that English is enough, especially since Russian is now generally available in translation.

I am old-fashioned enough to prefer the original, but in any case I said we need to go back in time and this means reading whatever languages were then used. For instance, I myself learnt at one time Danish, which I have now forgotten, but which enabled me to read an important article I might otherwise have missed, on the addition of convex curves, and also the book of Hjemslev on a geometry in which lines have thickness. This last idea comes from Felix Klein, it was one of the many things that led me to generalised curves. Another was, of course, Hilbert's famous statement that every problem of the Calculus of Variations has a solution, if the word 'solution' is suitable interpreted. I might say that Danish is a difficult language to learn: Danes insist on speaking English to the English-speaking! Fortunately my landlady was of Swedish origin ... In order to communicate with her, I had at first to phone her son-in-law, manager of a big factory ... it was less trouble all round to teach me Danish ...

Learning Russian was more straightforward: I do not speak it, but I can read fairly fluently. I learnt this after spending 3 weeks in deciphering a Russian article by the Pole T. Ważewski, even though, to understand that article mathematically took me months ... Other languages that I speak, not all of them too well, I learnt when I was very young with no difficulty at all, so that, for instance, on my first day in Munich, I was told by a fellow student straight from Berlin who had been conversing with me most of the day,

> You Munich people have a queer way of speaking.

Similarly, I learnt Italian at the age of 8. For visiting Italy I had taken, I thought, every precaution. I had been at pains to learn properly the crucial "I don't understand Italian",

> Non capisco Italiano.

I had followed this up with the polite Italian leave-taking,

A revederci e tanti saluti a casa.

Left alone in a Viareggion bathing-hut, I was unworried when a workman came there to speak to my parents. Need I describe the fury of words, aroused by my 2 sentences followed by my shutting the door? That is when I realized that I must learn more than the 2 sentences: I did. And that is how I as able so easily to read, some 10 or 12 years later, the famous two volumes of Tonelli's Fondamenti. By then I had met Tonelli, and my mother and I had had long talks with him. He was still young, he had a apartment in Bologna, and his books were packed in the basement: he was expecting a call to Rome, which never came ... Eventually, as Bianchi's successor, he went to Pisa, and made of it the mathematical centre many of my friends know well.

There are perhaps disadvantages to learning very young various foreign languages: I was not conscious of them, nor of those of having mathematical parents, though I now realize that people may have wondered what part of my early work was really my own. There is, however, the obvious advantage that one acquires a certain facility rather easily. Nevertheless, for those who find mathematics or languages difficult, let me say at once that lack of facility does not imply, and should be confused with, last of ability. Newton, for two whole years at Cambridge, could make no sense of Euclid's Elements. Hilbert too, as he tells us, found mathematics very difficult -- it was only after long and hard thinking that what seemed difficult was really rather simple. And Wittgenstein stood out in a class, according the Bertrand Russell, because he was the only one who looked puzzled. What we call ability generally requires very hard work. It takes time, the "midnight oil". Insight comes suddenly, but after much groping in the dark; often amongst a wealth of new ideas ... The work, the thankless work, we all know; but the ideas are like the views we get occasionally, on a long hard mountain climb. We have what T. Rado used to call, our lucid moments. They make seem worthwhile the thankless work, the tiring climb.

The Calculus of Variations, Optimal Control if you prefer, is a rich source of such ideas, and we can look at them from many sides. For instance, Jacobi's conjugate points were later used by Marston Morse's theory in the Large, to derive topological and other new results. The truth is that certain features of the landscape below, are most visible from mountains above. Thus Minkowski's beautiful Geometry of Numbers came into being thanks to the ideas of duality and convexity, which really originated in the Calculus of Variations with Hamilton and with Weierstrass, except that a simpler form of duality, analogous to that of poles and polars with respect to the unit circle, occurs in effect in Functional Analysis in the work of Euler, in what was long known as the 'Fundamental Lemma of the Calculus of Variations'. The latter, as I mention in my Lectures on the Calculus of Variations, leads in a natural way to the theory of distributions of Laurent Schwartz, which revolutionized much of classical analysis.

Conversely of course, ideas from outside have had quite an influence on the Calculus of Variations. Semi-continuity is absolutely essential in Tonelli's approach to existence theory: it was my father who suggested that he use this idea, which had previously not had any great impact on classical problems. There had been the well-known example, due to H. A. Schwarz, of arbitrarily finely inscribed polyhedra to a circular cylinder S, whose areas could be made to approximate any number not less than the area of S, or even to tend to infinity; but this was generally regarded, like non-differentiable functions at the time, as one of those hideous things, from which Hermite wrote,

Je me détourne avec horreur ...

At least Hermite was honest, more honest than many of us who prefer not to see whatever offends our prejudices ...

In mathematics, as in life, there is the problem: to hold fast to what we think right, to what surely ought to be? Or to seek in our ways an error, perhaps so ingrained, that in us everything cries out in its support? Did not the great Ptolemy think it obvious that the tiny stars, and the Sun and Moon, must move around our huge Earth? Did he not also take for granted that they must move in circles, since nothing so unsymmetrical as an ellipse -- a mere toy of pure mathematicians -- could possibly be of use to nature? We smile, but we too, not very long ago, had prejudices. Anything not in accord with them, was at best, controversial. We had our Set Theory controversy. Hilbert declared Cantor the greatest mathematician living.
He said,

> They shall not expel us from the Paradise
> that Georg Cantor has created for us!

Goettingen also stood by Einstein and Relativity -- controversial too, because, as Hilbert put it, Physics was too hard for physicists! The Goettingen tradition was for mathematicians to retain an interest in Physics, but not all physicists appreciate this! When the book of Courant-Hilbert appeared, "it was another book by mathematicians who believed it would be of use to physicists". That was years later, and they soon had to change their tune, but it shows how easily people are swayed by what we later regard as prejudices, and how free from such prejudices Goettingen once was.

There my parents lived for 8 years, and I and most of my brothers and sisters were born. That was also where Carathéodory first made his name with his famous 1905 doctoral thesis on Discontinuous Solutions in the Calculus of Variations. I studied it, and all his papers and books, particularly those relating to the Calculus of Variations, in which I was specially interested. It was he who took me under his wing when I came to Munich in 1924. Perron and he were succeeding the finally retiring Pringsheim and Lindemann. Hartogs and Tietze were associate professors. What was missing, the retirees having not been precisely inspiring, was even one promising student, so I was particularly welcome. However I was welcome too in Theoretical Physics, where Sommerfeld had a great school.

Carathéodory had been in Goettingen the successor of Felix Klein, but it was in the 1st War and things were unsatisfactory. A year later he accepted a call to Berlin, then he was wanted and made much of in his own country, Greece, where he was chosen as one of the signatories at the Versailles treaty and eventually as rector for the University of Smyrna, from which he barely escaped when the Turks marched in. He was glad to accept a Chair at Munich.

I was put in charge of the Carathéodory-Perron Seminar, that year devoted to Julia's little book. You doubtless remember the pretty pictures of zero-angled triangles made up of circular arcs, all derived from one another by reflections, and how together they filled a disc or a half-plane. They illustrated the bilinear maps associated with the famous theorem of Picard in many books on function-theory. Julia's little book was about extensions of that theorem, both more precise forms due to Schottky, Landau and Carathéodory himself, and analogues in an angle. It made extensive use of Montel's Normal Families, which were really methods from functional analysis. Interestingly enough, the groups of the above bilinear maps, which are still being studied, for instance in the books by Gelfand and his associates, are simple transforms of the relativistic groups which largely determine elementary particles, in the theory being developed by my friend P. Nowosad, a theory which also uses the ideas of my generalised curves. The unity of mathematics, the strong connections between one topic and another, are once again illustrated.

In the Munich Seminar I spoke of, I had, in effect, to coach all the other speakers, and to correct them when their presentation was -- as it generally was -- inadequate. My reward -- if indeed one can speak of rewarding what is already an honour -- was that I spent with Carathéodory every Wednesday afternoon, and that I learnt in this way of his many sided mathematical contributions. Needless to say, I also attended his lectures, which that year were on function-theory and conformal maps. He started off with a 3 or 4 weeks proof of Jordan's theorem, for which he made each time on the blackboard a marvellous drawing in coloured chalk, to make plausible that the theorem might be false. Did any of the 100 or so prospective school-teachers, attending the course along with me, understand any of this? I doubt it. But they were carried away by his enthusiasm, they could not help feeling the drama of it all, as if they were attending a play in a theatre.

I attended other lectures and Seminars, notably in theoretical Physics. I had a place to work at that Institut, among Sommerfeld's brilliant doctoral students and assistants. Heisenberg had left for Leipzig, but I heard him when he came back to give the first account of his Matrix Mechanics. I was there again the following year when Schroedinger spoke of his Wave Mechanics: I was now an undergraduate at Trinity College, Cambridge, but the academic year then finished there in May with Tripos Examinations, which in mathematics had been early. I was thus able to attend a further Summer Semester in Munich.

I was on the whole reasonably well prepared for this advanced work. In the two years since finishing early -- by special permission -- at an excellent classical Swiss gymnasium, I had been somewhat held back by illness, but I had at least read extensively, if not very critically. In mathematics this had been all the more pleasurable, by giving meaning to terms I had heard used many times by my parents. Of course I lacked the intensive examination preparation provided in English schools: I was shocked to find I could do in 3 hours only half a question in a past Tripos (Part I)! However I soon put this right: I was as good a parrot as anyone after a few weeks, when I had set myself daily 3 such Tripos papers from previous years. In the actual Part I Tripos, I was quite able to write without thought and very fast, the solutions of all 16 questions in each paper, easily within the 3 hours alloted. My father, who until the results were out, was sure I had failed, as indeed I would have if I had not prepared myself, wrote to me that I was too good for the English system and so forth, all in the belief that I had failed! I think he was almost disappointed that I had done well ... I did well again the following year in Part II Schedule A, and of course in Schedule B which was more advanced. After that I was deluged with Scholarships, Studentships, Grants, until I became a Fellow: it made up for the lean years at the beginning. Perhaps after all the intensive examination preparation provided by English schools would have been less successful in the long run? Or perhaps the quality of a system of education is mainly dependent on that of those in charge of it?

One course I heard Carathéodory give, was on the Calculus of Variations. In this I was particularly interested, and therefore, as I read and studied his books and papers, I gave special attention to any relating to it. I also read and studied Hilbert's papers and mimeographed lectures connected with it, for instance lectures on the foundations of physics, and I reread many times Tonelli's Fundamenti which I mentioned before. The Calculus of Variations was also one of the topics Carathéodory often talked about on my Wednesday afternoon visits, and I well remember his drawing my attention to peculiar features of problems with differential equations as subsidiary relations.

One such feature, which is known as rigidity, shows that the name of Calculus of Variations is hardly suitable, since we are unable to vary anything! This occurs, for instance, in (t,x,y)-space, for the differential equation $\dot{y} = \sqrt{1+\dot{x}^2}$, so that the difference of y-coordinates at the ends of a curve satisfying this, is the length of the curve's projection on the (t,x)-plane. If then this projection is a straight segment, there is no other curve subject to our differential equation with the same ends.

Another feature of the problems is the role played by the smallest convex set of directions which contains those subject to the given differential equations at a given point. Carathéodory drew my attention, in this connection, to a paper of his on Fourier Series, in which he studied the smallest convex set containing a given set of points. The important thing is that he realised that what mattered was not so much the given differential equations, but rather an appropriate inclusion in a certain convex set of directions. In the case in which this convex set is the interior and boundary of a quadratic cone, a similar inclusion occurs in Relativity, and plays an important part in Hilbert's lectures. However the general case requires much of what we have since been doing in Optimal Control, for instance the general solution of the differential equation $\dot{x} = f(t,x)$, as given in Carathéodory's Real Functions, where f is measurable in t, continuous in x; it also requires a measurability theorem such as that of McShane and Warfield, not to speak of the notion of generalised curve. Carathéodory was thus a real pioneer of our modern Optimal Control. I would also like to mention a passage in one of his papers, in the Mathematische Annalen, in which he speaks of Hilbert's construction: he says this leads to a curve which consists, in the non-regular case, partly of extremal arcs and partly of pieces on the singular set $\Omega = 0$; but in fact, in the modern terminology, it leads to a generalised solution, a generalised curve.

I actually arrived at this notion differently, from Number Theory, from the famous Kronecker-Weyl theorem. It asserts that if θ is irrational, then the fractional parts of its integer multiples are uniformly distributed on the unit interval. Probability distributions can thus be arrived at as a sort of generalised limit of sequences, for instance the generalised limit of the sequence

$$1, 0, 1, 0, 1, 0, \ldots$$

would be a probability $1/2$ at ech of the values 0 and -1, and not, as Euler would have it, the value $1/2$, which is the value he assigns as the sum of the series

$$1 - 1 + 1 - 1 + 1 - \ldots$$

which has the above partial sums.

Doubtless, I was influenced by many things subconsciously, both within the Calculus of Variations and without, I have mentioned Hjemslev and Felix Klein, I might have mentioned too the whole trend of modern Physics, which I had continued to follow when I cam to Cambridge. But the decisive idea came from Number Theory, which at the time was rather looked down upon, as a sort of luxury which could have no sort of practical use! We know better: in our age of computers and space exploration, such subjects as Number Theory and Mathematical Logic can save much energy, time and money. For instance the existence of periodic orbits in the restricted 3 body problem was proved by Arenstorff and Conley, using Number Theory techniques.

Generalised curves first appeared in the paper "Approximation by polygons in the Calculus of Variations" in the Royal Society Proceedings of

1933. Hardy advised me not to publish there again: my father had often published there, but since then the place of mathematics had become impossible. Subsequently this has become no better: the number of new Fellows allotted yearly to each science increased by 50%, which in fact means that the number allotted to mathematics remained one. Nevertheless, true scientists, many of whom are mathematicians in all but the name, more than ever feel the need for more mathematics, particularly Optimal Control. In a local chess tournament, my first opponent, an undergraduate in economics, turned out to be reading my book on this topic. That was the last year before my so-called retirement, and my lectures based on this book were attended mostly by students and professors from all manner of departments. One of those attending, and not missing a single lecture, was an engineering student, who had been specifically advised not to attend, since he was completing his thesis. I was asked to the thesis presentation: my engineering colleagues, while praising the thesis, all said they did not follow the details, but, they continued, Professor Young understands! (I wasn't so sure ...)

Generalised curves appeared again under the title "Generalised curves and the existence of an attained absolute minimum in the Calculus of Variations: in the Comptes Rendus de la Société des Sciences et Lettres de Varsovie in 1937. I had tried following my father's publishing in the Royal Society, but now I wanted to please my Polish friends, Saks and Zygmund, and generally to show that I shared my parents' admiration of Poland and its people.

In the paper, I had extended Hilbert's existence theorem, and it now made sense to tackle afresh the general theory. This was done by McShane in a series of papers. We saw much of each other at the International Congress at Oslo in 1936, and I of course told him of my forthcoming Polish paper with its automatic existence theory. (I was working on other things too, of which abstracts appeared, inequalities whose final form, some 30 years later, is connected with other quantities besides Stieltjes integrals, and depends on a trick taken from Number Theory -- the supposedly useless subject! This now has to do with stochastic integrals, and is therefore again related with corresponding work of McShane.)

The Congress of 1936 laid much stress on the minimal surface problem, so that I naturally went on to higher dimensions, to generalised surfaces and higher varieties. Let me just say that this really involved the collective work of a number of us on such concepts as area, Hausdorff measure, prime ends, ending up with the existence theorem of Danskin,

Sigalov and Cesari for a parametric surface of topological type not exceeding N, which solves a corresponding regular problem a theorem which Fleming and I extended to the non-regular case. Let me mention also the rediscovery of generalised curves under various names, in von Neumann's theory of games and in Wald's decision theory, subjects which are really variants of Optimal Control -- so much so that, for instance, Fleming, who was my former pupil, was able to solve overnight at the Rand Corporation, their problem for the year!

The next Congress, the Harvard Congress of 1950, laid similar stress on Laurent Schwartz's distributions. These, as I mentioned earlier, were to some extent implied in Euler's Fundamental Lemma, but I did not say that to see this was easy! In point of fact, we require for this a duality of function-spaces, and it is this duality that led Fleming and myself to a dual definition of my generalised curves, surfaces and higher varieties which does away with many classical paradoxes. All this is close to the de Rham currents and so to the beautiful later joint work of Federer and Fleming -- which in turn has had applications right into Number Theory. Ah, but generalised surfaces have had applications too, right into phase transitions in solids and in the theory of crystals! As for generalised curves, the basic idea is the same as that of relaxed controls in Optimal Control, or of Young measures in Elasticity, in Electricity and Magnetism, and generally in (ordinary or partial) Non-Linear Differential Equations. If you think generalised curves are sophisticated or unreal, you have only to bear in mind that the infinitesimal zigzags of light in a curved flexible tube of negligeable thickness constitute such a generalised curve of light, and that a medical man, with the most elementary notion of mathematics, may be using it to locate the ailing part of an organ or of an artery, and so to save some life! The zigzags result from corresponding reflections at the successive tangent planes, they are no more unnatural than the dribbling of a football ...

What is to many of us unnatural, magical I almost say, is what I have called duality. You younger people already take it for granted, like the telephone, or like travel by air. It is at the root of what I have spoken of as streamlining. How well I remember the first time I realised its power! I had had only a few sessions of an hour or so with my pupil Fleming, and I thought it time to tell him what I was now working on. I spent the whole session covering the board with the chalk of my lemmas, and finally the proof of my main theorem was complete. I was exhausted, but triumphant! That was when Fleming quietly took out of his pocket a half-page proof ... Need I add that it used duality?

The duality that leads to generalised curves, surfaces, and so on, is as I said, a duality of function spaces, and so is, with a rather sophisticated convergence partly reminiscent of Chittenden's, that leading to Schwartz distributions. However the magic already appears in the classical duality of Lagrangian and Hamiltonian, even though it was somewhat obscured by a profusion of formulae. As I say in my book, Tonelli and Marston Morse did not use it. Carathéodory, on the other hand, used it whenever possible: he had learnt from his friend and teacher, the great Minkowski, to interpret it as a duality of convex figures, and how this leads to Minkowski's famous Number Theory result about lattice points. Strictly this duality of convex figures corresponds in the Calculus of Variations, only in the parametric form to a Hamiltonian duality: in the non-parametric case, we have dual convex functions of W. H. Young and W. Fenchel.

The Hamiltonian duality is at the heart of Optimal Control: it is the starting point of the famous Maximum Principle of Pontrjagin. Let me only say, first that equivalent, but less illuminating, versions of this condition, were found earlier by Hestenes and others; and further that the Principle, as a necessary condition, would be valueless if we did not have an existence theory too. We have such a theory thanks to relaxed controls, or what comes to the same, thanks to generalised curves. This theory is due to J. Warga, to Filippov, and others; as I point out in my book, it rests on delicate measurability theorems of such people as McShane and Warfield, and ultimately, in its general form, on the continuum hypothesis. It also necessitates a most unsymmetrical form of the existence of a solution of a differential equation, which appeared in Carathéodory's Real Functions. No one imagined at the time that the additional generality provided by this lack of symmetry would ever be important!

I have said nothing of the developments subsequent to this, in a number of directions, for instance in Stochastic Control which is evidently the proper setting for relaxed controls in as much as they already involve probability. But I have said enough to make clear, as I have emphasized throughout, that we need constantly to go back to the old ideas, and to ideas outside our narrow field, to find in them the source of really important new methods in our own problems.

Index